Global Navigation Satellite Systems,
Inertial Navigation, and Integration

Global Navigation Satellite Systems, Inertial Navigation, and Integration

Mohinder S. Grewal
California State University at Fullerton
Fullerton, California

Angus P. Andrews
Rockwell Science Center (retired)
Thousand Oaks, California

Chris G. Bartone
Ohio University
Athens, Ohio

Fourth Edition

This fourth edition first published 2020

Edition History
Wiley-Interscience; 1st edition 2006
Wiley-Interscience; 2nd edition 2011
Wiley-Interscience; 3rd edition 2013

Registered Office
John Wiley & Sons, Inc., 111 River Street, Hoboken, NJ 07030, USA

Editorial Office
111 River Street, Hoboken, NJ 07030, USA

For details of our global editorial offices, customer services, and more information about Wiley products visit us at www.wiley.com.

Wiley also publishes its books in a variety of electronic formats and by print-on-demand. Some content that appears in standard print versions of this book may not be available in other formats.

Library of Congress Cataloging-in-Publication Data

Names: Grewal, Mohinder S., author. | Andrews, Angus P., author. | Bartone, Chris G., author. | John Wiley & Sons.
Title: Global navigation satellite systems, inertial navigation, and integration / Mohinder S. Grewal, California State University at Fullerton, Angus P. Andrews, Chris G. Bartone.
Description: Fourth Edition. | Hoboken : Wiley, 2020. | Third edition published 2013. | Includes bibliographical references and index.
Identifiers: LCCN 2019038560 (print) | LCCN 2019038561 (ebook) | ISBN 9781119547839 (Hardback) | ISBN 9781119547846 (Adobe PDF) | ISBN 9781119547815 (ePub)
Subjects: LCSH: Global Positioning System. | Inertial navigation. | Kalman filtering.
Classification: LCC G109.5 .G74 2020 (print) | LCC G109.5 (ebook) | DDC 910.285–dc23
LC record available at https://lccn.loc.gov/2019038560
LC ebook record available at https://lccn.loc.gov/2019038561

Cover design by Wiley
Cover image: © Busakorn Pongparnit/Getty Images

Set in 10/12pt WarnockPro by SPi Global, Chennai, India

10 9 8 7 6 5 4 3 2 1

M.S.G. dedicates this book to the memory of his parents, Livlin Kaur and Sardar Sahib Sardar Karam Singh Grewal.

A.P.A. dedicates his contributions to his wife Jeri, without whom it never would have happened.

C.G.B. dedicates this work to his wife Shirley and two sons, Christopher and Stephen, for their never-ending support over the years.

Contents

Preface to the Fourth Edition

This book is intended for people who need a working knowledge of Global Navigation Satellite Systems (GNSS), Inertial Navigation Systems (INS), and the Kalman filtering models and methods used in their integration. The book is designed to provide a useable, working familiarity with both the *theoretical* and *practical* aspects of these subjects. For that purpose we have included "real-world" problems from practice as illustrative examples. We also cover the more practical aspects of implementation: how to represent problems in a mathematical model, analyze performance as a function of model parameters, implement the mechanization equations in numerically stable algorithms, assess its computational requirements, test the validity of results, and monitor performance in operation with sensor data from GPS and INS. These important attributes, often overlooked in theoretical treatments, are essential for effective application of theory to real-world problems.

The companion Wiley website (www.wiley.com/go/grewal/gnss) contains MATLAB® m-files to demonstrate the workings of the navigation solutions involved. It includes Kalman filter algorithms with GNSS and INS data sets, so that the reader can better discover how the Kalman filter works by observing it in action with GNSS and INS. The implementation of GNSS, INS, and Kalman filtering on computers also illuminates some of the practical considerations of finite-word-length arithmetic and the need for alternative algorithms to preserve the accuracy of the results. If the student wishes to apply what she or he learns, then it is essential that she or he experience its workings and failings – and learn to recognize the difference.

The book is organized for use as a text for an introductory course in GNSS technology at the senior level or as a first-year graduate level course in GNSS, INS, and Kalman filtering theory and applications. It could also be used for self-instruction or review by practicing engineers and scientists in these fields.

This fourth edition has been updated to include advancements in GNSS/INS technology since the third edition in 2013, as well as many improvements suggested by reviewers and readers of the second edition. Changes in this fourth edition include the following:

1. Updates on the significant upgrades in existing GNSS systems and on other systems currently under advanced development.
2. Expanded coverage of basic principles of antenna design and practical antenna design solutions are included.
3. Expanded coverage of basic principles of receiver design, and an update of the foundations for code and carrier acquisition and tracking within a GNSS receiver.
4. Examples demonstrating independence of Kalman filtering from probability density functions of error sources beyond their means and covariances, and how this breaks down with nonlinearities.
5. Updated coverage of inertial navigation to cover recent technology developments and the mathematical models and methods used in its implementation.
6. Updated dynamic models for the propagation of inertial navigation errors, including the effects of drifting sensor compensation parameters and nonlinearities.
7. Greatly expanded coverage of GNSS/INS integration, including derivation of a unified GNSS/INS integration model, its MATLAB implementations, and performance evaluation under simulated dynamic conditions.

The companion Wiley website has also been augmented to include updated background material and additional MATLAB scripts for simulating GNSS-only and integrated GNSS/INS navigation. The companion website (www.wiley.com/go/grewal/gnss) includes satellite position determination, calculation of ionospheric delays, and dilution of precision.

Chapter 1 provides an overview of navigation in general, and GNSS and inertial navigation in particular. These overviews include fairly detailed descriptions of their respective histories, technologies, different implementation strategies, and applications.

Chapter 2 covers the fundamental attributes of satellite navigation systems in general, the technologies involved, how the navigation solution is implemented, and how satellite geometries influence errors in the solution.

Chapter 3 covers the fundamentals of inertial navigation, starting with its nomenclature, and continuing through to practical implementation methods, error sources, performance attributes, and development strategies.

Chapters 4–9 cover basic theory of GNSS for a senior-level class in geomatics, electrical engineering, systems engineering, and computer science. Subjects covered in detail include basic GNSS satellite signal structures, practical receiver antenna designs, receiver implementation structures, error sources, signal processing methods for eliminating or reducing recognized error sources, and system augmentation methods for improving system integrity and security.

Chapter 10 covers the fundamental aspects of Kalman filtering essential for GNSS/INS integration: its mathematical foundations and basic implementation methods, its application to sensor integration in general, and to GNSS navigation in particular. It also covers how the implementation includes its own performance evaluation, and how this can be used in performance-predictive design of sensor systems.

Chapter 11 covers the basic errors sources and models for inertial navigation, including the effects of sensor noise and errors due to drifting inertial sensor error characteristics, how the resulting navigation errors evolve over time, and the resulting models that enable INS integration with other sensor systems.

Chapter 12 covers the essential mathematical foundations for GNSS/INS integration, including a unified navigation model, its implementation in MATLAB, evaluations of the resulting unified system performance under simulated dynamic conditions, and demonstration of the navigation performance improvement attainable through integrated navigation.

Appendix A contains brief descriptions of the MATLAB® software, including formulas implementing the models developed in MATLAB® different chapters and used for demonstrating how they work. Appendix B and Appendix C (www.wiley.com/go/grewal/gnss) contains background material on coordinate systems and transformations implemented in the software, including derivations of the rotational dynamics used in navigation error modeling and GNSS/INS integration.

For instructors that wish to cover the fundamental aspects of GNSS, Chapters 1–2 and 4–9 are recommended. Instructors for a course covering the fundamental concepts of inertial navigation can cover Chapters 1, 3, 10, and 11. A follow-on class or a more advanced course in GNSS and INS integration should include Chapter 12 as well as significant utilization of the software routines provided for computer-based GNSS/INS integration projects.

October 2019

Mohinder S. Grewal, Ph.D., P.E.
California State University at Fullerton
Fullerton, California

Angus P. Andrews, Ph.D.
Rockwell Science Center (retired)
Thousand Oaks, California

Chris G. Bartone, Ph.D., P.E.
Ohio University
Athens, Ohio

Acknowledgments

We acknowledge Professor John Angus, Jay A. Farrell, and Richard B. Langley for assistance and inspiration on the outline of this edition. We acknowledge the assistance of Mrs. Laura A. Cheung of the Raytheon Company for her expert assistance in reviewing Chapter 8 (Differential GNSS) and with the MATLAB® programs. Special thanks goes to Dr. Larry Weill for his contribution to Chapter 7 on multipath mitigation algorithms.

A. P. A. thanks Andrey Podkorytov at the Moscow Aviation Institute for corrections to the Schmidt–Kalman filter; Randall Corey from Northrop Grumman and Michael Ash from C. S. Draper Laboratory for access to the developing Draft IEEE Standard for Inertial Sensor Technology; Dr. Michael Braasch at GPSoft, Inc. for providing evaluation copies of the GPSoft INS and GPS MATLAB Toolboxes; Drs. Jeff Schmidt and Robert F. Nease, former Vice President of Engineering and Chief Scientist at Autonetics, respectively, for information on the early history of inertial navigation; and Edward H. Martin, member of the GPS development team awarded the 1992 Robert J. Collier Trophy by the National Aeronautics Association, and winner of the 2009 Captain P.V.H. Weems Award presented by the Institute of Navigation for his role in GPS receiver development, for information on the very early history of GPS/INS integration.

C. G. B. would like to thank Ohio University and many of its fine faculty, staff, and students that I have had the pleasure to interact with in my research and teaching over the years. Such a rich environment has enabled me to develop a wide variety of classes and research efforts that these writings draw upon. Thanks also goes to Neil Gerein and Jerry Freestone from NovAtel, Dave Brooks from Sensor Systems, James Horne from Roke, and Herbert Blaser from u-blox for providing antenna information.

About the Authors

Mohinder S. Grewal, Ph.D., P.E., is well known for his innovative application of Kalman filtering techniques to real world modeling problems and his ability to communicate this complex subject to his students. His original research appears in IEEE and ION refereed journals and proceedings. He holds patents in GUS clock steering and L_1/L_5 differential bias estimation. Dr. Grewal is Professor of Electrical Engineering at California State University, Fullerton, which awarded him its 2008-2009 Outstanding Professor Award. His consulting associations include Raytheon Systems, Boeing Company, Lockheed-Martin, University of California, Riverside, staff of the US Department of the Interior, Geodetics, and Northrop. He is a Senior Member of IEEE and member of the Institute of Navigation. His Ph.D. in Control Systems and Computers is from University of Southern California.

Angus P. Andrews derived the first electrostatic bearing torque parametric models for calibrating electrostatic gyroscopes in 1967 at the Autonetics Division of Rockwell International, and then saw its development through two generations of strapdown inertial navigation systems to the N73 competitor for the US Air Force Standard Navigator. His career in inertial navigation also included derivations of new square root filtering formulas. His undergraduate degree is from MIT and his Ph.D. in mathematics is from University of California, Los Angeles.

Chris G. Bartone, Ph.D., P.E., is a professor at Ohio University with over 35 years experience in communications, navigation, and surveillance systems. He received his Ph.D., E.E. from Ohio University, M.S.E.E. from the Naval Postgraduate School, and B.S. E.E. from The Pennsylvania State University. Dr. Bartone has developed and teaches a number of GNSS, antenna, and microwave classes. He is a recipient of the RTCA William E. Jackson award, the ION Captain P.V.H. Weems award, and is a Fellow of the ION. His research concentrates on all aspects of navigation systems.

Acronyms

A/D	analog-to-digital (conversion)
ADC	analog-to-digital converter
ADR	accumulated delta range
ADS	automatic dependent surveillance
AGC	automatic gain control
AHRS	attitude and heading reference system
AIC	Akaike information-theoretic criterion
AIRS	advanced inertial reference sphere
ALF	atmospheric loss factor
ALS	autonomous landing system
altBOC	alternate binary offset carrier
AODE	age of data word, ephemeris
AOR-E	Atlantic Ocean Region East (WAAS)
AOR-W	Atlantic Ocean Region West (WAAS)
AR	autoregressive or axial ratio
ARMA	autoregressive moving average
ARNS	aeronautical radio navigation services
ASD	amplitude spectral density
ASIC	application-specific integrated circuit
ASQF	application-specific qualification facility (EGNOS)
A-S	antispoofing
ATC	air traffic control
BD	BeiDou
bps	bits per second
BOC	binary offset carrier
BPSK	binary phase-shift keying
BS	base station
C	civil
C/A	coarse acquisition (channel or code)
C&V	correction and verification (WAAS)
CDM	code-division multiplexing

CDMA	code-division multiple access
CEM	computational electromagnetic model
cps	chips per second
CEP	circle error probable
CL	code long
CM	code moderate
CNAV	civil navigation
CNMP	code noise and multipath
CONUS	conterminous United States, also continental United States
CORS	continuously operating reference station
COSPAS	Cosmicheskaya Sistyema Poiska Avariynich Sudov
CBOC	combined BOC
C/NAV	commercial navigation
CRC	cyclic redundancy check
CRPA	controlled reception pattern antenna
CWAAS	Canadian WAAS
DGNSS	differential GNSS
DGPS	differential GPS
DME	distance measurement equipment
DOD	Department of Defense (USA)
DOP	dilution of precision
E	eccentric anomaly
ECEF	Earth-centered, Earth-fixed (coordinates)
ECI	Earth-centered inertial (coordinates)
EGNOS	European Geostationary Navigation Overlay System
EIRP	effective isotropic radiated power
EKF	extended Kalman filter
EMA	electromagnetic accelerator or electromagnetic accelerometer
ENU	east–north–up (coordinates)
ESA	European Space Agency
ESG	electrostatic gyroscope
ESGN	electrostatically supported gyro navigator (US Navy)
EU	European Union
EWAN	EGNOS wide-area (communication) network
FAA	federal aviation administration (USA)
FDMA	frequency division multiple access
FEC	forward error correction
FLL	frequency-lock loop
FM	frequency modulation
FOG	fiber optic gyroscope
FPE	final prediction error (Akaike's)
FSLF	free-space loss factor
F/NAV	free navigation

FT	feet
GAGAN	GPS and GEO augmented navigation (India)
GBAS	ground-based augmentation system
GCCS	GEO communication and control segment
GDOP	geometric dilution of precision
GEO	geostationary Earth orbit
GES	GPS Earth station COMSAT
GIC	GPS integrity channel
GIPSY	GPS infrared positioning system
GIS	geographic information system(s)
GIVE	grid ionosphere vertical error
GLONASS	global orbiting navigation satellite system
GNSS	global navigation satellite system
GOA	GIPSY/OASIS analysis
GPS	global positioning system
GUS	GEO uplink subsystem
GUST	GEO uplink subsystem type 1
HDOP	horizontal dilution of precision
HEO	highly inclined elliptical orbit or high earth orbit
HMI	hazardously misleading information
HOW	handover word
HRG	hemispheric resonator gyroscope
ICAO	International Civil Aviation Organization
ICC	ionospheric correction computation
ICD	interface control document
IDV	independent data verification (of WAAS)
IF	intermediate frequency
IFOG	integrating or interferometric fiber optic gyroscope
IGP	ionospheric grid point (for WAAS)
IGS	international GNSS service
ILS	instrument landing system
IMU	inertial measurement unit
Inmarsat	international mobile (originally "Maritime") satellite organization
I/NAV	integrity navigation
INS	inertial navigation system
IODC	issue of data, clock
IODE	issue of data, ephemeris
IONO	ionosphere, ionospheric
IOT	in-orbit test
IR U	inertial reference unit
IS	interface specification
ISA	inertial sensor assembly

ITRF	International Terrestrial Reference Frame
JPALS	Joint Precision Approach and Landing System
JT1DS	Joint Tactical Information Distribution System
LAAS	local-area augmentation system
LADGPS	local-area differential GPS
LAMBDA	least-squares ambiguity decorrelation adjustment
LD	location determination
LLMSE	linear least mean squares estimator
LHCP	left-hand circularly polarized
LORAN	long-range navigation
LOS	line of sight
LPV	lateral positioning with vertical guidance
LSB	least significant bit
LTP	local tangent plane
M	mean anomaly, meter or military
MBOC	modified BOC
MCC	mission/master control center (EGNOS)
MCPS	million chips per second
MEDLL	multipath-estimating delay-lock loop
MEMS	microelectromechanical system(s)
MEO	medium Earth orbit
MS	mobile station (i.e. cell phone)
MMSE	minimum mean-squared error (estimator)
MMT	multipath mitigation technology
MOPS	minimum operational performance standards
MSAS	MTSAT satellite-based augmentation system (Japan)
MSB	most significant bit
MTSAT	multifunctional transport satellite (Japan)
MVDR	minimum variance distortionless response
MVUE	minimum-variance unbiased estimator
MWG	momentum wheel gyroscope
NAS	National Airspace System
NAVSTAR	navigation system with time and ranging
NCO	numerically controlled oscillator
NED	north–east–down (coordinates)
NGS	National Geodetic Survey (USA)
NLES	navigation land Earth station(s) (EGNOS)
NPA	nonprecision approach
NSRS	National Spatial Reference System
NSTB	National Satellite Test Bed
OASIS	orbit analysis simulation software
OBAD	old but active data
OD	orbit determination

OPUS	online positioning user service (of NGS)
OS	open service (of Galileo)
PA	precision approach
PACF	performance assessment and checkout facility (EGNOS)
P-code	precision code
PDF	probability density function
pdf	portable document format
PDI	pre-detection integration
PDOP	position dilution of precision
PI	proportional and integral (controller)
PID	process input data (of WAAS) or proportional, integral, and differential (control)
PIGA	pendulous integrating gyroscopic accelerometer
PLL	phase-lock loop
PLRS	position location and reporting system (US Army)
PN	pseudorandom noise
POR	pacific ocean region
PPS	precise positioning service or pulse(s) per second
PR	pseudorange
PRN	pseudorandom noise or pseudorandom number (=SVN for GPS)
PRS	public regulated service (of Galileo)
PSD	power spectral density
QZS	Quasi-Zenith Satellite
QZSS	Quasi-Zenith Satellite System
RAAN	right ascension of ascending node
RAG	receiver antenna gain (relative to isotropic)
RAIM	receiver autonomous integrity monitoring
RF	radiofrequency
RHCP	right-hand circularly polarized
RIMS	ranging and integrity monitoring station(s) (EGNOS)
RINEX	receiver independent exchange format (for GPS data)
RLG	ring laser gyroscope
RM A	reliability, maintainability, availability
RMS	root-mean-squared or reference monitoring station
RNSS	radio navigation satellite services
RPY	roll–pitch–yaw (coordinates)
RTCA	radio technical commission for aeronautics
RTCM	radio technical commission for maritime service
RTOS	real-time operating system
RVCG	rotational vibratory coriolis gyroscope
s	second
SAP	space adaptive processing

SAR	synthetic aperture radar, or search and rescue (Galileo service)
SARP	standards and recommended practices (Japan)
SARSAT	search and rescue satellite-aided tracking
SAW	surface acoustic wave
SBAS	space-based augmentation system
SBIRLEO	space-based infrared low Earth orbit
SCOUT	scripps coordinate update tool
SCP	Satellite Correction Processing (of WAAS)
SDR	software defined radio
SF	scale factor
SI	system international (metric)
SIS	signal in space
SM	solar magnetic
SNAS	Satellite Navigation Augmentation System (China)
SNR	signal-to-noise ratio
SOL	safety of life service (of Galileo)
SPS	standard positioning service (GPS)
sps	symbols per second
SSBN	ship submersible ballistic nuclear (USA)
STAP	space–time adaptive processing
STF	signal task force (of Galileo)
SV	space vehicle
SVN	space vehicle number (= PRN for GPS)
SWR	standing wave ratio
TCS	Terrestrial Communications Subsystem (for WAAS)
TCXO	temperature-compensated Xtal (crystal) oscillator
TDOA	time difference of arrival
TDOP	time dilution of precision
TEC	total electron content
TECU	total electron content units
3GPP	3rd generation partnership project
TLM	telemetry word
TMBOC	time-multiplexed BOC
TOA	time of arrival
TOW	time of week
TTA	time to alarm
TTFF	time to first fix
UDRE	user differential range error
UERE	user-equivalent range error
UKF	unscented Kalman filter
URE	user range error
USAF	United States Air Force
USN	United States Navy

UTC	universal time, coordinated (or coordinated universal time)
UTM	universal transverse mercator
VAL	vertical alert limit
VCG	vibratory coriolis gyroscope
VDOP	vertical dilution of precision
VHF	very high frequency (30–300 MHz)
VOR	VHF omnirange (radionavigation aid)
VRW	velocity random walk
WAAS	wide-area augmentation system (USA)
WADGPS	wide-area differential GPS
WGS	world geodetic system
WMS	wide-area master station
WN	week number
WNT	WAAS network time
WRE	wide-area reference equipment
WRS	wide-area reference station

About the Companion Website

This book is accompanied by a companion website:

www.wiley.com/go/grewal/gnss

The website includes:

- Solution Manual for Instructors only
- MATLAB files for selected chapters
- Appendices B and C

1

Introduction

A book on navigation? Fine reading for a child of six![1]

1.1 Navigation

During the European Age of Discovery, in the fifteenth to seventeenth centuries, the word *navigation* was synthesized from the Latin noun *navis* (ship) and the Latin verb stem *agare* (to do, drive, or lead) to designate the operation of a ship on a voyage from A to B – or the art thereof.

In this context, the word *art* is used in the sense of a *skill, craft, method,* or *practice.* The Greek word for it is $\tau\epsilon\chi\nu\upsilon$, with which the Greek suffix -$\lambda o\gamma\iota\alpha$ (the study thereof) gives us the word *technology.*

1.1.1 Navigation-Related Technologies

In current engineering usage, the art of getting from A to B is commonly divided into three interrelated technologies:

- *Navigation* refers to the art of determining the current location of an object – usually a vehicle of some sort, which could be in space, in the air, on land, on or under the surface of a body of water, or underground. It could also be a comet, a projectile, a drill bit, or anything else we would like to locate and track. In modern usage, A and B may refer to the object's current and intended dynamic *state,* which can also include its velocity, attitude, or attitude rate relative to other objects. The practical implementation of navigation generally requires observations, measurements, or sensors to measure relevant variables, and methods of estimating the state of the object from the measured values.

1 Source: Truant officer Agatha Morgan, played by Sara Haden in the 1936 film *Captain January,* starring Shirley Temple and produced by Daryl F. Zanuck for 20th Century Fox Studios.

Global Navigation Satellite Systems, Inertial Navigation, and Integration,
Fourth Edition. Mohinder S. Grewal, Angus P. Andrews, and Chris G. Bartone.

Companion website: www.wiley.com/go/grewal/gnss

- *Guidance* refers to the art of determining a suitable trajectory for getting the object to a desired *state*, which may include position, velocity, attitude, or attitude rate. What would be considered a "suitable" trajectory may involve such factors as cost, consumables and/or time required, risks involved, or constraints imposed by existing transportation corridors and geopolitical boundaries.
- *Control* refers to the art of determining what actions (e.g. applied forces or torques) may be required for getting the object to follow the desired trajectory.

These distinctions can become blurred – especially in applications when they share hardware and software. This has happened in missile guidance [1], where the focus is on getting to B, which may be implemented without requiring the intermediate locations. The distinctions are clearer in what is called "Global Positioning System (GPS) navigation" for highway vehicles:

- *Navigation* is implemented by the GPS receiver, which gives the user an estimate of the current location (A) of the vehicle.
- *Guidance* is implemented as *route planning*, which finds a route (trajectory) from A to the intended destination B, using the connecting road system and applying user-specified measures of route suitability (e.g. travel distance or total time).
- *Control* is implemented as a sequence of requested driver actions to follow the planned route.

1.1.2 Navigation Modes

From time immemorial, we have had to solve the problem of getting from A to B, and many solution methods have evolved. Solutions are commonly grouped into five basic navigation modes, listed here in their approximate chronological order of discovery:

- *Pilotage* essentially relies on recognizing your surroundings to know where you are (A) and how you are oriented relative to where you want to be (B). It is older than human kind.
- *Celestial navigation* uses relevant angles between local vertical and celestial objects (e.g. the Sun, planets, moons, stars) with known directions to estimate orientation, and possibly location on the surface of the Earth. Some birds have been using celestial navigation in some form for millions of years. Because the Earth and these celestial objects are moving with respect to one another, accurate celestial navigation requires some method for estimating time. By the early eighteenth century, it was recognized that estimating longitude with comparable accuracy to that of latitude (around half a degree at that time) would require clocks accurate to a few minutes over long sea voyages. The requisite clock technology was not developed until the middle of

the eighteenth century, by John Harrison (1693–1776). The development of atomic clocks in the twentieth century would also play a major role in the development of satellite-based navigation.
- *Dead reckoning* relies on knowing where you started from, plus some form of heading information and some estimate of speed and elapsed time to determine the distance traveled. Heading may be determined from celestial observations or by using a magnetic compass. Dead reckoning is generally implemented by plotting lines connecting successive locations on a chart, a practice at least as old as the works of Claudius Ptolemy (~85–168 CE).
- *Radio navigation* relies on radio-frequency sources with known locations, suitable receiver technologies, signal structure at the transmitter, and signal availability at the receiver. Radio navigation technology using land-fixed transmitters has been evolving for about a century. Radio navigation technologies using satellites began soon after the first artificial satellite was launched.
- *Inertial navigation* is much like an automated form of dead reckoning. It relies on knowing your initial position, velocity, and attitude, and thereafter measuring and integrating your accelerations and attitude rates to maintain an estimate of velocity, position, and attitude. Because it is self-contained and does not rely on external sources, it has the potential for secure and stealthy navigation in military applications. However, the sensor accuracy requirements for these applications can be extremely demanding [2]. Adequate sensor technologies were not developed until the middle of the twentieth century, and early systems tended to be rather expensive.

These modes of navigation can be used in combination, as well. The subject of this book is a combination of the last two modes of navigation: global navigation satellite system (GNSS) as a form of radio navigation combined with inertial navigation. The key integration technology is Kalman filtering, which also played a major role in the development of both navigation modes.

The pace of technological innovation in navigation has been accelerating for decades. Over the last few decades, navigation accuracies improved dramatically and user costs have fallen by orders of magnitude. As a consequence, the number of marketable applications has been growing phenomenally. From the standpoint of navigation technology, we are living in interesting times.

1.2 GNSS Overview

Satellite navigation development began in 1957 with the work of William W. Guier (1926–2011) and George C. Weiffenbach (1921–2003) at the Applied Physics Laboratory of Johns Hopkins University [3], resulting in the US Navy Transit GNSS [4]. Transit became operational in the mid-1960s, achieving navigational accuracies in the order of 200 m and remained operational until it was superseded by the US Air Force GPS 28 years later. The Transit navigation solution is based on the Doppler history of the received satellite signal as the satellite passed overhead from horizon to horizon – a period of about a quarter of an hour. The US Navy also developed the TIMATION (TIMe/navigATION) in the mid-1960s to explore the performance of highly accurate space-based clocks for precise satellite-based positioning. While Transit and TIMATION were "carrier-phase" only-based systems, the US Air Force 621B experimental program validates the use of ranging codes for a global satellite-based precision navigation system. These programs were instrumental in the concepts and techniques in the development of GPS as well as other satellite-based GNSS that we know today.

Currently there are several GNSS in various stages of operation and development. This section provides a brief overview of these systems, where a more detailed discussion is given in Chapter 4.

1.2.1 GPS

The GPS is part of a satellite-based navigation system developed by the US Department of Defense under its NAVSTAR satellite program [5–16].

1.2.1.1 GPS Orbits

The fully populated GPS constellation includes 31 active satellites with additional operational spares, in six operational planes. The satellites are in circular orbits with four or more satellites in each orbital plane. The orbital planes are each inclined at an angle of 55° relative to the equator and are separated from each other by multiples of 60° right ascension. Each satellite is in a medium Earth orbit (MEO), is nongeostationary, and is approximately circular, with radii of 26 560 km, with orbital period of one-half sidereal day ($\approx$11.967 hours). Four or more GPS satellites will always be visible from any point on the Earth's surface, where the GPS satellites can be used to determine an observer's position, velocity, and time (PVT) anywhere on the Earth's surface 24 h/d.

1.2.1.2 Legacy GPS Signals

Each GPS satellite carries a cesium and/or rubidium atomic clock (i.e. frequency reference oscillator) to provide timing information for the signals transmitted by the satellites. While each satellite carries several internal clock, all navigation signals are generated from one clock. Satellite clock corrections are provided to the users in the signals broadcast by each satellite, with the aid of the GPS Ground Control Segment. The legacy GPS satellite transmits two L-band spread spectrum navigation signals on – an L1 signal with carrier frequency $f_1 = 1575.42\,\text{MHz}$ and an L2 signal with carrier frequency $f_2 = 1227.6\,\text{MHz}$. These two frequencies are integral multiples $f_1 = 154 f_0$ and $f_2 = 120 f_0$ of a base frequency $f_0 = 10.23\,\text{MHz}$. The L1 signal from each satellite is *binary phase-shift keying* (BPSK) modulated by two *pseudorandom noise* (PRN) codes in phase quadrature, designated as the C/A-code and P(Y)-code. The L2 signal from each satellite is BPSK modulated by only the P(Y)-code. A brief description of the nature of these PRN codes follows, with greater detail given in Chapter 4.

Compensating for ionosphere propagation delays. The time delay from when a navigation signal is transmitted, to when the signal is received, is used to eventually estimate the distance between the satellite and the user. This signal propagation delay is affected by the atmosphere. As the signals pass through the ionosphere, the delay chances with frequency. This is one motivation for use of two different carrier signals, L1 and L2. Because delay through the ionosphere varies approximately as the inverse square of signal frequency f (delay $\propto f^{-2}$), the measurable differential delay between the two carrier frequencies can be used to compensate for the delay in each carrier (see Ref. [16] for details).

Code-division multiplexing. Knowledge of the PRN codes allows users independent access to multiple GPS satellite signals on the same carrier frequency. The signal transmitted by a particular GPS signal can be selected by generating and matching, or correlating, the PRN code for that particular satellite. All PRN codes are known and are generated or stored in GPS satellite signal receivers. For legacy GPS there are two PRN codes transmitted from each satellite. The first PRN code from each GPS satellite, sometimes referred to as a *precision code* or *P-code,* is a relatively long, fine-grained code having an associated clock or chip rate of $f_0 = 10.23$ MHz. A second PRN code from each GPS satellite, sometimes referred to as a *clear* or *coarse acquisition code* or *C/A-code,* is intended to facilitate rapid satellite signal acquisition and handover to the P-code. It is a relatively short, coarser-grained code having an associated clock or chip rate of $f_0 = 1.023$ MHz. The C/A-code for any GPS satellite has a length of 1023 chips or time increments before it repeats. The full P-code has a length of 259 days, during which each satellite transmits a unique portion of the full P-code. The portion of P-code used for a given GPS satellite has a length of precisely one week (seven days) before this code portion repeats. Accepted methods for generating the C/A-code and P-code were established by the satellite developer (Satellite Systems Division of Rockwell International Corporation) in 1991 [17].

Navigation signal. The GPS satellite bit stream includes navigational information on the ephemeris of the transmitting GPS satellite and an almanac for all GPS satellites, with parameters providing approximate corrections for ionospheric signal propagation delays suitable for single-frequency receivers and for an offset time between satellite clock time and true GPS time. The legacy navigational information is transmitted at a rate of 50 baud. Further discussion of the GPS and techniques for obtaining position information from satellite signals can be found in chapter 4 of Ref. [18].

Precise positioning service (PPS). Formal, proprietary service PPS is the full-accuracy, single-receiver GPS positioning service provided to the United States and its allied military organizations and other selected agencies. This service includes access to the encrypted P(Y)-code.

Standard positioning service (SPS). SPS provides GPS single-receiver (stand-alone) positioning service to any user on a continuous, worldwide basis. SPS is intended to provide access only to the C/A-code and the L1 carrier.

1.2.1.3 Modernization of GPS

GPS IIF, GPS IIR–M, and GPS III provide the legacy and new modernized signals. These may include L2 civil (L2C) signal and the L5 signal (at 1176.45 MHz) modulated by a new code structure, as well as, the M and L1C codes. These modernized GPS signals improve the ionospheric delay calculation, ranging performance, ambiguity resolution, and overall PVT accuracy.

The GPS Ground Control Segment monitors the GPS signals in space, interfaces with the US Naval Observatory for timing information, and has remote monitor/uplink transmitter sites throughout the globe. Over the years, the GPS GCS has been upgraded and the Next-Generation Operational Control System (OCX) will monitor all legacy and modernized GPS signals to provide for enhanced PVT solutions for the user segment. See Sections 4.2.8 and 10.5.5.5 and Ref. 18, Chapter 4.

1.2.2 Global Orbiting Navigation Satellite System (GLONASS)

A second system for global positioning is the Global Orbiting Navigation Satellite System (GLONASS), placed in orbit by the former Soviet Union and now operated and maintained by the Russian Republic [19, 20].

1.2.2.1 GLONASS Orbits

GLONASS has 24 satellites, distributed approximately uniformly in three orbital planes (as opposed to six for GPS) of 8 satellites each. Each orbital plane has a nominal inclination of 64.8° relative to the equator, and the three orbital planes are separated from each other by multiples of 120° right ascension. GLONASS orbits have smaller radii than GPS orbits, about 25 510 km, and a satellite period of revolution of approximately 8/17 of a sidereal day.

1.2.2.2 GLONASS Signals

The legacy GLONASS system uses frequency-division multiplexing of independent satellite signals. Each GLONASS satellite transmits two navigation signals in the L1 and L2 frequency bands, corresponding to $f_1 = (1.602 + 9k/16)$ GHz and $f_2 = (1.246 + 7k/16)$ GHz, where $k = -7, -6, \ldots 5, 6$ is the satellite number. These frequencies lie in two bands at 1.598–1.605 GHz (L1) and 1.242–1.248 GHz (L2). The L1 code is modulated by a C/A-code (chip rate = 0.511 MHz) and by a P-code (chip rate = 5.11 MHz). The L2 code is presently modulated only by the P-code. The GLONASS satellites also transmit navigational data at a rate of 50 baud. Because the satellite frequencies are distinguishable from each other, the P-code and the C/A-code are the same for each satellite. The methods for receiving and analyzing GLONASS signals are similar to the methods used for GPS signals. Further details can be found in the patent by Janky [21].

1.2.2.3 Modernized GLONASS

The first of next-generation GLONASS-K satellites was first launched on 26 February 2011 and continues to undergo flight tests. This satellite transmits the legacy FDMA (frequency division multiple access) GLONASS signals and a L3OC code-division multiple access (CDMA) signal at a frequency of 1202 MHz. Other GLONASS CDMA signals are under development within the legacy L1 (L1OC signal) and L2 (L2OC signal) bands.

1.2.3 Galileo

The Galileo system is satellite-based navigation system currently under development by the European Union (EU). This development has completed definition and development phases and is nearly complete with launching operational satellites to achieve a 30 satellite constellation. Galileo operates in the L-band with MEO satellites at height slightly above the GPS MEO satellites (23 222 km for Galileo versus 20 180 km for GPS). Galileo satellites operate in three orbital planes at an inclination angle similar to GPS. Galileo operates in three spectral bands known as E1 (1559–1592 MHz), E5 (1164–1215 MHz), and E6 (1260–1300 MHz).

1.2.3.1 Galileo Navigation Services

The EU intends the Galileo system to provide various levels of services.

Open service (OS). The OS provides signals for positioning and timing, is free of direct user charge, and is accessible to any user equipped with a suitable receiver, with no authorization required. The OS provides dual-frequency operation in the L1/E1 and L5/E5 frequency bands. The Galileo E1 L1C signal centered at 1575.42 MHz is compatible with the modernized GPS L1C signal transmitted by GPS III satellites. The Galileo E5a signal at 1176.45 MHz is part of a combined AltBOC signal. Modernized GNSS receiver equipment may use a combination of Galileo and GPS signals, thereby improving performance in severe environments such as urban canyons and heavy vegetation.

Commercial service (CS). The CS service is intended for applications requiring performance higher than that offered by the OS. Users of this service pay a fee for the added value. CS is implemented by adding two additional signals to the OS signal suite. The additional signals are protected by commercial encryption, and access protection keys are used in the receiver to decrypt the signals. Typical value-added services include service guarantees, precise timing, multifrequency ionospheric delay measurements, local differential correction signals for very high-accuracy positioning applications, and other specialized requirements. These services will be developed by service providers, which will buy the right to use the multifrequency commercial signals from the Galileo operator.

Public regulated service (PRS). The PRS is an access-controlled service for government-authorized applications. It is expected to be used by groups such as police, coast guards, and customs. The signals will be encrypted, and access by region or user group will follow the security policy rules applicable in Europe. The PRS will be operational at all times and in all circumstances, including periods of crisis. A major feature of PRS is the robustness of its signal, which protects it against jamming and spoofing.

Search and rescue (SAR). The SAR service is Europe's contribution to the international cooperative effort on humanitarian SAR. It will feature near-real-time reception of distress messages from anywhere on Earth, precise location of alerts (within a few meters), multiple satellite detection to overcome terrain blockage, and augmentation by the four low Earth orbit (LEO) satellites and the three geostationary satellites in the current Cosmicheskaya Sistema Poiska Avariynyh Sudov-Search and Rescue Satellite Aided Tracking (COSPAS-SARSAT) system.

1.2.3.2 Galileo Signal Characteristics

Galileo will provide 10 right-hand circularly polarized navigation signals in three frequency bands. The various Galileo navigation signals will use four different navigation (NAV) data formats to support the various service supported by Galileo: F/NAV (Free NAV), I/NAV (Integrity NAV), C/NAV, and G/NAV (Galileo NAV). The I/NAV signals contain integrity information, while the F/NAV signals do not. The C/NAV signals are used by the CS, and the G/NAV signals are used by the PRS.

E5a–E5b Band This band, which spans the frequency range from 1164 to 1214 MHz, contains two signals, denoted E5a and E5b, which are centered at 1176.45 and 1207.140 MHz, respectively. Each signal has an in-phase component and a quadrature component. Both components use spreading codes with a chipping rate of 10.23 Mcps (million chips per second). The in-phase components are modulated by navigation data, while the quadrature components, called *pilot signals*, are data-free. The data-free pilot signals permit arbitrarily long coherent processing, thereby greatly improving detection and tracking sensitivity. A major feature of the E5a and E5b signals is that they can be treated as either separate signals or a single wide-band signal. Low-cost receivers can use either signal, but the E5a signal might be preferred, since it is centered at the same frequency as the modernized GPS L5 signal and would enable the simultaneous reception of E5a and L5 signals by a relatively simple receiver without the need for reception on two separate frequencies. Receivers with sufficient bandwidth to receive the combined E5a and E5b signals would have the advantage of greater ranging accuracy and better multipath performance.

Even though the E5a and E5b signals can be received separately, they actually are two spectral components produced by a single modulation called alternate *binary offset carrier* (AltBOC) modulation. This form of modulation retains the simplicity of standard binary offset carrier (BOC) modulation (used in the modernized GPS M-code military signals) and has a constant envelope while permitting receivers to differentiate the two spectral lobes.

The in-phase component of the E5a signal is modulated with 50 sps (symbols per second) navigation data without integrity information, and the in-phase component of the E5b signal is modulated with 250 sps data with integrity information. Both the E5a and E5b signals are available to the OS and CS services.

E6 Band This band spans the frequency range from 1260 to 1300 MHz and contains a C/NAV signal and a G/NAV signal, each centered at 1278.75 MHz. The C/NAV signal is used by the CS service and has both an in-phase and a quadrature pilot component using a BPSK spreading code modulation of 5×1.023 Mcps. The in-phase component contains 1000-sps data modulation, and the pilot component is data-free. The G/NAV signal is used by the PRS service and has only an in-phase component modulated by a BOC(10,5) spreading code and data modulation with a symbol rate that is to be determined.

L1/E1 Band The L1/E1 band (sometimes denoted as L1 for convenience) spans the frequency range from 1559 to 1591 MHz and contains a G/NAV signal used by the PRS service and an I/NAV signal used by the OS and CS services. The G/NAV signal has only an in-phase component with a BOC spreading code and data modulation. The I/NAV signal has an in-phase and quadrature component. The in-phase component contains 250-sps data modulation with a BOC(1,1) spreading code. The quadrature component is data-free and utilizes a combined BOC signal.

1.2.4 BeiDou

The BeiDou Navigation Satellite System (BDS) is being developed by the People's Republic of China (PRC), starting with regional services and expanding to global services. Phase I was established in 2000. Phase II (BDS-2) provides service for areas in China and its surrounding areas. Phase III (i.e. BDS-3) is being deployed to provide global service.

1.2.4.1 BeiDou Satellites

BeiDou will consist of 27 MEO satellites, including 5 geostationary Earth orbit (GEO) satellites and 3 inclined geosynchronous orbit (IGSO) satellites. The GEO and IGSO satellites will be at longitudes to support the China and the surround areas of Southeast Asia.

1.2.4.2 Frequency

The BDS-2 and BDS-3 operate on various frequencies in the L1 (BDS-3 B1C signal at 1575.42 MHz), E6 (BDS B3I signal at 1268.5 MHz), E5 (BDS-2 and BDS-3 at 1207.14 MHz; and BDS-3 (B2a) at 1176.45 MHz). These signals use various navigation data formats from the BDS MEO, GEO, or IGSO satellites to support global and regional civil services. Details of this section are given in Chapter 4.

1.2.5 Regional Satellite Systems

There are several regional satellite systems that provide regional navigation and/or augmentation service.

1.2.5.1 QZSS

Quasi-Zenith Satellite System (QZSS) is satellite-based navigation system being developed by the Japanese government. QZSS has a constellation of four IGSO satellites that provide navigation and augmentation to GPS over Japan and Southeast Asia. The system transmits GPS-type signals: L1 (L1 C/A and L1C), L2C, and L5, as well as, augmentation signals to support submeter and centimeter level services. Details of this section are given in Chapter 4.

1.2.5.2 NAVIC

The Indian Regional Navigation Satellite Systems, operationally known as NAVIC, is a regional satellite navigation system developed by the Indian Space Research Organization (ISRO). NAVIC constellation consists of eight satellites operation in GEO- and IGSO-type orbits, where satellites transmit navigation signal in the L5 and S-band. Details of this section are given in Chapter 4.

1.3 Inertial Navigation Overview

The following is a more-or-less heuristic overview of inertial navigation technology. Chapter 3 has the essential technical details about hardware and software used for inertial navigation, and Chapter 11 is about analytical methods for statistical characterization of navigation performance.

1.3.1 History

Although the theoretical foundations for inertial navigation have been around since the time of Isaac Newton (1643–1727), the technology for reducing it to practice would not become available until the twentieth century. This history can be found in the accounts by Draper [22], Gibson [23], Hellman [24], Mackenzie [2], Mueller [18], Wagner [25], and Wrigley [26]. For an account of the related computer hardware and software developments through that period, see McMurran [27].

1.3.1.1 Theoretical Foundations

It has been called "Newtonian navigation" [28] because its theoretical foundations have been known since the time of Newton.[2]

Given the position $\boldsymbol{x}(t_0)$ and velocity $\boldsymbol{v}(t_0)$ of a vehicle at time t_0, and its acceleration $\boldsymbol{a}(s)$ for times $s > t_0$, then its velocity $\boldsymbol{v}(t)$ and position $\boldsymbol{x}(t)$ for all time $t > t_0$ can be defined as

$$\boldsymbol{v}(t) = \boldsymbol{v}(t_0) + \int_{t_0}^{t} \boldsymbol{a}(s)ds \tag{1.1}$$

$$\boldsymbol{x}(t) = \boldsymbol{x}(t_0) + \int_{t_0}^{t} \boldsymbol{v}(s)ds \tag{1.2}$$

It follows that given the initial position $\boldsymbol{x}(t_0)$ and velocity $\boldsymbol{v}(t_0)$ of a vehicle or vessel, its subsequent position depends only on its subsequent accelerations. If these accelerations could be measured and integrated, this would provide a navigation solution.

Indications that this can be done can be found in nature:

1) *Halters* are an extra set of modified wings on some flying insects. During flight, these function as *Coriolis vibratory gyroscopes* (CVGs) to provide rotation rate feedback in attitude control. Their function was not known for a long time because they are too tiny to observe.
2) *Vestibular systems* include ensembles of gyroscopes (*semicircular canals*) and accelerometers (*saccule* and *utricle*) located in the bony mass behind each of your ears. Each of these is a complete 3D *inertial measurement unit* (IMU), used primarily to aid your vision system during rotations of the head but also useful for short-term navigation in total darkness. These have been evolving since the time your ancestors were fish [29].

The development of man-made solutions for inertial navigation came to have the same essential parts, except that they are not biological (yet).

---

2 Newton called derivatives "fluxions" and integrals "fluents" and had his own unique notation. Modern notation used here has evolved from that of Gottfried Wilhelm Leibniz (1646–1716) and others.

1.3.1.2 Development Challenges: Then and Now

The technology of Newton's time was not adequate for practical implementation. What was missing included the following:

1) Methods for measuring three components of the acceleration vector $\boldsymbol{a}(t)$ in Eq. (1.1) with sufficient accuracy for the navigation problem.
2) Methods for maintaining the representation of vectors $\boldsymbol{a}(t)$ and $\boldsymbol{v}(t)$ in a common inertial coordinate frame for integration.
3) Methods for integrating $\boldsymbol{a}(t)$ and $\boldsymbol{v}(t)$ in *real time* (i.e. fast enough for the demands of the application).
4) Methods for initializing $\boldsymbol{v}(t)$ and $\boldsymbol{x}(t)$ in the common inertial coordinate frame.
5) Applications to justify the investments in technology required for developing the necessary solutions. It could not be justified for transportation at the pace of a sailing ship or a horse, but it could (and would) be justified by the military activities of World War II and the Cold War.

These challenges for inertial navigation system (INS) development were met and overcome during the great arms races of the mid-to-late twentieth century, the same time period that gave us GNSS. Like GNSS, INS technology first evolved for military applications and then became available for consumer applications as well. Today, the same concerns that have given us chip-level GNSS receivers are giving us chip-level inertial navigation technology. Chapter 3 describes the principles behind modern-day inertial sensors, often using macro-scale sensor models to illustrate what may be less obvious at the micro-scale. The following subsections provide a more heuristic overview.

1.3.2 Development Results

1.3.2.1 Inertial Sensors

Rotation sensors are used in INSs for keeping track of the directions of the sensitive (input) axes of the acceleration sensors – so that sensed accelerations can be integrated in a fixed coordinate frame. Sensors for measuring rotation or rotation rates are collectively called *gyroscopes,* a term coined by Jean Bernard Léon Foucault (1819–1868). Foucault measured the rotation rate of Earth using two types of gyroscopes:

1) What is now called a *momentum wheel gyroscope* (MWG), which – if no torques are applied – tends to keep its spin axis in an inertially fixed direction while Earth rotates under it.
2) The Foucault pendulum, which might now be called a *vibrating Coriolis gyroscope* (vibrating at about 0.06 Hz). It depends on sinusoidal motion of a proof mass and the Coriolis[3] effect, an important design principle for miniature gyroscopes, as well.

---

3 Named after Gaspard-Gustave de Coriolis (1792–1843), who remodeled Newtonian mechanics in rotating coordinates.

Table 1.1 A sampling of inertial sensor types.

What it measures	Sensor types	Physical phenomena	Implementation methods
Rotation (gyroscope)	Momentum wheel gyroscope (MWG)	Conservation of angular momentum	Angle displacement
			Torque rebalance
	Coriolis vibratory gyroscope (CVG)	Coriolis effect and vibration	Balanced (tuning fork)
			Wine glass resonance
	Optical gyroscope	Sagnac effect	Fiber optic gyroscope (FOG)
			Ring laser gyroscope (RLG)
Acceleration (accelerometer)	Mass spring (fish example)	Stress in support	Piezoresistive
			Piezoelectric
			Surface acoustic wave
			Vibrating wire in tension
	Electromagnetic	Induction	Drag cup
		Electromagnetic force	Force rebalance
	Pendulous integrating gyroscopic accelerometer (PIGA)	Gyroscopic precession	Angular displacement
			Torque rebalance
	Electrostatic	Electrostatic force	Force rebalance

Some of the physical phenomena used in the design of gyroscopes are listed in Table 1.1. Further explanation of the terminology is provided in Chapter 3.

What we call *acceleration sensors* or *accelerometers* actually measure *specific force*, equal to the physical force applied to a mass divided by the mass, solving $f = ma$ for a, given f (the sensor input) and m (a known constant). *Accelerometers do not measure gravitational acceleration*, but can measure the force applied to counter gravity. A spring scale used in a fish market for measuring the mass m of a fish, for example, is actually measuring the spring force f applied to the fish by the spring for countering gravity, solving $f = ma$ for m, given f (spring strain, proportional to force) and a (local gravity being countered).

The essential element of an accelerometers is its *proof mass* m, which is a known mass suspended within an enclosure with limited degrees of travel, with the enclosure affixed to the item to be navigated. However, there are other means for measuring the force required to keep the proof mass centered in its enclosure, some of which are listed in Table 1.1. The terminology used there will be elaborated upon in Chapter 3.

The *relative* input axis directions of all the inertial sensors in an INS is controlled by hard-mounting them all to a common base, forming what is called an *inertial sensor assembly* (ISA) or IMU.[4] This way, all the sensor input axes will have a fixed direction relative to one another. The ISA is essentially the "holy point" of inertial navigation, just as the receiver antenna is the holy point for GNSS navigation. It is the reference point for the navigation solution.

1.3.2.2 Sensor Attitude Control

Methods by which the information from gyroscopes is used for resolving acceleration inputs into suitable coordinates for integration can be divided into two general approaches:

1) *Inertial stabilization of the accelerometer input axes, using the gyroscopes.* The first precise and reliable integrating accelerometer was invented in Germany by Fritz Mueller (1907–2001) in the 1930s [18]. Its proof mass is a MWG mounted such that its center of mass is offset axially from its center of support within its enclosure, causing the gyroscope to precess about the input acceleration axis at an angular rate proportional to specific force applied orthogonal to its spin axis, and accumulating a net precession angle proportional to velocity change in that direction. However, because it contains a MWG, it is also sensitive to rotation. That error source could be eliminated by maintaining Mueller's gyroscope in a fixed inertial direction, and the German visionary Johannes Boykow (1879–1935) played a significant role in developing an approach by which the outputs of rotation sensors could be fed back in servo control loops to gimbals nulling the inertial rotations of the ISA [18]. The result is now called an *inertially stabilized platform* or *inertial platform*, and the resulting INS is called "gimbaled."
2) *Strapdown systems* use software to replace hardware (gimbals), processing the gyro outputs to maintain the coordinate transformation between accelerometer-fixed coordinates and (essentially) inertial platform coordinates. The accelerometer outputs can then be transformed to what would have been inertial platform coordinates and processed just as they had been with a gimbaled system – without requiring an inertial platform. Getting rid of gimbals reduces INS weight and cost, but it also increases computer requirements. As a consequence, strapdown system development had to

4 These terms are often used interchangeably, although an IMU usually refers to a functioning ISA capable of measuring rotation and acceleration in three dimensions, whereas an ISA may not necessarily contain a complete sensor suite.

await the essential computer development, which did not happen until the mid-twentieth century. However, as silicon-based technologies advanced to produce chip-based sensors and computers, only high-end inertial systems would require gimbals.

1.3.2.3 Initialization

The initial solution to the inertial navigation problem is the starting position and orientation of the ISA. The starting position must usually be entered from other sources (including GNSS). If the ISA is sufficiently stationary with respect to the Earth, its starting orientation with respect to local vertical can be determined from the accelerometer outputs. If it is not at the poles and its rotation sensors are sufficiently accurate, its orientation with respect to North, East, South, and West can be determined by sensing the direction of Earth rotation. Otherwise, if the local Earth magnetic field is not too close to being vertical, it can be oriented using magnetometers. Other alternatives are mentioned in Chapter 3.

1.3.2.4 Integrating Acceleration and Velocity

Integration of gyro outputs (to maintain the attitude solution) and accelerometer outputs (to maintain the velocity and position solution) is done digitally. This is relatively straightforward for integrating accelerations and velocities in inertially stabilized Cartesian coordinates for gimbaled systems, and gimbal servo control takes care of the attitude solution. Otherwise, more sophisticated integration methods are required for maintaining the required coordinate transformation matrices in strapdown systems.

Because sensor output errors are being integrated, the resulting navigation solution errors tend to accumulate over time. Performance of INSs is generally specified in terms of navigation error growth rates.

1.3.2.5 Accounting for Gravity

Because accelerometers cannot measure gravitational accelerations, these must be calculated using the position and attitude solutions and added to the sensed acceleration. Otherwise, an INS resting on the surface of the Earth would measure only the upward support force countering gravity and think it was being accelerated upward.

The need to calculate gravity causes vertical dynamic instability of the navigation solution, because gravitational acceleration decreases with altitude (Newton's law of universal gravitation). Therefore, a navigation error in altitude leads to an error in calculating the unsensed downward gravitational acceleration. Upward navigational errors cause decreased downward gravity estimates, leading to a net upward acceleration error and to exponential growth of vertical velocity error and even faster growth in altitude error. This may not be a problem for ships at sea level, but it is serious enough otherwise that inertial navigators require auxiliary sensors (e.g. altimeters or GNSS) to stabilize altitude errors.

Gravity modeling also alters the dynamics of horizontal errors in position and velocity, but this effect is not unstable. It causes horizontal errors to behave like a pendulum with a period of about 84.4 minutes, something called the Schuler period.[5] When Coriolis force is taken into account, this error behaves like a Foucault pendulum with that period.

1.4 GNSS/INS Integration Overview

1.4.1 The Role of Kalman Filtering

The Kalman filter has been called "navigation's integration workhorse," [30] for the essential role it has played in navigation, and especially for integrating different navigation modes. Ever since its introduction in 1960 [31], the Kalman filter has played a major role in the design and implementation of most new navigation systems, as a statistically optimal method for estimating position using noisy measurements. Because the filter also produces an estimate of its own accuracy, it has also become an essential part of a methodology for the optimal design of navigation systems. The Kalman filter has been essential for the design and implementation of every GNSS.

Using the Kalman filter, navigation systems designers have been able to exploit a powerful synergism between GNSSs and INSs, which is possible because they have very complementary error characteristics:

- Short-term position errors from the INS are relatively small, but they degrade significantly over time.
- GNSS position accuracies, on the other hand, are not as good over the short term, but they do not degrade with time.

The Kalman filter takes advantage of these characteristics to provide a common, integrated navigation implementation with performance superior to that of either subsystem (GNSS or INS). By using statistical information about the errors in both systems, it is able to combine a system with tens of meters position uncertainty (GNSS) with another system whose position uncertainty degrades at kilometers per hour (INS) and achieve bounded position uncertainties in the order of centimeters (with differential GNSS) to meters.

5 Named after Maximilian Schuler (1882–1972), a German engineer who discovered the phenomenon while analyzing the error characteristics of his cousin Hermann Anschütz-Kaempfe's gyrocompass.

1.4.2 Implementation

The Kalman filter solves for the solution with the least mean-squared error by using data-weighting proportional to statistical information content (the inverse of uncertainty) in the measured data. It combines GNSS and INS information to the following:

1) Track drifting parameters of the sensors in the INS, so that INS performance does not degrade with time when GNSS is available.
2) Improve overall performance even when there are insufficient satellite signals for obtaining a complete GNSS solution.
3) Allow the INS to navigate with improved initial error whenever GNSS signals become unavailable.
4) Improve GNSS signal reacquisition when GNSS signals become available again by providing better navigation solutions (based on INS data).
5) Use acceleration and attitude rate information from the INS for reducing the signal phase-tracking filter lags in the GNSS receiver, which can significantly improve GNSS reliability during periods of high maneuvering, jamming, or reduced signal availability.

The more intimate levels of GNSS/INS integration necessarily penetrate deeply into each of the subsystems, in that it makes use of partial results that are not ordinarily accessible to users. To take full advantage of the offered integration potential, we must delve into technical details of the designs of both types of systems.

Problems

1.1 How many satellites and orbit planes exist for GPS, GLONASS, and Galileo? What are the respective orbit plane inclinations?

1.2 List the differences in signal characteristics between GPS, GLONASS, and Galileo.

1.3 What are the reference points for GNSS and INS navigators? That is, when one of these produces a position estimate, what part of the respective system is that the position of?

1.4 Would an error-free accelerometer attached to a GNSS satellite orbiting the Earth have any output? Why or why not?

1.5 Does the same fish, weighed on the same spring scale, appear to weigh more (as indicated by the stretching of the spring) at sea level on the equator or at the North Pole? Justify your answer.

References

1 Biezad, D.J. (1999). *Integrated Navigation and Guidance Systems.* New York: American Institute of Aeronautics and Astronautics.
2 Mackenzie, D. (2001). *Inventing Accuracy: A Historical Sociology of Nuclear Missile Guidance.* Cambridge, MA: MIT Press.
3 Guier, W.H. and Weiffenbach, G.C. (1997). Genesis of satellite navigation. *Johns Hopkins APL Technical Digest* 18 (2): 178–181.
4 Stansell, T. (1978). *The Transit Navigation Satellite System: Status, Theory, Performance, Applications.* Magnavox.
5 Global Positioning System (1999). *Selected Papers on Satellite Based Augmentation Systems (SBASs) ("Redbook"),* vol. VI. Alexandria, VA: ION.
6 Herring, T.A. (1996). The global positioning system. *Scientific American* 274 (2): 44–50.
7 Hofmann-Wellenhof, B., Lichtenegger, H., and Collins, J. (1997). *GPS: Theory and Practice.* Vienna: Springer-Verlag.
8 Institute of Navigation (1980). *Monographs of the Global Positioning System: Papers Published in Navigation ("Redbook"),* vol. I. Alexandria, VA: ION.
9 Institute of Navigation (1984). *Monographs of the Global Positioning System: Papers Published in Navigation ("Redbook"),* vol. II. Alexandria, VA: ION.
10 Institute of Navigation (1986). *Monographs of the Global Positioning System: Papers Published in Navigation ("Redbook"),* with Overview by R. Kalafus, vol. III. Alexandria, VA: ION.
11 Institute of Navigation (1993). *Monographs of the Global Positioning System: Papers Published in Navigation ("Redbook"),* with Overview by R. Hatch, vol. IV. Alexandria, VA: ION.
12 Institute of Navigation (1998). *Monographs of the Global Positioning System: Papers Published in Navigation ("Redbook"),* vol. V. Alexandria, VA: ION.
13 Logsdon, T. (1992). *The NAVSTAR Global Positioning System.* New York: Van Nostrand Reinhold.
14 Parkinson, B.W. and Spilker, J.J. Jr., (eds.) (1996). *Global Positioning System: Theory and Applications,* Progress in Astronautics and Aeronautics, vol. 1. Washington, DC: American Institute of Aeronautics and Astronautics.
15 Parkinson, B.W. and Spilker, J.J. Jr., (eds.) (1996). *Global Positioning System: Theory and Applications,* Progress in Astronautics and Aeronautics, vol. 2. Washington, DC: American Institute of Aeronautics and Astronautics.
16 Parkinson, B.W., O'Connor, M.L., and Fitzgibbon, K.T. (1995). Aircraft automatic approach and landing using GPS. In: *Global Positioning System: Theory & Applications,* Progress in Astronautics and Aeronautics, Chapter 14, vols. II and 164 (ed. B.W. Parkinson, J.J. Spilker Jr., and editor-in-chief P. Zarchan), 397–425. Washington, DC: American Institute of Aeronautics and Astronautics.

17 Rockwell International Corporation, Satellite Systems Division, Revision B (1991). GPS Interface Control Document ICD-GPS-200, July 3, 1991.
18 Mueller, F.K. (1985). A history of inertial navigation. *Journal of the British Interplanetary Society* 38: 180–192.
19 Kayton, M. and Fried, W.L. (1997). *Avionics Navigation Systems*, 2e. New York: Wiley.
20 Leick, A. (1995). *GPS: Satellite Surveying*, 2e, 534–537. New York: Wiley.
21 Janky, J.M. (1997). Clandestine location reporting by a missing vehicle. US Patent 5, 629, 693, 13 May 1997.
22 Draper, C.S. (1981). Origins of inertial navigation. *AIAA Journal of Guidance and Control* 4 (5): 449–456.
23 Gibson, J.N. (1996). *The Navaho Missile Project: The Story of the "Know-How" Missile of American Rocketry*. Atglen, PA: Schiffer Military/Aviation History.
24 Hellman, H. (1962). The development of inertial navigation. *NAVIGATION, Journal of the Institute of Navigation* 9 (2): 82–94.
25 Wagner, J.F. (2005). From Bohnenberger's machine to integrated navigation systems, 200 years of inertial navigation. In: *Photogrammetric Week 05* (ed. D. Fritsch), 123–134. Heidelberg: Wichmann Verlag.
26 Wrigley, W. (1977). History of inertial navigation. *NAVIGATION, Journal of the Institute of Navigation* 24: 1–6.
27 McMurran, M.W. (2008). *Achieving Accuracy: A Legacy of Computers and Missiles*. Bloomington, IN: Xlibris.
28 Slater, J.M. (1967). *Newtonian Navigation*, 2e. Anaheim, CA: Autonetics Division of Rockwell International.
29 Shubin, N. (2009). *Your Inner Fish: A Journey into the 3.5-Billion-Year History of the Human Body*. New York: Random House.
30 Levy, J.J. (1997). The Kalman filter: navigation's integration workhorse. *GPS World* (September 1997), pp. 65–71.
31 Kalman, R.E. (1960). A new approach to linear filtering and prediction problems. *ASME Transactions, Series D: Journal of Basic Engineering* 82: 35–45.

2

Fundamentals of Satellite Navigation Systems

2.1 Chapter Focus

This chapter presents a concise system-level overview of constellations, operational configurations, and signaling characteristics of a global navigation satellite system (GNSS). Various GNSSs have been developed over the decades to fulfill user position, velocity, and timing (PVT) application requirements. More detailed coverage of the essential technical details of various GNSSs is presented in Chapters 4–9.

2.2 Satellite Navigation Systems Considerations

Since the early days of satellite-based navigation systems, they have operated independently and have been integrated with other radionavigation systems, as well as internal systems to increase the robustness of the PVT solution. While radionavigation systems, including GNSS, provide good accuracy, these types of systems may be vulnerable to interference and periods of unavailability. On the other hand, while inertial navigation systems are self-contained, their accuracy degrades over time. The integration of these two complementary sensor systems enables the position navigation timing (PNT) engineer to implement a robust PVT solution for a variety of applications.

2.2.1 Systems Other than GNSS

In many parts of the Globe, GNSS signals have replaced other terrestrial radionavigation systems, such as long-range navigation-version C (LORAN-C) signals produced by three or more LORAN signal sources positioned at fixed, known terrestrial locations for outside-the-building location determination. A LORAN-C system relies on a plurality of ground-based signal towers, preferably spaced 100–300 km apart, that transmit distinguishable electromagnetic

Global Navigation Satellite Systems, Inertial Navigation, and Integration,
Fourth Edition. Mohinder S. Grewal, Angus P. Andrews, and Chris G. Bartone.

Companion website: www.wiley.com/go/grewal/gnss

signals that are received and processed by an LORAN receiver system. A representative LORAN-C system is discussed in the US DOT LORAN-C User Handbook [1]. LORAN-C signals use carrier frequencies of the order of 100 kHz and have maximum reception distances of hundreds of kilometers. An enhanced version of LORAN (eLORAN) provides ranging from each independent station and enhanced data. The low frequency and high transmitter power in LORAN provides for a diverse PNT capability. Although LORAN has been decommissioned in the United States, other counties and regions continue and operate LORAN for an alternative to other radionavigation systems.

There are other ground-based radiowave signal systems suitable for PVT applications. These include Multi-lateral DME, TACAN, US Air Force Joint Tactical Information Distribution System Relative Navigation (JTIDS Relnav), US Army Position Location and Reporting System (PLRS) (see summaries in [2], pp. 6, 7 and 35–60), and cellular radiolocation services.

Cellular radiolocation services provide for enhanced 911 (E911) and location-based services (LBSs). These PVT solutions may be based on GNSS or range and bearing determination using the actual cell towers. Two-way ranging (e.g. round trip timing [RTT]), and bearing determination may be determined by antenna sector or advanced phase angle-of-arrival (AOA) determination. Several standards in the 3GPP and Userplane community govern the implementation (http://www.3gpp.org/specifications and https://www.omaspecworks.org/).

2.2.2 Comparison Criteria

The following criteria may be used in selecting navigation systems appropriate for a given application system:

1. Navigation method(s) used
2. System reliability/integrity
3. Navigational accuracy
4. Region(s) of coverage/availability
5. Required transmission frequencies and bands of operation
6. Navigation fix update rate
7. User set cost
8. Status of system development and readiness.

2.3 Satellite Navigation

Early satellite-based navigation systems c. 1960–1970, such as TRANSIT, utilized a low Earth orbit (LEO) to obtain sufficient Doppler on the carrier phase for positioning. In 1973, satellite navigation in the form of the global position system (GPS) took hold with the formulation of the GPS Program Office, which helps consolidate knowledge from the TRANSIT program, the Navy's Timation Satellites, and the USAF Project 621B satellite pseudoranging program.

Today, GNSSs use medium Earth orbit (MEO) for good visibility/availably of the satellite and still provide moderate Doppler for PVT determination. Some GNSSs also supplement the GNSS constellation with geostationary (GEO), geosynchronous, or inclined geosynchronous (IGSO). These GEO and/or IGSO satellites may provide additional ranging and a data link augmentation for the GNSS constellation.

The various governing bodies for satellite navigation publish technical interface control documents and specification on their respective GNSS. For GPS, the US Government publishes various Interface Specifications and performance specifications at GPS.gov [Ref. [2] in Chapter 4]. For the Russian GLONASS, a multitude of performance and system characterizations are published by the Russian Information and Analysis Center for Positioning, Navigation and Timing. For the European Galileo GNSS, these types of documents are published by the European Global Navigation Satellite Systems Agency [Ref. [26] in Chapter 4]. For the Chinese BeiDou Navigation Satellite System, information and official documents are published (Ref. [27] in Chapter 4). Other regional type satellite navigation systems cover a specific region of Globe such as the Japanese Quazi-Satellite System (QZSS) (Ref. [29] in Chapter 4) and the Indian Regional Navigation Satellite System (IRNSS): NAVIC (Ref. [30] in Chapter 4) provide service and/or augmentation to a specific geographical region of the Globe.

2.3.1 GNSS Orbits

Figure 2.1 illustrates the key parameters that may be used to describe a GNSS satellite that orbits the Earth.

Various configurations may be used to complete a GNSS constellation. Various design parameters may be considered regarding the number of satellites, their location in the constellation, location on ground infrastructure components, and user segment service area.

GPS satellites occupy six orbital planes in an MEO that are inclined 55° from the equatorial plane, as illustrated in Figure 2.2. Each of the six orbit planes in Figure 2.2 contains four or more satellites. The European Galileo and Chinese BeiDou use three orbital planes separated in longitude by 120°. Galileo has plans for 10 MEO satellites in each orbital plane. Glonass also used three orbital planes while maintaining a nominal 24 satellite constellation. GNSS augmentation or regional satellite-based navigation systems are most often placed in GEO or IGSO over the region to be services; these satellites will have an orbital period of 24 hours.

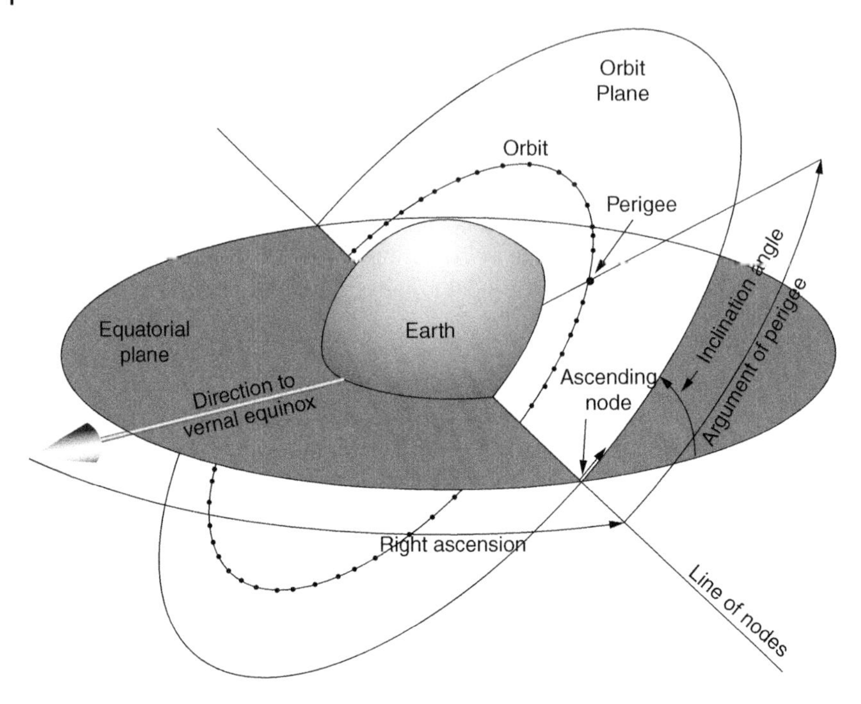

Figure 2.1 Parameters defining satellite orbit geometry.

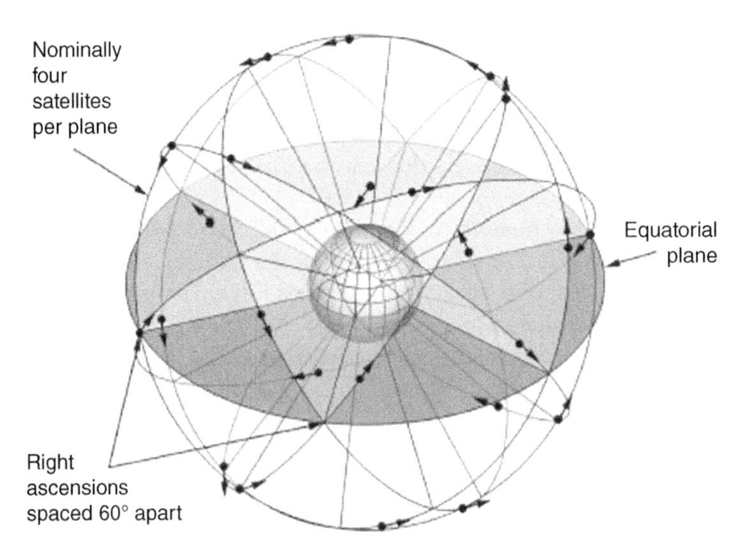

Figure 2.2 Six GPS orbit planes inclined 55° from the equatorial plane.

Regardless of the type of radionavigation (satellite-based or terrestrial-based) system, most often the solution method requires the determination of a unknown user antenna location with known transmitter locations and measured range (or pseudorange) observations.

2.3.2 Navigation Solution (Two-Dimensional Example)

Antenna location in two dimensions can be calculated by using range measurements [3].

2.3.2.1 Symmetric Solution Using Two Transmitters on Land

In this case, the receiver and two transmitters are located in the same plane, as shown in Figure 2.3, with known positions of the two transmitters: x_1, y_1 and x_2, y_2. Ranges R_1 and R_2 from the two transmitters to the user position are calculated as

$$R_1 = c\Delta T_1 \tag{2.1}$$

$$R_2 = c\Delta T_2 \tag{2.2}$$

where

c = speed of light (0.299 792 458 m/ns)

ΔT_1 = time taken for the radiowave to travel from transmitter 1 to the user (ns)

ΔT_2 = time taken for the radiowave to travel from transmitter 2 to the user (ns)

X, Y = unknown user position to be solved for (m)

The range to each transmitter can be written as

$$R_1 = [(X - x_1)^2 + (Y - y_1)^2]^{1/2} \tag{2.3}$$

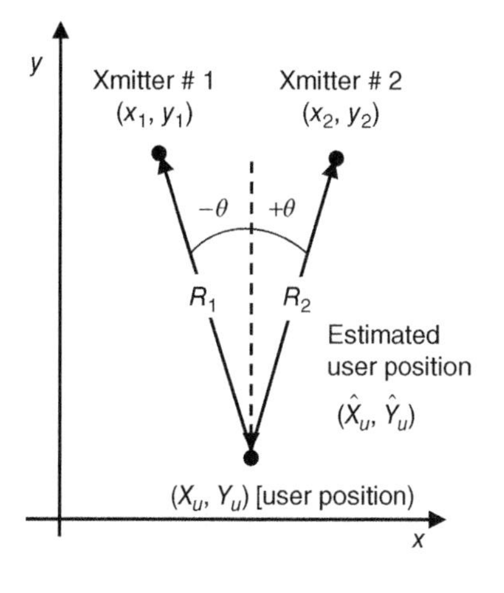

Figure 2.3 Two transmitters with known 2D positions.

$$R_2 = [(X - x_2)^2 + (Y - y_2)^2]^{1/2} \tag{2.4}$$

Expanding R_1 and R_2 in Taylor series expansion with small perturbation in X by Δx and Y by Δy yields

$$\Delta R_1 = \frac{\partial R_1}{\partial X}\Delta x + \frac{\partial R_1}{\partial Y}\Delta y + u_1 \tag{2.5}$$

$$\Delta R_2 = \frac{\partial R_2}{\partial X}\Delta x + \frac{\partial R_2}{\partial Y}\Delta y + u_2 \tag{2.6}$$

where u_1 and u_2 are higher-order terms. The derivatives of Eqs. (2.3) and (2.4) with respect to X, Y are substituted into Eqs. (2.5) and (2.6), respectively.

Thus, for the symmetric case, we obtain

$$\begin{aligned}\Delta R_1 &= \frac{X - x_1}{[(X - x_1)^2 + (Y - y_1)^2]^{\frac{1}{2}}}\Delta x + \frac{Y - y_1}{[(X - x_1)^2 + (Y - y_1)^2]^{\frac{1}{2}}}\Delta y + u_1 \\ &= \sin\theta\Delta x + \cos\theta\Delta y + u_1 \end{aligned} \tag{2.7}$$

$$\Delta R_2 = -\sin\theta\Delta x + \cos\theta\Delta y + u_2 \tag{2.8}$$

To obtain the least-squares estimate of (X, Y), we need to minimize the quantity

$$J = u_1^2 + u_2^2 \tag{2.9}$$

which is

$$J = \left(\underbrace{\Delta R_1 - \sin\theta\Delta x - \cos\theta\Delta y}_{u_1}\right)^2 + \left(\underbrace{\Delta R_2 + \sin\theta\Delta x - \cos\theta\Delta y}_{u_2}\right)^2 \tag{2.10}$$

The solution for the minimum can be found by setting $\partial J/\partial\Delta x = 0 = \partial J/\partial\Delta y$, then solving for Δx and Δy:

$$\begin{aligned} 0 &= \frac{\partial J}{\partial \Delta x} \\ &= 2(\Delta R_1 - \sin\theta\Delta x - \cos\theta\Delta y)(-\sin\theta) \\ &\quad + 2(\Delta R_2 + \sin\theta\Delta x - \cos\theta\Delta y)(\sin\theta) \end{aligned} \tag{2.11}$$

$$= \Delta R_2 - \Delta R_1 + 2\sin\theta\Delta x \tag{2.12}$$

with solution

$$\Delta x = \frac{\Delta R_1 - \Delta R_2}{2\sin\theta} \tag{2.13}$$

The solution for Δy may be found in similar fashion as

$$\Delta y = \frac{\Delta R_1 + \Delta R_2}{2\cos\theta} \tag{2.14}$$

2.3.2.2 Navigation Solution Procedure

Transmitter positions x_1, y_1, x_2, y_2 are known; the transmitter locations are typically provided by the GNSS satellites or known to be at a fixed surveyed location for a terrestrial source. Signal travel times ΔT_1, ΔT_2 are measured by the system. Initially, the user positions $\widehat{X}_u$, $\widehat{Y}_u$ are assumed. Set position coordinates X, Y equal to their initial estimates:

$$X = \widehat{X}_u, Y = \widehat{Y}_u$$

Compute the range errors:

$$\Delta R_2 = \overbrace{[(\widehat{X}_u - x_1)^2 + (\widehat{Y}_u - y_1)^2]^{1/2}}^{\text{Geometrical ranges}} - \overbrace{c\Delta T_1}^{\text{Measured pseudoranges}} \tag{2.15}$$

$$\Delta R_2 = [(\widehat{X}_u - x_1)^2 + (\widehat{Y}_u - y_1)^2]^{1/2} - c\Delta T_1 \tag{2.16}$$

Compute the θ angle (see Figure 2.3):

$$\begin{aligned}\theta &= \tan^{-1}\frac{\widehat{X}_u - x_1}{\widehat{Y}_u - y_1}\\ &= \sin^{-1}\frac{\widehat{X}_u - x_1}{\sqrt{(\widehat{X}_u - x_1)^2 + (\widehat{Y}_u - y_1)^2}}\end{aligned} \tag{2.17}$$

Compute update to user position:

$$\Delta x = \frac{1}{2\sin\theta}(\Delta R_1 - \Delta R_2) \tag{2.18}$$

$$\Delta y = \frac{1}{2\cos\theta}(\Delta R_1 + \Delta R_2) \tag{2.19}$$

Compute a new estimate of position using the update:

$$\widehat{X}_u = \widehat{X}_u + \Delta x, \quad \widehat{Y}_u = \widehat{Y}_u + \Delta y \tag{2.20}$$

Continue to compute θ, ΔR_1, and ΔR_2 from these equations with new values of $\widehat{X}_u$ and $\widehat{Y}_u$. Iterate Eqs. (2.15)–(2.20) and stop when Δx and Δy become less than the desired accuracy:

$$\begin{aligned}\Delta X_{\text{best}} &= \overbrace{\frac{1}{2\sin\theta}(\Delta R_1 - \Delta R_2)}^{\text{Correction equations}}, \quad \overbrace{X_{\text{new}} = X_{\text{old}} + X_{\text{best}}}^{\text{Iteration equations}}\\ \Delta Y_{\text{best}} &= \frac{1}{2\sin\theta}(\Delta R_1 - \Delta R_2), \quad Y_{\text{new}} = Y_{\text{old}} + Y_{\text{best}}\end{aligned}$$

2.3.3 User Solution and Dilution of Precision (DOP)

Just as in a land-based system, better accuracy is obtained by using reference points (i.e. ranging sources) well separated in space. For example, the range measurements made to four reference points clustered together will yield nearly equal values. Position calculations involve range differences, and where the ranges are nearly equal, small relative errors are greatly magnified in the difference. This effect, brought about as a result of satellite geometry, is known as *dilution of precision* (DOP). This means that range errors that occur from other independent effects such as multipath, atmospheric delays, and/or satellite clock errors are also magnified by the geometric effect.

The DOP can be interpreted as the dilution of the precision from the measurement (i.e. range) domain to the solution (i.e. PVT) domain.

The observation measurement equations in three dimensions for each satellite with known coordinates (x_i, y_i, z_i) and unknown user coordinates (X, Y, Z) are given by

$$Z_\rho^i = \rho^i = \sqrt{(x_i - X)^2 + (y_i - Y)^2 + (z_i - Z)^2} + C_b \tag{2.21}$$

where ρ^i is the pseudorange to the ith satellite and C_b is the residual satellite clock bias and receiver clock bias, where an estimate of the satellite clock error from the GNSS ground control segment has been removed from C_b.

These are nonlinear equations that can be linearized using the Taylor series (see, e.g., chapter 5 of Ref. [4]). The satellite positions may be converted to east–north–up (ENU) from Earth-centered, Earth-fixed (ECEF) coordinates (see Appendix B).

Let the vector of ranges be $Z_\rho = \mathbf{h}(\mathbf{x})$, a nonlinear function $\mathbf{h}(\mathbf{x})$ of the four-dimensional vector $\mathbf{x}$ representing user solution for position and receiver clock bias and expand the left-hand side of this equation in a Taylor series about some nominal solution $\mathbf{x}^{\text{nom}}$ for the unknown vector

$$x = [X, Y, Z, C_b]^T \tag{2.22}$$

of variables,

X = east component of the user's antenna location (m)

Y = north component of the user's antenna location (m)

Z = upward vertical component of the user's antenna location (m)

C_b = receiver clock bias (m)

for which

$$\begin{aligned} Z_\rho &= \mathbf{h}(\mathbf{x}) = \mathbf{h}(\mathbf{x}^{\text{nom}}) + \left.\frac{\partial \mathbf{h}(\mathbf{x})}{\partial \mathbf{x}}\right|_{\mathbf{x}=\mathbf{x}^{\text{nom}}} \delta\mathbf{x} + \text{ HOT} \\ \delta\mathbf{x} &= \mathbf{x} - \mathbf{x}^{\text{nom}}, \quad \delta Z_\rho = \mathbf{h}(\mathbf{x}) - \mathbf{h}(\mathbf{x}^{\text{nom}}) \end{aligned} \tag{2.23}$$

where HOT stands for "higher-order terms."

These equations become

$$Z_\rho = \frac{\partial \mathbf{h}(\mathbf{x})}{\partial \mathbf{x}} \Big|_{\mathbf{x}=\mathbf{x}^{\text{nom}}} \partial \mathbf{x} = H^{[1]} \delta \mathbf{x}$$

$$\delta \mathbf{x} = X - X_{\text{nom}}, \quad \delta \mathbf{y} = Y - Y_{\text{nom}}, \quad \delta \mathbf{z} = Z - Z_{\text{nom}} \tag{2.24}$$

where $H^{[1]}$ is the first-order term in the Taylor series expansion:

$$\begin{aligned} \delta Z_\rho &= \rho_r(X, Y, Z) - \rho_r(X_{\text{nom}}, Y_{\text{nom}}, Z_{\text{nom}}) \\ &\approx \underbrace{\frac{\partial \rho_r}{\partial X}\Big|_{X_{\text{nom}}, Y_{\text{nom}}, Z_{\text{nom}}}}_{H^{[1]}} \delta \mathbf{x} + v_\rho \end{aligned} \tag{2.25}$$

for v_ρ is the noise in receiver measurements. This vector equation can be written in scalar form where i is the satellite number as

$$\left.\begin{aligned} \frac{\partial \rho_r^i}{\partial X} &= \frac{-(x_i - X)}{\sqrt{(x_i - X)^2 + (y_i - Y)^2 + (z_i - Z)^2}}\Bigg|_{X_{\text{nom}}, Y_{\text{nom}}, Z_{\text{nom}}} \\ &= \frac{-(x_i - X_{\text{nom}})}{\sqrt{(x_i - X_{\text{nom}})^2 + (y_i - Y_{\text{nom}})^2 + (z_i - Z_{\text{nom}})^2}} \\ \frac{\partial \rho_r^i}{\partial Y} &= \frac{-(y_i - Y_{\text{nom}})}{\sqrt{(x_i - X_{\text{nom}})^2 + (y_i - Y_{\text{nom}})^2 + (z_i - Z_{\text{nom}})^2}} \\ \frac{\partial \rho_r^i}{\partial Z} &= \frac{-(z_i - Z_{\text{nom}})}{\sqrt{(x_i - X_{\text{nom}})^2 + (y_i - Y_{\text{nom}})^2 + (z_i - Z_{\text{nom}})^2}} \end{aligned}\right\} \tag{2.26a}$$

for $i = 1, 2, 3, 4$ (i.e. four satellites).

We can combine Eqs. (2.25) and (2.26a) into the matrix equation with measurements as

$$\underbrace{\begin{bmatrix} \partial z_\rho^1 \\ \partial z_\rho^2 \\ \partial z_\rho^3 \\ \partial z_\rho^4 \end{bmatrix}}_{4\times 1} = \underbrace{\begin{bmatrix} \frac{\partial \rho_r^1}{\partial X} & \frac{\partial \rho_r^1}{\partial Y} & \frac{\partial \rho_r^1}{\partial Z} & 1 \\ \frac{\partial \rho_r^2}{\partial X} & \frac{\partial \rho_r^2}{\partial Y} & \frac{\partial \rho_r^2}{\partial Z} & 1 \\ \frac{\partial \rho_r^3}{\partial X} & \frac{\partial \rho_r^3}{\partial Y} & \frac{\partial \rho_r^3}{\partial Z} & 1 \\ \frac{\partial \rho_r^4}{\partial X} & \frac{\partial \rho_r^4}{\partial Y} & \frac{\partial \rho_r^4}{\partial Z} & 1 \end{bmatrix}}_{4\times 4} \underbrace{\begin{bmatrix} \delta_x \\ \delta_y \\ \delta_z \\ C_b \end{bmatrix}}_{4\times 1} + \underbrace{\begin{bmatrix} v_\rho^1 \\ v_\rho^2 \\ v_\rho^3 \\ v_\rho^4 \end{bmatrix}}_{4\times 1} \tag{2.26b}$$

which we can write in symbolic form as

$$\overbrace{\delta Z_\rho}^{4\times1} = \overbrace{H^{[1]}}^{4\times4} \overbrace{\delta \mathbf{x}}^{4\times1} + \overbrace{v_\rho}^{4\times1} \tag{2.27}$$

(see table 5.3 in Ref. [4]).

To calculate $H^{[1]}$, one needs satellite positions and the nominal value of the user's position in ENU coordinate frames.

To calculate the geometric dilution of precision (GDOP) (approximately), we obtain

$$\overbrace{\delta Z_\rho}^{4\times1} = \overbrace{H^{[1]}}^{4\times1} \overbrace{\delta \mathbf{x}}^{4\times1} \tag{2.28}$$

Known are δZ_ρ and $H^{[1]}$ from the pseudorange, satellite position, and nominal value of the user's position. The correction $\delta \mathbf{x}$ is the unknown vector.

If we premultiply both sides of Eq. (2.28) by $H^{[1]T}$, the result will be

$$H^{[1]T}\delta Z_\rho = \underbrace{\overbrace{H^{[1]T}}^{4\times4} \overbrace{H^{[1]}}^{4\times4}}_{4\times4} \delta \mathbf{x} \tag{2.29}$$

Then, we premultiply Eq. (2.29) by $(H^{[1]T}\, H^{[1]})^{-1}$:

$$\delta \mathbf{x} = (H^{[1]T} H^{[1]})^{-1} H^{[1]T} \delta Z_\rho \tag{2.30}$$

If $\delta \mathbf{x}$ and δZ_ρ are assumed random with zero mean, the error covariance (E = expected value)

$$\begin{aligned} E\langle(\delta \mathbf{x})(\delta \mathbf{x})^T\rangle &= E\langle(H^{[1]T} H^{[1]})^{-1} H^{[1]T} \delta Z_\rho [(H^{[1]T} H^{[1]})^{-1} H^{[1]T} \delta Z_\rho]^T\rangle \\ &= (H^{[1]T} H^{[1]})^{-1} H^{[1]T} \underbrace{E\langle \delta Z_\rho \delta Z_\rho^T\rangle} H^{[1]} (H^{[1]T} H^{[1]})^{-1} \end{aligned} \tag{2.31}$$

The pseudorange measurement covariance is assumed uncorrelated satellite to satellite with variance σ^2:

$$E\langle \delta Z_\rho \delta Z_\rho^T\rangle = \sigma^2 \mathbf{I} \tag{2.32}$$

Substituting Eq. (2.32) into Eq. (2.31) gives

$$\begin{aligned} E[\delta \mathbf{x}(\delta \mathbf{x})^T] &= \sigma^2 (H^{[1]T} H^{[1]})^{-1} \underbrace{(H^{[1]T} H^{[1]})(H^{[1]T} H^{[1]})^{-1}}_{\mathbf{I}} \\ &= \sigma^2 (H^{[1]T} H^{[1]})^{-1} \end{aligned} \tag{2.33}$$

for

$$\overbrace{\delta\mathbf{x}}^{4\times 1} = \begin{bmatrix} \Delta E \\ \Delta N \\ \Delta U \\ C_{\mathrm{b}} \end{bmatrix}$$

and

$$\left.\begin{aligned} \Delta E &= \text{east error} \\ \Delta N &= \text{north error} \\ \Delta U &= \text{up error} \end{aligned}\right\} \begin{pmatrix} \text{locally} \\ \text{level} \\ \text{coordinate} \\ \text{frame} \end{pmatrix}$$

and the covariance matrix becomes

$$E[\delta\mathbf{x}(\delta\mathbf{x})^T] = \begin{bmatrix} E\langle\Delta E^2\rangle & E\langle\Delta E\Delta N\rangle & E\langle\Delta E\Delta U\rangle & E\langle\Delta E\Delta C_{\mathrm{b}}\rangle \\ E\langle\Delta N\Delta E\rangle & E\langle\Delta N^2\rangle & E\langle\Delta N\Delta U\rangle & E\langle\Delta N\Delta C_{\mathrm{b}}\rangle \\ E\langle\Delta U\Delta E\rangle & E\langle\Delta U\Delta N\rangle & E\langle\Delta U^2\rangle & E\langle\Delta U\Delta C_{\mathrm{b}}\rangle \\ E\langle\Delta C_{\mathrm{b}}\Delta E\rangle & E\langle\Delta C_{\mathrm{b}}\Delta N\rangle & E\langle\Delta C_{\mathrm{b}}\Delta U\rangle & E\langle C_{\mathrm{b}}^2\rangle \end{bmatrix} \tag{2.34}$$

We are principally interested in the diagonal elements of

$$(H^{[1]T}H^{[1]})^{-1} = \begin{bmatrix} A_{11} & A_{12} & A_{13} & A_{14} \\ A_{21} & A_{22} & A_{23} & A_{24} \\ A_{31} & A_{32} & A_{33} & A_{34} \\ A_{41} & A_{42} & A_{43} & A_{44} \end{bmatrix} \tag{2.35}$$

that represent the DOP of range measurement error to the user solution error (see Figure 2.4):

$$\begin{aligned} \text{Geometric DOP(GDOP)} &= \sqrt{A_{11} + A_{22} + A_{33} + A_{44}} \\ \text{Position DOP(PDOP)} &= \sqrt{A_{11} + A_{22} + A_{33}} \\ \text{Horizontal DOP(HDOP)} &= \sqrt{A_{11} + A_{22}} \\ \text{Vertical DOP(VDOP)} &= \sqrt{A_{33}} \\ \text{Time DOP(TDOP)} &= \sqrt{A_{44}} \end{aligned}$$

Hence, all DOPs represent the sensitivities of user solution error to pseudorange errors. Figure 2.4 illustrates the relationship between the various DOP terms.

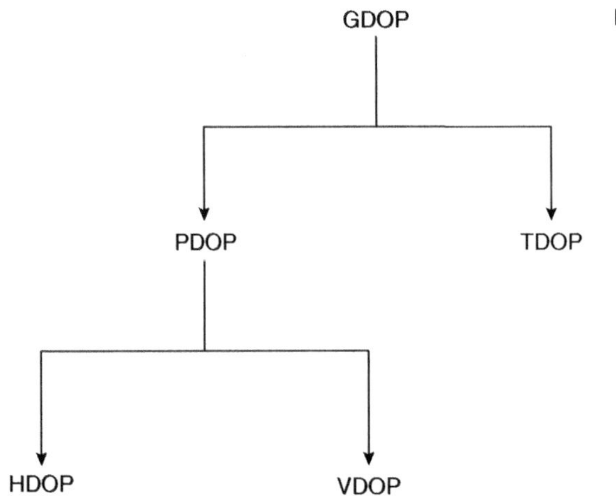

Figure 2.4 DOP hierarchy.

2.3.4 Example Calculation of DOPs

2.3.4.1 Four Satellites

For simplicity, consider four satellite measurements. The best accuracy is found with three satellites equally spaced on the horizon, at minimum elevation angle, with the fourth satellite directly overhead, as listed in Table 2.1.

The diagonal of the unscaled covariance matrix $(H^{[1]T}\ H^{[1]})^{-1}$ then has the terms

$$\begin{bmatrix} (\text{east DOP})^2 & & & (\text{Cross-terms}) \\ & (\text{north DOP})^2 & & \\ & & (\text{vertical DOP})^2 & \\ (\text{Cross-terms}) & & & (\text{time DOP})^2 \end{bmatrix}$$

where

$$\text{GDOP} = \sqrt{\text{trace}(H^{[1]T}H^{[1]})^{-1}},\ H^{[1]} = \left.\frac{\partial \rho}{\partial x}\right|_{X_{\text{nom}}, Y_{\text{nom}}, Z_{\text{nom}}}$$

Table 2.1 Example with four satellites.

	Satellite location			
	1	**2**	**3**	**4**
Elevation (°)	5	5	5	90
Azimuth (°)	0	120	240	0

Typical example values of $H^{[1]}$ for this geometry are

$$H^{[1]} = \begin{bmatrix} 0.000 & 0.996 & 0.087 & 1.000 \\ 0.863 & -0.498 & 0.087 & 1.000 \\ -0.863 & -0.498 & 0.087 & 1.000 \\ 0.000 & 0.000 & 1.000 & 1.000 \end{bmatrix}$$

The GDOP calculations for this example are

$$(H^{[1]T}H^{[1]})^{-1} = \begin{bmatrix} 0.672 & 0.000 & 0.000 & 0.000 \\ 0.000 & 0.672 & 0.000 & 0.000 \\ 0.000 & 0.000 & 1.600 & -.0505 \\ 0.000 & 0.000 & -0.505 & 0.409 \end{bmatrix}$$

$$\text{GDOP} = \sqrt{0.672 + 0.672 + 1.6 + 0.409} = 1.83$$

$$\text{PDOP} = 1.72$$

$$\text{HDOP} = 1.16$$

$$\text{VDOP} = 1.26$$

$$\text{TDOP} = 0.64$$

Gdop.m calculates the GDOP for the chosen constellation for GPS_perf.m by calculating $H^{[1]}$ matrix calcH. See Appendix A on www.wiley.com/go/grewal/gnss.

2.4 Time and GPS

2.4.1 Coordinated Universal Time (UTC) Generation

Coordinated universal time (UTC) is the timescale based on the atomic second but is occasionally corrected by the insertion of leap seconds so as to keep it approximately synchronized with the Earth's rotation. The leap second adjustments keep UTC within 0.9 seconds of UT1, which is a timescale based on the Earth's axial spin. UT1 is a measure of the true angular orientation of the Earth in space. Because the Earth does not spin at exactly a constant rate, UT1 is not a uniform timescale [5].

2.4.2 GPS System Time

The timescale to which GPS signals are referenced is referred to as *GPS time.* GPS time is derived from a composite or "paper" clock that consists of all operational monitor station and satellite atomic clocks. Over the long run, it is steered to keep it within about 90 nanoseconds (1σ) of UTC, as maintained by the master clock at the US Naval Observatory, ignoring the UTC leap seconds. At the integer second level, GPS time equaled UTC in 1980. However, due to the leap seconds that have been inserted into UTC, GPS time was ahead of UTC by 18 seconds after September 2019.

2.4.3 Receiver Computation of UTC

The parameters needed to calculate UTC from GPS time are found in subframe 4 of the navigation data message. These data include a notice to the user regarding the scheduled future or recent past (relative to the navigation message upload) value of the delta time due to leap seconds Δt_{LFS}, together with the week number WN_{LFS} and the day number DN, at the end of which the leap second becomes effective. The latter two quantities are known as the *effectivity time* of the leap second. "Day one" is defined as the first day relative to the end/start of a week, and the WN_{LFS} value consists of the eight least significant bits (LSBs) of the full week number.

Three different UTC/GPS time relationships exist, depending on the relationship of the effectivity time to the user's current GPS time:

1. *First case.* Whenever the effectivity time indicated by the WN_{LFS} and WN values is not in the past relative to the user's present GPS time, ***and*** the user's present time does not fall in the time span starting at DN + 3/4 and ending at DN + 5/4, the UTC time is calculated as

$$t_{\text{UTC}} = t_E - \Delta t_{\text{UTC}} \quad (\text{modulo } 86\,400) \ \ (\text{s})$$

where t_{UTC} is in seconds; 86 400 is the number of seconds per day; and

$$\Delta t_{\text{UTC}} = \Delta T_{\text{LS}} + A_0 + A_1[t_E - t_{0t} + 604\,800(\text{WN} - \text{WN}_t)] \ \ (\text{s})$$

where 604 800 is the number of seconds per week and

t_E = user GPS time from start of week (seconds)
ΔT_{LS} = delta time due to leap seconds
A_0 = a constant polynomial term from the ephemeris message
A_1 = a first-order polynomial term from the ephemeris message
t_{0t} = reference time for UTC date
WN = current week number derived from subframe 1
WN_t = UTC reference week number

The user GPS time t_E is in seconds relative to the end/start of the week, and the reference time t_{0t} for UTC data is referenced to the start of that week, whose number WN_t is given in word 8 of page 18 in subframe 4. The WN_t value consists of the eight LSBs of the full week number. Thus, the user must account for the truncated nature of this parameter as well as truncation of WN, WN_t, and WN_{LFS} due to the rollover of the full week number. These parameters are managed by the GPS control segment so that the absolute value of the difference between the untruncated WN and WN_t values does not exceed 127.

2. *Second case.* Whenever the user's current GPS time falls within the time span from DN +3/4 to DN +5/4, proper accommodation of the leap second event with a possible week number transition is provided by the following expression for UTC:

$$t_{\mathrm{UTC}} = W[\mathrm{modulo}(86\,400 + \Delta t_{\mathrm{LSF}} \Delta t_{\mathrm{LS}})]\ (\mathrm{s})$$

where

$$W = (t_E - \Delta t_{\mathrm{UTC}} - 43\,200)(\mathrm{modulo}\ 86\,400) + 43\,200\ (\mathrm{s})$$

and the definition of Δt_{UTC} given previously applies throughout the transition period.

3. *Third case.* Whenever the effectivity time of the leap second event, as indicated by the WNLFS and DN values, is in the past relative to the user's current GPS time, the expression given for t_{UTC} in the first case earlier is valid except that the value of Δt_{LFS} is used instead of Δt_{LS}. The GPS control segment coordinates the update of UTC parameters at a future upload in order to maintain a proper continuity of the t_{UTC} timescale.

2.5 Example: User Position Calculations with No Errors

2.5.1 User Position Calculations

This section demonstrates how to go about calculating the user position, given ranges (pseudoranges) to satellites, the known positions of the satellites, and ignoring the effects of clock errors, receiver errors, propagation errors, and so on.

Then, the pseudoranges will be used to calculate the user's antenna location.

2.5.1.1 Position Calculations

Neglecting clock errors, let us first determine the position calculation with no errors:

$$\begin{aligned} \rho_r &= \text{pseudorange (known)} \\ x, y, z &= \text{satellite position coordinates (known), in ECEF} \\ X, Y, Z &= \text{user position coordinates (unknown)} \end{aligned}$$

where x, y, z, X, Y, Z are in the ECEF coordinate system. (It can be converted to ENU.)

Position calculation with no errors is

$$\rho_r = \sqrt{(x - X)^2 + (y - Y)^2 + (z - Z)^2} \tag{2.36}$$

Squaring both sides yields

$$\begin{aligned} \rho_r^2 &= (x - X)^2 + (y - Y)^2 + (z - Z)^2 \\ &= \underbrace{X^2 + Y^2 + Z^2}_{r^2 + C_b} + x^2 + y^2 + z^2 - 2Xx - 2Yz - 2Zz \end{aligned} \tag{2.37}$$

$$\rho_r^2 - (x^2 + y^2 + z^2) - r^2 = C_b - 2Xx - 2Yz - 2Zz \tag{2.38}$$

where r equals the radius of the Earth and C_b is the clock bias correction. The four unknowns are (X, Y, Z, C_b). Satellite position (x, y, z) is calculated from ephemeris data. For four satellites, Eq. (2.38) becomes

$$\begin{aligned} \rho_{r_1}^2 - (x_1^2 + y_1^2 + z_1^2) - r^2 &= C_b - 2Xx_1 - 2Yy_1 - 2Zz_1 \\ \rho_{r_2}^2 - (x_2^2 + y_2^2 + z_2^2) - r^2 &= C_b - 2Xx_2 - 2Yy_2 - 2Zz_2 \\ \rho_{r_3}^2 - (x_3^2 + y_3^2 + z_3^2) - r^2 &= C_b - 2Xx_3 - 2Yy_3 - 2Zz_3 \\ \rho_{r_4}^2 - (x_4^2 + y_4^2 + z_4^2) - r^2 &= C_b - 2Xx_4 - 2Yy_4 - 2Zz_4 \end{aligned} \tag{2.39}$$

with unknown 4×1 state vector

$$\begin{bmatrix} X \\ Y \\ Z \\ C_b \end{bmatrix}$$

We can rewrite the four equations in matrix form as

$$\begin{bmatrix} \rho_{r_1}^2 - (x_1^2 + y_1^2 + z_1^2) - r^2 \\ \rho_{r_2}^2 - (x_2^2 + y_2^2 + z_2^2) - r^2 \\ \rho_{r_3}^2 - (x_3^2 + y_3^2 + z_3^2) - r^2 \\ \rho_{r_4}^2 - (x_4^2 + y_4^2 + z_4^2) - r^2 \end{bmatrix} = \begin{bmatrix} -2x_1 - 2y_1 - 2z_1 1 \\ -2x_2 - 2y_2 - 2z_2 1 \\ -2x_3 - 2y_3 - 2z_3 1 \\ -2x_4 - 2y_4 - 2z_4 1 \end{bmatrix} = \begin{bmatrix} X \\ Y \\ Z \\ C_b \end{bmatrix}$$

or

$$\overbrace{Y}^{4\times1} = \overbrace{M}^{4\times4} \overbrace{X_\rho}^{4\times1} \tag{2.40}$$

where

Y = vector (known)

M = matrix (known)

X_ρ = vector (unknown)

Then, we premultiply both sides of Eq. (2.40) by M^{-1}:

$$\begin{aligned} M^{-1}Y &= M^{-1}MX_\rho \\ &= X_\rho \\ &= \begin{bmatrix} X \\ Y \\ Z \\ C_b \end{bmatrix} \end{aligned}$$

If the rank of M, the number of linear independent columns of the matrix M, is less than 4, then M will not be invertible.

2.5.2 User Velocity Calculations

Differentiate Eq. (2.21) with respect to time without C_b.

$$\dot{Z}_{\rho_r} = \dot{\rho}_r = \frac{(x-X)(\dot{x}-\dot{X}) + (y-Y)(\dot{y}-\dot{Y}) + (z-Z)(\dot{z}-\dot{Z})}{\rho_r} \tag{2.41}$$

Differentiate Eq. (2.41) with respect to $\dot{X}$, $\dot{Y}$, $\dot{Z}$

$$\underbrace{\begin{bmatrix} \partial \dot{Z}^1_\rho \\ \partial \dot{Z}^2_\rho \\ \partial \dot{Z}^3_\rho \end{bmatrix}}_{3\times1} = \underbrace{\begin{bmatrix} \frac{\partial \rho^1_r}{\partial x} & \frac{\partial \rho^1_r}{\partial y} & \frac{\partial \rho^1_r}{\partial z} \\ \frac{\partial \rho^2_r}{\partial x} & \frac{\partial \rho^2_r}{\partial y} & \frac{\partial \rho^2_r}{\partial z} \\ \frac{\partial \rho^3_r}{\partial x} & \frac{\partial \rho^3_r}{\partial y} & \frac{\partial \rho^3_r}{\partial z} \end{bmatrix}}_{3\times3} \underbrace{\begin{bmatrix} \delta\dot{x} \\ \delta\dot{y} \\ \delta\dot{z} \end{bmatrix}}_{3\times1} \tag{2.42}$$

where $\delta\dot{x} = \dot{X} - \dot{X}_{\text{nom}}$, $\delta\dot{y} = \dot{Y} - \dot{Y}_{\text{nom}}$, $\delta\dot{z} = \dot{Z} - \dot{Z}_{\text{nom}}$.

In classical navigation geometry, the components (3×3) of this unit vector are often called direction cosine. It is interesting to note that these components are the same as the position linearization shown in Eqs. (2.26a) and (2.26b).

Equations (2.42) and (2.26b) will be used in GPS/INS tightly coupled implementation as measurement equations for pseudoranges and/or delta pseudoranges in chapters 11 and 12 in the extended Kalman filters. Equation (2.27) will be used in integrity determination of GNSS satellites in Chapter 9 and from Eq. (2.41),

$$-\dot{\rho}_r + \frac{1}{\rho_r}[\dot{x}(x-X) + \dot{y}(y-Y) + \dot{z}(z-Z)] = \left(\frac{x-X}{\rho_r}\dot{X} + \frac{y-Y}{\rho_r}\dot{Y} + \frac{z-Z}{\rho_r}\dot{Z}\right) \tag{2.43}$$

where

$$\dot{\rho}_r = \text{range rate (known)}$$
$$\rho_r = \text{range (known)}$$
$$(x, y, z) = \text{satellite positions (known)}$$
$$(\dot{x}, \dot{y}, \dot{z}) = \text{satellite rates (known)}$$
$$X, Y, Z = \text{user position (known from position calculations)}$$
$$(\dot{X}, \dot{Y}, \dot{Z}) = \text{user velocity (unknown)}$$

For three satellites, Eq. (2.43) becomes

$$\begin{bmatrix} -\dot{\rho}_{r_1} + \frac{1}{\rho_{r_1}}[\dot{x}_1(x_1 - X) + \dot{y}_1(y_1 - Y) + \dot{z}_1(z_1 - Z)] \\ -\dot{\rho}_{r_1} + \frac{1}{\rho_{r_2}}[\dot{x}_2(x_2 - X) + \dot{y}_2(y_2 - Y) + \dot{z}_2(z_2 - Z)] \\ -\dot{\rho}_{r_1} + \frac{1}{\rho_{r_3}}[\dot{x}_3(x_3 - X) + \dot{y}_3(y_3 - Y) + \dot{z}_3(z_3 - Z)] \end{bmatrix} = \begin{bmatrix} \frac{(x_1 - X)}{\rho_{r_1}} & \frac{(y_1 - Y)}{\rho_{r_1}} & \frac{(z_1 - Z)}{\rho_{r_1}} \\ \frac{(x_2 - X)}{\rho_{r_2}} & \frac{(y_2 - Y)}{\rho_{r_2}} & \frac{(z_2 - Z)}{\rho_{r_2}} \\ \frac{(x_3 - X)}{\rho_{r_3}} & \frac{(y_3 - Y)}{\rho_{r_3}} & \frac{(z_3 - Z)}{\rho_{r_3}} \end{bmatrix} \begin{bmatrix} \dot{X} \\ \dot{Y} \\ \dot{Z} \end{bmatrix} \tag{2.44}$$

Equation (2.44) becomes

$$\overbrace{D}^{3\times 1} = \overbrace{N}^{3\times 3}\overbrace{U_v}^{3\times 1} \tag{2.45}$$

$$\overbrace{U_v}^{3\times 1} = N^{-1}D \tag{2.46}$$

where

$$D = \text{known vector}$$
$$N = \text{known matrix}$$
$$U_v = \text{unknown user velocity vector}$$

However, if the rank of N is <3, N will not be invertible.

Problems

Refer to Appendix B for coordinate system definitions and to Section B.3.10 for satellite orbit equations.

2.1 Which of the following coordinate systems is not rotating?
(a) North–east–down (NED)
(b) East–north–up (ENU)
(c) Earth-centered, Earth-fixed (ECEF)
(d) Earth-centered inertial (ECI)
(e) Moon-centered, moon fixed

2.2 Show that $C_{\text{ENU}}^{\text{ECEF}} \times C_{\text{ECEF}}^{\text{ENU}} = I$, the 3 × 3 identity matrix. (*Hint*: $\left(C_{\text{ENU}}^{\text{ECEF}} = \left[C_{\text{ECEF}}^{\text{ENU}}\right]^T\right)$).

2.3 Rank VDOP, HDOP, and PDOP from smallest (best) to largest (worst) under normal conditions:
(a) VDOP ≤ HDOP ≤ PDOP
(b) VDOP ≤ PDOP ≤ HDOP
(c) HDOP ≤ VDOP≤PDOP
(d) HDOP ≤ PDOP ≤ VDOP
(e) PDOP ≤ HDOP ≤ VDOP
(f) PDOP ≤ VDOP ≤ HDOP

2.4 UTC time and the GPS time are offset by an integer number of seconds (e.g. 16 seconds as of June 2012) as well as a fraction of a second. The fractional part is approximately.
(a) 0.1–0.5 s
(b) 1–2 ms
(c) 100–200 ns
(d) 10–20 ns

2.5 Derive equations (2.41) and (2.42).

2.6 For the following GPS satellites, find the satellite position in ECEF coordinates at t = 3 seconds. (*Hint*: See Appendix B.) Ω_0 and θ_0 are given below at time $t_0 = 0$:

	Ω_0 (°)	θ_0 (°)
(a)	326	68
(b)	26	34

2.7 Using the results of the previous problem, find the satellite positions in the local reference frame. Reference should be to the COMSAT facility in Santa Paula, California, located at 32.4° latitude, −119.2° longitude. Use coordinate shift matrix $S = 0$. (Refer to Section B.3.10.)

2.8 Given the following GPS satellite coordinates and pseudoranges:

Satellite	Ω_0 (°)	θ_0 (°)	ρ (m)
1	326	68	2.324×10^7
2	26	340	2.0755×10^7
3	146	198	2.1103×10^7
4	86	271	2.3491×10^7

(a) Find the user's antenna position in ECEF coordinates.
(b) Find the user's antenna position in locally level coordinates referenced to 0° latitude, 0° longitude. Coordinate shift matrix $S = 0$.
(c) Find the various DOPs.

2.9 Given two satellites in north and east coordinates

$$x(1) = 6.1464 \times 10^6, \;\; y(1) = 2.0172 \times 10^7 \text{ (m)}$$
$$x(2) = 6.2579 \times 10^6, y(2) = -7.4412 \times 10^6 \text{ (m)}$$

with pseudoranges

$$c\Delta t(1) = \rho_r(1) = 2.324 \times 10^7 \text{ (m)}$$
$$c\Delta t(2) = \rho_r(2) = 2.0755 \times 10^7 \text{ (m)}$$

and starting with an initial guess of x_{est}, y_{est}, find the user's antenna position.

2.10 A satellite position at time $t = 0$ is specified by its orbital parameters as $\Omega_0 = 92.847°$, $\theta_0 = 135.226°$, $\alpha = 55°$, $R = 26\,560\,000$ m.
(a) Find the satellite position at one second, in ECEF coordinates.
(b) Convert the satellite position from (a) with user at

$$\begin{bmatrix} X_u \\ Y_u \\ Z_u \end{bmatrix}_{\text{ECEF}} = \begin{bmatrix} -2.430601 \\ -4.702442 \\ -3.546587 \end{bmatrix} \times \; 10^6 \text{ m}$$

from WGS84 (ECEF) to ENU coordinates with origin at

$$\theta = \text{local reference longitude} = 32.4°$$
$$\phi = \text{local reference latitude} = -119.2°$$

References

1 Department of Transportation (1990). *LORAN-C User's Handbook*, Commandant Instruction M12562.3. Washington, DC: U.S. Coast Guard.
2 Logsdon, T. (1992). *The NAVSTAR Global Positioning System*. New York: Van Nostrand Reinhold.
3 Grewal, M.S. and Andrews, A.P. (2019). *Application of Kalman Filtering to GPS, INS, & Navigation, Short Course Notes*. Anaheim, CA: Kalman Filtering Consultant Associates.
4 Grewal, M.S. and Andrews, A.P. (2015). *Kalman Filtering: Theory and Practice Using MATLAB®*, 4e. New York: Wiley.
5 Allan, D.W., Ashby, N., and Hodge, C.C. (1997). *The Science of Timekeeping*, Hewlett Packard, Application Note 1289. Palo Alto, CA: Hewlett-Packard.

3

Fundamentals of Inertial Navigation

An inertial system does for geometry…what a watch does for time.[1]
Charles Stark Draper (1901–1987)

Charles Stark Draper was the American pioneer in inertial navigation who founded the Instrumentation Laboratory at MIT in 1932 to develop aircraft instrumentation technologies. In his analogy previously quoted, watches keep track of time by being set to the correct time, then incrementing that time according to the inputs from a "time sensor" (a frequency source) to update that initial value.

An inertial navigation system (INS) does something similar, only with different variables – and it increments doubly. An INS needs to be set to the correct position and velocity. Thereafter, they use measured accelerations to increment that initial velocity, and use the resulting velocities to increment position.

3.1 Chapter Focus

The overview of inertial navigation in Section 1.3 alluded to the history and terminology of the technology. The focus here is on how inertial sensors function and how they are integrated into navigation systems, including the following:

1. Terminology for the phenomenology and apparatus of inertial navigation
2. Technologies used for sensing rotation and acceleration
3. Error characteristics of inertial sensors
4. Sensor error compensation methods
5. How to compensate for unsensed gravitational accelerations
6. Initializing and propagating navigation solutions for attitude (rotational orientation), velocity, and position

1 Quoted by author Tom Pickens in "Doc Gyro and His Wonderful 'Where Am I?' Machine," *American Way Magazine*, 1972.

Global Navigation Satellite Systems, Inertial Navigation, and Integration,
Fourth Edition. Mohinder S. Grewal, Angus P. Andrews, and Chris G. Bartone.

Companion website: www.wiley.com/go/grewal/gnss

7. Carouseling and indexing as methods for mitigating the effects of sensor errors
8. System-level testing and evaluation
9. INS performance metrics and standards

How this all affects navigation performance is discussed in Chapter 11.

Scope. The technology of inertial navigation has been evolving for nearly a century, its diversity and sophistication have grown enormously, and the scale of inertial systems has shrunk by orders of magnitude – from the unbearable to the wearable. As a result, we cannot cover every aspect of every technical approach to every application in full detail in this one chapter. The focus here will be limited to just those INS applications involving global navigation satellite system (GNSS) navigation, which limits it to the terrestrial environment with GNSS availability at least part of the time, and to those aspects of the application necessary for GNSS/INS integration. We will endeavor to cover here the essential design and implementation issues involved in these applications.

3.2 Terminology

Much of the terminology for inertial navigation evolved when the technology was highly classified and being developed by independent design teams, the result of which has been considerable diversity. The terminology used throughout the book, listed in the following text, generally follows a standardized terminology for inertial sensors [1] and systems [2].

Inertia is the propensity of bodies to maintain constant translational and rotational velocity, unless disturbed by forces or torques, respectively (Newton's first law or motion).

Inertial reference frames are coordinate frames in which Newton's laws of motion are valid. They cannot be rotating or accelerating. They are not necessarily the same as the *navigation coordinates*, which are typically dictated by the navigation problem at hand. We live in a rotating and accelerating environment here on Earth, and that defines an Earth-fixed locally level coordinate system we already feel comfortable with – even though it is accelerating (to counter gravity) and rotating. These rotations and accelerations must be taken into account in the practical implementation of inertial navigation.

Navigation coordinates are those used for representing the position of the inertial sensors with respect to its environment. In GNSS/INS integration, this will generally be the same as that used by the GNSS, representing the near-Earth environment. See Appendix B (www.wiley.com/go/grewal/gnss) for descriptions of navigation coordinates and the transformations involved.

The navigation solution for inertial navigation includes the instantaneous values of position, velocity, and rotational orientation of the inertial sensors with respect to navigation coordinates. It must be sufficient for propagating the solution forward in time, given the inertial sensor outputs.

Inertial sensors measure inertial accelerations and rotations, both of which are vector-valued variables.

Accelerometers measure *specific force*, the point being that *accelerometers do not measure gravitational acceleration.* Specific force is modeled by Newton's second law as $\mathbf{a} = \mathbf{F}/\mathrm{m}$, where $\mathbf{F}$ is the physically applied force (not including gravity) and $\mathbf{m}$ is the mass it is applied to. Specific force is the force per unit mass, $\mathbf{F}/\mathrm{m}$, and accelerometers are sometimes called **specific force receivers**. SI units for specific force are meters per second per second.

Gyroscopes (often shortened to "gyros") are sensors for measuring rotation.

Displacement gyros (also called **whole-angle gyros**) measure accumulated rotation angle, in angular units (e.g. radians or degrees).

Rate gyros measure rotation rates in angular rate units (e.g. radians per second, degrees per hour, etc.).
Inertial navigation depends on gyros for maintaining knowledge of how the accelerometers are oriented in inertial and navigational coordinates.

Input axes of an inertial sensor define which vector component(s) of acceleration, rotation, or rotation rate it measures. These are illustrated by arrows in Figure 3.1, with rotation arrows wrapped around the input axes of gyroscopes to indicate the direction of rotation. Multi-axis sensors measure more than one component.

Calibration is a process for characterizing sensor input–output behavior from a set of observed input–output pairs. The objective of sensor calibration is to be able to determine its inputs, given its outputs.

Scale factor and bias are the most common sensor error characteristics determined by calibration.

Scale factor is the ratio of sensor output variation to sensor input variation.

Bias is the sensor output with zero input.

Inertial sensor assemblies (ISAs) are ensembles of inertial sensors rigidly mounted to a common base to maintain the same relative orientations, as illustrated in Figure 3.1.
ISAs used in inertial navigation usually contain three accelerometers and three gyroscopes, represented in the figure by lettered blocks with arrows representing their respective input axes, or an equivalent configuration using

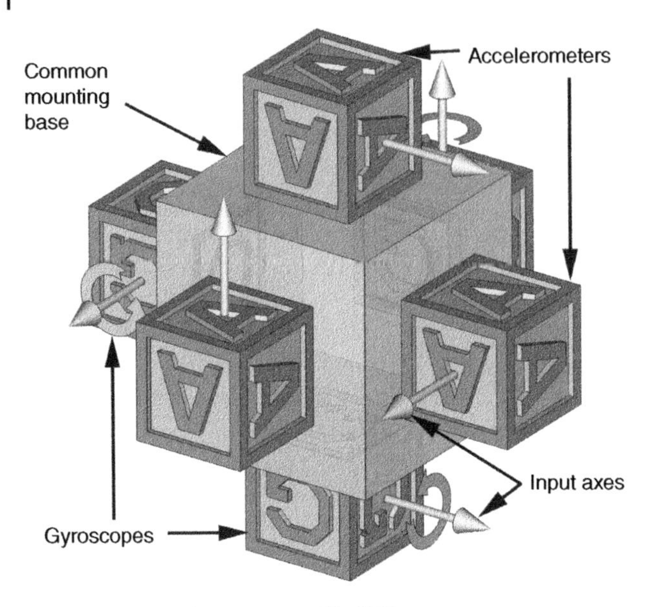

Figure 3.1 Inertial sensor assembly (ISA) components.

multi-axis sensors. However, ISAs used for some other purposes (e.g. dynamic control applications such as autopilots or automotive steering augmentation) may not need as many sensors, and some designs use redundant sensors. Other terms used for the ISA are **instrument cluster** and (for inertially stabilized implementations) **stable element** or **stable platform**.

Inertial reference unit (IRU) is a term commonly used for inertial sensor system for attitude information only (i.e. using only gyroscopes). Space-based telescopes, for example, do not generally need acccelerometers, but they do need gyroscopes to keeping track of orientation.

Inertial measurement units (IMUs) include ISAs and associated support electronics for calibration and control of the ISA. Support electronics may also include thermal control or compensation, signal conditioning and input–output control. An IMU may also include an IMU processor, and – for inertially stabilized systems – the gimbal control electronics.

Inertial navigation systems (INS) measure rotation rates and accelerations, and calculate attitude, velocity, and position. Its subsystems include:

IMUs, already mentioned earlier.
Navigation computers (one or more) to calculate the gravitational acceleration (not measured by accelerometers) and process the outputs of the accelerometers and gyroscopes from the IMU to maintain an estimate of the position of

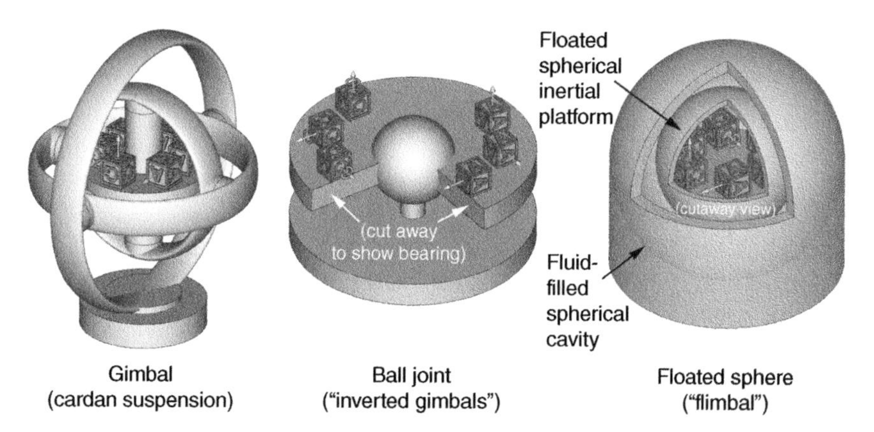

Figure 3.2 Inertially stabilized IMU alternatives.

the IMU. Intermediate results of the implementation method usually include estimates of velocity, attitude, and attitude rates of the IMU.

User interfaces, such as display consoles for human operators and analog and/or digital data interfaces for vehicle guidance and control functions.

Power supplies and/or raw power conditioning for the complete INS.

Implementations of inertial navigation systems include two general types:

Strapdown systems do nothing to physically control the orientation of the ISA, but they do process the gyroscope outputs to keep track of its orientation with respect to navigation coordinates. A strapdown system is usually attached to its host vehicle so it can keep track of its host vehicle orientation with respect to navigation coordinates.
Inertially stabilized systems use their gyroscopes for controlling ISA attitude, as illustrated in Fig 3.2a–c. This shows three alternative structures that have been tried at different times:

Gimbals, also called a Cardan[2] suspension. This was the most popular implementation using hardware to solve the attitude problem. The US Navy's Electrostatically Supported Gyro Navigator (ESGN) is a gimbaled system, and possibly the most accurate INS for long-term inertial navigation.

Ball-joint, which Fritz Mueller called "inverted gimbals" [3]. It is only useful for applications with limited rotational freedom in pitch and roll, such as for ships

2 Named after the Italian physician, inventor, and polymath Girolamo Cardano (1501–1576), who also invented what Americans call a "universal joint" and Europeans call a "Cardan shaft."

in lesser sea states. It has not become popular, perhaps because of the difficulties of applying controlled torques about the spherical bearing to stabilize the ISA.

Floated systems, a configuration also called a "FLIMBAL" system, an acronym for floated inertial measurement ball. The advanced inertial reference sphere (AIRS) inertial navigator for the US Air Force LGM-30G Minuteman III ICBM is a floated system. Despite the difficulties of transferring power, signals, heat, torque, and relative attitude between the housing and the inner spherical ISA, AIRS is probably the most accurate (and expensive) inertial navigator for high-g rocket booster applications.

In all three cases, the rotation-isolated ISA is also called an *inertial platform, stable platform,* or *stable element.* The IMU in this case may include the ISA, the gimbal/float structure and all associated electronics (e.g. gimbal wiring, rotary slip rings, gimbal bearing angle encoders, signal conditioning, gimbal bearing torque motors, and thermal control).

Commonly used inertially stabilized ISA orientations in terrestrial applications include:

Inertially fixed (non-rotating), a common orientation for operations in space. In this case, the ISA may include one or more star trackers to correct for any gyroscope errors. However, locally level implementations may also use star trackers for the same purpose.

Locally level, a common orientation for terrestrial navigation. In this case, the ISA rotates with the Earth, and keeps two of its reference axes locally level during horizontal motion over the surface. Some early systems aligned the gyro and accelerometer input axes with the local directions of north, east, and down, because the gimbal angles could then represent the Euler angles for heading (yaw), pitch, and roll of the vehicle. However, there are also advantages in allowing the locally stabilized element to physically rotate about the local vertical direction.

Inertially stabilized systems are generally more expensive than strapdown systems, but their performance is usually better. This is due, in part, to the fact that their gyroscopes and accelerometers are not required to endure high rotation rates.

"Host vehicle" refers to the transportation system using INS for navigation. It could be a spacecraft, aircraft, surface ship, submarine, land vehicle, or pack animal (including humans).

Shock and vibration isolation. High-frequency dynamic forces acting on the host vehicle (e.g. from propulsion noise, bumpy terrain, turbulence, or impacts) can excite elastic waves and vibrations in the host vehicle that are transmitted through the vehicle frame to the INS through its mounting hardware. The resulting zero-mean high-frequency inputs to the inertial sensors

should not influence the navigation solution significantly, but they can create numerical errors in the real-time computer methods used for integrating attitude rates and acceleration, and they can damage the sensors used. These effects can be mitigated at the interface between INS and host by using shock and vibration isolators (generally made from "lossy" elastomers) to dampen the high-frequency components of contact forces.
Because inertial navigation systems perform integrals of acceleration and attitude rates, these integrals need initial values.

Initialization is a procedure for obtaining an initial value of the navigation solution.

Rotational orientation or *attitude* refers to the angular pose of a rigid object in three-dimensional space relative to the axes of a coordinate system.

Alignment is a procedure used for establishing the initial value of the rotational orientation of the ISA relative to navigation coordinates. Inertial systems with sufficiently accurate sensors can perform **self-alignment** when the system is sufficiently stationary with respect to the Earth. In that case, the implementation can be divided into two parts:
Leveling uses the accelerometers to measure the upward acceleration required to counter gravity, from which the system can determine the orientation of its ISA relative to local vertical. For inertially stabilized systems, the stable element (ISA) is physically leveled during this process (hence the name).

Gyrocompassing is a procedure for estimating the direction of the Earth's rotation axes with respect to ISA coordinates, using its gyroscopes. This and the direction of the local vertical then determines the north–south direction, so long as the stationary location is not in the vicinity of the poles. Given these two directions, the INS can orient itself relative to its location on the Earth. The term *gyrocompassing* is a reference to the gyrocompass, an instrument introduced toward the end of the nineteenth century to replace the magnetic compass on iron ships. The gyrocompass uses mechanical means to orient itself relative to north, whereas the INS requires a computer. For some inertially stabilized systems, gyrocompassing physically aligns the ISA with its level sensor axes pointing north and east.

Transfer alignment uses an independent navigation solution for a host vehicle to initialize the navigation solution (including alignment) in another vehicle carried by the host vehicle. This was originally developed for using the INS in a host vehicle to initialize an INS in guided munitions, and it usually requires some amount of maneuvering of the host vehicle to attain observability of the required alignment variables. A version of this makes use of the INS in an aircraft carrier, the roll and pitch of its deck, and the direction of the launch catapult as inputs for aligning the aircraft INS during takeoff.

Magnetic alignment uses the directions of sensed acceleration (from countering gravity) and the local magnetic field to orient itself. This does not work where the magnetic field is close to vertical (near the magnetic poles), and it can be compromised by magnetic materials warping the local magnetic field.

3.3 Inertial Sensor Technologies

3.3.1 Gyroscopes

The French physicist Léon Foucault (1819–1868) gave them this name (from the Greek for "rotation sensor" or something like that) in the mid-nineteenth century. The technology has developed considerably since then. The following is but a sampling from a vast reservoir of technological approaches to inertial sensing.

3.3.1.1 Momentum Wheel Gyroscopes (MWGs)

Foucault used one to measure the rotation rate of the Earth[3] in 1852, this one featuring a spinning brass wheel passively isolated from disturbing torques by using two nested gimbals.

Bearing Technologies A limiting design factor in momentum wheel gyroscope (MWG) performance has been bearing torque, which has been addressed by going from sleeve bearings to jewel bearings, to ball bearings, to air bearings, and to electrostatic bearings. Even though electrostatic suspension is inherently unstable, it can achieve very low bearing torques. Perhaps the most accurate momentum wheel gyroscopes to date were the superconducting electrostatic gyroscopes used in a theoretical physics experiment named "Gravity Probe B" [4], a NASA-funded program to resolve two fine points of Einstein's theory. It was able to achieve drift rate accuracies in the order of 10^{-9} deg/h, but only in a zero-g environment, and at enormous cost. Unfortunately, scaling down the size of momentum wheel gyroscopes tends to scale up the ratio of surface area to angular momentum, which scales up angular drift rates due to bearing torques.

Whole-angle Gyroscopes Foucault's 1852 gyroscope was mounted inside two sets of gimbals that freed its spin axis to remain pointing in a fixed inertial direction, held there by the conservation of angular momentum in inertial coordinates. Electrostatic gyroscopes have spherical rotors supported by

---

3 The inertial rotation rate of the Earth was already quite well known, thanks to astronomers. Foucault was only using it to demonstrate gyroscopic physics.

electrostatic forces inside spherical suspension cavities, which gives their spin axes freedom to remain in fixed inertial directions. These are called *whole-angle* gyroscopes, and they can use the directions of their spin axes as inertial reference directions.

Rate Gyroscopes These use torques applied to the spinning rotor to keep its spin axis aligned with its enclosure. The spin axis rotational slewing rate is then proportional to the applied torque. There are also rate gyroscopes that do not use momentum wheels.

Axial Mass Unbalance Torques If the center of mass of the rotor of a momentum wheel gyroscope is not concentric with its center of support, then the offset between the downward gravitational force on its mass and the upward force supporting it will create a torque. The component of that torque perpendicular to the spin axis of the rotor will then cause the rotor angular momentum to precess about the applied vertical force. It is an acceleration-sensitive error torque due to axial mass unbalance that is difficult to avoid within manufacturing tolerances. It is commonly mitigated by calibrating its magnitude and compensating for it during operation.

3.3.1.2 Coriolis Vibratory Gyroscopes (CVGs)

Tuning Fork Gyroscopes The tuning fork is the paradigmatic Coriolis vibratory gyroscope, long used as voice-band frequency source before being pressed into service as a gyroscope, and it has since inspired miniaturized MESG designs using the same general principles. These mechanical principles are illustrated in Figure 3.3, showing how the normally counterbalanced synchronized motions

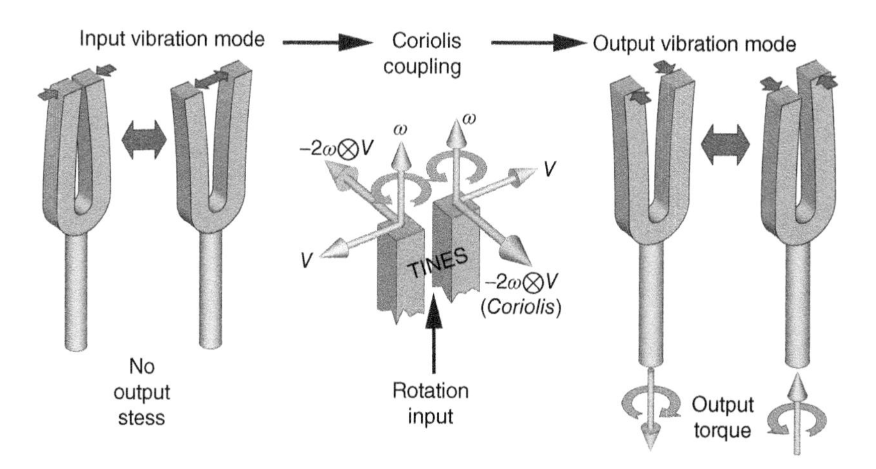

Figure 3.3 Tuning fork gyroscope.

of the two tines are coupled through the Coriolis effect due to rotations about the handle into an unbalanced vibratory torsion mode that is output through the handle. Counterbalanced momentum of the tines is an essential design feature for Coriolis vibratory gyroscopes (CVGs), here using two masses traveling in opposite directions to mitigate the effects of Newton's third law. Vibrations tend to launch elastic waves into the surrounding materials, so control of the distribution of vibration is a critical design issue for all CVGs – keeping it where it belongs and eliminating it where it does not belong. Operation requires methods for controlling the tuning fork motion and sensors for detecting the output mode. Tuning fork gyroscopes fabricated from quartz can perform both functions using the piezoelectric properties of quartz. Tuning forks fabricated in silicon are discussed in the following text.

MEMS Tuning Fork Gyroscope A design originally developed at the Charles Stark Draper Laboratories in the 1980s and 1990s [5] does not physically resemble a tuning fork, but it uses the same principle of two masses resonating in the plane of their silicon substrate, and in synchrony to maintain zero net momentum. The input rotation axis is in the plane of the substrate, as illustrated in Figure 3.4. Vibratory motion is controlled by electrostatic "comb drives" (interdigitated electrodes) developed at the University of California at Berkeley. The input axis is in the plane of the substrate and the output vibration mode is normal to the substrate surface. Many improvements have been made in this original design. There are also devices using rotational vibrations coupled with Coriolis effects.

Hemispherical Resonator Gyroscopes These are also called *wine glass gyroscopes*, referring to resonant modes of wine glass rims, the nodes of which move when the wine glass is rotated about its stem – caused by Coriolis coupling. These

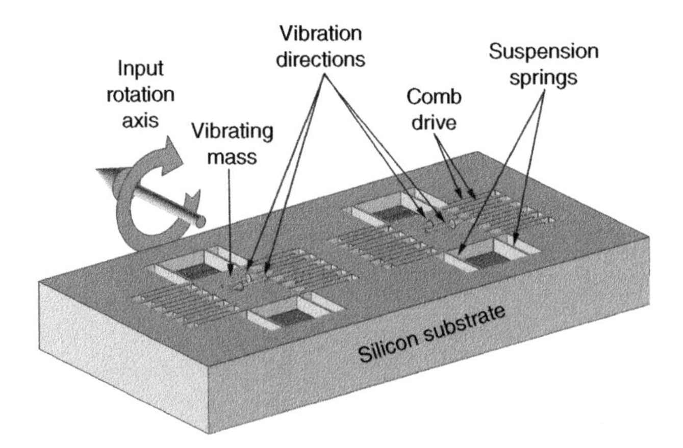

Figure 3.4 MEMS tuning fork gyroscope.

devices can continue to operate through short-term radiation events when electronic devices are shut down.

3.3.1.3 Optical Gyroscopes (RLGs and FOGs)

There are two essential designs for optical gyroscopes, both of which depend on the *Sagnac effect*, a phenomenon studied by Franz Harress in 1911 [6] and Georges Sagnac in 1913. The effect has to do with the relative delay of two light beams from the same source traveling in opposite directions around the same closed loop, and how their relative delay depends on the rotation rate of the apparatus in the plane of the loop. The effect has been named for Sagnac, who showed that the delay difference was proportional to the rotation rate, and the effect scaled as the area of the loop. The effect was not used for sensing rotation until after a working laser was demonstrated in 1960, first with the lasing cavity in the closed optical path – the ring laser gyroscope (RLG) – and later using a kilometers-long coil of optical fiber – the fiber optic gyroscope (FOG).

Ring laser gyroscopes are rate integrating gyroscopes. Their output interferometric phase rate is proportional to rotation rate, so each output phase shift represents an incremental inertial angular rotation angle. To minimize temperature and pressure sensitivities, their closed-loop optical paths are typically machined into very stable materials. Early designs exhibited a "lock-in" problem near zero input rates, due to backscatter off the mirrors. Later designs avoided this by using out-of-plane optical paths and multi-frequency lasing cavities.

Fiber optic gyroscopes were first developed after single-mode optical fibers became available, about a decade after the first laser. Unlike RLGs, FOGs are rate gyros. Their output is proportional to the input rotation rate, and must be integrated to get rotation angles. The optical loop in this case is a very long coil of optical fiber with an external laser source.

3.3.2 Accelerometers

Accelerometers used in inertial navigation measure the force required to keep a proof mass stationary with respect to its enclosure, which is called *specific force* to distinguish it from unsensed gravitational accelerations. Accelerometer designs differ in how that force is measured, and how that force is distributed. Examples of these different design approaches are given in the following text.

3.3.2.1 Mass-spring Designs

These measure stress in the material connecting the proof mass to its enclosure. A spring scale measures the strain in a spring, which is proportional to the force (stress) applied. Similar designs use other means to measure that stress:

Beam accelerometers sense the surface tension on a supporting beam surface to measure the load normal to the beam surface due to applied specific force.

Piezoresistive accelerometers use the change in resistance in the stressed support material. MEMS accelerometers used for automotive air bag deployment have used piezoresistance.

Piezoelectric accelerometers have long been used for measuring vibrational acceleration, and have been used as essentially DC sensors in piezoelectric capacitors on MEMS beam accelerometers.

SAW (surface acoustic wave) accelerometers use strain-induced shifts in the frequency of surface SAW resonators as a measure of the strain in the support material (a beam, for example).

Vibrating wire accelerometers use the frequency change in vibrating support wires due to changes in tension (stress) in the wires to measure the force applied by the wire to the supported proof mass. Because the fundamental vibration frequency of a wire under tension varies as the square root of tension, these are not linear sensors.
… and there are many more.

3.3.2.2 Pendulous Integrating Gyroscopic Accelerometers (PIGA)

This was the first "inertial grade" accelerometer (see Section 1.3.2.2), and is still in use today for high-end applications such as ICBM navigation during launch.

3.3.2.3 Electromagnetic

Force-rebalance. A common design for electromagnetic accelerometers uses permanent magnets as part of the proof mass, surrounded by a voice coil used to keep the magnet in a fixed position. The current required in the coil to keep it there will then be proportional to the force applied.

Inductive designs. A *drag cup* is a non-magnetic conducting cylindrical sleeve with a rotating bar magnet inside, so that the axial torque on the drag cup will be proportional to the rotation rate of the magnet. Analog automobile tachometers and speedometers use them with a torsion spring on the drag cup to indicate rpms or speed. They have also been used with mass-unbalanced drag cups such that the magnet rotation rate required to keep the drag cup stationary in an accelerating environment is proportional to acceleration and each revolution of the magnet represents an increment in velocity – making it an integrating accelerometer. Two of these can also be concatenated together in series such that they also integrate velocity to get position. Although they have performed well as tachometers and speedometers, they have not yet been sufficiently accurate for inertial navigation.

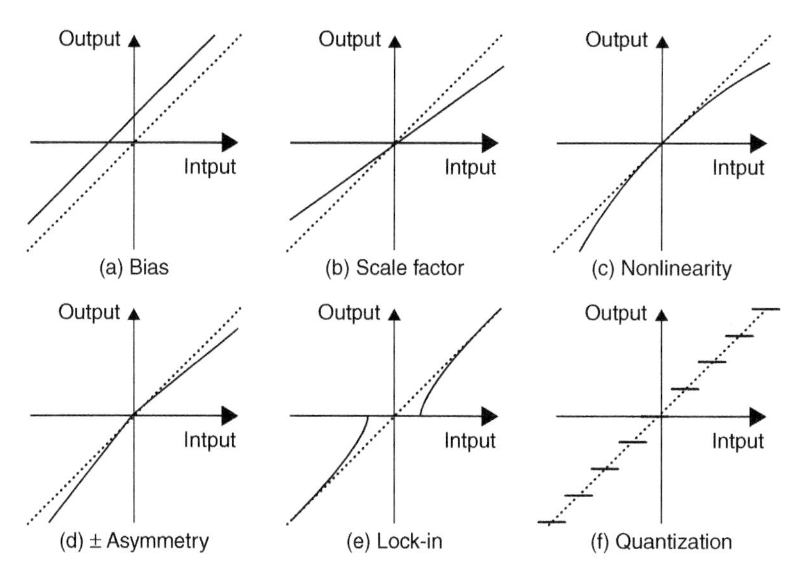

Figure 3.5 Common input–output error types.

3.3.2.4 Electrostatic

The surface-to-mass ratios of MESG-scale devices are such that surface electrostatic forces eventually come to dominate acceleration-induced forces. Electrical signals can then be used to keep a thin proof mass centered in its enclosure during accelerations.

3.3.3 Sensor Errors

3.3.3.1 Additive Output Noise

Sensor noise is most commonly modeled as zero-mean additive random noise. As a rule, sensor calibration removes all but the zero-mean noise component. Models and methods for dealing with various forms of zero-mean random additive noise using Kalman filtering are discussed in Chapter 10.

3.3.3.2 Input–output Errors

The ideal sensor input–output function for rotation and acceleration sensors is linear and *unbiased*, meaning that the sensor output is zero when the sensor input is zero.

These are repeatable sensor output errors, unlike the zero-mean random noise considered earlier. The same types of models apply to accelerometers and gyroscopes. Some of the more common types of sensor input–output errors are illustrated in Figure 3.5. These are listed for the specific panels:

(a) bias, which is any nonzero sensor output when the input is zero;
(b) scale factor error, usually due to manufacturing tolerances;

(c) nonlinearity, which is present in most sensors to some degree;
(d) scale factor sign asymmetry (often from mismatched push–pull amplifiers);
(e) lock-in, often due to mechanical stiction or (for ring laser gyroscopes) mirror backscatter; and
(f) quantization error, inherent in all digitized systems.

Theoretically, one can recover the sensor input from the sensor output so long as the input–output relationship is known and invertible. Lock-in (or "dead zone") errors and quantization errors are the only ones shown with this problem. The cumulative effects of both types (lock-in and quantization) often benefit from zero-mean input noise or dithering. Also, not all digitization methods have equal cumulative effects. Cumulative quantization errors for sensors with frequency outputs are bounded by $\pm$ one-half least significant bit (LSB) of the digitized output, but the variance of cumulative errors from independent sample-to-sample A/D conversion errors can grow linearly with time.

In inertial navigation, integration turns white noise into random walks.

3.3.3.3 Error Compensation

The accuracy demands on sensors used in inertial navigation cannot always be met within the tolerance limits of manufacturing, but can often be met by calibrating those errors after manufacture and using the results to compensate them during operation. Calibration is the process of characterizing the sensor output, given its input. Sensor error compensation is the process of determining the sensor input, given its output. Sensor design is all about making that process easier. Another problem is that any apparatus using physical phenomena that might be used to sense rotation or acceleration may also be sensitive to other phenomena, as well. Many sensors also function as thermometers, for example.

Figure 3.6 is a schematic of such an error compensation procedure, using the example of a gyroscope that is also sensitive to acceleration and temperature (not an unusual situation). The first problem is to determine the input–output function

$$\boldsymbol{\omega}_{\text{out}} = \mathbf{f}\ (\boldsymbol{\omega}_{\text{in}}, \mathbf{a}_{\text{in}}, T, \ldots),$$

where the ellipsis ", …)" allows for the effects of more variables to be compensated. The functional characterization is usually done using a set of controlled input values and measured output values. The next problem is to determine its inverse,

$$\boldsymbol{\omega}_{\text{in}} \approx \mathbf{f}^{-1}\ (\boldsymbol{\omega}_{\text{out}}, \mathbf{a}_{\text{in}}, T, \ldots)$$

and use it with independently sensed values for the variables involved – $\boldsymbol{\omega}_{\text{out}}$ (sensor output), $\mathbf{a}_{\text{in}}$ (compensated accelerometer output) and T (temperature) in this example.

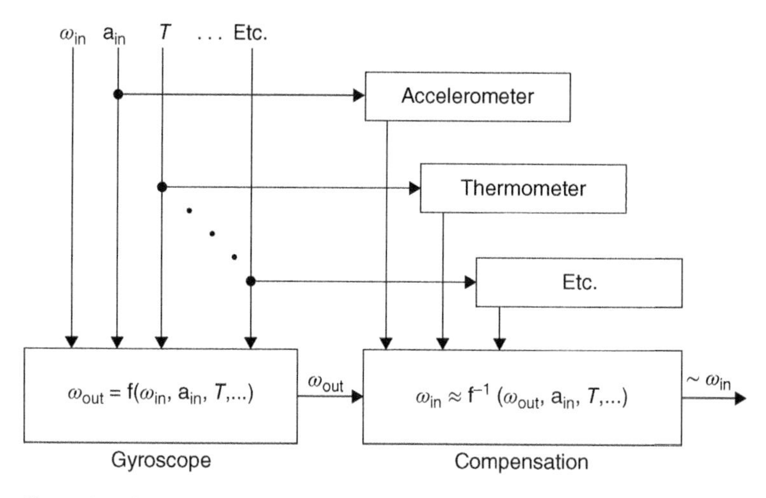

Figure 3.6 Gyro error compensation example.

If the input–output function $\boldsymbol{\omega}_{\text{out}} = \mathbf{f}(\boldsymbol{\omega}_{\text{in}}, \mathbf{a}_{\text{in}}, T, \ldots)$ is common to all sensors of the same design, then this only has to be done once. Otherwise, it can become expensive.

There are also methods using nonlinear Kalman filtering and auxiliary sensor aiding for tracking and updating compensation parameters that may drift over time.

3.3.4 Inertial Sensor Assembly (ISA) Calibration

The individual sensor input axes within an inertial sensor assembly (ISA) must be aligned to a common reference frame, and this can be combined with sensor-level calibration of all sensor compensation parameters, as illustrated in Figure 3.5. Figure 3.7 illustrates how input axis misalignments and scale factors at the ISA level affect sensor outputs, in terms of how they are related to the linear input–output model,

$$\mathbf{z}_{\text{output}} = \mathbf{M}(\mathbf{z}_{\text{input}} + \mathbf{b}_z) \tag{3.1}$$

$$\mathbf{M} = \begin{bmatrix} m_{11} & m_{12} & m_{13} \\ m_{21} & m_{22} & m_{23} \\ m_{31} & m_{32} & m_{33} \end{bmatrix}, \tag{3.2}$$

where $\mathbf{z}_{\text{input}}$ is a vector representing the inputs (accelerations or rotation rates) to three inertial sensors with nominally orthogonal input axes, $\mathbf{z}_{\text{output}}$ is a vector representing the corresponding outputs, $\mathbf{b}_z$ is a vector of sensor output biases, and the corresponding elements of $\mathbf{M}$ are labeled in Figure 3.7.

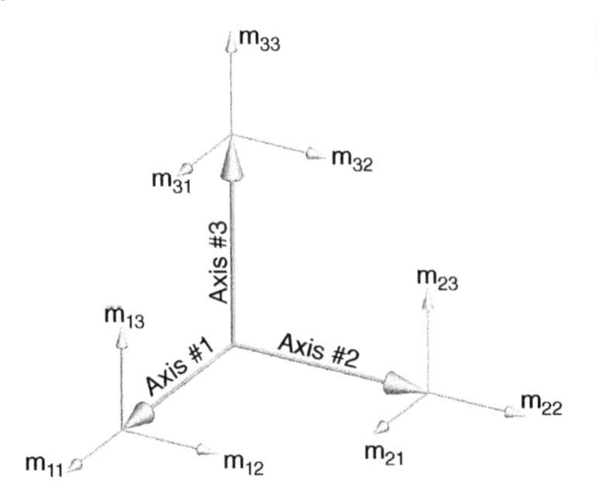

Figure 3.7 Directions of modeled sensor cluster errors.

3.3.4.1 ISA Calibration Parameters

The parameters m_{ij} and $\mathbf{b}_z$ of this model can be estimated from observations of sensor outputs when the inputs are known, the process called **calibration.**

The purpose of calibration is **sensor compensation**, which is essentially inverting the input-output of Equation 3.1 to obtain

$$\mathbf{z}_{\text{input}} = \mathbf{M}^{-1}\mathbf{z}_{\text{output}} - \mathbf{b}_z, \tag{3.3}$$

the sensor inputs compensated for scale factor, misalignment, and bias errors.

This result can be generalized for a cluster of $N \geq 3$ gyroscopes or accelerometers, the effects of individual **biases, scale factors**, and **input axis misalignments** can be modeled by an equation of the form

$$\underbrace{\mathbf{z}_{\text{input}}}_{3\times 1} = \underbrace{\mathbf{M}^{\dagger}_{\text{scale factor \& misalignment}}}_{3\times N} \underbrace{\mathbf{z}_{\text{output}}}_{N\times 1} - \underbrace{\mathbf{b}_z}_{3\times 1}, \tag{3.4}$$

where $\mathbf{M}^{\dagger}$ is the Moore–Penrose pseudoinverse of the corresponding $\mathbf{M}$, which can be determined by calibration.

Compensation In this case, calibration amounts to estimating the values of $\mathbf{M}^{\dagger}$ and $\mathbf{b}_z$, given input–output pairs $[\mathbf{z}_{\text{input},\,k}, \mathbf{z}_{\text{output},\,k}]$, where $\mathbf{z}_{\text{input},\,k}$ is known from controlled calibration conditions and $\mathbf{z}_{\text{output},\,k}$ is recorded under these conditions. For accelerometers, controlled conditions may include the direction and magnitude of gravity, conditions on a shake table, or those on a centrifuge. For gyroscopes, controlled conditions may include the relative direction of the rotation axis of Earth (e.g. with sensors mounted on a two-axis indexed rotary table), or controlled conditions on a rate table.

The full set of input–output pairs under J sets of calibration conditions yields a system of $3J$ linear equations

$$\underbrace{\begin{bmatrix} z_{1,\text{ input},\ 1} \\ z_{2,\text{ input},\ 1} \\ z_{3,\text{ input},\ 1} \\ \vdots \\ z_{3,\text{ input},\ J} \end{bmatrix}}_{3J\text{ knowns}} = \underbrace{\begin{bmatrix} z_{1,\text{ output},\ 1} & z_{2,\text{ output},\ 1} & z_{3,\text{ output},\ 1} & \cdots & 0 \\ 0 & 0 & 0 & \cdots & 0 \\ 0 & 0 & 0 & \cdots & 1 \\ \vdots & \vdots & \vdots & \ddots & \vdots \\ 0 & 0 & 0 & \cdots & 1 \end{bmatrix}}_{\mathbf{Z},\text{ a } 3J\times(3N+3)\text{ matrix of knowns}} \underbrace{\begin{bmatrix} m_{1,\ 1} \\ m_{1,\ 2} \\ m_{1,\ 3} \\ \vdots \\ b_{3,\ z} \end{bmatrix}}_{3N+3\text{ unknowns}} \tag{3.5}$$

in the $3N$ unknown parameters $m_{i,\ j}$ (the elements of the matrix $\mathbf{M}^{\dagger}$) and 3 unknown parameters $b_{i,\ z}$ (rows of the 3-vector $\mathbf{b}_z$), which will be overdetermined for $J > N + 1$. In that case, the system of linear equations may be solvable for the $3(N + 1)$ calibration parameters by using the method of least-squares,

$$\begin{bmatrix} m_{1,\ 1} \\ m_{1,\ 2} \\ m_{1,\ 3} \\ \vdots \\ b_{3,\ z} \end{bmatrix} = [\mathbf{Z}^{\mathrm{T}}\mathbf{Z}]^{-1}\,\mathbf{Z}^{\mathrm{T}} \begin{bmatrix} z_{1,\text{ input},\ 1} \\ z_{2,\text{ input},\ 1} \\ z_{3,\text{ input},\ 1} \\ \vdots \\ z_{3,\text{ input},\ J} \end{bmatrix} \tag{3.6}$$

provided that the matrix $\mathbf{Z}^{\mathrm{T}}\mathbf{Z}$ is nonsingular.

The values of $\mathbf{M}^{\dagger}$ and $\mathbf{b}_z$ determined in this way are called **calibration parameters.**

Estimation of the calibration parameters can also be done using Kalman filtering, a by-product of which would be the covariance matrix of calibration parameter uncertainty. This covariance matrix is also useful in modeling system-level performance.

3.3.4.2 Calibration Parameter Drift

INS calibration parameters may not be exactly constant over time. Their values may change significantly over the operational life of the INS. Specifications for calibration stability generally divide these calibration parameter variations into two categories:

Turn-on to turn-on changes that occur between a system shut-down and the next start-up. They may be caused by temperature transients or power turn-on effects during shut-downs and turn-ons, and may represent stress relief mechanisms within materials and between assembled parts. They are generally considered to be independent from turn-on to turn-on, so the model for the covariance of calibration errors for the kth turn-on would be of the form

$$\mathbf{P}_{\text{calib.},\ k} = \mathbf{P}_{\text{calib.},\ k-1} + \mathbf{\Delta P}_{\text{calib.}}, \tag{3.7}$$

where $\Delta \mathbf{P}_{\text{calib.}}$ is the covariance of turn-on-to-turn-on parameter changes. The initial value $\mathbf{P}_{\text{calib., 0}}$ at the end of calibration is usually determinable from error covariance analysis of the calibration process. Note that this is the covariance model for a random walk, the covariance of which grows without bound.

Long-term drift, sometimes called "aging," which has been attributed to such long-term phenomena as material migration within solids or ion diffusion within crystalline materials. Its calibration parameter uncertainty covariance equation has the same form as Eq. (3.7), but with $\Delta \mathbf{P}_{\text{calib.}}$ now representing the calibration parameter drift in the time interval $\Delta t = t_k - t_{k-1}$ between successive discrete times within an operational period.

Predicting Incipient System or Sensor Failures Incipient sensor failures can possibly be predicted by observing over time the rate of change of sensor calibration parameters, depending on experience over the lifetimes of inertial navigators of the same type. One of the advantages of tightly coupled GNSS/INS integration is that INS sensors can be continuously calibrated all the time that GNSS data is available. System health monitoring can then include tests for the trends of sensor calibration parameters, setting threshold conditions for failing the INS system, and isolating a likely set of causes for the observed trends.

3.3.5 Carouseling and Indexing

These are methods used for cancelling the effect of some sensor output bias errors on navigation errors by rotating the sensor input axis either continuously (carouseling) or discretely (indexing), effectively zeroing the inertial integral of the sensor-fixed errors.

Carouseling was first introduced at the Delco Division of General Motors in the early 1970s for gimbaled systems, using a vertical gimbal axis for the rotation. It has also been applied to strapdown systems, using ISA rotation about a nominally vertical yaw axis, and to strapdown MEMS systems for navigation [7], and evaluated for MEMS gyrocompasses [8].

Indexing (also called "gimbal flipping") has been used quite extensively in the US Navy's electrostatically supported gyro navigator (ESGN), using independent indexing around and end-to-end on the spin axes of its electrostatic gyroscopes. This indexing does not cancel the effects of rotor axial mass unbalance, but it cancels just about everything else.

3.4 Inertial Navigation Models

Besides models for mitigating sensor errors, inertial navigation systems require internal models for the external world in which they must navigate. These models define the navigation coordinates to be used, including any departures from inertial coordinates. For terrestrial navigation, those departures include gravity, rotation, and the non-rectilinear shapes of Earth-fixed coordinates.

3.4.1 Geoid Models

Carl Friedrich Gauss introduced the idea of a *geoid* in 1828 as the "mathematical figure of the Earth," defining its shape as that of a reference equipotential surface (e.g. mean sea level), and defining locations in the near-Earth environment in terms of longitude, latitude and altitude with respect to that geoid shape.

Geoid models began as oblate spheroids – something predicted by Isaac Newton as the real shape of the Earth – with the Earth's center of mass at its center, the Earth rotation axis along its short axis, and the Airy Transit Circle at the Royal Observatory in Greenwich at its prime meridian.

Later refinements included undulations of the equipotential surface above and below the reference ellipsoid, represented in terms of spherical harmonics. That basic concept is still in use today, and it is used for GNSS navigation. Geoid models are also used for inertial navigation. Besides providing a common grid solution for position, geoids also provide models for the near-Earth gravitational field.

There are different geoids for different purposes. Those used for surveying generally include some model for the local vertical direction as a function of location, because the local vertical is used as a reference direction in surveying. Those used in hydrology need to include something equivalent to gravitational potential. Those used in inertial navigation need to include the local gravitational acceleration magnitude and direction.

There have been many geoid models over the years, and multiple international efforts to establish a common one. The WGS84 reference geoid was established by the 1984 World Geodetic Survey, and it has become widely used in GNSS and inertial navigation.

3.4.2 Terrestrial Navigation Coordinates

Descriptions of the major coordinates used in inertial navigation and GNSS/INS integration are described in Appendix B (www.wiley.com/go/grewal/gnss). These include coordinate systems used for representing the trajectories of GNSS satellites and user vehicles in the near-Earth environment and for representing the attitudes of host vehicles relative to locally level coordinates, including the following:

1. **Inertial coordinates**
   (a) Earth-centered inertial (ECI), with origin at the center of mass of the Earth and principal axes in the directions of the vernal equinox and the rotation axis of the Earth.
   (b) Satellite orbital coordinates, used in GNSS ephemerides.

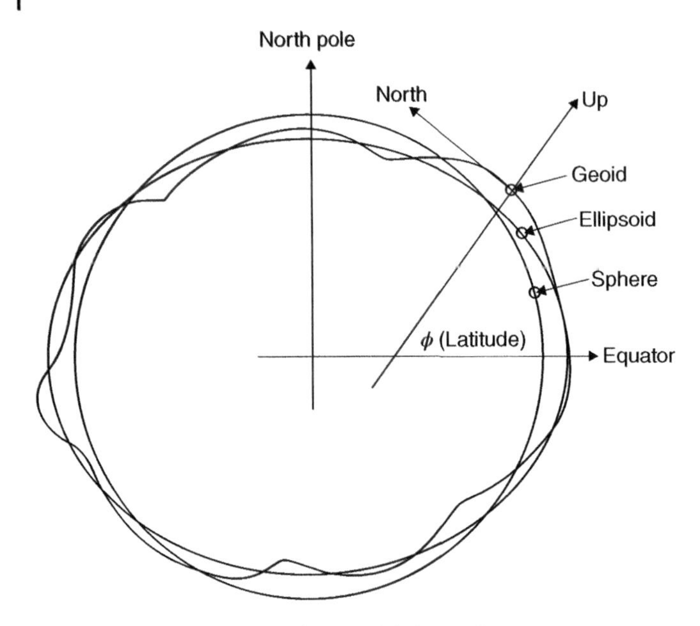

Figure 3.8 Equipotential surface models for Earth.

2. **Earth-fixed coordinates**
   (a) Earth-centered, Earth-fixed (ECEF), with origin at the center of mass of the Earth and principal axes in the directions of the prime meridian at the equator and the rotation axis of the Earth.
   (b) Geodetic coordinates based on a reference geoid. Geodetic latitude, defined as the angle between the equatorial plane and the local vertical, as illustrated in Figure 3.8. It can differ from geocentric latitude by as much as 12 arc minutes, equivalent to about 20 km of northing distance.
   (c) Local tangent plane (LTP) coordinates, also called "locally level coordinates," essentially representing the Earth as being locally flat. These coordinates are particularly useful from a human factors standpoint for representing the attitude of the host vehicle and for representing local directions. They include
      i. East–north–up (ENU), shown in Figure B.7;
      ii. North–east–down (NED), which can be simpler to relate to vehicle coordinates; and
      iii. Alpha wander or azimuth wander, rotated from ENU coordinates through an angle α about the local vertical, as shown in Figure B.8.
3. **Vehicle-fixed coordinates**
   (a) Roll–pitch–yaw (RPY), as shown in Figure B.9.

Transformations between these different coordinate systems are important for representing vehicle attitudes, for resolving inertial sensor outputs into inertial navigation coordinates, and for GNSS/INS integration. Methods used for representing and implementing coordinate transformations are also presented in Section B.4.

3.4.3 Earth Rotation Model

Our Earth is the mother of all clocks. It has given us the time units of days, hours, minutes, and seconds that govern our lives. Not until the development of atomic clocks in the mid-twentieth century were we able to observe the imperfections in our Earth-clock due to shifting mass distributions caused by tectonic events, the transfers of angular momentum with the atmosphere and hydrosphere, and the gradual slowdown from the transfer of energy and angular momentum within the Earth–moon system due to tides.[4] Despite these, we continue to use Earth rotation as our primary time reference, adding or subtracting leap seconds to atomic clocks to keep them synchronized to the rotation of the Earth. These time variations are significant for GNSS navigation, but not for inertial navigation.

The value of earthrate in the World Geodetic System 1984 (WGS 84) Earth model is $7\ 292\ 115\ 167 \times 10^{-14}$ rad/s, or about 15.04109 deg/h. This is its *sidereal* rotation rate with respect to distant stars. Its mean rotation rate with respect to the nearest star (our Sun), as viewed from the rotating Earth, is 15 deg/h, averaged over one year.

3.4.4 Gravity Models

Because an INS operates in a world with gravitational accelerations it is unable to sense and unable to ignore, it must use a reasonably faithful model of gravity.

Gravity models for the Earth include centrifugal acceleration due to rotation of the Earth as well as true gravitational accelerations due to the mass distribution of the Earth, but they do not generally include oscillatory effects such as tidal variations.

3.4.4.1 Gravitational Potential

Gravitational potential of a unit of mass is defined to be zero at a point infinitely distant from all massive bodies and to decrease toward massive bodies such as the Earth. That is, a point at infinity is the reference point for gravitational potential.

In effect, the gravitational potential at a point in or near the Earth is defined by the potential energy lost per unit of mass falling to that point from infinite

---

4 An effect discovered by George Darwin, second son of Charles Darwin.

altitude. In falling from infinity, potential energy is converted to kinetic energy, $\mathbf{m}\mathbf{v}_{\text{escape}}^2/2$, where $\mathbf{v}_{\text{escape}}$ is the *escape velocity*. Escape velocity at the surface of the Earth is about 11 km/s.

3.4.4.2 Gravitational Acceleration

Gravitational acceleration is the negative gradient of gravitational potential. Potential is a scalar function, and its gradient is a vector. Because gravitational potential increases with altitude, its gradient points upward and the negative gradient points downward.

3.4.4.3 Equipotential Surfaces

An equipotential surface is a surface of constant gravitational potential. If the ocean and atmosphere were not moving, then the surface of the ocean at static equilibrium would be an equipotential surface. *Mean sea level* is a theoretical equipotential surface obtained by time-averaging the dynamic effects. *Orthometric altitude* is measured along the (curved) plumbline.

WGS84 Ellipsoid The WGS84 Earth model approximates mean sea level (an equipotential surface) by an ellipsoid of revolution with its rotation axis coincident with the rotation axis of the Earth, its center at the center of mass of the Earth, and its prime meridian through Greenwich. Its semimajor axis (equatorial radius) is defined to be 6 378 137 m, and its semiminor axis (polar radius) is defined to be 6 356 752.3142 m.

Geoid Models Geoids are approximations of mean sea-level orthometric height with respect to a reference ellipsoid. Geoids are defined by additional higher-order shapes, commonly modeled by spherical harmonics of height deviations from an ellipsoid, as illustrated in Figure 3.8. There are many geoid models based on different data, but the more recent, most accurate models depend heavily on GPS data. Geoid heights deviate from reference ellipsoids by tens of meters, typically.

The WGS84 geoid heights vary about ± 100 m from the reference ellipsoid. As a rule, oceans tend to have lower geoid heights and continents tend to have higher geoid heights. Coarse 20-m contour intervals are plotted versus longitude and latitude in Figure 3.9, with geoid regions above the ellipsoid shaded gray.

3.4.4.4 Longitude and Latitude Rates

The second integral of acceleration in locally level coordinates should result in the estimated vehicle position. This integral is somewhat less than straightforward when longitude and latitude are the preferred horizontal location variables.

The rate of change of vehicle altitude equals its vertical velocity, which is the first integral of net (i.e. including gravity) vertical acceleration. The rates of change of vehicle longitude and latitude depend on the horizontal components of vehicle velocity, but in a less direct manner. The relationship between longitude and latitude rates and east and north velocities is further complicated by the oblate shape of the Earth.

The rates at which these angular coordinates change as the vehicle moves tangent to the surface will depend upon the radius of curvature of the reference surface model. Radius of curvature can depend on the direction of travel, and for an ellipsoidal model there is one radius of curvature for north–south motion and another radius of curvature for east–west motion.

Meridional Radius of Curvature The radius of curvature for north–south motion is called the "meridional" radius of curvature, because north–south travel is along a meridian (i.e. line of constant longitude). For an ellipsoid of revolution, all meridians have the same shape, which is that of the ellipse that was rotated to produce the ellipsoidal surface model. The tangent circle with the same radius of curvature as the ellipse is called the "*osculating circle*" (osculating

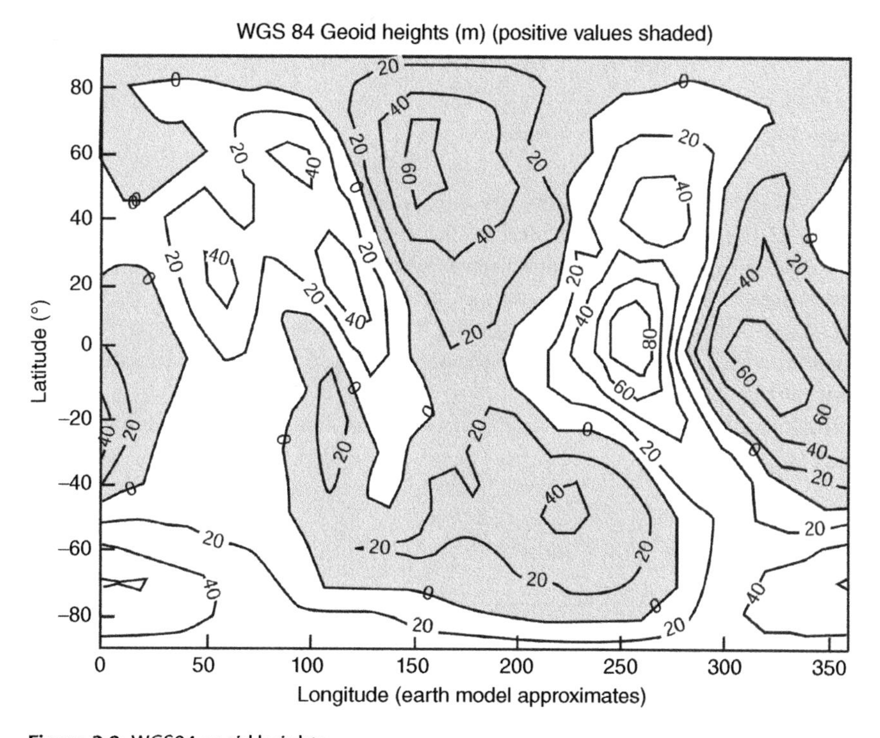

Figure 3.9 WGS84 geoid heights.

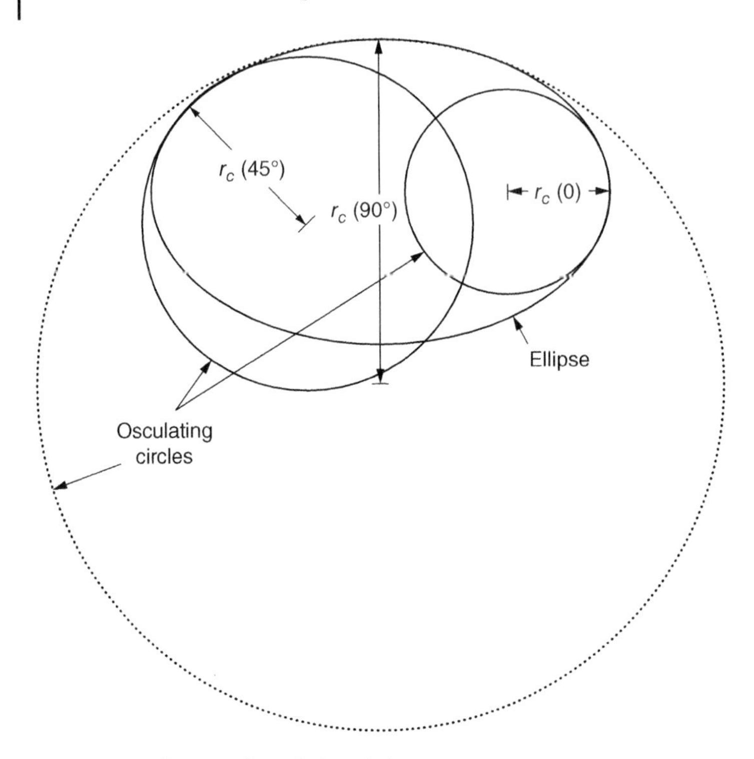

Figure 3.10 Ellipse and osculating circles.

means "kissing"). As illustrated in Figure 3.10 for an oblate Earth model, the radius of the meridional osculating circle is smallest where the geocentric radius is largest (at the equator), and the radius of the osculating circle is largest where the geocentric radius is smallest (at the poles). The osculating circle lies inside or on the ellipsoid at the equator and outside or on the ellipsoid at the poles and passes through the ellipsoid surface for latitudes in between.

The formula for meridional radius of curvature as a function of geodetic latitude (ϕ_{geodetic}) is

$$r_M = \frac{b^2}{a[1 - e^2 \sin^2(\phi_{\text{geodetic}})]^{3/2}} = \frac{a(1 - e^2)}{[1 - e^2 \sin^2(\phi_{\text{geodetic}})]^{3/2}} \tag{3.8}$$

where a is the semimajor axis of the ellipse, b is the semiminor axis, and $e^2 = (a^2 - b^2)/a^2$ is the eccentricity squared.

Geodetic Latitude Rate The rate of change of geodetic latitude as a function of north velocity is then

$$\frac{d\phi_{\text{geodetic}}}{dt} = \frac{v_N}{r_M + h} \tag{3.9}$$

and geodetic latitude can be maintained as the integral

$$\phi_{\text{geodetic}}(t_{\text{now}}) = \phi_{\text{geodetic}}(t_{\text{start}}) + \int_{t_{\text{start}}}^{t_{\text{now}}} \frac{v_{N(t)}\, dt}{a(1-e^2)/\{1-e^2\sin^2[\phi_{\text{geodetic}}(t)]\}^{3/2} + h(t)]} \tag{3.10}$$

where $h(t)$ is height above (+) or below (−) the ellipsoid surface and $\phi_{\text{geodetic}}(t)$ will be in radians if $v_N(t)$ is in meters per second and $r_M(t)$ and $h(t)$ are in meters.

Transverse Radius of Curvature The radius of curvature of the reference ellipsoid surface in the east–west direction (i.e. orthogonal to the direction in which the meridional radius of curvature is measured) is called the *transverse radius of curvature.* It is the radius of the osculating circle in the local east–up plane, as illustrated in Figure 3.11, where the arrows at the point of tangency of the transverse osculating circle are in the local ENU coordinate directions. As this figure illustrates, on an oblate Earth, the plane of a transverse osculating circle does not pass through the center of the Earth, except when the point of osculation is at the equator. (All osculating circles at the poles are in meridional planes.) Also, unlike meridional osculating circles, transverse osculating circles generally lie outside the ellipsoidal surface, except at the point of tangency and at the equator, where the transverse osculating circle *is* the equator.

Figure 3.11 Transverse osculating circle.

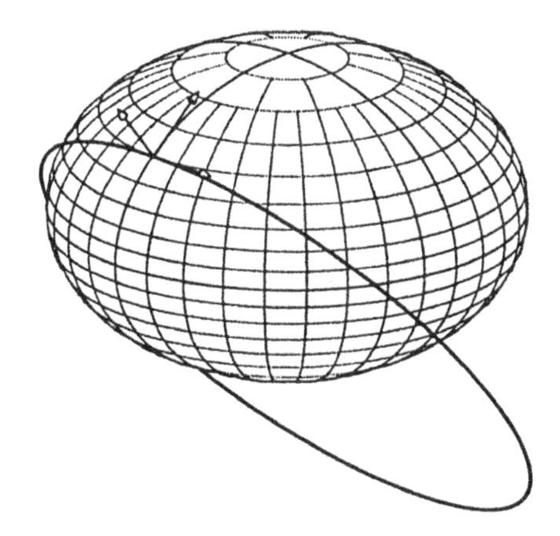

The formula for the transverse radius of curvature on an ellipsoid of revolution is

$$r_T = \frac{a}{\sqrt{1 - e^2 \sin^2(\phi_{\text{geodetic}})}} \tag{3.11}$$

where a is the semimajor axis of the generating ellipse and e is its eccentricity.

Longitude Rate The rate of change of longitude as a function of east velocity is then

$$\frac{d\theta}{dt} = \frac{v_E}{\cos(\phi_{\text{geodetic}})(r_T + h)} \tag{3.12}$$

and longitude can be maintained by the integral

$$\theta(t_{\text{now}}) = \theta(t_{\text{start}}) + \int_{t_{\text{start}}}^{t_{\text{now}}} \frac{v_E(t)\, dt}{\cos[\phi_{\text{geodetic}}(t)](a/\sqrt{1 - e^2 \sin^2(\phi_{\text{geodetic}}(t))} + h(t))} \tag{3.13}$$

where $h(t)$ is height above (+) or below (−) the ellipsoid surface and θ will be in radians if $v_E(t)$ is in meters per second and $r_T(t)$ and $h(t)$ are in meters. Note that this formula has a singularity at the poles, where $\cos(\phi_{\text{geodetic}}) = 0$, a consequence of using latitude and longitude as location variables.

WGS84 Reference Surface Curvatures The apparent variations in meridional radius of curvature in Figure 3.10 are rather large because the ellipse used in generating Figure 3.10 has an eccentricity of about 0.75. The WGS84 ellipse has an eccentricity of about 0.08, with geocentric, meridional, and transverse radius of curvature as plotted in Figure 3.12 versus geodetic latitude. For the WGS84 model,

- Mean geocentric radius is about 6371 km, from which it varies by −14.3 km (−0.22%) to +7.1 km (+0.11%).
- Mean meridional radius of curvature is about 6357 km, from which it varies by −21.3 km (−0.33%) to 42.8 km (+0.67%).
- Mean transverse radius of curvature is about 6385 km, from which it varies by −7.1 km (−0.11%) to +14.3 km (+0.22%).

Because these vary by several parts per thousand, one must take radius of curvature into account when integrating horizontal velocity increments to obtain longitude and latitude.

3.4.5 Attitude Models

Attitude models for inertial navigation represent

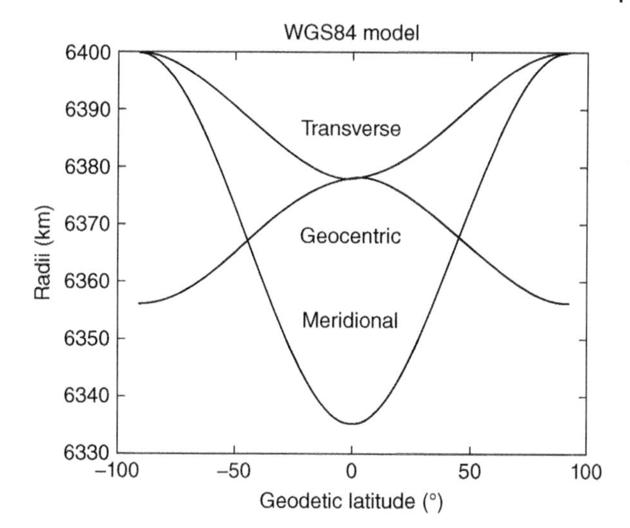

Figure 3.12 Radii of WGS84 reference ellipsoid.

1) The relative rotational orientations of two coordinate systems, usually represented by coordinate transformation matrices but also represented in terms of rotation vectors.
2) Attitude dynamics, usually represented in terms of three-dimensional rotation rate vectors but also represented by four-dimensional quaternion.

3.4.5.1 Coordinate Transformation Matrices and Rotation Vectors

Appendix B on www.wiley.com/go/grewal/gnss is all about the coordinates, coordinate transformation matrices, and rotation vectors used in inertial navigation.

3.4.5.2 Attitude Dynamics

Rate gyroscopes used in inertial navigation measure components of a rotation rate vector $\vec{\omega}_{\mathrm{ISA}}$ in ISA-fixed coordinates, which had historically been used to express its effect on a coordinate transformation matrix $\mathbf{C}_{\mathrm{INT}}^{\mathrm{ISA}}(t)$ from ISA coordinates to the coordinates for integration in terms of a linear differential equation of the sort

$$\begin{aligned}\frac{d}{dt}\mathbf{C}_{\mathrm{INT}}^{\mathrm{ISA}}(t) &= [\vec{\omega}_{\mathrm{ISA}}\otimes]\mathbf{C}_{\mathrm{INT}}^{\mathrm{ISA}}(t)\\ &= \begin{bmatrix} 0 & -\omega_3 & \omega_2 \\ \omega_3 & 0 & -\omega_1 \\ -\omega_2 & \omega_1 & 0 \end{bmatrix}\mathbf{C}_{\mathrm{INT}}^{\mathrm{ISA}}(t), \end{aligned} \tag{3.14}$$

which is not particularly well-conditioned for numerical integration.

The alternative representation in terms of quaternions is described in Section 3.6.1.2 and (in greater detail) in Appendix B.

3.5 Initializing The Navigation Solution

3.5.1 Initialization from an Earth-fixed Stationary State

3.5.1.1 Accelerometer Recalibration

This is only possible with gimbaled systems and it adds to the start-up time, but it can improve performance. Navigation accuracy is very sensitive to accelerometer biases, which can shift due to thermal transients in turn-on/turn-off cycles, and can also drift randomly over time for some accelerometer design. Fortunately, gimbals can be used to calibrate accelerometer biases in a stationary 1g environment. Bias and scale factors can both be determined by using the gimbals to point each accelerometer input axes straight up and then straight down (by nulling the horizontal accelerometer outputs). Then each accelerometer's bias is the average of the up and down outputs and scale factor is half the difference divided by the local gravitational acceleration.

3.5.1.2 Initializing Position and Velocity

Velocity. Zero, by definition.

Position. From GNSS, if available, otherwise from local sources – e.g. signs or local auxiliary sensors (see Section 3.5.1.3).

3.5.1.3 Initializing ISA Attitude

Gyrocompassing. Gyrocompass alignment of stationary vehicles uses the sensed direction of acceleration to determine the local vertical and the sensed direction of rotation to determine north, as illustrated in Figure 3.13.

Using auxiliary sensors. Non-gyroscopic attitude sensors can also be used as aids in alignment. These include the following:

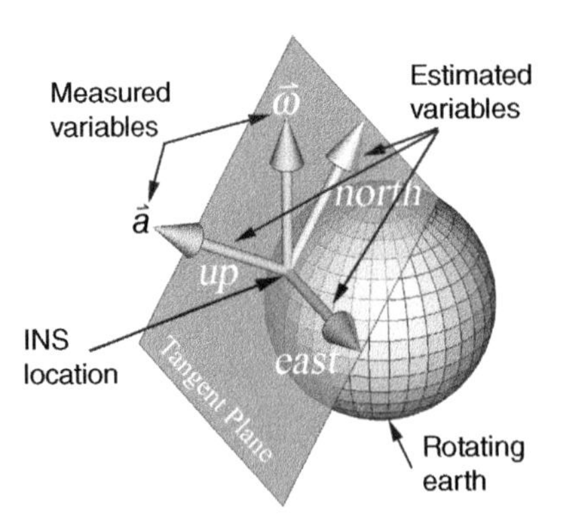

Figure 3.13 Gyrocompassing determines sensor orientations with respect to east, north, and up.

Magnetic sensors are used primarily for coarse heading alignment, to speed up INS alignment.

Star trackers are used primarily for space-based or near-space applications. The Snark cruise missile and the U-2 spy plane used inertial-platform-mounted star trackers to maintain INS alignment on long flights, an idea attributed to Northrop Aviation.

Optical alignment systems have been used on some systems prior to launch. Some use Porro prisms mounted on the inertial platform to maintain optical line-of-sight reference through ground-based theodolites to reference directions at the launch complex.

Quasi magnetostatic sensors have been used in virtual reality systems for determining the attitude and position of the headset relative to its environment. These use three orthogonal and independently coded low-to-medium frequency magnetic dipole sources to illuminate the local area where rotational and locational orientation data is required, and three-axis AC magnetic sensors as receivers.

3.5.1.4 Gyrocompass Alignment Accuracy

Gyrocompass alignment is not necessary for integrated GNSS/INS navigation, although many INSs may already be configured for it.

Accuracy A rough rule-of-thumb for gyrocompass alignment accuracy is

$$\sigma^2_{\text{gyrocompass}} > \sigma^2_{\text{acc}} + \frac{\sigma^2_{\text{gyro}}}{15^2\cos^2(\phi_{\text{geodetic}})} \tag{3.15}$$

where

$\sigma_{\text{gyrocompass}}$ is the minimum achievable root-mean-square (RMS) alignment error in radians,
σ_{acc} is the RMS accelerometer accuracy in g's,
σ_{gyro} is the RMS gyroscope accuracy in degrees per hour,
15 deg/h is the rotation rate of the Earth, and
ϕ_{geodetic} is the latitude at which gyrocompassing is performed.

Alignment accuracy is also a function of the time allotted for it, and the time required to achieve a specified accuracy is generally a function of sensor error magnitudes (including noise) and the degree to which the vehicle remains stationary.

Gimbaled implementation. Gyrocompass alignment for gimbaled systems is a process for aligning the inertial platform axes with the navigation coordinates using only the sensor outputs while the host vehicle is essentially stationary.

For systems using ENU navigation coordinates, for example, the platform can be tilted until two of its accelerometer inputs are zero, at which time both input axes will be horizontal. In this locally leveled orientation, the sensed rotation axis will be in the north–up plane, and the platform can be slewed about the vertical axis to null the input of one of its horizontal gyroscopes, at which time that gyroscope input axis will point east–west. That is the basic concept used for gyrocompass alignment, but practical implementation requires filtering[5] to reduce the effects of sensor noise and unpredictable zero-mean vehicle disturbances due to loading activities and/or wind gusts.

Strapdown implementation. Gyrocompass alignment for strapdown systems is a process for "virtual alignment" by determining the sensor cluster attitude with respect to navigation coordinates using only the sensor outputs while the system is essentially stationary.

Error-free implementation. If the sensor cluster could be firmly affixed to the Earth and there were no sensor errors, then the sensed acceleration vector $\mathbf{a}_{\text{output}}$ in sensor coordinates would be in the direction of the local vertical, the sensed rotation vector $\boldsymbol{\omega}_{\text{output}}$ would be in the direction of the Earth rotation axis, and the unit column vectors

$$\mathbf{1}_U = \frac{\mathbf{a}_{\text{output}}}{|\mathbf{a}_{\text{output}}|} \tag{3.16}$$

$$\mathbf{1}_N = \frac{w_{\text{output}} - (\mathbf{1}_U^{\mathrm{T}} w_{\text{output}})\mathbf{1}_U}{|w_{\text{output}} - (\mathbf{1}_U^{\mathrm{T}} w_{\text{output}})\mathbf{1}_U|} \tag{3.17}$$

$$\mathbf{1}_E = \mathbf{1}_N \otimes \mathbf{1}_U \tag{3.18}$$

would define the initial value of the coordinate transformation matrix from sensor-fixed coordinates to ENU coordinates:

$$\mathbf{C}_{\text{ENU}}^{\text{sensor}} = [\mathbf{1}_E | \mathbf{1}_N | \mathbf{1}_U]^{\mathrm{T}} \tag{3.19}$$

Practical implementation. In practice, the sensor cluster is usually mounted in a vehicle that is not moving over the surface of the Earth, but may be buffeted by wind gusts or disturbed during fueling and loading operations. Gyrocompassing then requires some amount of filtering (Kalman filtering, as a rule) to reduce the effects of vehicle buffeting and sensor noise. The gyrocompass filtering period is typically on the order of several minutes for a medium-accuracy INS but may continue for hours, days, or continuously for high-accuracy systems.

5 The vehicle dynamic model used for gyrocompass alignment filtering can be "tuned" to include the major resonance modes of the vehicle suspension.

3.5.2 Initialization on the Move

3.5.2.1 Transfer Alignment

This method is generally faster than gyrocompass alignment, but it requires another INS on the host vehicle and it may require special maneuvering of the host vehicle to attain observability of the alignment variables. It is commonly used for in-air INS alignment for missiles launched from aircraft and for on-deck INS alignment for aircraft launched from carriers. Alignment of carrier-launched aircraft may also use the direction of the velocity impulse imparted by the steam catapult.

3.5.2.2 Initializing Using GNSS

This is an issue in GNSS/INS integration, which is covered in Chapter 12. In this case it must also estimate the INS orientation and velocity, the observability of which generally depends on the host vehicle trajectory.

3.6 Propagating The Navigation Solution

3.6.1 Attitude Propagation

Knowing the instantaneous rotational orientations of the inertial sensor input axes with respect to navigational coordinates is essential for inertial navigation to work. The integration of accelerations for maintaining the navigation solution for velocity and position depends on it.

3.6.1.1 Strapdown Attitude Propagation

Strapdown Attitude Problems Early on, strapdown systems technology had an "attitude problem," which was the problem of representing attitude rate in a format amenable to accurate computer integration over high dynamic ranges. The eventual solution was to represent attitude in different mathematical formats as it is processed from raw gyro outputs to the matrices used for transforming sensed acceleration to inertial coordinates for integration.

Figure 3.14 illustrates the resulting major gyro signal processing operations, and the formats of the data used for representing attitude information. The processing starts with gyro outputs and ends with a coordinate transformation matrix from sensor coordinates to the coordinates used for integrating the sensed accelerations.

Coning Motion This type of motion is a problem for attitude integration when the frequency of motion is near or above the sampling frequency. It is usually a consequence of host vehicle frame vibration modes or resonances in the INS mounting, and INS shock and vibration isolation is often designed to eliminate or substantially reduce this type of rotational vibration.

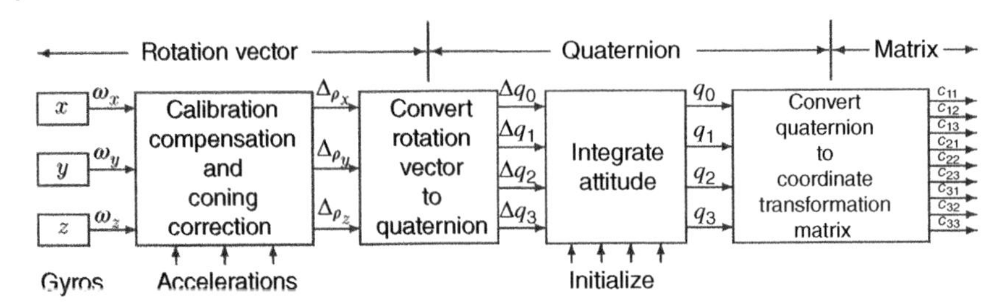

Figure 3.14 Strapdown attitude representations.

Coning motion is an example of an attitude trajectory (i.e. attitude as a function of time) for which the integral of attitude rates does *not* equal the attitude change. An example trajectory would be

$$\boldsymbol{\rho}(t) = \theta_{\text{cone}} \begin{bmatrix} \cos(\Omega_{\text{coning}} t) \\ \sin(\Omega_{\text{coning}} t) \\ 0 \end{bmatrix} \tag{3.20}$$

$$\dot{\boldsymbol{\rho}}(t) = \theta_{\text{cone}} \Omega_{\text{coning}} \begin{bmatrix} -\sin(\Omega_{\text{coning}} t) \\ \cos(\Omega_{\text{coning}} t) \\ 0 \end{bmatrix}, \tag{3.21}$$

where

$\boldsymbol{\rho}$ is the rotation vector,
θ_{cone} is called the *cone angle* of the motion,
Ω_{coning} is the *coning frequency* of the motion,

as illustrated in Figure 3.15.

The coordinate transformation matrix from body coordinates to inertial coordinates will be

$$\begin{aligned} \mathbf{C}^{\text{body}}_{\text{inertial}}(\boldsymbol{\rho}) &= \cos\theta \mathbf{I} + (1 - \cos\theta) \\ &\begin{bmatrix} \cos(\Omega_{\text{coning}} t)^2 & \sin(\Omega_{\text{coning}} t)\cos(\Omega_{\text{coning}} t) & 0 \\ \sin(\Omega_{\text{coning}} t)\cos(\Omega_{\text{coning}} t) & \sin(\Omega_{\text{coning}} t)^2 & 0 \\ 0 & 0 & 0 \end{bmatrix} \\ &+ \sin\theta \begin{bmatrix} 0 & 0 & \sin(\Omega_{\text{coning}} t) \\ 0 & 0 & -\cos(\Omega_{\text{coning}} t) \\ -\sin(\Omega_{\text{coning}} t) & \cos(\Omega_{\text{coning}} t) & 0 \end{bmatrix}, \end{aligned} \tag{3.22}$$

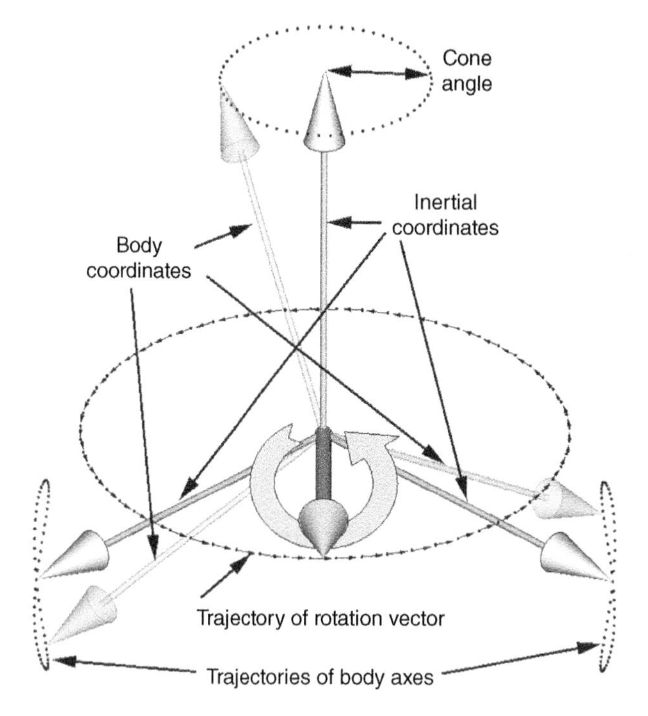

Figure 3.15 Coning motion.

and the measured inertial rotation rates in body coordinates will be

$$\begin{aligned}
\boldsymbol{\omega}_{\text{body}} &= \mathbf{C}_{\text{body}}^{\text{inertial}} \dot{\boldsymbol{\rho}}_{\text{inertial}} \\
&= \theta_{\text{cone}} \Omega_{\text{coning}} \, [\mathbf{C}_{\text{inertial}}^{\text{body}}]^{\text{T}} \begin{bmatrix} -\sin(\Omega_{\text{coning}} t) \\ \cos(\Omega_{\text{coning}} t) \\ 0 \end{bmatrix} \\
&= \begin{bmatrix} -\theta_{\text{cone}} \ \Omega_{\text{coning}} \ \sin(\Omega_{\text{coning}} \ t) \cos(\theta_{\text{cone}}) \\ \theta_{\text{cone}} \ \Omega_{\text{coning}} \ \cos(\Omega_{\text{coning}} \ t) \cos(\theta_{\text{cone}}) \\ -\sin(\theta_{\text{cone}}) \theta_{\text{cone}} \ \Omega_{\text{coning}} \end{bmatrix}.
\end{aligned} \tag{3.23}$$

The integral of $\boldsymbol{\omega}_{\text{body}}$

$$\int_{s=0}^{t} \boldsymbol{\omega}_{\text{body}}(s) \, ds = \begin{bmatrix} -\theta_{\text{cone}} \ \cos(\theta_{\text{cone}})[1 - \cos(\Omega_{\text{coning}} \ t)] \\ \theta_{\text{cone}} \ \cos(\theta_{\text{cone}}) \sin(\Omega_{\text{coning}} \ t) \\ -\sin(\theta_{\text{cone}}) \theta_{\text{cone}} \ \Omega_{\text{coning}} \ t \end{bmatrix}, \tag{3.24}$$

which is what a rate integrating gyroscope would measure.

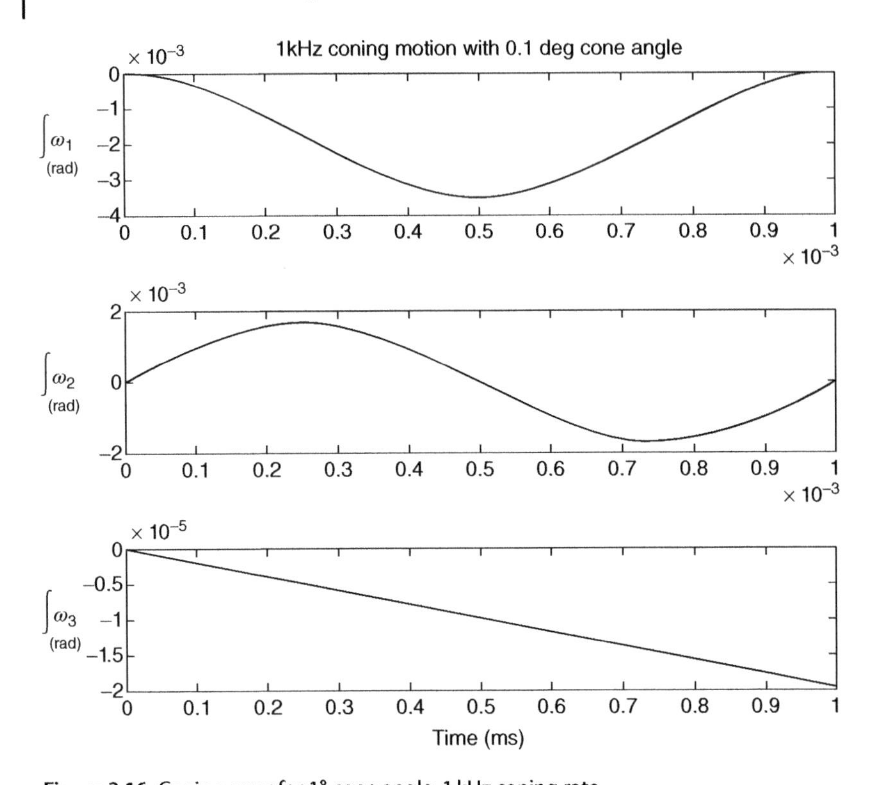

Figure 3.16 Coning error for 1° cone angle, 1 kHz coning rate.

The solutions for $\theta_{\text{cone}} = 0.1°$ and $\Omega_{\text{coning}} = 1\text{kHz}$ are plotted over one cycle (1 ms) in Figure 3.16. The first two components are cyclical, but the third component accumulates linearly over time at about -1.9×10^{-5} radians in one millisecond, which is a bit more than -1 deg/s. *This is why coning error compensation is important.*

Rotation Vector Implementation This implementation is primarily used at a faster sampling rate than the nominal sampling rate (i.e. that required for resolving measured accelerations into navigation coordinates). It is used to remove the nonlinear effects of coning and skulling motion that would otherwise corrupt the accumulated angle rates over the nominal intersample period. This implementation is also called a "coning correction."

Bortz Model for Attitude Dynamics This exact model for attitude integration based on measured rotation rates and rotation vectors was developed by John Bortz (1935–2013) [9]. It represents ISA attitude with respect to the reference inertial coordinate frame in terms of the rotation vector $\boldsymbol{\rho}$ required to rotate

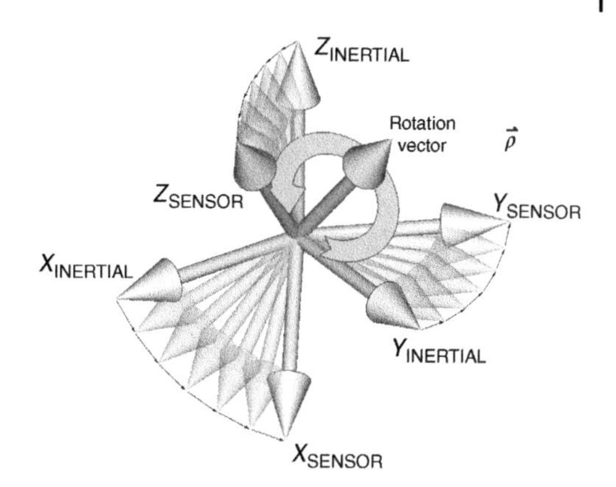

Figure 3.17 Rotation vector representing coordinate transformation.

the reference inertial coordinate frame into coincidence with the sensor-fixed coordinate frame, as illustrated in Figure 3.17.

The Bortz dynamic model for attitude then has the form

$$\dot{\rho} = \boldsymbol{\omega} + \mathbf{f}_{\text{Bortz}}(\boldsymbol{\omega}, \boldsymbol{\rho}), \tag{3.25}$$

where $\boldsymbol{\omega}$ is the vector of measured rotation rates. The Bortz "noncommutative rate vector"

$$\begin{aligned} \mathbf{f}_{\text{Bortz}}(\boldsymbol{\omega}, \boldsymbol{\rho}) &= \frac{1}{2}\boldsymbol{\rho} \otimes \boldsymbol{\omega} \\ &+ \frac{1}{\| \boldsymbol{\rho}\|^2}\left\{1 - \frac{\| \boldsymbol{\rho} \| \ \sin(\| \boldsymbol{\rho} \|)}{2\,[1 - \cos(\| \boldsymbol{\rho} \|)]}\right\} \boldsymbol{\rho} \otimes (\boldsymbol{\rho} \otimes \boldsymbol{\omega}) \end{aligned} \tag{3.26}$$

$$|\boldsymbol{\rho}| < \frac{\pi}{2}. \tag{3.27}$$

Equation (3.25) represents the rate of change of attitude as a nonlinear differential equation that is linear in the measured instantaneous body rates $\boldsymbol{\omega}$. Therefore, by integrating this equation over the nominal intersample period $[0,\ \Delta t]$ with initial value $\boldsymbol{\rho}(0) = 0$, an exact solution of the body attitude change over that period can be obtained in terms of the net rotation vector

$$\Delta\boldsymbol{\rho}(\Delta t) = \int_0^{\Delta t} \dot{\boldsymbol{\rho}}(\boldsymbol{\rho}(s),\ \boldsymbol{\omega}(s))\, ds \tag{3.28}$$

that avoids all the noncommutativity errors, and satisfies the constraint of Eq. (3.27) so long as the body cannot turn 180° in one sample interval Δt. In practice, the integral is done numerically with the gyro outputs ω_1, ω_2, ω_3 sampled at intervals $\delta t \ll \Delta t$. The choice of δt is usually made by analyzing the gyro outputs under operating conditions (including vibration isolation), and selecting a sampling frequency $1/\delta t$ well above the Nyquist frequency for the observed attitude rate spectrum. The frequency response of the gyros also enters into this design analysis.

The MATLAB® function `fBortz.m` on www.wiley.com/go/grewal/gnss calculates $\mathbf{f}_{\text{Bortz}}(\boldsymbol{\omega})$ defined by Eq. (3.26).

3.6.1.2 Quaternion Implementation

The quaternion representation of vehicle attitude is the most reliable, and it is used as the "holy point" of attitude representation. Its value is maintained using the incremental rotations $\Delta\boldsymbol{\rho}$ from the rotation vector representation, and the resulting values are used to generate the coordinate transformation matrix for accumulating velocity changes in inertial coordinates.

Quaternions represent three-dimensional attitude on the three-dimensional surface of the four-dimensional sphere, much like two-dimensional directions can be represented on the two-dimensional surface of the three-dimensional sphere.

Converting Incremental Rotations to Incremental Quaternions An incremental rotation vector $\Delta\boldsymbol{\rho}$ from the Bortz coning correction implementation of Eq. (3.28) can be converted to an equivalent incremental quaternion $\Delta\mathbf{q}$ by the operations

$$\theta = |\Delta\boldsymbol{\rho}| \text{ (rotation angle in radians)} \tag{3.29}$$

$$\begin{aligned}\mathbf{u} &= \frac{1}{\theta}\Delta\boldsymbol{\rho} \\ &= \begin{bmatrix} u_1 \\ u_2 \\ u_3 \end{bmatrix} \text{ (unit vector)}\end{aligned} \tag{3.30}$$

$$\begin{aligned}\Delta\mathbf{q} &= \begin{bmatrix} \cos\left(\frac{\theta}{2}\right) \\ u_1 \sin\left(\frac{\theta}{2}\right) \\ u_2 \sin\left(\frac{\theta}{2}\right) \\ u_3 \sin\left(\frac{\theta}{2}\right) \end{bmatrix} \\ &= \begin{bmatrix} \Delta q_0 \\ \Delta q_1 \\ \Delta q_2 \\ \Delta q_3 \end{bmatrix} \text{ (unit quaternion).}\end{aligned} \tag{3.31}$$

Quaternion implementation of attitude integration

If $\mathbf{q}_{k-1}$ is the quaternion representing the prior value of attitude,
$\Delta\mathbf{q}$ is the quaternion representing the change in attitude, and
$\mathbf{q}_k$ is the quaternion representing the updated value of attitude,
then the update equation for quaternion representation of attitude is

$$\mathbf{q}_k = \Delta\mathbf{q} \times \mathbf{q}_{k-1} \times \Delta\mathbf{q}^{\star} \tag{3.32}$$

here "× represents quaternion multiplication (defined in Appendix B) and the superscript ⋆ represents the conjugate of a quaternion,

$$\begin{bmatrix} q_1 \\ q_2 \\ q_3 \\ q_4 \end{bmatrix}^{\star} \stackrel{\text{def}}{=} \begin{bmatrix} q_1 \\ -q_2 \\ -q_3 \\ -q_4 \end{bmatrix} \tag{3.33}$$

3.6.1.3 Direction Cosines Implementation

The coordinate transformation matrix $\mathbf{C}^{\text{body}}_{\text{inertial}}$ from body-fixed coordinates to inertial coordinates is needed for transforming discretized velocity changes measured by accelerometers into inertial coordinates for integration. The quaternion representation of attitude is used for computing $\mathbf{C}^{\text{body}}_{\text{inertial}}$.

Quaternions to Direction Cosines Matrices The direction cosines matrix $\mathbf{C}^{\text{body}}_{\text{inertial}}$ from body-fixed coordinates to inertial coordinates can be computed from its equivalent unit quaternion representation

$$\mathbf{q}^{\text{body}}_{\text{inertial}} = \begin{bmatrix} q_0 \\ q_1 \\ q_2 \\ q_3 \end{bmatrix} \tag{3.34}$$

as

$$\begin{aligned} \mathbf{C}^{\text{body}}_{\text{inertial}} &= (2\,q_0^2 - 1)\mathbf{I}_3 + 2\begin{bmatrix} q_1 \\ q_2 \\ q_3 \end{bmatrix} \times \begin{bmatrix} q_1 \\ q_2 \\ q_3 \end{bmatrix}^{\mathrm{T}} - 2\,q_0 \begin{bmatrix} q_1 \\ q_2 \\ q_3 \end{bmatrix} \otimes \\ &= \begin{bmatrix} (2\,q_0^2 - 1 + 2\,q_1^2) & (2\,q_1q_2 + 2\,q_0q_3) & (2\,q_1q_3 - 2\,q_0q_2) \\ (2\,q_1q_2 - 2\,q_0q_3) & (2\,q_0^2 - 1 + 2\,q_2^2) & (2\,q_2^2 + 2\,q_0q_1) \\ (2\,q_1q_3 + 2\,q_0q_2) & (2\,q_2^2 - 2\,q_0q_1) & (2\,q_0^2 - 1 + 2\,q_3^2) \end{bmatrix} \end{aligned} \tag{3.35}$$

Strapdown with Whole-angle Gyroscopes These are momentum wheel gyroscopes whose spin axes are unconstrained, such as they are for Foucault's gimbaled gyroscope or electrostatic gyroscopes. In either case, the direction of the spin axis serves as an inertial reference – like a star. Two of these with quasi-orthogonal spin axes provide a complete inertial reference system. The only attitude propagation required in this case is to compensate for the spin axis drifts due to bearing torques and rotor axial mass unbalance (an acceleration-sensitive effect). The different coordinate systems involved are illustrated in Figure 3.18.

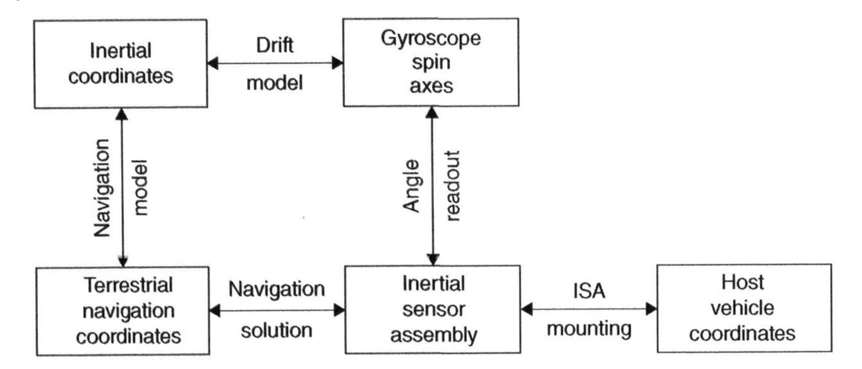

Figure 3.18 Coordinates for strapdown navigation with whole-angle gyroscopes.

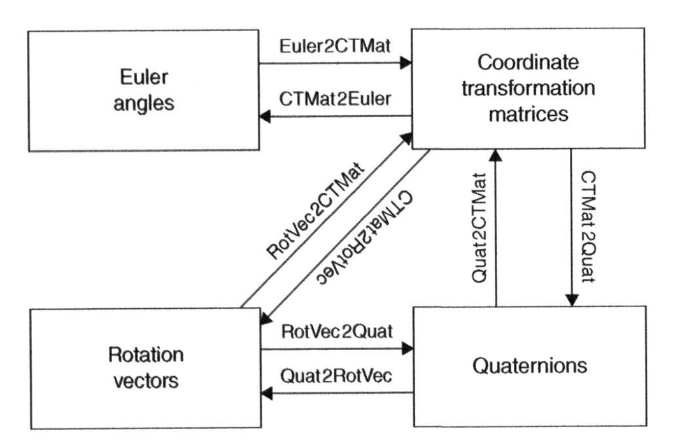

Figure 3.19 Attitude representation formats and MATLAB® transformations.

3.6.1.4 MATLAB® Implementations

Figure 3.19 shows four different representations used for relative rotational orientations, and the names of the MATLAB® script m-files (i.e. with the added ending `.m`) on www.wiley.com/go/grewal/gnss for transforming from one representation to another.

3.6.1.5 Gimbal Attitude Implementations

The primary function of gimbals is to isolate the ISA from vehicle rotations, but they are also used for other INS functions.

Vehicle Attitude Determination The gimbal angles determine the vehicle attitude with respect to the ISA, which has a controlled orientation with respect to navigation coordinates. Each gimbal angle encoder output determines the relative rotation of the structure outside gimbal axis relative to the structure inside the gimbal axis, the effect of each rotation can be represented by a 3×3 rotation

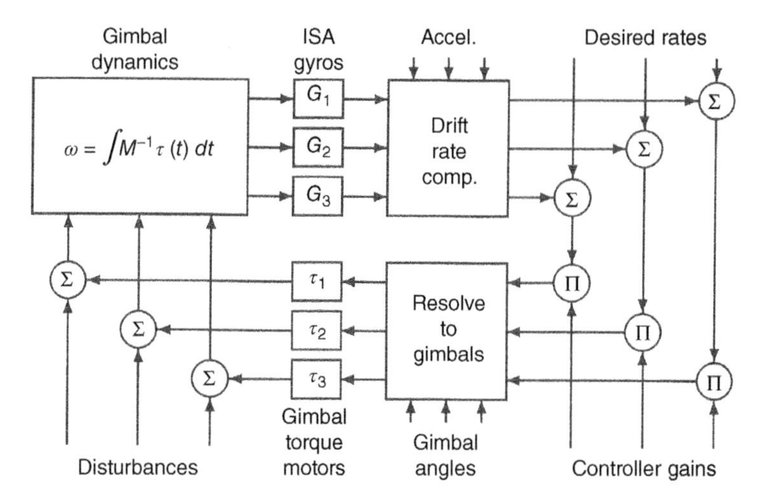

Figure 3.20 Simplified control flow diagram for three gimbals.

matrix, and the coordinate transformation matrix representing the attitude of vehicle with respect to the ISA will be the ordered product of these matrices.

ISA Attitude Control Gimbals control ISA orientation. This is a three-degree-of-freedom problem, and the solution is unique for three gimbals. That is, there are three attitude-control loops with (at least) three sensors (the gyroscopes) and three torquers. Each control loop can use a PID controller, with the commanded torque distributed to the three torquers according to the direction of the torquer/gimbal axis with respect to the gyro input axis, somewhat as illustrated in Figure 3.20, where

Disturbances includes the sum of all torque disturbances on the individual gimbals and the ISA, including those due to ISA mass unbalance and acceleration, rotations of the host vehicle, air currents, torque motor errors, etc.

Gimbal dynamics is actually quite a bit more complicated than the rigid-body torque equation

$$\boldsymbol{\tau} = \mathbf{M}_{\text{inertia}}\dot{\boldsymbol{\omega}}$$

which is the torque analog of $\mathbf{F} = m\mathbf{a}$, where $\mathbf{M}_{\text{inertia}}$ is the moment of inertia matrix. The IMU is not a rigid body, and the gimbal torque motors apply torques *between* the gimbal elements (i.e. ISA, gimbal rings, and host vehicle).

Desired rates refers to the rates required to keep the ISA aligned to a moving coordinate frame (e.g. locally level).

Resolve to gimbals is where the required torques are apportioned among the individual torquer motors on the gimbal axes.

The actual control loop is more complicated than that shown in the figure, but it does illustrate in general terms how the sensors and actuators are used.

For systems using four gimbals to avoid gimbal lock, the added gimbal adds another degree of freedom to be controlled. In this case, the control law usually adds a fourth constraint (e.g. maximize the minimum angle between gimbal axes) to avoid gimbal lock.

3.6.2 Position and Velocity Propagation

3.6.2.1 Vertical Channel Instability

The INS navigation solution for altitude and altitude rate is called its *vertical channel*. It might have become the Achilles heel of inertial navigation if it had not been recognized (by physicist George Gamow [10]) and resolved (by Charles Stark Draper and others) early on.

The reason for this is that the vertical gradient of the gravitational acceleration is negative. Because accelerometers cannot sense gravitational accelerations, the INS must rely on Newton's universal law of gravitation to take them into account in the navigation solution. Newton's law has the downward gravitational acceleration inversely proportional to the square of the radius from the Earth's center, which then falls off with increasing altitude. Therefore an INS resting stationary on the surface of the Earth with an upward navigational error in altitude would compute a downward gravitational acceleration smaller that the (measured) upward specific force countering gravity, which would result in an upward navigational acceleration error, which only makes matters worse. This would not be a problem for surface ships, it might have been a problem for aircraft if they did not already use barometric altimeters, and similarly for submarines if they did not already use depth sensors. It became an early example of sensor integration successfully applied to inertial navigation.

This is no longer a serious issue, now that we have chip-scale barometric altimeters.

3.6.2.2 Strapdown Navigation Propagation

The basic signal processing functions for strapdown INS navigation are illustrated in Figure 3.21, where

G is the estimated gravitational acceleration, computed as a function of estimated position.

POS_{NAV} is the estimated position of the host vehicle in navigation coordinates.

VEL_{NAV} is the estimated velocity of the host vehicle in navigation coordinates.

ACC_{NAV} is the estimated acceleration of the host vehicle in navigation coordinates, which may be used for trajectory control (i.e. vehicle guidance).

$\text{ACC}_{\text{SENSOR}}$ is the estimated acceleration of the host vehicle in sensor-fixed coordinates, which may be used for vehicle steering stabilization and control.

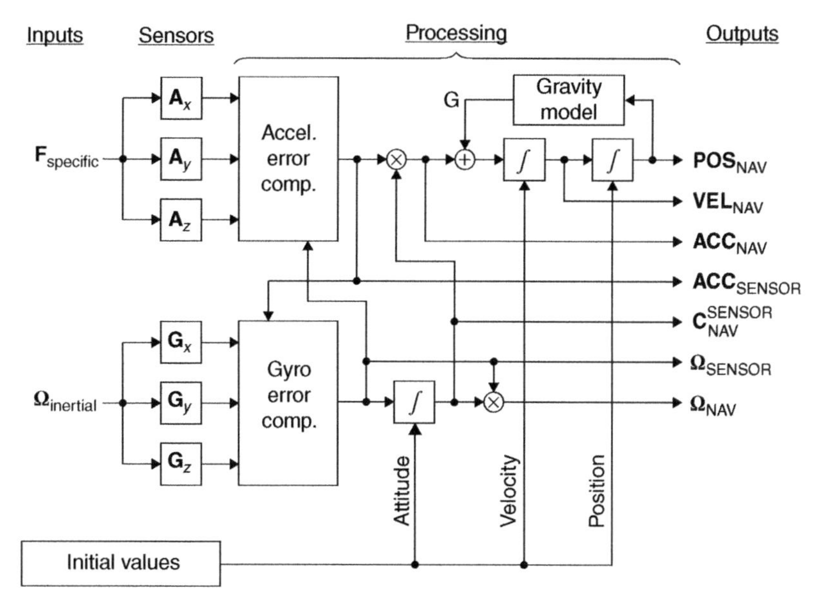

Figure 3.21 Essential navigation signal processing for strapdown INS.

C_{NAV}^{SENSOR} is the 3 × 3 coordinate transformation matrix from sensor-fixed coordinates to navigation coordinates, representing the attitude of the sensors in navigation coordinates.

ω_{SENSOR} is the estimated angular velocity of the host vehicle in sensor-fixed (ISA) coordinates, which may be used for vehicle attitude stabilization and control.

ω_{NAV} is the estimated angular velocity of the host vehicle in navigation coordinates, which may be used in a vehicle pointing and attitude control loop.

The essential processing functions include double integration (represented by boxes containing integration symbols) of acceleration to obtain position, and computation of (unsensed) gravitational acceleration as a function of position. The sensed angular rates also need to be integrated to maintain the knowledge of sensor attitudes. The initial values of all the integrals (i.e. position, velocity, and attitude) must also be known before integration can begin.

The position vector POS_{NAV} is the essential navigation solution. The other outputs shown are not needed for all applications, but most of them (except ω_{NAV}) are intermediate results that are available "for free" (i.e. without requiring further processing). The velocity vector VEL_{NAV}, for example, characterizes speed and heading, which are also useful for correcting the course of the host vehicle to bring it to a desired location. Most of the other outputs shown would be required for implementing control of an unmanned or autonomous host vehicle to follow a desired trajectory and/or to bring the host vehicle to a desired final position.

Navigation functions that are not shown in Figure 3.21 include:

1. How initialization of the integrals for position, velocity, and attitude is implemented. Initial position and velocity can be input from other sources (GNSS, for example), and attitude can be inferred from some form of trajectory matching (using GPS, for example) or by *gyrocompassing*.
2. How attitude rates are integrated to obtain attitude, described in Section 3.6.1.1.
3. For the case that navigation coordinates are Earth-fixed, the computation of navigational coordinate rotation due to earthrate as a function of position, and its summation with sensed rates before integration.
4. For the case that navigation coordinates are locally-level, the computation includes the rotation rate of navigation coordinates due to vehicle horizontal velocity and its summation with sensed rates before integration.
5. Calibration of the sensors for error compensation. If the errors are sufficiently stable, it needs to be done only once. Otherwise, it can be implemented using the GNSS/INS integration techniques discussed in Chapter 12.

Figure 3.22 is a process flow diagram for the same implementation, arranged such that the variables available for other functions is around the periphery. These are the sorts of variables that might be needed for driving cockpit displays, antennas, weaponry, sensors, or other surveillance assets.

3.6.2.3 Gimbaled Navigation Propagation

The signal flowchart in Figure 3.23 shows the essential navigation signal processing functions for a gimbaled INS with inertial sensor axes aligned to locally level coordinates, where

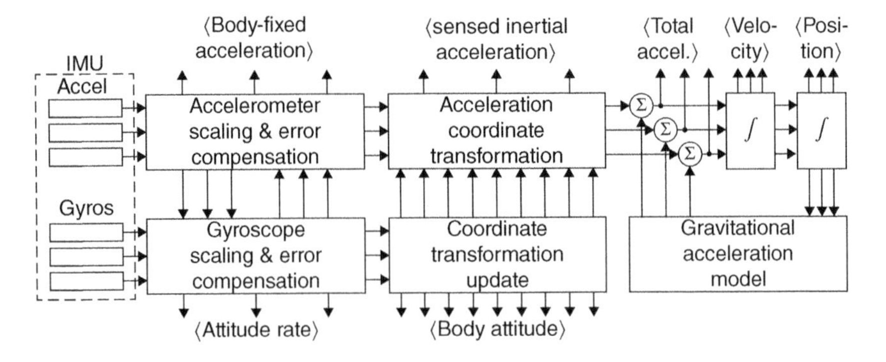

Figure 3.22 Outputs (in angular brackets) of simple strapdown INS.

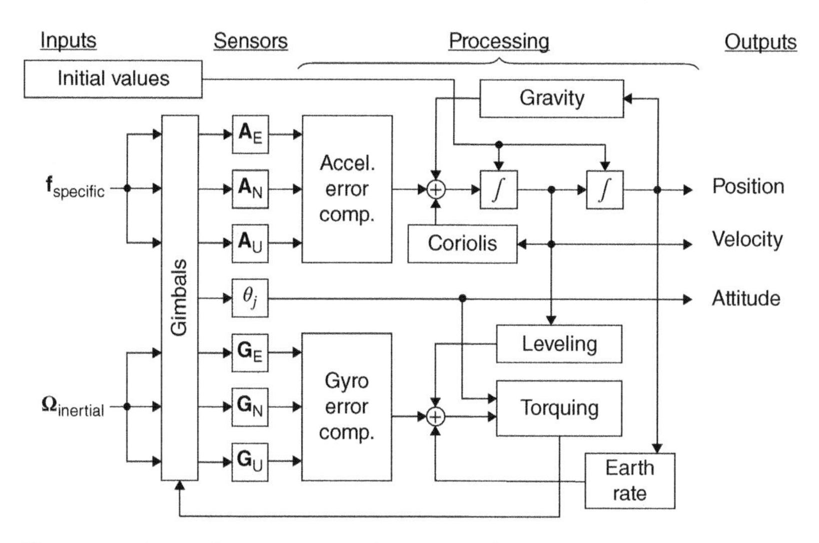

Figure 3.23 Essential navigation signal processing for gimbaled INS.

$\mathbf{f}_{\text{specific}}$ is the **specific force** (i.e. the sensible acceleration, which does not include gravitational acceleration) applied to the host vehicle.

$\boldsymbol{\omega}_{\text{inertial}}$ is the instantaneous inertial rotation rate vector of the host vehicle.

A denotes a specific force sensor (accelerometer).

$\boldsymbol{\theta}_j$ denotes the ensemble of gimbal angle encoders, one for each gimbal angle. There are several possible formats for the gimbal angles, including digitized angles, three-wire synchros signals, or sin / cos pairs.

G denotes an inertial rotation sensor (gyroscope).

Position is the estimated position of the host vehicle in navigation coordinates (e.g. longitude, latitude, and altitude relative to sea level).

Velocity is the estimated velocity of the host vehicle in navigation coordinates (e.g. east, north, and vertical).

Attitude is the estimated attitude of the host vehicle relative to locally level coordinates. For some three-gimbal systems, the gimbal angles are the Euler angles representing vehicle heading (with respect to north), pitch (with respect to horizontal), and roll. Output attitude may also be used to drive cockpit displays such as compass cards or artificial horizon indicators.

Accelerometer Error Compensation and Gyroscope Error Compensation denote the calibrated corrections for sensor errors. These generally include corrections for scale factor variations, output biases, and input axis misalignments for both types of sensors, and acceleration-dependent errors for gyroscopes.

Gravity denotes the gravity model used to compute the acceleration due to gravity as a function of position.

Coriolis denotes the acceleration correction for Coriolis effect in rotating coordinates.
Leveling denotes the rotation rate correction to maintain locally level coordinates while moving over the surface of the Earth.
Earth Rate denotes the model used to calculate the Earth rotation rate in locally level INS coordinates.
Torquing denotes the servo loop gain computations used in stabilizing the INS in locally level coordinates.

Not shown in the figure is the input altitude reference (e.g. barometric altimeter or GNSS) required for vertical channel (altitude) stabilization.

3.7 Testing and Evaluation

The final stage of the development cycle is testing and performance evaluation. For standalone inertial systems, this usually proceeds from the laboratory to a succession of host vehicles, depending on the application.

3.7.1 Laboratory Testing

Laboratory testing is used to evaluate sensors before and after their installation in the ISA, and then to evaluate the system implementation during operation. The navigation solution from a stationary system should remain stationary, and any deviation is due to navigation errors. Testing with the system stationary can also be used to verify that position errors due to intentional initial velocity errors follow a path predicted by Schuler oscillations ($\approx$84.4-minute period, described in Chapter 11) and the Coriolis effect. If not, there is an implementation error. Other laboratory testing may include controlled tilts and rotations to verify the attitude estimation implementations, and detect any un-compensated sensitivities to rotation and acceleration.

Additional laboratory testing may be required for specific applications. Systems designed to operate aboard Navy ships, for example, may be required to meet their performance requirements under dynamic disturbances at least as bad as those to be expected aboard ships under the worst sea conditions. This may include what is known as a "Scoresby test," used at the US Naval Observatory in the early twentieth century for testing gyrocompasses. Test conditions may include roll angles of $\pm 80^\circ$ and pitch angles of $\pm 15^\circ$, at varying periods in the order of a second.

Drop tests (for handling verification) and shake-table or centrifuge tests (for assessing acceleration capabilities) can also be done in the laboratory.

3.7.2 Field Testing

After laboratory testing, systems are commonly evaluated next in highway testing.

Systems designed for tactical aircraft must be designed to meet their performance specifications under the expected peak dynamic loading, which is generally determined by the pilot's limitations.

Systems designed for rockets must be tested under conditions expected during launch, sometimes as a "piggyback" payload during the launch of a rocket for another purpose. Accelerations can reach around $3g$ for manned launch vehicles, and much higher for unmanned launch vehicles.

In all cases, GNSS has become an important part of field instrumentation. The Central Inertial Guidance Test Facility at Holoman AFB was once equipped with elaborate range instrumentation for this purpose, which is now performed at much lower cost using GNSS.

3.7.3 Performance Qualification Testing

The essential performance metric for any navigation system is accuracy, commonly specified in terms of position accuracy. However, because inertial navigation also provides attitude and velocity information, those may also be factors in performance assessments. In the case of integrated GNSS/INS systems, there may be an additional performance metric related to how fast navigational accuracy degrades when GNSS signal reception is lost. INS-only navigation is called *free inertia navigation,* and we will start there. Integrated GNSS/INS performance is addressed in Chapter 12.

3.7.3.1 CEP and Nautical Miles

Here, "CEP" stands for "*circular error probable*" or "*circle of equal probability,*" specified as the radius of the circle about the navigation solution with a radius such that the true solution is equally likely to be inside or outside of that circle. That radius is usually specified in units of nautical miles (1852 m in SI units[6]).

3.7.3.2 Free Inertial Performance

"Free" here means unaided by external sensors – except for altimeter aiding required to stabilize altitude errors.

Free Inertial Error Heuristics Unaided inertial navigators are essentially integrators, integrating sensed accelerations (including sensor noise) to get velocity and integrating velocity to get position. Unfortunately, integration does bad things to zero-mean random noise. The integral of zero-mean additive uncorrelated random noise on an accelerometer output is a random walk process,

6 The nautical mile was originally defined in the seventeenth century as the surface distance covered by one arc-minute of latitude at sea level. However, because the Earth is not exactly spherical and latitude is measured with respect to the local vertical, this number varies from equator to pole from about 1843 m to about 1861 m.

the variance of which grows linearly over time. Integrated twice to get position, we might expect variance to grow quadratically with time, in which case its standard deviation would grow linearly with time. As a consequence, one might expect INS performance to be characterized by how fast an error distance statistic grows linearly with time.

CEP Rates A common INS performance metric used in military applications is *CEP rate*, generally calculated over typical INS mission times (a few hours, generally) and originally measured (before GNSS availability) during INS performance qualification trials aboard different military aircraft or surface vehicles on suitably instrumented test ranges. That can now be done using GNSS as the independent position solution for CEP rate determinations.

INS Performance Categories In the 1970s, before GPS became a reality, the US Department of Defense established the following categories of INS performance:

High accuracy systems have free inertial CEP rates in the order of 0.1 nautical miles per hour (~185 m/h) or better.

Medium accuracy systems have free inertial CEP rates in the order of 1 nautical mile per hour (~1.85 km/h). This was the level of accuracy deemed sufficient for most military and commercial aircraft [11].

Low accuracy systems have free inertial CEP rates in the order of 10 nautical miles per hour (~18.5 km/h) or worse. Sometimes called *tactical grade* INS performance, this range covered requirements for many short-range standoff weapons such as guided artillery or tactical missiles.

Comparable Sensor Performance Ranges Order-of-magnitude ranges for the inertial sensor errors in these INS performance categories are summarized in Table 3.1. Inertial sensors below tactical grade are sometimes called *commercial grade* or *consumer grade*.

Table 3.1 INS and inertial sensor performance ranges.

	Performance ranges			
System or sensor	**High**	**Medium**	**Low**	**Units**
INS	$\leq 10^{-1}$	~ 1	≥ 10	NMi/h CEP rate
Gyroscopes	$\leq 10^{-3}$	$\sim 10^{-2}$	$\geq 10^{-1}$	deg/h drift rate
Accelerometers	$\leq 10^{-7}$	$\sim 10^{-6}$	$\geq 10^{-5}$	g (9.8 m/s^2) bias

CEP versus RMS Unfortunately, the probability-based CEP statistic is not exactly compatible with the RMS statistics used in the Kalman filter Riccati equations[7] for characterizing INS performance. One must assume some standard probability density function (e.g. Gaussian) to convert mean-squared horizontal position errors to CEPs.

3.8 Summary

1. Inertial navigation accuracy is mostly limited by inertial sensor accuracy.
2. The accuracy requirements for inertial sensors cannot always be met within manufacturing tolerances. Some form of calibration is usually required for compensating the residual errors.
3. INS accuracy degrades over time, and the most accurate systems generally have shortest mission times. For example, ICBMs only need their inertial systems for a few minutes.
4. Performance of inertial systems is commonly specified in terms of CEP rate.
5. Accelerometers cannot measure gravitational acceleration.
6. Both inertial and satellite navigation require accurate models of the Earth's gravitational field.
7. Both navigation modes also require an accurate model of the shape of the Earth.
8. The first successful navigation systems were gimbaled, in part because the computer technology required for strapdown implementations was decades away. That has not been a problem for about four decades.
9. Gimbaled systems tend to be more accurate and more expensive than strapdown systems.
10. The more reliable attitude implementations for strapdown systems use quaternions to represent attitude.
11. Systems traditionally go through a testing and evaluation process to verify performance.
12. Before testing and evaluation of an INS, its expected performance is commonly evaluated using the analytical models of Chapter 11.

3.8.1 Further Reading

Inertial navigation has a rich and growing technology base – more than can be covered in a single book, and certainly not in one chapter – but there is some good open-source literature on the subject:

1. Titterton and Weston [12] is a good source for additional information on strapdown hardware and software.

---

7 Defined in Chapter 10.

2. Paul Savage's two volume tome [13] on strapdown system implementations is also rather thorough.
3. Chapter 5 of [14] and the references therein include some recent developments.
4. Journals of the IEEE, IEE, Institute of Navigation, and other professional engineering societies generally have the latest developments on inertial sensors and systems.
5. The Mathworks file exchange at https://www.mathworks.com/matlabcentral/fileexchange/ includes many m-file implementations of navigation procedures, including a complete WGS84 geoid model.
6. In addition, the World Wide Web includes many surveys and reports on inertial sensors and systems.

Problems

Refer to Appendix B for coordinate system definitions, and satellite orbit equations.

3.1 Which, if any, of the following coordinate systems is not rotating?
(a) Northeast–down (NED)
(b) East–north–up (ENU)
(c) Earth-centered Earth-fixed (ECEF)
(d) Earth-centered inertial (ECI)
(e) Moon-centered moon-fixed

3.2 What is the minimum number of two-axis gyroscopes (i.e. gyroscopes with two, independent, orthogonal input axes) required for inertial navigation?
(a) 1
(b) 2
(c) 3
(d) Not determined.

3.3 What is the minimum number of gimbal axes required for gimbaled inertial navigators in fully maneuverable host vehicles? Explain your answer.
(a) 1
(b) 2
(c) 3
(d) 4

3.4 Define *specific force*.

3.5 An inertial sensor assembly (ISA) operating at a fixed location on the surface of the Earth would measure
(a) No acceleration
(b) 1g acceleration downward
(c) 1g acceleration upward

3.6 Explain why an inertial navigation system is not a good altimeter.

3.7 The *inertial* rotation rate of the Earth is
(a) 1 revolution per day
(b) 15 deg/h
(c) 15 arc-seconds per second
(d) None of the above

3.8 Define CEP and CEP rate for an INS.

3.9 The CEP rate for a *medium accuracy* INS is in the order of
(a) 2 m/s
(b) 200 m/h
(c) 2000 m/h
(d) 20 km/h

3.10 Derive the equivalent formulas in terms of Y (yaw angle), P (pitch angle), and R (roll angle) for unit vectors $\mathbf{1}_R$, $\mathbf{1}_P$, $\mathbf{1}_Y$ in NED coordinates and $\mathbf{1}_N$, $\mathbf{1}_E$, $\mathbf{1}_D$ in RPY coordinates.

3.11 Explain why accelerometers cannot sense gravitational accelerations.

3.12 Show that the matrix $\mathbf{C}^{\text{body}}_{\text{inertial}}$ defined in Eq. (3.35) is orthogonal by showing that $\mathbf{C}^{\text{body}}_{\text{inertial}} \times \mathbf{C}^{\text{body}\,\text{T}}_{\text{inertial}} = \mathbf{I}$, the identity matrix. (*Hint*: Use $q_0^2 + q_1^2 + q_2^2 + q_3^2 = 1$.)

3.13 Calculate the numbers of computer multiplies and adds required for
(a) gyroscope scale factor/misalignment/bias compensation (Eq. (3.4 with $N = 3$)
(b) accelerometer scale factor/misalignment/bias compensation (Eq. (3.4 with $N = 3$) and
(c) transformation of accelerations to navigation coordinates (Figure 3.22) using quaternion rotations (see Appendix B on quaternion algebra)

If the INS performs these 100 times per second, how many operations per second will be required?

References

1 IEEE Standard 528-2001 (2001). IEEE Standard for Inertial Sensor Terminology. New York: Institute of Electrical and Electronics Engineers.
2 IEEE Standard 1559-2009 (2009). IEEE Standard for Inertial System Terminology. New York: Institute of Electrical and Electronics Engineers.
3 Mueller, F.K. (1985). A history of inertial navigation. *Journal of the British Interplanetary Society* 38: 180–192.
4 Everitt, C.W.F., DeBra, D.B., Parkinson, B.W. et al. (2011). Gravity probe B: final results of a space experiment to test general relativity. *Physical Review Letters* 106: 221101.
5 Bernstein, J., Cho, S., King, A.T. et al. (1993). A micromachined comb-drive tuning fork rate gyroscope. Proceedings IEEE Micro Electro Mechanical Systems. IEEE, pp. 143–148.
6 von Laue, M. (1920). Zum Versuch von F. Harress. *Annalen der Physik* 367 (13): 448–463.
7 Collin, J., Kirkko-Jaakkola, M., and Takala, J. (2015). Effect of carouseling on angular rate sensor error processes. *IEEE Transactions on Instrumentation and Measurement* 64 (1): 230–240.
8 Renkoski, B.M. (2008). The effect of carouseling on MEMS IMU performance for gyrocompassing applications. MS thesis. Massachusetts Institute of Technology.
9 Bortz, J.E. (1971). A new mathematical formulation for strapdown inertial navigation. *IEEE Transactions on Aerospace and Electronic Systems* AES-7: 61–66.
10 Draper, C.S. (1981). Origins of inertial navigation *AIAA Journal of Guidance and Control* 4 (5): 449–456.
11 Chairman of Joint Chiefs of Staff, US Department of Defense (2003). 2003 CJCS Master Positioning, Navigation and Timing Plan. Rept. CJCSI 6130.01C.
12 Titterton, D.H. and Weston, J.L. (2004). *Strapdown Inertial Navigation Technology*, 2e. Stevenage, UK: Institution of Electrical Engineers.
13 Savage, P.G. (1996). *Introduction to Strapdown Inertial Navigation Systems*, Vols. 1 & 2. Maple Plain, MN: Strapdown Associates.
14 Groves, P.D. (2013). *Principles of GNSS, Inertial, and Multisensor Integrated Navigation Systems*, 2e. Artech House.

4

GNSS Signal Structure, Characteristics, and Information Utilization

Why are the global navigation satellite system (GNSS) signals so complex? GNSSs are designed to be readily accessible to millions of military and civilian users. The GNSSs are receive-only passive systems that enable a very large number of users to simultaneously use the system. Because there are many functions that must be performed, the GNSS signals have a rather complex structure. As a consequence, there is a correspondingly complex sequence of operations that a GNSS receiver must carry out in order to extract and utilize the desired information from the signals in space. Modernized Global Positioning System (GPS) and other GNSSs have improved upon the legacy GPS waveform in terms of multipath, correlation, and overall accuracy and availability performance. In this chapter, we characterize the signal mathematically, describe the purposes and properties of the important signal components, and discuss generic methods for extracting information from these GNSS navigation signals.

This chapter provides details of the legacy GPS signals, followed by the modernized GPS signals. Discussions of the Global Orbiting Navigation Satellite System (GLONASS), Galileo, BeiDou Navigation Satellite Systems (BDS), Quasi-Zenith Satellite System (QZSS), and the Indian Regional Navigation Satellite Systems (IRNSS)/NAVIC will follow.

4.1 Legacy GPS Signal Components, Purposes, and Properties

The GPS is a code-division multiple access (CDMA) satellite-based ranging system. The system is considered a spread-spectrum system where the radio frequency (RF) bandwidth that is used is much wider than as required to transmit the underlying navigation data. The GPS has two basic services that are provided by the legacy GPS: the standard positioning service (SPS) and the precise positioning service (PPS), with documented performance standards [1].

Global Navigation Satellite Systems, Inertial Navigation, and Integration,
Fourth Edition. Mohinder S. Grewal, Angus P. Andrews, and Chris G. Bartone.

Companion website: www.wiley.com/go/grewal/gnss

The legacy GPS signals in space are well documented by the US Department of Defense in the form of an interface specification (IS). This IS provides basic characteristics of the signal in space and information on how user equipment should interface to and process the navigation signals [2].

The interface between the legacy GPS space and user segments consists of two RF links, L1 and L2. The carriers of the L-band links can be modulated by two navigation data bit streams, each of which normally is a composite generated by the modulo-2 addition of a pseudorandom noise (PRN) ranging code and the downlink system navigation (NAV) data. The legacy 50-bps (bits per second) NAV message is now referred to specifically as the legacy navigation (LNAV), and further designated as LNAV-L (lower) for PRNs 1–32 and LNAV-U (upper) for PRNs 33–63. For simplicity here, we will refer to it as just the NAV message. Utilizing these links, the space vehicles (SVs) of the GPS space segment can provide continuous Earth coverage of navigation signals that provide the user segment ranging PRN code phase, carrier phase, and system data needed for the user to calculate a position, velocity, and time (PVT) solution. These signals are available to a suitably equipped GPS user with RF visibility to the SVs.

4.1.1 Signal Models for the Legacy GPS Signals

Each GPS satellite simultaneously transmits on two of the legacy L-band frequencies denoted by L1 and L2, which are 1575.42 and 1227.60 MHz, respectively. The carrier of the L1 signal consists of two signal components that are orthogonal (i.e. cos and sin functions in quadrature). The first component is biphase modulated (i.e. binary phase shift keying [BPSK]) by a 50-bps navigation data stream and a PRN spreading code, called the *C/A-code*, consisting of a 1023-chip sequence that has a period of 1 ms and a chipping rate of 1.023 MHz. The second component is also biphase modulated by the same 50-bps data stream but with a different PRN code called the *P(Y)-code*, which has a 10.23-MHz chipping rate and has a one-week period. While the native P-code can be transmitted without data encryption, when the P-code is encrypted, it is referred to as the P(Y)-code. The mathematical model of the GPS navigation signal on the L1 frequency is shown in Eq. (4.1) for each SV i:

$$s_i(t) = \sqrt{2P_{\mathrm{C/A}}}d(t)c_i(t)\cos(\omega t + \theta_i) + \sqrt{2P_{\mathrm{P(Y)}}}d(t)p_i(t)\sin(\omega t + \theta_i) \tag{4.1}$$

where

$P_{\mathrm{C/A}}$ = power in C/A-encoded component

$P_{\mathrm{P(Y)}}$ = power in P(Y)-encoded component

$d(t)$ = navigation date at 50-bps rate

$c_i(t)$ = C/A PRN code at 1.023-MHz rate for SV i

$p_i(t)$ = P PRN code at 10.23-MHz rate for SV i

ω = carrier frequency (rad/s)

In Eq. (4.1), $P_{C/A}$ and $P_{P(Y)}$ are the respective carrier powers for the C/A- and P(Y)-encoded carrier components. The data, $d(t)$, is the 50-bps navigation data modulation; $c(t)$ and $p(t)$ are the respective C/A and P PRN code waveforms; ω is the L1 carrier frequency in radian per second; and θ is a common phase shift in radians at the respective carrier frequency. The C/A-code carrier component lags the P-code carrier component by 90° when both data chips are 0 (i.e. −1). The carrier power for the P-code carrier is approximately 3 dB less than the power in the C/A-code carrier [2].

In contrast to the L1 signal, the L2 signal is modulated with the navigation data at the 50-bps rate and the P(Y)-code, although there is the option of not transmitting the 50-bps data stream. The mathematical model of the L2 waveform is shown in Eq. (4.2) for each SV i:

$$s_i(t) = \sqrt{2P_{P(Y)}}d(t)p_i(t)\sin(\omega t + \theta_i) \tag{4.2}$$

where

$P_{P(Y)}$ = power in P(Y)-encoded component

$d(t)$ = navigation date at 50-bps rate

$p_i(t)$ = P PRN code at 10.23-MHz rate for SV i

ω = carrier frequency (rad/s)

Figure 4.1 illustrates a function block diagram of how the legacy GPS signals are generated at L1 and L2. Both of the output signals go to their respective power amplifiers and are then combined for transmission out of the GPS helix antenna array. The gain adjustments depicted in Figure 4.1 as negative values would be implemented in the generation of the signals themselves and not as actual attenuation. It should also be noted that although there were provisions to transmit the C/A code on L2 in early block GPS SVs, this functionality is superseded with the L2 civil (L2C) signal, as will be discussed later in this chapter.

It is worthy to note that all signals generated are from the same reference oscillator, and this reference oscillator is divided by integer numbers for the C/A-code clock, P(Y)-code clock, and navigation data register clock and multiplied up by an integer number for the L1 and L2 carrier signals. This helps

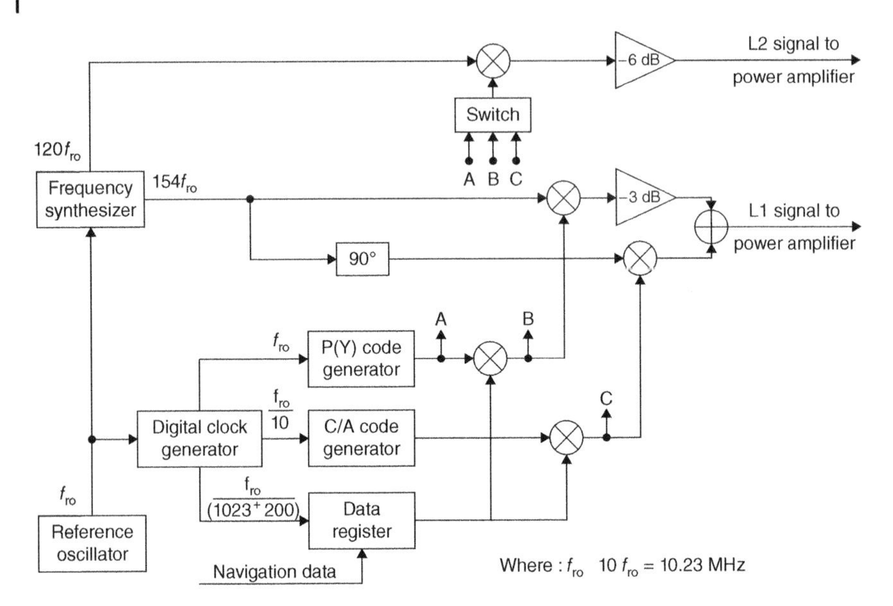

Figure 4.1 Block diagram of GPS signal generation at L1 and L2 frequencies. *Note*: All signals coherent with reference oscillator.

in the synthesizer and clock designs as well as in maintaining code to carrier coherency.

Figure 4.2 illustrated the C/A-encoded structure on the L1 carrier signal. The 50-bps navigation data bit boundaries always occur at an epoch of the C/A-code. The C/A-code epochs mark the beginning of each period of the C/A-code, and there are precisely 20 code epochs per data bit, which simply repeats 20 times. Within each C/A-code chip, there are precisely 1540 L1 carrier cycles. The navigation data and code are multiplied together (implemented digitally with a modulo addition) and are then multiplied by the carrier signal. When $(d(t) \oplus c(t))$ is a "1," simple the phase of the carrier is unaffected, but when $(d(t) \oplus c(t))$ is a "−1," the phase of the carrier is inverted by 180° to produce the BPSK modulation.

Figure 4.3 illustrated the P(Y)-encoded structure on the L1 carrier, with the navigation data bits. This modulation process is similar to the C/A signal encoding with some key differences. The carrier competent for the P(Y)-code signal is in quadrature (i.e. orthogonal) to that used for the C/A-code generation. Also, the P(Y) is more random and does not repeat within a navigation data bit interval, and there is a total of 204 600 P(Y)-code chips within a navigation data bit of duration 20 ms; again, the P(Y) is in alignment with the data bit boundaries. The P(Y)-code runs at a 10 times faster rate than that of the C/A-code. Hence, there

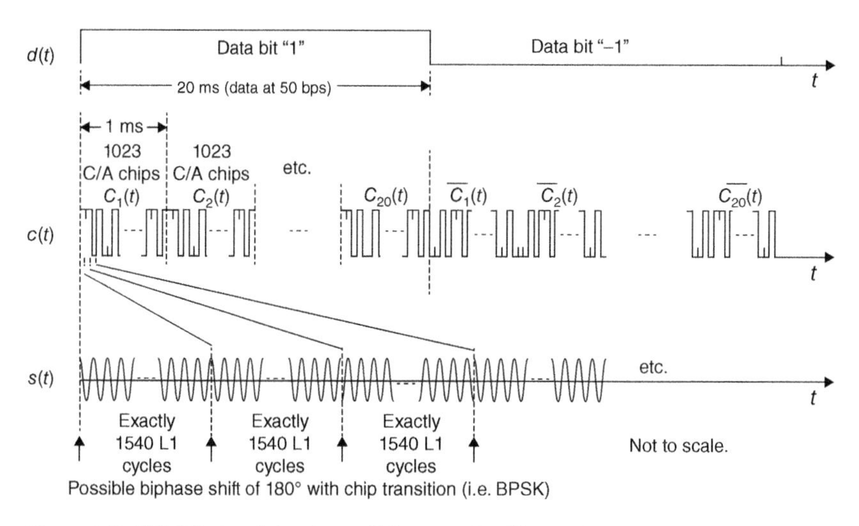

Figure 4.2 GPS C/A signal structure at L1 generation illustration.

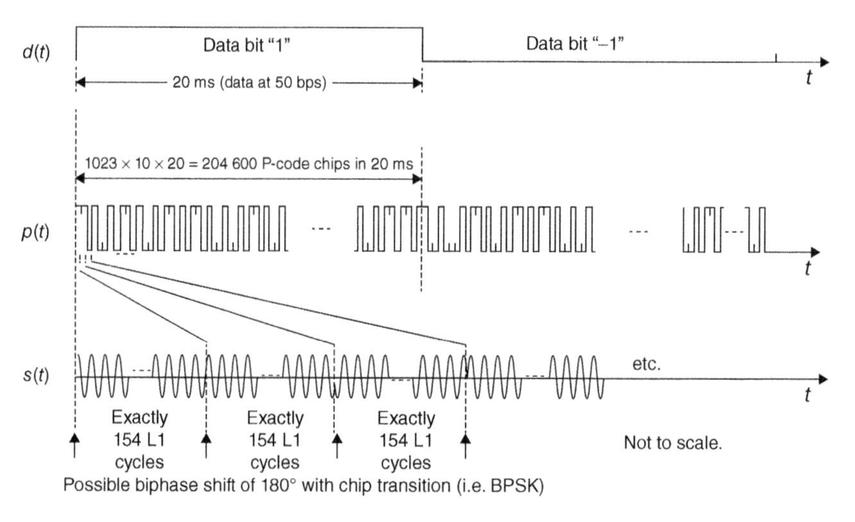

Figure 4.3 GPS P(Y) signal structure at L1 generation illustration.

are 10 times less carrier cycles per chip in the P(Y) than the C/A-code. For the GPS L1 P(Y) signal, there are exactly 154 L1 carrier cycles per P(Y) chip. BPSK modulation is again performed.

When the C/A-encoded signal is combined with the P(Y)-encoded signal, there are two synchronized BPSK signals transmitted in quadrature. This feature allows these two signals to be separated in the GPS receiver.

4.1.2 Navigation Data Format

The navigation data bit stream running at a 50-bps rate conveys the *navigation message* that is determined by the GPS Ground Control Segment and uploaded to each satellite. This navigation information is then encoded onto the navigation signals and transmitted to the users. The navigation message includes, but is not limited to, the following information:

1. *Satellite almanac data.* Each satellite transmits orbital data called the *almanac,* which enables the user to calculate the approximate location of every satellite in the GPS constellation at any given time. Almanac data are not accurate enough to determine an accurate user position but can be stored in a receiver where they remain valid for many months, for their intended purpose. They are used primarily to determine which satellites are visible at an estimated location so that the receiver can search for those satellites when it is first turned on. They can also be used to determine the approximate expected signal Doppler shift to aid in rapid acquisition of the satellite signals. Every GPS satellite transmits almanac data about the GPS constellation (i.e. about all the other satellites, including itself). Almanac data include the course satellite orbital (i.e. position) and course satellite clock error data.
2. *Satellite ephemeris data.* Ephemeris data are similar to almanac data but enable a much more accurate determination of satellite position and clock error needed to convert signal propagation delay into an estimate of the user's PVT. Satellite ephemeris data include details on the position of the satellite (i.e. orbital ephemeris data) and details on the satellite's clock error (i.e. clock ephemeris data). In contrast to almanac data, ephemeris data for a particular satellite are broadcast only by that satellite and the data are typically valid for many hours (e.g. 0–24 hours).
3. *Signal timing data.* The navigation data stream includes time tagging, which is used to establish a time of transmission (TOT) at specific points in the GPS signal. This information is needed to determine the satellite-to-user propagation delay used for ranging.
4. *Ionosphere delay data.* Ranging errors due to the signal propagation delay through the Earth's ionosphere can affect GPS positioning. While these effects can be measured directly for multifrequency GPS users, they most often are modeled for single-frequency users. To help mitigate these effects, ionosphere error perditions are encoded into the navigation message that can be decoded and applied by the user to partially cancel these errors [2, 3].
5. *Satellite health message.* The navigation data stream also contains information regarding the current health of the satellite, so that the receiver can ignore that satellite if it is not operating properly.

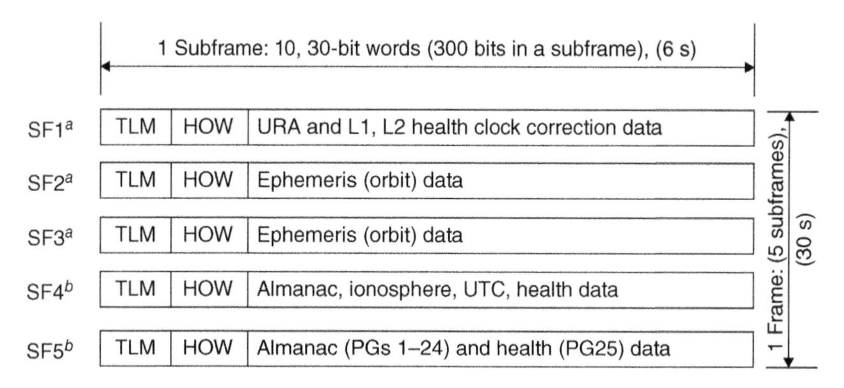

Figure 4.4 GPS navigation data format (legacy) frame structure.[a] Same data transmitted every frame, by each SV (until updated by CS). [b] Data commutated in each frame (i.e., page), common to all SVs (until updated by CS). *Note*: With multichannel receiver, data are decoded simultaneously from SVs.

The format of the navigation message has a basic frame structure as shown in Figure 4.4. The format consists of 25 frames, where a frame contains 1500 bits. Each *frame* is subdivided into five 300-bit subframes, and each *subframe* consists of 10 words of 30 bits each, with the most significant bit (MSB) of the word transmitted first. Thus, at the 50-bps rate, it takes 6 seconds to transmit a subframe and 30 seconds to complete one frame. Transmission of the complete 25-frame navigation message requires 750 seconds, or 12.5 minutes. Except for occasional updating, subframes 1, 2, and 3 are constant (i.e. repeat) with each frame at the 30-seconds frame repetition rate. On the other hand, subframes 4 and 5 are each subcommutated 25 times. The 25 versions of subframes 4 and 5 are referred to as *pages* 1–25. Hence, except for occasional updating, each of these pages repeats every 750 seconds, or 12.5 minutes.

A detailed description of all information contained in the navigation message is beyond the scope of this text but can be found in Ref. [2]. Therefore, we give only an overview of the fundamental elements. Each subframe begins with a *telemetry word* (TLM). The first 8 bits of the TLM is a preamble that enables the receiver to determine when a subframe begins and for data recovery purposes. The remainder of the TLM contains parity bits and a telemetry message that is available only to authorized users. The second word of each subframe is called the *handover word* (HOW).

4.1.2.1 *Z*-Count

Information contained in the HOW is derived from a 29-bit quantity called the *Z-count*. The *Z*-count is not transmitted as a single word, but part of it is transmitted within the HOW. The Z-count counts *epochs* generated by the X1 register of the P-code generator in the satellite, which occur every 1.5 seconds.

The 19 least significant bits (LSBs) of the Z-count is called the *time of week* (TOW) count and indicate the number of X1 epochs that have occurred since the start of the current week. The start of the current week occurs at the X1 epoch, which occurs at approximately midnight of Saturday night/Sunday morning. The TOW count increases from 0 at the start of the week to 403 199 and then resets to 0 again at the start of the following week. A TOW count of 0 always occurs at the beginning of subframe 1 of the first frame (the frame containing page 1 of subcommutated subframes 4 and 5) [2].

A truncated version of the TOW count, containing its 17 MSBs, constitutes the first 17 bits of the HOW. Multiplication of this truncated count by 4 gives the TOW count at the start of the following subframe. Since the receiver can use the TLM preamble to determine precisely the time at which each subframe begins, a method for determining the TOT of any part of the GPS signal is thereby provided. The time projection to the next subframe beginning can also be used to rapidly acquire the P(Y)-code, which has a week-long period. The relationship between the HOW counts and TOW counts is shown in Figure 4.5.

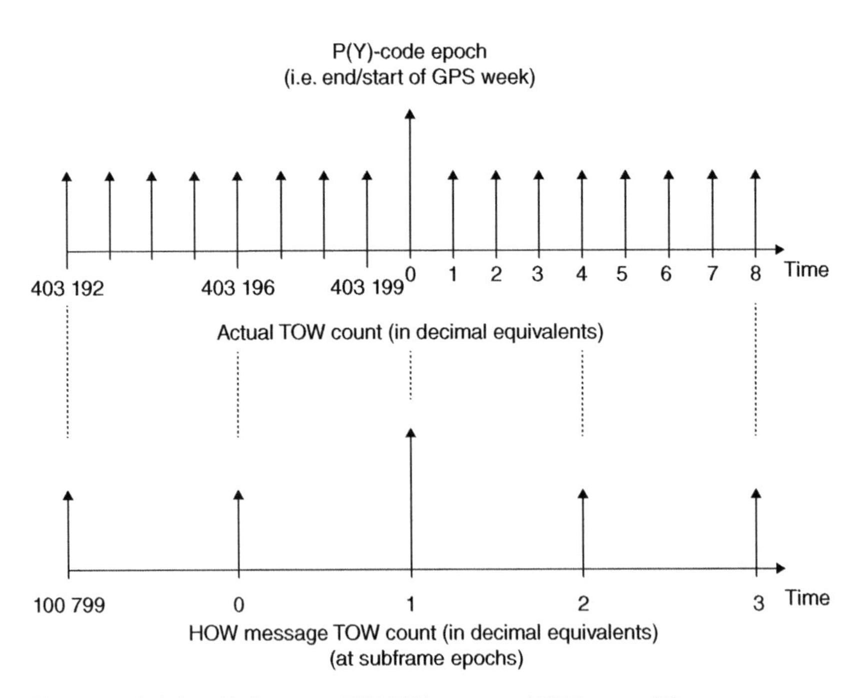

Figure 4.5 Relationship between GPS HOW counts and TOW counts [2].

4.1.2.2 GPS Week Number (WN)

The 10 MSBs of the Z-count contain the GPS *week number* (WN), which is a modulo-1024 week count. The *zero state* is defined to be the week that started with the X1 epoch occurring at approximately midnight on the night of 5 January 1980/morning of 6 January 1980, where GPS time began. Because WN is a modulo-1024 count, an event called the *week rollover* occurs every 1024 weeks (a few months short of 20 years), and GPS receivers must be designed to accommodate. The first GPS WN rollover occurred at GPS time zero on 22 August 1999 with few difficulties. The second GPS week rollover occurred GPS time zero on 7 April 2019. The WN is not part of the HOW but instead appears as the first 10 bits of the third word in subframe 1.

Frame and Subframe Identification Three bits of the HOW are used to identify which of the five subframes is being transmitted. The frame being transmitted (corresponding to a page number from 1 to 25) can readily be identified from the TOW count computed from the HOW of subframe 5. This TOW count is the TOW at the start of the next frame. Since there are 6 TOW counts per frame, the frame number of that frame is simply (TOW/6) (mod 25).

4.1.2.3 Information by Subframe

In addition to the TLM and HOW, which occur in every subframe, the following information is contained within the remaining eight words of subframes 1–5 (only fundamental information is described):

1. *Subframe 1.* The WN portion of the Z-count is part of word 3 in this subframe. Subframe 1 also contains GPS clock correction data for the satellite in the form of polynomial coefficients defining how the correction varies with time. Time defined by the clocks in the satellite is commonly called *SV time*; the time after corrections have been applied is called *GPS time.* Thus, even though individual satellites may not have perfectly synchronized SV times, they do share a common GPS time. Additional information in subframe 1 includes the quantities t_{oc}, T_{GD}, and issue of data clock (IODC). The clock reference time, t_{oc}, is used as a time origin to calculate satellite clock error. The T_{GD} term is used to correct for satellite group delay errors for single-frequency GPS users, relative to an L1 and L2 dual-frequency P(Y) user solution. The IODC indicates the issue number of the clock data intended to alert users to changes in clock error parameters provided by the navigation message.
2. *Subframes 2 and 3.* These subframes contain the ephemeris data, which are used to determine the precise satellite position and velocity required by the user solution. Unlike the almanac data, these data are very precise, are valid over a relatively short period of time (e.g. 0–24 hours), and apply only to the satellite transmitting it. The issue of data ephemeris (IODE) informs users

when changes in ephemeris parameters have occurred within the navigation data. Each time new parameters are uploaded from the GPS Ground Control Segment, the IODE number changes.

3. *Subframe 4*. The 25 pages of this subframe contain the almanac for satellites with PRN numbers 25 and higher, as well as special messages, ionosphere correction terms, and coefficients to convert GPS time to coordinated universal time (UTC). There are also spare words for possible future applications. The components of an almanac (for orbit and clock) are very similar to those of the ephemeris but contain less bits. The satellite positions and SV clock error can be calculated in a similar fashion as is done using the accurate ephemeris data.
4. *Subframe 5*. The 25 pages of this subframe include the almanac for satellites with PRN numbers from 1 to 24 and SV health information.

It should be noted that since each satellite transmits all 25 pages, almanac data for all satellites are transmitted by every satellite. Unlike ephemeris data, almanac data remain valid for long periods (e.g. months) but are much less precise. Additional data contained in the navigation message is the user range error (URE), which estimate the range error due to satellite ephemeris and timing errors (but no errors due to propagation) on a per SV i basis.

4.1.3 GPS Satellite Position Calculations

The precise positions of the GPS satellites are calculated by the GPS Ground Control Segment and then modeled to describe the orbit. The parameters that describe the orbit of each satellite are encoded, in terms of Keplerian orbital parameters in the navigation data bit stream, which is uploaded to the satellite for broadcast to the user. The ephemeris parameters describe the orbit during an interval of time (at least one hour) include primary parameters (i.e. semi-major axis, longitude of ascending node, etc.), as well as secondary parameters, such as various higher-order correction terms (i.e. the C' terms listed in Table 4.1). The complete list and definitions of these parameters are given in Table 4.1.

After the GPS receiver has decoded these data, the precise positions of the satellites needs to be calculated so the user solution for PVT can be determined. GPS uses an Earth-centered, Earth-fixed (ECEF) coordinate system with respect to a World Geodetic System 1984 (WGS84) datum using GPS time as established by the US Naval Observatory (USNO). The SV positions and hence the user solutions are with respect to these system references.

The Keplerian and other GPS system parameter information decoded from the navigation data stream can be used by the GPS user to calculate the satellite position with the algorithm outlined in Table 4.2. Overall, the user GPS receiver SV position calculation algorithm uses the data provided in the navigation signal broadcast and calculates the position of the SV in the orbital plane (with

Table 4.1 Components of ephemeris data.

Term	Description	Unit
t_{0e}	Reference time of ephemeris	s
$\sqrt{a}$	Square root of semimajor axis	$\sqrt{\text{m}}$
e	Eccentricity	Dimensionless
i_0	Inclination angle (at time t_{0e})	Semicircle
Ω_0	Longitude of the ascending node (at weekly epoch)	Semicircle
ω	Argument of perigee (at time t_{0e})	Semicircle
M_0	Mean anomaly (at time t_{0e})	Semicircle
IDOT	Rate of change of inclination angle (i.e. di/dt)	Semicircle/s
$\dot{\Omega}$	Rate of change of longitude of the ascending node	Semicircle/s
Δn	Mean motion correction	Semicircle/s
C_{uc}	Amplitude of cosine correction to argument of latitude	rad
C_{us}	Amplitude of sine correction to argument of latitude	rad
C_{rc}	Amplitude of cosine correction to orbital radius	m
C_{rs}	Amplitude of sine correction to orbital radius	m
C_{ic}	Amplitude of cosine correction to inclination angle	rad
C_{is}	Amplitude of sine correction to inclination angle	rad

respect to an Earth-centered inertial [ECI] coordinate system, lines 15 and 16 in Table 4.2) and then rotates this SV position into the ECEF coordinate frame, lines 17–19 in Table 4.2. While the long list of equations in Table 4.2 looks a bit daunting at first glance, it is fairly straightforward, but a couple of items deserve additional discussion.

Calculation of the satellite's ECEF antenna phase center position is very sensitive to small perturbations in most ephemeris parameters. The sensitivity of position to the parameters $\sqrt{a}$, C_{rc}, and C_{rs} is about 1 m/m. The sensitivity to angular parameters is on the order of 10^8 m/semicircle and to the angular rate parameters on the order of 10^{12} m/semicircle/s. Because of this extreme sensitivity, when performing SV calculations in any GNSS system, it is very important to utilize the fixed parameters defined for the system and full precision (usually double precision) in the computation engine used. Some of these parameters are identified at the beginning of Table 4.2. Attention to units and scale factors of the raw data bits transmitted in the navigation message should be exercised [2].

4.1.3.1 Ephemeris Data Reference Time Step and Transit Time Correction

The reference time for the ephemeris orbit data, t_{0e}, represents the middle of the orbit fit, for which the ephemeris data are valid. When propagating

Table 4.2 Algorithm for computing satellite position.

	$\mu = 3.986\,005 \times 10^{14}\ \text{m}^3/\text{s}^2$	WGS84 value of Earth's universal gravitational parameter
	$c = 2.997\,924\,58 \times 10^{8}\ \text{m/s}$	GPS value for speed of light
	$\dot{\Omega}_e = 7.292\,115\,146\,7 \times 10^{-5}\ \text{rad/s}$	WGS84 value of Earth's rotation rate
	$\pi = 3.141\,592\,653\,589\,8$	GPS value for ratio of circumference to radius of circle
1	$a = (\sqrt{a})^2$	Semimajor axis
2	$n = \sqrt{\frac{\mu}{a^3}} + \Delta n$	Corrected mean motion (rad/s)
3	$t_k = t - t_{0e}$	Time from ephemeris epoch
4	$M_k = M_0 + (n)(t_k)$	Mean anomaly
5	$M_k = E_k - e \sin E_k$	Eccentric anomaly (must be solved iteratively for E_k)
6	$\sin \nu_k = \frac{\sqrt{1-e^2} \sin E_k}{1 - e \cos E_k}$ $\cos \nu_k = \frac{\cos E_k - e}{1 - e \cos E_k}$ $\nu_k = a \tan 2 \left[\frac{\sin \nu_k}{\cos \nu_k}\right]$	True anomaly (solve for in each quadrant)
7	$\phi_k = \nu_k + \omega$	Argument of latitude
8	$\delta\phi_k = C_{\text{us}} \sin(2\phi_k) + C_{\text{uc}} \cos(2\phi_k)$	Argument of latitude correction
9	$\delta r_k = C_{\text{rs}} \sin(2\phi_k) + C_{\text{rc}} \cos(2\phi_k)$	Radius correction
10	$\delta i_k = C_{\text{is}} \sin(2\phi_k) + C_{\text{ic}} \cos(2\phi_k)$	Inclination correction
11	$u_k = \phi_k + \delta\phi_k$	Corrected argument of latitude
12	$r_k = a(1 - e \cos E_k) + \delta r_k$	Corrected radius
13	$i_k = i_0 + (di/dt)t_k + \delta i_k$	Corrected inclination
14	$\Omega_k = \Omega_0 + (\dot{\Omega} - \dot{\Omega}_e)(t_k) - \dot{\Omega}_e t_{0e}$	Corrected longitude of ascending node
15	$x_k' = r_k \cos u_k$	In-plane x position (ECI frame)
16	$y_k' = r_k \sin u_k$	In-plane y position (ECI frame)
17	$x_k = x_k' \cos \Omega_k - y_k' \cos i_k \sin \Omega_k$	ECEF x coordinate
18	$y_k = x_k' \sin \Omega_k - y_k' \cos i_k \sin \Omega_k$	ECEF y coordinate
19	$z_k = y_k' \sin i_k$	ECEF z coordinate

the GPS SV position at a particular GPS time (t), the time index t_k is used. Thus, t_k can be positive or negative. With a time corrected GPS time t, the SV position would be the position of the SV at the TOT. Additionally, t_k must account for beginning- or end-of-week crossovers. Thus, if t_k is greater than 302 400 seconds, subtract 604 800 seconds from t_k; if t_k is less than −302 400 seconds, add 604 800 seconds to t_k.

A compensation that must be made by the user to provide an accurate user solution is compensation for the transit time (t_{transit}), which is the finite amount of time it takes the SV signal to propagate from the satellite antenna to the user antenna. First, the transit time must be estimated. This can be done by the GPS receiver using the adjusted pseudorange measurement (removing the transmitter clock error and T_{GD} for single frequency solution) and dividing by the speed of light. (Atmospheric delay should not be removed in this adjusted pseudorange measurement, when calculating the transit time.) Propagation times are typically 68–83 ms, depending upon the aspect angle to the satellite. With the transit time, SV position compensations can be done by subtracting an additional rotation term to the corrected longitude of ascending node term of Table 4.2 by the amount of $\dot{\Omega}_e t_{\text{transit}}$, to compensate for the transmit time. It should be noted that an alternative, and equivalent approach, is to calculate the SV position at the time of reception, and then perform a coordinate rotation, about the ECEF Z axis, using the angular rotation rate of the Earth and the transmit time calculate, i.e. $\dot{\Omega}_e t_{\text{transit}}$.

4.1.3.2 True, Eccentric, and Mean Anomaly

With the GPS satellite in a "circular orbit," they are reality in a slightly elliptical orbit that cannot be ignored. (A perfect circular orbit has an eccentricity of 0.0, whereas GPS orbits typically have eccentricities on the order of 0.001 or so.) Orbit phase variables used for determining the position of a satellite in its elliptical orbit are illustrated in Figure 4.6. The variable v in Figure 4.6 is called the

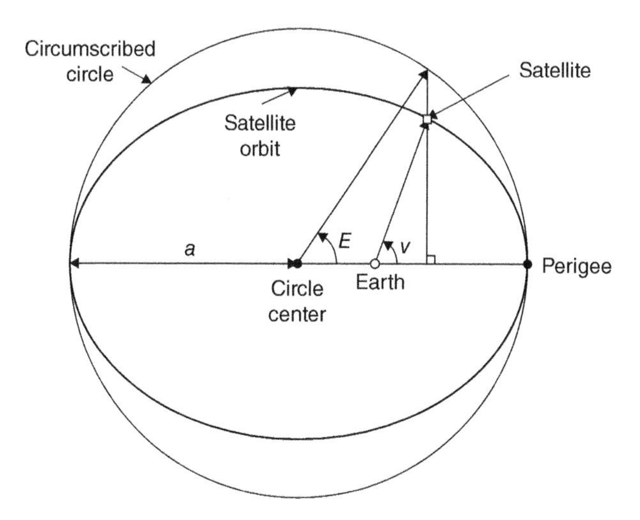

Figure 4.6 Geometric relationship between true anomaly v and eccentric anomaly E.

true anomaly in orbit mechanics. The problem of determining satellite position from these data and equations is called the *Kepler problem*. The hardest part of the Kepler problem is determining true anomaly as a function of time, which is nonlinear with respect to time. This problem was eventually solved by introducing two intermediate "anomaly" variables.

The *eccentric anomaly* (E) is defined as a geometric function of true anomaly, as shown in Figure 4.6. Eccentric anomaly E is defined by projecting the satellite position on the elliptical orbit out perpendicular to the semimajor axis a and onto the circumscribed circle. Eccentric anomaly is then defined as the central angle to this projection point on the circle, as shown in Figure 4.6.

The *mean anomaly* (M) can be described as the arc length (in radians) the satellite would have traveled since the perigee passage if it were moving in the circumscribed circle at the mean average angular velocity n. (Perigee is the point where the satellite is closest to the Earth.) The mean anomaly can then be defined as shown in Eq. (4.3), which is a linear function of time:

$$M(t) = \frac{2\pi(t - t_{\text{perigee}})}{T_{\text{period}}} \tag{4.3}$$

In Eq. (4.3), t is the time in seconds where the true anomaly is to be determined, t_{perigee} is the time when the satellite was at its perigee, and T_{period} is the period of the satellite orbit, in seconds to complete one full orbit. From this equation, calculating mean anomaly as a function of time is relatively easy, but the solution for true anomaly as a function of eccentric anomaly is much more difficult.

Note that if the orbit was perfectly circular, then the *eccentric anomaly* (E), *mean anomaly* (M), and *true anomaly* (ν) would all be equal; however, for the low eccentricities of GPS orbits, the numerical values of the true, eccentric, and mean anomalies are quite close together. The precision required in calculating true anomaly requires they be treated as separate variables.

4.1.3.3 Kepler's Equation for the Eccentric Anomaly

One equation in Table 4.2, shown in Eq. (4.4), is called *Kepler's equation*. It relates the eccentric anomaly E_k for the satellite at the step time t_k to its mean anomaly, M_k, and the orbit eccentricity e. This equation is the most difficult of all the equations in Table 4.2 to solve for E_k as a function of M_k:

$$M_k = E_k - e \sin E_k \tag{4.4}$$

The solution of Kepler's equation in Eq. (4.4) includes a transcendental function of eccentric anomaly E_k. It is impractical to solve for E_k in any way except by approximation. Standard practice is to solve the true anomaly equation iteratively for E_k, using the second-order Newton–Raphson method to solve for a residual term expressed in Eq. (4.5), and to make it arbitrarily small,

$$\varepsilon_k \equiv -M_k + E_k - e \sin E_k = 0 \tag{4.5}$$

Then, use the resulting value of E_k to calculate true anomaly; this process could continue at a given time step t_k until an arbitrarily small value is obtained. It starts by assigning an initial guess to E_k, say, $E_k[n]$ equal to $M_k[n]$ for $n = 0$; see line 4 in Table 4.2. Since the GPS orbits are fairly circular to start with, the mean anomaly is a fair starting point. Then, form successively better estimates $E_k[n+1]$ by the second-order Newton–Raphson formula illustrated in Eq. (4.6):

$$E_k[n+1] = E_k[n] - \left[\frac{\varepsilon_k(E_k[n])}{\left.\frac{\partial \varepsilon_k}{E_k}\right|_{E_k=E_k[n]} - \left[\frac{\left.\frac{\partial^2 \varepsilon_k}{E_k^2}\right|_{E_k=E_k[n]} \varepsilon_k(E_k[n])}{2 \left.\frac{\partial \varepsilon_k}{E_k}\right|_{E_k=E_k[n]}} \right]} \right] \tag{4.6}$$

The iteration of Eq. (4.6) can stop once the difference between the estimated $E_k[n+1]$ and the last value; that is, $E_k[n]$ is sufficiently small, for example, less than 10^{-6}. Usually, within about three to six iteration steps, at a particular t_k, the $E_k[n+1]$ value is within 10^{-6} of the $E_k[n]$ value, and iteration can stop.

4.1.3.4 Satellite Time Corrections

The GPS Ground Control Segment estimates the satellite transmitter clock error, with respect to GPS time (t), and encodes clock bias, velocity, and acceleration error corrections, as well as a clock reference time (t_{oc}) in the broadcast navigation data stream. The user must correct the satellite clock error in accordance with Eq. (4.7):

$$t = t_{\text{SV}i} - \Delta t_{\text{SV}i} \tag{4.7}$$

In Eq. (4.7), t is GPS system time in seconds, that is, corrected from, $t_{\text{SV}i}$, which is the effective SV PRN code phase time at message transmission time in seconds, and $\Delta t_{\text{SV}i}$ is the total SV PRN code phase time offset in seconds. The SV PRN code phase offset (i.e. transmitter clock error) is given by Eq. (4.8). The clock error is made up of a bias, velocity, acceleration, and a relativistic correction term:

$$\Delta t_{\text{SV}i} = a_{f0} + a_{f1}(t - t_{\text{oc}}) + a_{f2}(t - t_{\text{oc}})^2 + \Delta t_r \text{ , [s]} \tag{4.8}$$

where

a_{f0} = SV i clock bias error (seconds)

a_{f1} = SV i clock velocity error (s/s)

a_{f2} = SV i clock acceleration error (s/s^2)

t_{oc} = clock reference time for clock correction date (seconds)

Δt_r = SV i relativistic correction (seconds)

In Eq. (4.8), clock error polynomial coefficients, the clock reference time for the clock correction data, and information needed to calculate the relativistic correction, are provided in the broadcast navigation data.

The relativistic correction term, in seconds (in the range of sub-nanoseconds), is straightforward, as shown in Eq. (4.9), but needs the iterated eccentric anomaly E_k. In Eq. (4.9), the eccentricity and semimajor axis are as described in Table 4.2:

$$\Delta t_r = Fe\sqrt{a}\sin E_k \ , [\text{s}] \tag{4.9}$$

where

$$F = \frac{-2\sqrt{\mu}}{c^2} = -4.442\,807\,622 \times 10^{-10} \ , [\text{s}/\sqrt{\text{m}}]$$

Note that Eqs. (4.7) and (4.8) are coupled, but this will not cause problems. While the coefficients a_{f0}, a_{f1}, and a_{f2} are generated by using GPS time as indicated in Eq. (4.7), sensitivity of t, to the difference between $t_{\text{SV}i}$ and t is negligible, when calculating the transmitter clock error expressed in Eq. (4.8). This negligible sensitivity allows the user to approximate t by $t_{\text{SV}i}$, using Eq. (4.8). The value of t must account for beginning- or end-of-week crossovers. Thus, if the quantity $t - t_{\text{oc}}$ is greater than 302 400 seconds, subtract 604 800 seconds from t; if the quantity $t - t_{\text{oc}}$ is less than −302 400 seconds, then add 604 800 seconds to t.

The MATLAB® m-files (`ephemeris.m`, `GPS_position.m`, `GPS_position_3D.m`, `GPS_el_az.m`, `GPS_el_az_all.m`, `GPS_el_az_one_time.m`) on www.wiley.com/go/grewal/gnss calculates satellite positions for one set of ephemeris data and one time. Other programs calculate satellite positions for a range of time; see Appendix A.

4.1.4 C/A-Code and Its Properties

The C/A-code has the following functions:

1. *Enable accurate range measurements and resistance to errors caused by multipath.* To establish the position of a user to within reasonable accurate (e.g. less than 30 m) satellite-to-user range, estimates are needed. The estimates are made from measurements of signal propagation delay from the satellite to the user. To achieve the required accuracy in measuring signal delay, the GPS carrier must be modulated by a waveform having a relatively large bandwidth. The needed bandwidth is provided by the C/A-code modulation rate, which also permits the receiver to use correlation processing to effectively combat measurement errors. Because the C/A-code causes the bandwidth of the signal to be much greater than that needed to convey the 50-bps navigation data bit stream, the resulting signal is called a *spread-spectrum* signal. Using the C/A-code to increase the signal bandwidth also reduces errors

in measuring signal delay caused by multipath (the arrival of the signal via multiple paths such as reflections from objects near the receiver antenna) since the ability to separate the direct path signal from the reflected signal improves as the signal bandwidth is made larger. While the C/A-code does a good job at this, the higher-rate spreading codes can do better at multipath mitigation.

2. *Permits simultaneous range measurement from several satellites.* The use of a distinct C/A-code for each satellite permits all satellites to use the same frequencies without interfering with each other. This is possible because the signal from an individual satellite can be isolated by correlating it with a replica of its C/A-code in the receiver. This causes the C/A-code modulation from that satellite to be removed so that resulting signal bandwidth is narrowband; the signal part that remains is the low-rate navigation data. This process is called *despreading* of the signal. However, the correlation process does not cause the signals from other satellites to become narrowband because the codes from different satellites are nearly orthogonal. Therefore, the interfering signals from other satellites can be largely rejected by passing the desired despread signal through a narrowband filter, a bandwidth-sharing process called *code-division multiplexing* (CDM) or CDMA.
3. *Protection from interference/jamming.* The C/A-code also provides a measure of protection from intentional or unintentional interference or jamming of the received signal by another man-made signal. The correlation process that despreads the desired signal has the property of spreading other undesirable signals. Therefore, the signal power of any interfering signal, even if it is narrowband, will be spread over a large frequency band, and only that portion of the power lying in the narrowband filter will compete with the desired signal. The C/A-code provides about 20–30 dB of improvement in resistance to jamming from narrowband signals. Despite this distinct advantage, we must also remember that the GPS CDMA signals are relatively weak signals received on Earth.
4. *Short period for fast acquisition.* The GPS C/A-code has a period of only 1 ms. This short period helps the GPS receiver to rapidly acquire and track it. The short period and simplicity of its implementation also help reduce the cost, complexity, and power consumption in the GPS to support the C/A-code implementation.

We next detail important properties of the C/A-code.

4.1.4.1 Temporal Structure

Each satellite has a unique C/A-code identified by the PRN number, but all the codes consist of a repeating sequence of 1023 chips occurring at a rate of 1.023 MHz with a period of 1 ms, as previously illustrated in Figure 4.2. The leading edge of a specific chip in the sequence, called the *C/A-code epoch*, defines the beginning of a new period of the code. The code sequence can be generated digitally, with digital values, or converted into a bipolar signal with either positive or negative values without any loss of generality. The sequences of the individual 1023 chips appear to be random but are in fact generated by a deterministic algorithm implemented by a combination of shift registers (i.e. delay units) and combination logic. The algorithm produces two intermediate PRN code sequences (i.e. $G_1(t)$ and $G_2(t)$). Then $G_2(t)$ is delayed with respect to $G_1(t)$, and they are combined to produce the C/A *Gold codes*. The details of Gold codes can be found in Ref. [4]. Gold codes are numerous to provide enough codes in the Gold code family to support the GPS constellation and other GNSS signal sources. While Gold codes do not have perfect correlation performance, their correlation properties are well understood and bounded [4]. Thus, the GPS C/A-code (based on the Gold codes) has the property of low cross correlation between different codes (nearly orthogonal) as well as reasonably small autocorrelation sidelobes.

4.1.4.2 Autocorrelation Function

The autocorrelation of a code sequence refers to how well that code correlates to a delayed version of itself over time, theoretically over all time. An approximation of the autocorrelation function can be made over a limited observation time (T), for code $c(t)$ and a delayed version of that same code over a delay (τ) (i.e. $c(t-\tau)$), as shown in Eq. (4.10):

$$R_c(\tau) = \frac{1}{T}\int_0^T c(t)c(t-\tau)d\tau \tag{4.10}$$

The autocorrelation of the GPS C/A can then be calculated from Eq. (4.10). An illustration of a GPS PRN code autocorrelation function (i.e. C/A-code) is shown in Figure 4.7, where the width of one C/A-code chip is $t_c = 0.9775\ \mu s$. The shape of the code autocorrelation is basically a triangle with a base length of two chips and a peak located at $\tau = 0$, when the codes are lined up in time. The

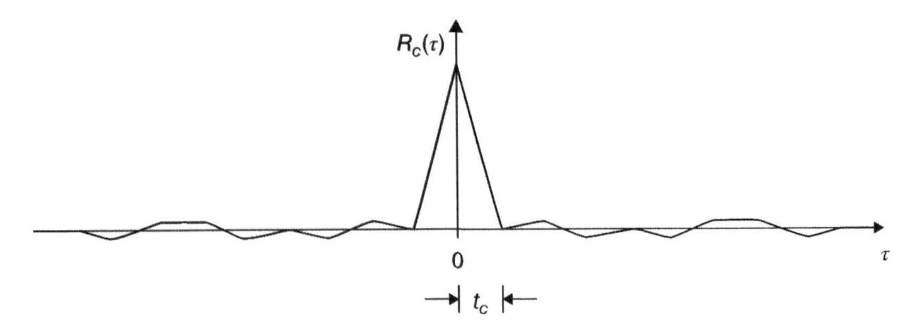

Figure 4.7 Illustration of autocorrelation functions of GPS PRN codes.

autocorrelation function contains small sidelobes outside the triangular region. In most benign cases, these sidelobes do not have a major impact on ranging performance.

The C/A-code autocorrelation function plays a substantial role in GPS receivers, inasmuch as it forms the basis for code tracking and accurate user-to-satellite range measurement estimation. In fact, the receiver continually computes values of this autocorrelation function in which the received signal code, $c(t)$, is correlated with a receiver-generated code (i.e. $c(t - \tau)$). Special hardware and software enable the receiver to adjust the reference waveform delay so that the value of τ is zero, thus enabling determination of the time of arrival of the received signal. As a final note, when the GPS signal is filtered in the GPS receiver, the autocorrelation function becomes rounded and accuracy is degraded.

4.1.4.3 Power Spectrum

The power spectrum of a spreading code describes how the power in the code is distributed in the frequency domain. It can be defined in terms of either a Fourier series expansion of the code waveform or, equivalently, the code autocorrelation function. The power spectral density is the Fourier transform of the autocorrelation function. Using the code autocorrelation function $R_c(\tau)$, the power spectral density of our spreading code can be described as shown in Eq. (4.11):

$$S_c(f) = \lim_{T\to\infty} \frac{1}{T} \int_{-T}^{T} R_c(\tau) e^{-2\pi f \tau} d\tau \tag{4.11}$$

A plot of $S_c(f)$ is shown as a smooth curve in Figure 4.8. The overall envelope of the power spectral density of our spreading code is a sinc^2 function (i.e. $\sin^2(x)/x^2$ shape), with its nulls (i.e. zero values) at multiples of the code rate away from the central peak. Approximately 90% of the signal power is located between the first two nulls, but the smaller portion lying outside the first nulls is very important for accurate ranging. As for the GPS C/A-code, the spectrum nulls are at multiples of the C/A-code rate of 1.023 MHz. The period nature of the C/A-code with 1 ms code epochs leads to spectral line components with 1-kHz spacing. Also shown in Figure 4.8, for comparative purposes, is a typical noise power spectral density found in a GPS receiver after frequency conversion of the signal to baseband (i.e. with carrier removed). It can be seen that the presence of the C/A-code causes the entire signal to lie well below the noise level because the signal power has been spread over a wide frequency range (approximately ± 1 MHz).

4.1.4.4 Despreading of the Signal Spectrum

Using the mathematical model of the signal modulated by the C/A-code presented in Eq. (4.1), within the GPS receiver, for a specific PRN code, the

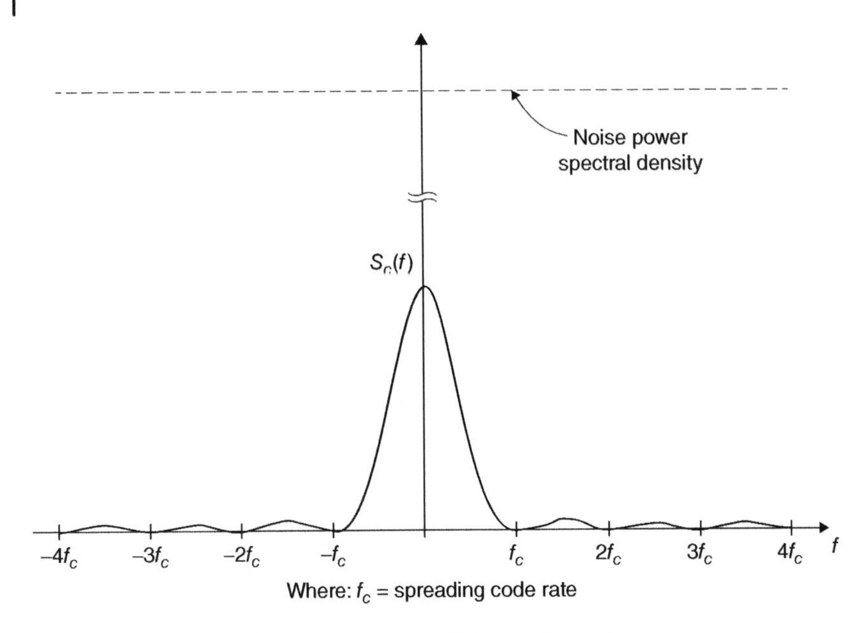

Figure 4.8 Illustration of power spectrum of GPS spreading codes.

received signal's carrier frequency may be tracked with a phase-lock loop (PLL) to remove the carrier frequency, and only the data modulation $d(t)$ and the spreading code $c(t)$ modulation remain. For a particular PRN C/A-code, the resulting signal (at baseband [bb]), $s_{\text{bb}}(t)$, in normalized form is shown in Eq. (4.12):

$$S_{\text{bb}}(t) = d(t)c(t) \tag{4.12}$$

The power spectrum of the signal $s_{\text{bb}}(t)$ is similar to that of the C/A-code illustrated in Figure 4.8. As previously mentioned, the signal in this form has a power spectrum lying below the receiver noise level. However, if this signal $s_{\text{bb}}(t)$ is multiplied by a replica of $c(t)$ in exact alignment with it (i.e. is punctual; we will call it $c_p(t)$), the resulting signal can then be fed through a signal recovery filter (e.g. low-pass filter [LPF]) function to produce an estimate of the original data, $\hat{d}(t)$, as represented in Eq. (4.13):

$$s_{\text{bb}}(t)c_p(t) = d(t)c(t)c_p(t) \xrightarrow[\text{LPF}]{} K\hat{d}(t) \tag{4.13}$$

As shown in Eq. (4.13), both of the coding sequences are considered to be of value ± 1. This multiplication and low-pass filtering operation is a correlation, and the proportionality constant K has no consequential effect on our ability to produce a good estimate of the original navigation data bit stream.

This procedure is called *code despreading*, which removes the C/A-code modulation from the signal and exposes the underlying navigation data. The resulting signal has a two-sided spectral width of approximately 100 Hz due to the 50-bps navigation data modulation. From the previously stated equation, it can be seen that the total signal power has not substantially changed in this process but is now contained in a much narrower bandwidth. Thus, the magnitude of the power spectrum is greatly increased, as indicated in Figure 4.9. In fact, it now exceeds that of the noise, and the signal can be recovered by passing it through a small-bandwidth signal recovery filter to remove the wideband noise, as shown in Figure 4.9.

4.1.4.5 Role of Despreading in Interference Suppression

At the same time that the spectrum of the desired GPS signal is narrowed by the despreading process, any interfering signal that is not modulated by the same C/A-code will instead have its spectrum *spread* to a width of at least 2 MHz so that only a small portion of the interfering power can pass through the signal recovery filter. The amount of interference suppression gained by using the C/A-code depends on the bandwidth of the recovery filter, the bandwidth of the interfering signal, and the bandwidth of the C/A-code. For a narrowband interfering source whose signal can be modeled by a nearly sinusoidal waveform and a signal recovery filter bandwidth of 1000 Hz or more, the amount of

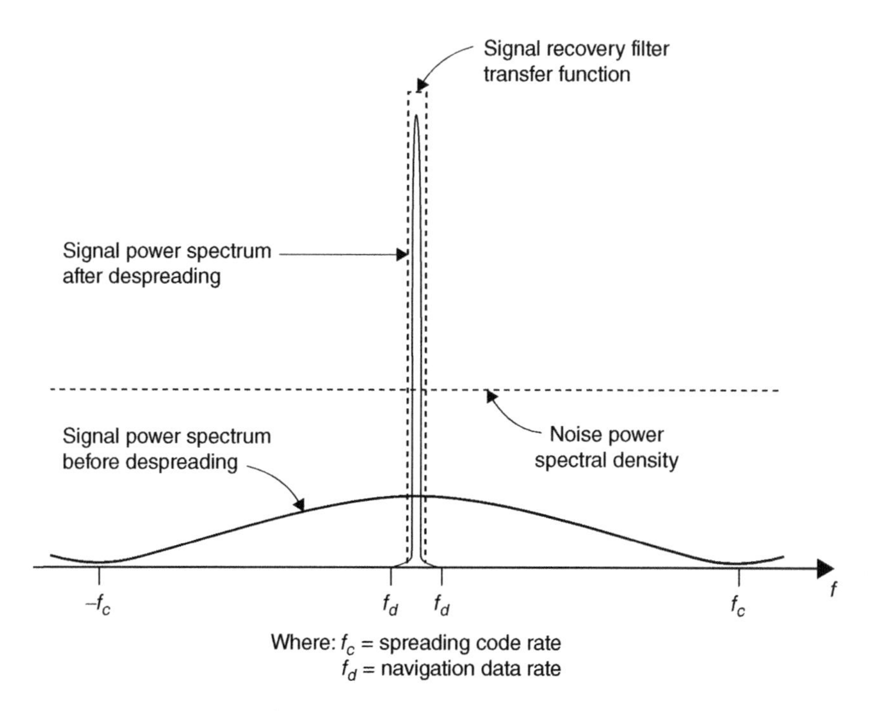

Figure 4.9 Despreading of the spreading code.

interference suppression in decibels can be expressed in Eq. (4.14):

$$\eta = 10\log_{10}\left(\frac{W_c}{W_f}\right), \quad [\text{dB}] \tag{4.14}$$

where

W_c = bandwidth of the spreading code

W_f = bandwidth of the filtering function

In Eq. (4.14), the null-to-null bandwidth of the C/A-code is 2.046 MHz; and if $W_f = 2000$ Hz, about 30 dB of suppression can be obtained for a narrowband interference source. When the signal recovery filter has a bandwidth smaller than 1000 Hz, the situation is more complicated since the despread interfering sinusoid will have discrete spectral components with 1000-Hz spacing. As the bandwidth of the interfering signal increases, the C/A-code despreading process provides a decreasing amount of interference suppression. For interference sources having a bandwidth greater than that of the signal recovery filter, the amount of suppression in decibels provided by the spreading code can be approximated as shown in Eq. (4.15):

$$\eta = 10\log_{10}\left(\frac{W_I + W_c}{W_I}\right), \quad [\text{dB}] \tag{4.15}$$

where

W_c = bandwidth of the spreading code

W_I = bandwidth of the interference source

4.1.4.6 Cross-correlation Function

The cross correlation of a code sequence refers to how well that code correlates to a delayed version of a *different* code over time, theoretically over all time. An approximation of the cross-correlation function can be made over a limited observation time (T), for code $c_1(t)$ and a delayed version of a different code (i.e. $c_2(t-\tau)$) as shown in Eq. (4.16):

$$R_{c_1c_2}(\tau) = \frac{1}{T}\int_0^T c_1(t)c_2(t-\tau)d\tau \tag{4.16}$$

Thus, when a selected satellite signal is encoded with a given spreading code, and it is despread using a replica of that same code, the signals from other satellites (encoded with different spreading codes) look like wideband interference sources that are below the noise level. This permits a GPS receiver to extract a multiplicity of individual satellite signals and process them individually, even though all signals are transmitted at the same frequency. This is why this process is called CDM and allows for access to all of the GPS signals (i.e. CDMA).

4.1.5 P(Y)-Code and Its Properties

The GPS P(Y)-code is a higher-rate code, is encrypted, and appears on both of the GPS L1 and L2 frequencies. It is used by military, authorized, and civil users to produce "semicodeless" measurements. The P(Y)-code has the following functions:

1. *Increased jamming protection.* Because the bandwidth of the P-code is 10 times greater than that of the C/A-code, it offers approximately 10 dB more protection from narrowband interference. In military applications, the interference is likely to be a deliberate attempt to jam (render useless) the received GPS signal.
2. *Provision for antispoofing.* In addition to jamming, another military tactic that an enemy can employ is to radiate a signal that appears to be a GPS signal (*spoofing*) but, in reality, is designed to confuse the GPS receiver. This is prevented by encrypting the P-code. The would-be spoofer cannot know the encryption process and cannot make the contending signal look like a properly encrypted signal. Thus, the receiver can reject the false signal and decrypt the desired one.
3. *Denial of P-code use.* The structure of the P-code is published in the open literature, so that anyone may generate it as a reference code for despreading the signal and for making range measurements. However, encryption of the P-code by the military will deny its use by unauthorized parties.
4. *Increased code range measurement accuracy.* All other parameters being equal, accuracy in range measurement improves as the signal bandwidth increases. Thus, the P-code provides improved range measurement accuracy as compared with the C/A-code. Simultaneous range measurements using both codes are even better. Because of its increased bandwidth, the P-code is also more resistant to range errors caused by multipath.

4.1.5.1 P-Code Characteristics

Unlike the C/A-code, the P-code modulates both the L1 and L2 carriers. Its chipping rate is 10.23 MHz, which is precisely 10 times the C/A rate, and it has a period of one week. It is transmitted synchronously with the C/A-code in the sense that each chip transition of the C/A-code always corresponds to a chip transition in the P-code. Like the C/A-code, the P-code autocorrelation function has a triangular central peak centered at $\tau = 0$, but with one-tenth the base width, as shown in Figure 4.7, where the width of one P(Y)-code chip is $t_c = 0.097\,75\,\mu s$. The power spectrum also has a $\sin^2(x)/x^2$ characteristic, but with 10 times the bandwidth, as indicated in Figure 4.8, where f_c is 10.23 MHz. Because the period of the P-code is so long, the power spectrum can be regarded as continuous envelope for practical purposes. Each satellite broadcasts a unique P-code. The technique used to generate it is similar to that of the C/A-code but somewhat more complicated. The details of the P-code generation can be found in Ref. [2].

4.1.5.2 Y-Code

The encrypted form of the P-code used for antispoofing and denial of the P-code to unauthorized users is called the *Y-code*. The Y-code is formed by multiplying the P-code by an encrypting code called the *W-code*. The W-code is a random-looking sequence of chips that occur at a 511.5-kHz rate. Thus there are 20 P-code chips for every W-code chip. Since both the P-code and the W-code have chip values of ± 1, the resulting P(Y)-code has the same appearance as the P-code; that is, it also has a 10.23-MHz chipping rate. However, the Y-code cannot be despread by a receiver replica P-code unless it is decrypted. Decryption consists of multiplying the Y-code by a receiver-generated replica of the W-code that is made available only to authorized users. Since the encrypting W-code is also not known by the creators of spoofing signals, it is easy to verify that such signals are not legitimate.

4.1.6 L1 and L2 Carriers

The L1 (or L2) carrier is used for the following purposes:

1. *Propagate the GPS signal from the satellite to the user in the service area provided by the system.*
2. *To provide very accurate relative range measurements for precision applications using carrier phase.*
3. *To provide accurate Doppler measurements.* The phase rate of the received carrier can be used for accurate determination of user velocity. The integrated Doppler, which can be obtained by counting the cycles of the received carrier, is often used as a precise delta range observable that can materially aid the performance of code tracking loops. The integrated Doppler history is also used as part of the carrier phase ambiguity resolution process.

4.1.6.1 Dual-Frequency Operation

The use of *both* the L1 and L2 frequencies provides the following benefits:

1. *Provides accurate measurement of ionosphere signal delay.* A major source of ranging error is caused by changes in both the phase velocity and group velocity of the signal as it passes through the ionosphere. Range errors of 10–20 m are commonplace and can be much larger (e.g. 100 m). Because the delay induced by the ionosphere is known to be inversely proportional to the square of frequency, ionosphere range error can be estimated accurately by comparing the times of arrival of the L1 and L2 signals. Details on the calculations appear in Chapter 7.

2. *Facilitates carrier phase ambiguity resolution.* In high-accuracy GPS differential positioning, the range estimates using carrier phase measurements are precise but highly ambiguous due to the periodic structure of the carrier. The ambiguity is more easily resolved (by various methods) as the carrier frequency decreases. By using L1 and L2 carrier frequencies, the ambiguity resolution can be based on their frequency difference (1575.42, 1227.6 MHz), to produce various code and carrier combinations. These combinations provide various tradeoffs with respect to faster and more reliable ambiguity resolution performance versus accuracy.
3. *Provides system redundancy (primarily for the military user).*

4.1.7 Transmitted Power Levels

The GPS signals are transmitted at a minimum power level on the order of 478-W (26.8-dBW) effective isotropic radiated power (EIRP), which means that the minimum received power is the same as would be obtained if the satellite radiated 478 W from an isotropic antenna. This effective power level is reached by radiating a smaller total power into a beam approximately 30° wide toward the Earth. The radiated power level was chosen to provide a signal-to-noise ratio (SNR) sufficient for tracking of the signal by a receiver on the Earth with an unobstructed view of the satellite and to not interfere with other terrestrial communication systems. The low signal power level provides a challenge to meet the need to operate GPS receivers under less desirable conditions, such as in heavy vegetation, in urban canyons, or indoor environments where considerable signal attenuation often occurs. These low signal power levels can also pose problems when attempting to operate a GPS receiver in a jamming or interference environment. The GPS P(Y) signal is transmitted at a power level that is 3 dB less on L1 and 6 dB less on L2 than the C/A-encoded signal on L1 as illustrated in Figure 4.1.

4.1.8 Free Space and Other Loss Factors

As the signal propagates toward the Earth, it loses power density due to spherical spreading. The loss is accounted for by a quantity called the *free space loss factor* (FSLF), given in Eq. (4.17):

$$\text{FSLF} = \left(\frac{\lambda}{4\pi R}\right)^2 \tag{4.17}$$

With the height of a GPS satellite in a medium Earth orbit (MEO) of approximately 20 000 km, and an L1 wavelength of 0.19 m, the FSLF is approximately 5.7×10^{-19} or −182.4 dB.

Other loss factors include an atmospheric loss factor (ALF) of 2.0 dB, which is allocated to account for any signal attenuation by the atmosphere. Additional provisions can be allocated for any antenna mismatch losses on the order of 2.4 dB.

A typical GPS antenna with right-hand circular polarization and a hemispherical pattern has about 3.0 dBi of gain relative to an isotropic antenna at zenith and decrease as the elevation angle to the satellite decreases. The actual GPS signal power levels received are specified for a 3-dBi gain linearly polarized antenna with optimum orientation. With this linearly polarized antenna optimally placed with respect to the orientation of the incoming GPS signal, this antenna will only receive one-half of the power from the signal, so receiver antenna gain (G_r) of 0 dBic (dBs relative to a right-handed circular polarized signal) can be used. Additionally, there may be antenna orientation and mismatch losses that can be grouped and allocated.

4.1.9 Received Signal Power

When the EIRP is used, all of the other loss factors are considered, and applying the Friis transmission equation, the power received at the user antenna output can be calculated in dB format as EIRP + G_r + FSLF − ALF − Mismatch = 26.8 + 0 + (−182.4) − 2.0 − 2.4 = −156.0 dBW. Now the actual GPS signal power levels vary somewhat from this calculated value based on transmitter and user received antenna pattern variations, frequency, code type, and user latitude as documented in Ref. [2].

4.2 Modernization of GPS

Since GPS was declared with a full operational capability (FOC) in April 1995, applications and the use of GPS have evolved rapidly, especially in the civil sector. As a result, radically improved levels of performance have been reached in positioning, navigation, and time transfer. However, the availability of GPS has also spawned new and demanding applications that reveal certain shortcomings of the legacy system. Therefore, since the mid-1990s, numerous governmental and civilian committees have investigated these shortcomings and requirements for a more modernized GPS.

The modernization of GPS is a complex task that requires tradeoffs in many areas. Major issues include spectrum needs and availability, military and civil performance, signal integrity and availability, financing and cost containment, and potential competing interests by other GNSSs and developing countries. Major decisions have been made for the incorporation of modernized civil and military signals that have enhanced the overall performance of the GPS.

4.2.1 Benefits from GPS Modernization

The benefit from a modernized GPS are the following:

1. *Robust dual-frequency ionosphere correction capability for civil users.* Since only the encrypted P(Y)-code appears on the L2 frequency, civil users benefit from a robust dual-frequency ionosphere error correction capability. Civil users had to rely on semicodeless tracking of the GPS L2 signal, which is not as robust as access to a full-strength unencrypted signal. While civil users could employ a differential technique, this adds complexity to an ionosphere free user solution.
2. *A better civil code.* While the GPS C/A-code is a good, simple spreading code, a better civil code would provide better correlation performance. Rather than just turning on the C/A-code on the L2 frequency, a more advanced L2C spreading code provides robust ranging and ionosphere error correction capability.
3. *Improved ambiguities resolution with phase measurements.* High-accuracy differential positioning at the centimeter level by civil users requires rapid and reliable resolution of ambiguities using phase measurements. Ambiguity resolution with single-frequency (L1) receivers generally requires advanced error mitigation strategies and a sufficient length of time for the satellite geometry to change significantly. Performance is improved with robust dual-frequency receivers. However, the effective SNR of the legacy P(Y) encrypted signal is dramatically reduced because the encrypted P(Y)-code cannot be despread by the civil user. A high CNR signal on a second frequency (i.e. L2C or L5) helps here.
4. *Dual-frequency navigation signals in the aeronautical radio navigation service (ARNS) band.* The ARNS band of frequencies is federally protected and can be used for safety of life (SOL) applications. The GPS L1 band is in an ARNS band, but the GPS L2 band is not. (The L2 band is in the radio navigation satellite service [RNSS] band that has a substantial amount of uncontrolled signals in it.) In applications involving public safety, the integrity of the current system can be improved with a robust dual-frequency capability where both GPS signals are within the ARNS bands. This is particularly true in aviation landing systems that demand the presence of an adequate number of high-integrity satellite signals and functional cross-checks during precision approaches. The L5 signal at 1176.45 MHz is within the ARNS 960–1215 MHz band.

5. *Improvement is needed in multipath mitigation capability.* Multipath remains a dominant source of GPS positioning error and cannot be removed by differential techniques. Although certain mitigation techniques, such as multipath mitigation technology (MMT), approach theoretical performance limits for in-receiver processing, the required processing adds to receiver costs. In contrast, improved multipath performance can be achieved with signal design and higher chipping rate codes.
6. *Military requirements in a jamming environment.* The feature of selective availability (SA) was suspended at 8 p.m. EDT on 1 May 2000. SA was the degradation in the autonomous positioning performance of GPS, which was a concern in many civil applications requiring the full accuracy of which GPS is capable. If the GPS C/A-code was ever interfered with, the accuracies would degrade or not be available. Because the P(Y)-code has an extremely long period (seven days), it is difficult to acquire unless some knowledge of the code timing is known. P(Y) timing information is supplied by the GPS HOW at the beginning of every subframe. However, to read the HOW, the C/A-code must first be acquired to gain access to the navigation message. Unfortunately, the C/A-code is relatively susceptible to jamming, which would seriously impair the ability of a military receiver to acquire the P(Y)-code. Direct P(Y) acquisition techniques are possible, but these techniques still require information about the satellite position, user position, and clock errors to be successful. Furthermore interference on the C/A-coded signal would also affect the P(Y)-coded signal in the same frequency band.
   Additional power received from the satellite would also help military users operate more effectively. However, if the power is increased for only some of the GNSS signals in a region, the undesirable signal-to-signal correlation could be a concern.
7. *Compatibility and operability with other GNSSs.* With the advances in other GNSSs by other nations, the requirement exists for international cooperation in the development of new GNSS signals to ensure they do not interfere with each other and potentially provide an interoperable combined GNSS service.

4.2.2 Elements of the Modernized GPS

Table 4.3 provides summary of the legacy and modernized navigation signals with Figure 4.10 illustrating the modernized GPS signal spectrum. The legacy GPS signals (L1 C/A and P(Y) and L2 P(Y)) as well as the additional modernized signals (L2C, L5, GPS L1 and L2 M-code, and L1C) are illustrated in Figure 4.10. The L2C signal was introduced to the GPS constellation with the launch of the first Block IIR-M satellite on 21 September 2005. One of the last GPS Block

Table 4.3 Key GPS signals and parameters.

GPS band	Frequency (MHz)	Bandwidth (MHz)	Code type	Component	Modulation	Data type (rate, sps)	Service
L1	**1575.42**	20.460					
			C/A	I	BPSK(1)	LNAV (50 bps)	SPS
			P(Y)	Q	BPSK(10)	LNAV (50 bps)	PPS
			M		BOC(10,5)		PPS
		30.690	L1C	L1C_P	TMBOC	CNAV-2	L1CP
				L1C_D	BOC(1,1)	Pilot (0)	L1CP
					BOC(1,1)		L1CD
			Gold		BPSK(1)	SBAS (500)	WAAS
L2	**1227.60**	20.460					
			P(Y)	Q	BPSK(10)	LNAV (50 bps)	PPS
			L2C	L2C	BPSK(1)	CNAV (50)	SPS
			L2CM L2CL	L2CM and L2CL combo			
			M		BOC(10,5)		
L5	**1176.45**	20.460	L5				SPS
				I5	BPSK(10)	CNAV (100)	
				Q5	BPSK(10)	Pilot (0)	

Main frequency signals are bolded.

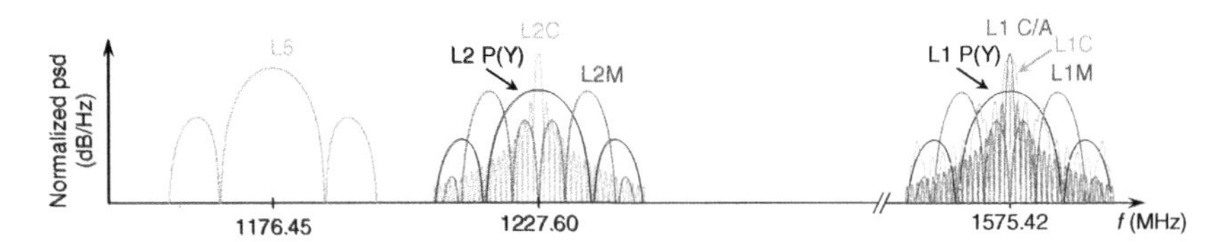

Figure 4.10 A modernized GPS signal spectrum. psd, power spectral density.

IIR-M SV49 added an L5 payload, the L5 signal was operationally introduced with Block IIF satellites that were first launched in 2010.

The bandwidth available for the modernized GPS signals is up to 24 MHz, but the compatibility and power levels relative to the other codes are design considerations. (GPS III SVs will have an increased bandwidth for the L1C signal.) Furthermore, assuming equal received power and filtered bandwidth, the ranging performance (with or without multipath) on a GPS signal is highly dependent upon the signal's spectral shape (or equivalently, the shape of the autocorrelation function). In this sense, the L1 C/A-coded and L2 civil signals are somewhat equivalent in scope as are the P(Y) and L5 civil signals (albeit with very different characteristics). As we will see, the military M-coded signal and other GNSS codes are different because they use different subcarrier frequencies and chipping rates. These different subcarriers, in essence, add an aspect known as frequency division multiplexing to the GPS spectrum.

Another important aspect of the modernized GPS signals is the navigation message formats, which moved away from the fixed frame, subframe navigation message format with 30 bit words, to a more conventional message type format with 8 bit bytes. Additional various message types can be transmitted at various rates to enable more efficient use of the data bandwidth.

Major elements of these modernized signals are as follows.

4.2.3 L2 Civil Signal (L2C)

This L2C signal has a code structure that has some performance advantages over the legacy C/A-code. The L2C signal offers civilian users the following improvements:

(a) *Robust dual-frequency ionosphere error correction.* The dispersive delay characteristic of the ionosphere proportional to $1/f^2$ can be estimated much more accurately with this full-strength signal on the L2 frequency. Thus, civil users can choose to use the semicodeless L2 P(Y) and L1 C/A signals or the L2C and L1 C/A signals to estimate the ionosphere delay.
(b) *Carrier phase ambiguity resolution is significantly improved.* The accessibility of the full-strength L1 and L2 signals provide robust "wide-lane" measurement combinations having ambiguities that are much easier to resolve.

(c) *The L2C signal improves robustness in acquisition and tracking.* The L2C spreading code provides for a more reliable acquisition and tracking performance.

Originally, modernization efforts considered turning on the C/A-code at the L2 carrier frequency (1227.60 MHz) to provide the civilian community with a robust ionosphere correction capability as well as additional flexibility and robustness. However, later in the planning process, it was realized that additional advantages could be obtained by implementing a new L2 civil (L2C) signal. The decision was made to use this new signal and its structure was made public early in 2001. Both the L2C and the new military M-code signal (described later in this chapter) appear on the L2 carrier orthogonal to the current P(Y).

Like the C/A-code, the C-code is a PRN code that runs at a 1.023×10^6 cps (chips per second) rate. However, it is generated by 2 : 1 time-division multiplexing of two independent subcodes, each having half the chipping rate, namely, 511.5×10^3 cps. Each of these subcodes is made available to the receiver by demultiplexing. These two subcodes have different periods before they repeat. The first subcode, the code moderate (CM), has a moderate length of 10 230 chips and a 20-ms period. The moderate length of this code permits relatively easy acquisition of the signal although the 2 : 1 multiplexing results in a 3-dB acquisition and data demodulation loss. The second subcode, the code long (CL), has a length of 707 250 chips, has a 1.5-seconds period, and is data-free. The CM and CL codes are combined to provide the C-code at the 1.023-Mcps rate. Navigation data can be modulated on the C-code. Provisions call for no data, legacy navigation data at a 50-bps rate, or new civil navigation (CNAV) data. The L2C CNAV data at a 25-bps rate would be encoded using a rate one-half convolutional encoding technique, to produce a 50 sps (symbols per second) data bit stream that could then be modulated onto the L2 frequency to form the L2C signal. With no data, the coherent processing time can be increased substantially, thereby permitting better code and carrier tracking performance, especially at low SNR. The relatively long CL code length also generates smaller correlation sidelobes as compared with the C/A-code. Details on the L2 civil signal are given by Fontana et al. [5] and are provided in Ref. [2].

4.2.4 L5 Signal

Although the use of the L1 and L2C signals can satisfy most civil users, there are concerns that the L2 frequency band may be subject to unacceptable levels of interference for applications involving public safety, such as aviation. The potential for interference arises because the International Telecommunications Union (ITU) has authorized the L2 band on a coprimary basis with radiolocation services, such as high-power radars. As a result of Federal Aviation Administration (FAA) requests, the Department of Transportation and Department of Defense have called for a new civil GPS frequency, called L5, at 1176.45 MHz in the ARNS band of 960–1215 MHz. To gain maximum performance, the L5 spread-spectrum codes were selected to have a higher chipping rate and longer period than do the C/A-codes to allow for better accuracy measurements. Additionally, the L5 signal has two signal components in phase quadrature, where one component has L5 CNAV data (i.e. I5) and the other component (i.e. Q5) has no navigation data. The IS details of the L5 signal can be found in Ref. [6], with benefits summarized as follows:

(a) *Ranging accuracy is improved.* Pseudorange errors due to random noise are reduced below levels obtainable with the C/A-codes due to the 10× larger bandwidth of the L5 codes. As a consequence, both code-based positioning accuracy and phase ambiguity resolution performance (i.e. more rapid ambiguity resolution) are improved over C/A-code tracking.
(b) *Errors due to multipath are reduced.* The larger bandwidth of the L5 code sharpens the peak of the code autocorrelation function, thereby reducing the shift in the peak due to multipath signal components. The realized multipath mitigation depends upon the receiver design and the delay of the multipath.
(c) *Carrier phase tracking is improved.* Weak-signal phase tracking performance of GPS receivers is severely limited by the necessity of using a Costas (or equivalent-type) PLL to remove carrier phase reversals of the data modulation. Such loops rapidly degrade below a certain threshold (about 25–30 dB-Hz) because truly coherent integration of the carrier phase is limited to the 20-ms data bit length. In contrast, the "data-free" quadrature component of the L5 signal permits coherent integration of the carrier for arbitrarily long periods, i.e. beyond the traditional 50-bps data bit edges, which permits better phase tracking accuracy and lower tracking thresholds.
(d) *Weak-signal code acquisition and tracking is enhanced.* The data-free component of the L5 signal permits enhanced levels of positioning capability with very weak signals. Acquisition is improved because fully coherent integration times longer than 20 ms are possible. Code tracking is also improved by virtue of better carrier phase tracking for the purpose of code rate aiding.
(e) *The L5 signal further supports rapid and reliable carrier phase ambiguity resolution.* The L5 signal is a full-strength, high-chipping rate code that provides high-quality code and carrier phase measurements. These can be used to support various code and carrier combinations for high-accuracy and more reliable carrier phase ambiguity resolution techniques.

(f) *The codes are better isolated from each other.* The longer length of the L5 codes reduces the cross correlation between codes from different satellites, thus minimizing the probability of locking onto the wrong code during acquisition, even at the increased power levels of the modernized signals.
(g) *Advanced navigation messaging.* The L5 signal structure has a new CNAV messaging structure that will allow for increased data integrity.

GPS modernization for the L5 signal calls for a completely new civil signal format (i.e. L5 code) at a carrier frequency of 1176.45 MHz (i.e. L5 carrier). The L5 signal is defined in a quadrature scene where the total signal power is divided equally between in-phase (I) and quadrature (Q) components. Each component is modulated with a different but synchronized 10 230-chip direct sequence L5 code transmitted at 10.23 Mcps (the same rate as the P(Y)-code), but with a 1-ms period (the same as the C/A-code period). The I channel is modulated with a 100-sps data stream, which is obtained by applying rate 1/2, constraint length 7, forward error correction (FEC) convolutional coding to a 50-bps navigation data message that contains a 24-bit cyclic redundancy check (CRC). The Q channel is unmodulated by navigation data. However, both channels are further modulated by Neumann Hoffman (NH) synchronization codes, which provide additional spectral spreading of narrowband interference, improve bit, and symbol synchronization, and also improve cross-correlation properties between signals from different GPS satellites. The L5 signal is shown in Figure 4.10 illustrating the modernized GPS (and legacy GPS) signal spectrum.

Compared with the C/A-code rate, the L5 code rate is 10-times as fast, provides lower autocorrelation sidelobes, substantially improves ranging accuracy and better interference protection, and substantially reduces multipath errors at longer path separations (i.e. long delay multipath). Additionally, these codes were selected to reduce, as much as possible, the cross correlation between satellite signals. The absence of data modulation on the Q channel permits longer coherent processing intervals in code and carrier tracking loops, with full-cycle carrier tracking in the latter. As a result, the tracking capability and phase ambiguity resolution become more robust.

Further details on the civil L5 signal can be found in Refs. [6–9].

4.2.5 M-Code

The GPS military (M) codes are transmitted on both the L1 and L2 carrier frequencies, beginning with the GPS Block IIR-M satellites in 2005. These M-codes are based on a family of split-spectrum GNSS codes for military and new GPS civil signals [10–12]. The M-code provides the following advantages to military users:

(a) *Direct acquisition of the M-codes.* The design of these codes eliminates the need to first acquire the L1 C/A-code with its relatively high vulnerability to jamming.
(b) *Better ranging accuracy.* As can be seen in Figure 4.10, the M-codes have significantly more energy near the edges of the bands, with a relatively small amount of energy near the band center. Since most of the C/A-code power is near the band center, potential interference between the codes is mitigated. The effective bandwidth of the M-codes is much larger than that of the P(Y)-codes, which concentrate most of their power near the L1 or L2 carrier. Because of the modulated subcarrier, the autocorrelation function of the M-codes has not just one peak but several peaks spaced one subcarrier period apart, with the largest at the center. The modulated subcarrier will cause the central peak to be significantly sharpened, significantly reducing pseudorange measurement error.
(c) *Reduced multipath error.* The sharp central peak of the M-code autocorrelation function is less susceptible to shifting in the presence of multipath correlation function components.

The M-coded signals transmitted on the L1 and L2 carriers have the capability of using different codes on the two frequencies. The M-codes are known as binary offset carrier (BOC) encoded signals where the notation of BOC(f_{sx}, f_{cx}) is used where f_{sx} represents the subcarrier multiplier and f_{cx} represents the code rate multiplier, with respect to a nominal code rate of 1.023 MHz. The M-code is a BOC(10,5) code in which a 5.115-Mcps chipping sequence modulates a 10.23-MHz square wave subcarrier. Each spreading chip subtends exactly two cycles of the subcarrier, with the rising edge of the first subcarrier cycle coincident with initiation of the spreading chip. The spectrum of the BOC(10,5) code has considerably more relative power near the edges of the signal bandwidth than any of the C/A, P(Y), L2C, and L5 coded signals. As a consequence, the M-coded signal has minimal spectral overlap with the other GPS transmitted signals, which permits transmission at higher power levels without mutual interference. The resulting spectrum has two lobes, one on each side of the band center, thereby producing the split-spectrum code. The M-code signals are illustrated in Figure 4.10. The M-code signal is transmitted in the same quadrature channel as the C/A-code (i.e. with the same carrier phase), that is, in phase quadrature with the P(Y)-code. The M-codes are encrypted and unavailable to unauthorized users. The nominal received power level is −158 dBW at Earth. Additional details on the BOC(10,5) code can be found in Refs. [10, 12].

4.2.6 L1C Signal

The L1 Civil (L1C) signal was introduced into the GPS constellation with the first GPS III satellite launched by the Falcon 9 vehicle on 23 December 2018.

Although the current C/A-code is planned to remain on the L1 frequency (1575.42 MHz), the additional L1C signal adds a higher-performance civil signal at the L1 frequency with interoperability with other GNSSs.

Like the L5 civil signal, the L1C signal has a data-free (i.e. pilot) quadrature component and an in-phase data component transmitted in quadrature. The L1C signal for GPS is based upon a time-multiplexed binary offset carrier (TMBOC) modulation technique that synchronously time-multiplexes the BOC(1,1) and BOC(6,1) spreading codes for the pilot component (designated as $L1C_p$), and a BOC(1,1) modulated signal with navigation data (designated as $L1C_D$). Both of the BOC codes are generated synchronously at a rate of 1.023 MHz and are based on the Legendre sequence called Weil code. These codes have a period of 10 ms, so 10 230 chips are within one period. Additionally, there is an overlay code that is encoded onto the $L1C_p$ pilot channel. One bit of the overlay code has duration of and is synchronized to the 10-ms period of the BOC code generators. The overlay code rate is 100 bps and has 1800 bits in an 18-seconds period.

To generate the TMBOC signal for the $L1C_p$ channel, the BOC(1,1) and BOC(6,1) spreading sequences are time-multiplexed. With 33 symbols of a BOC(1,1) sequence, four symbols are replaced with BOC(6,1) chips. These occur at symbols 0, 4, 6, and 29. Thus, with 75% of the power distributed in the pilot signal, there is 1/11 of the power in the BOC(6,1) component and 10/11 of the power in the BOC(1,1) component of the carrier.

The L1C signal also has a new navigation message structure, designated as CNAV-2, with three defined different subframe formats. Subframe 1 contains GPS time information (i.e. time of interval [TOI]). Subframe 2 contains ephemeris and clock correction data. Subframe 3 is commutated over various pages and provides less time-sensitive data such as almanac, UTC, and ionosphere that can be expended in the future. Advanced encoding is also implemented in the form of a Bose, Chaudhuri, and Hocquenghem (BCH) for subframe 1, and rate 1/2 low density parity check (LDPC) with interleaving in subframes 2 and 3.

The split-spectrum nature of the L1C encoded signal provides some frequency isolation from the L1 C/A-encoded signal. Each spreading chip subtends exactly one cycle of the subcarrier, with the rising edge of the first subcarrier half-cycle coincident with initiation of the spreading chip. The TMBOC codes provide a larger root mean square (RMS) bandwidth compared with pure BOC(1,1).

For many years now, cooperation at the international level has been ongoing to enable the L1C signal to be interoperable with other GNSSs. A combined interoperable signal will allow a user to ubiquitously use navigation signals from different GNSSs with known and specified performances attributes.

Additional details on the L1C signal can be found in Refs. [13–15]. As a final note, it is anticipated that the existing C/A-code at the L1 frequency will be retained for legacy purposes.

4.2.7 GPS Satellite Blocks

The families of GPS satellites launched prior to recent modernization efforts are referred to as Block I (1978–1985), Block II (1989–1990), and Block IIA (1990–1997); all of these satellites transmit the legacy GPS signals (i.e. L1 C/A and P(Y) and L2 P(Y)). (The USNO has an up-to-date listing of all of the GPS satellites in use today [16].)

In 1997 the Block IIR satellites began to replace the older Block II/IIA satellites. The Block IIR satellites have several improvements, including reprogrammable processors enabling problem fixes and upgrades in flight. Eight Block IIR satellites were modernized (designated as Block IIR-M) to include the new military M-code signals on both the L1 and L2 frequencies, as well as the new L2C signal on L2. The first Block IIR-M was launched in September 2005.

To help secure the L5 frequency utilization, one of the Block IIR-M satellites (GPS IIR-20(M)), SV49, was outfitted with a special L5 payload and was launched on 24 March 2009. This particular satellite has hardware configuration issues relating to the L5 payload installation and is transmitting a degraded signal. Since that time, the navigation signals have been set unhealthy in the broadcast navigation message.

The Block IIF (i.e. follow-on) family was the next generation of GPS satellites, retaining all the capabilities of the previous blocks, but with many improvements, including an extended design life of 12 years, faster processors with more memory, and the inclusion of the L2C and the L5 signal on a third, L5 frequency (1176.45 MHz). A total of 12 Block IIF satellites were launched between May 2010 and February 2016.

The next block of GPS satellites joining the GPS constellation, designated as Block III, began with the launch on 23 December 2018. The GPS III satellites added the L1C signal and enhanced signal generation capabilities. GPS III is planned to include all of the legacy and modernized GPS signal components, including the L1C signal, and to add specified signal integrity with enhanced ground monitoring. The added signal integrity planned for GPS III may be able to satisfy additional aviation requirements [17]. Improvements for military users include high-power spot beams for the L1 and L2 military M-code signals by providing 20-dB higher received power over the earlier M-code signals. However, in the fully modernized Block III satellites, the M-coded signal components are planned to be radiated as physically distinct signals from a separate antenna on the same satellite. This is done in order to enable optional transmission of a spot beam for greater anti-jam resistance within a selected local region on the Earth. There will be 10 GPS III satellites produced, followed by an additional 11, GPSIIIF follow-on satellites.

4.2.8 GPS Ground Control Segment

The GPS broadcast signals are monitored by a complex ground network of GPS receivers located throughout the globe. These ground reference receivers are networked to a master control station (MCS) at Schriever AFB, Colorado, with a backup MCS at Vandenberg AFB, California. The MCS interfaces to the USNO for GPS timing, manages the SV orbits and clocks, formats the GPS broadcast messages, and sends the formatted navigation messages to remote ground transmitter sites that uplink the broadcast message information so the broadcast navigation message can then be transmitted to the user segment over the L-band links.

Many enhancements to the ground control segment have occurred over the years that include expanding the USAF monitor station to a total of 16, under the Legacy Accuracy Improvement Initiative (L-AII) using National Geospatial-Intelligence Agency (NGA) monitoring stations. The Architecture Evolution Plan (AEP) added the capability to monitor the L2C and L5 CNAV messages, as well as, being used for early use of the M-code.

The Next-Generation Operation Control System (OCX) will be a major change in the AEP. OCX will replace the existing AEP and is being developed and implemented in stages (i.e. Blocks). Block 0 is used for the initial launch and control of GPS III satellites, while the Block 1 and 2 will provide full monitor and control of all legacy and modernized GPS signals [18, 19]. See Section 10.5.5.5, page 388.

4.3 GLONASS Signal Structure and Characteristics

The GLONASS is the Russian GNSS. GLONASS has similar operational requirements to GPS with some key differences in its configuration and signal structure. Like GPS, GLONASS is an all-weather, 24-hour satellite-based navigation system that has space, control, and user segments. The first GLONASS satellite was launched in 1982, and the GLONASS was declared an operational system on 24 September 1993.

The GLONASS satellite constellation is designed to operate with 24 satellites in three orbital planes at 19 100-km altitude (whereas GPS uses six planes at 20 180-km altitude). GLONASS calls for eight SVs equally spaced in each plane. The GLONASS orbital period is 11 h 15 min, which is slightly shorter than the 11-h 56-min 02-s orbital period for a GPS satellite. Because some areas of Russia are located at high latitudes, the orbital inclination of 64.8° is used as opposed to the inclination of 55° used for GPS.

Each GLONASS satellite transmits its own ephemeris and system almanac data. Via the GLONASS Ground Control Segment, each GLONASS satellite transmits its position, velocity, and lunar/solar acceleration effects in an ECEF coordinate frame (as opposed to GPS that encodes SV positions using Keplerian orbital parameters). GLONASS ECEF coordinates are with reference to the PZ-90.02 datum with time reference linked to their national reference of UTC (SU) (Soviet Union, now Russia).

4.3.1 Frequency Division Multiple Access (FDMA) Signals

GLONASS uses multiple frequencies in the L-band, separated by a substantial distance for ionosphere mitigation (i.e. L1 and L2), but these GLONASS L1 and L2 bands are slightly different from the GPS L1 and L2 frequencies. One significant difference between GLONASS and GPS is that legacy GLONASS uses a frequency division multiple access (FDMA) architecture in each band as opposed to the CDMA approach used by GPS.

4.3.1.1 Carrier Components

Legacy GLONASS uses two L-band frequencies, L1 and L2, as defined in Eq. (4.18). The channel numbers for GLONASS signal operation are

$$\begin{aligned} f_{K1} &= f_{01} + K\Delta f_1 \\ f_{K2} &= f_{02} + K\Delta f_2 \end{aligned} \tag{4.18}$$

where

$$\begin{aligned} K &= \text{channel number } (-7 \leq K \leq +6) \\ f_{01} &= 1602\,\text{MHz} \\ \Delta f_1 &= 562.5\,\text{MHz} \\ f_{02} &= 1246\,\text{MHz} \\ \Delta f_2 &= 437.5\,\text{MHz} \end{aligned}$$

4.3.1.2 Spreading Codes and Modulation

With the GLONASS signals isolated in frequency, an optimum maximal-length (m-sequence) spreading code can be used. GLONASS utilizes two such codes, one standard precision navigation signal (SPNS) at a 0.511-Mcps rate that repeats every 1 ms and a second high-precision navigation signal (HPNS) at a 5.11-Mcps rate that repeats every 1 second. Similar to GPS, the GLONASS signals utilize BPSK modulation and are transmitted out of a right-hand circularly polarized (RHCP) antenna.

4.3.1.3 Navigation Data Format

The format of the legacy GLONASS navigation data is similar to the legacy GPS navigation data format, with different names and content. The GLONASS navigation data format is organized as a superframe that is made up of frames, where frames are made up of strings. A superframe has a duration of 150 seconds and is made up of five frames, so each frame lasts 30 seconds. Each frame is made up of 15 strings, where a string has a duration of 2 seconds. GLONASS encodes satellite ephemeris data as immediate data and almanac data as nonimmediate data. There is a time mark in the GLONASS navigation data (last 0.3 seconds of a string) that is an encoded PRN sequence.

4.3.1.4 Satellite Families

While the first series of GLONASS satellites were launched from 1982 to 2003, the GLONASS-M satellites began launching in 2003. These GLONASS-M satellites had improved frequency plans and accessible signals that make up the bulk of the GLONASS constellation. GLONASS-K satellites represent a major modernization effort by the Russian government. These GLONASS-K satellites transmit the legacy GLONASS FDMA signals as well as a new CDMA format signals. Details on the signal structure of these new CDMA signals can be found in Ref. [17].

4.3.2 CDMA Modernization

One of the issues with an FDMA GNSS structure is the interchannel (i.e. interfrequency) biases that can arise within the FDMA GNSS receiver. If not properly addressed in the receiver design, these interchannel biases can be a significant error source in the user solutions. These error sources arise because the various navigation signals pass through the components within the receiver at slightly different frequencies. The group delay thought these components are noncommon, at the different frequencies, and they produced different delays on the various navigation signals, coming from different satellites. These interfrequency biases are substantially reduced (on a comparative basis) with CDMA-based navigation systems because all of the signals are transmitted at the same frequency. (The relatively small amounts of Doppler received from the various CDMA navigation signals are relatively minor when considering the group delay.)

An additional consideration with an FDMA GNSS signal structure is the amount of frequency bandwidth that is required to support the FDMA architecture. CDMA architecture typically has all the signals transmitted at the same carrier frequency for more efficient utilization of a given bandwidth.

GLONASS has established several separate versions of its GLONASS-K satellites. The first GLONASS-K1 satellite launched on 26 February 2011 carried the first GLONASS CDMA signal structure and has been successfully tracked on Earth [20]. The GLONASS-K1 satellite has transmitted a CDMA signal at a designated L3 frequency of 1202.025 MHz (test signal), as well as the legacy GLONASS FDMA signals at L1 and L2. The CDMA signal from the GLONASS-K1 satellite is considered a test signal. The follow-on generation of satellites is designated as the GLONASS-K2 satellites. A full constellation of legacy and new CDMA signals (using a base code generation rate of 1.023 MHz) are planned for these GLONASS-K2 satellites including plans to transmit its open services CDMA signal in the GLONASS L1 (L1OC centered at 1600.995 MHz), L2 (L2OC centered at 1248.06 MHz), and L3 (L3OC centered at 1202.025 MHz) frequency bands [21–26].

4.4 Galileo

Galileo is a GNSS being developed by the European Union and the European Space Agency (ESA). Like GPS and GLONASS, it is an all-weather, 24-hour satellite-based navigation system being designed to provide various services. The program has had three development phases: (i) definition (completed), (ii) development and launch of on-orbit validation satellites (completed), and (iii) launch of operational satellites, including additional development [27]. The first Galileo In-Orbit Validation Element (GIOVE) satellite, designated as GIOVE-A, was launched in 28 December 1995, followed by the GIOVE-B on 27 April 2008; both are now decommissioned. The next four Galileo in-orbit validation (IOV) satellites were launched in 2011/2012 to provide additional validation of Galileo. FOC satellites began to be launched in 2014. The constellation of Galileo satellites is nearly complete, and recent launches have put satellites into orbit four at a time with the Ariane 5 launch vehicle. Most of the Galileo satellites transmit their navigation signals using a passive hydrogen maser clock.

4.4.1 Constellation and Levels of Services

The full constellation of Galileo is planned to have 30 satellites in MEOs with an orbital radius of 23 222 km (similar to the GPS orbital height of 20 180 km). The inclination angle of the orbital plane is 56° (GPS is 55°), with three orbital planes (GPS has 6). This constellation will have 10 satellites in each orbital plane.

Various services are supported for Galileo, including open service (OS), commercial service (CS), public regulated service (PRS), and search and rescue (SAR). These services are supported with different signal structures and encoding formats tailored to support each particular service.

4.4.2 Navigation Data and Signals

Table 4.4 lists some of the key parameters for Galileo that will be discussed in this section. The table lists the Galileo signals, frequencies, identifiable signal

Table 4.4 Key Galileo signals and parameters.

Signal	Frequency (MHz)	Bandwidth (MHz)	Component	Data type (rate, sps)	Service
E1	**1575.420**	24.552			
			E1-B	I/NAV (250)	OS/CS
			E1-C	Pilot (0)	
E6	**1278.750**	40.920			
			E6-B	C/NAV (1000)	CS
			E6-C	Pilot (0)	
E5	**1191.795**	51.150			
E5a	1176.450	20.460	E5a-I	F/NAV (50)	OS
			E5a-Q	Pilot (0)	
E5b	1207.140	20.460	E5b-I	I/NAV (250)	OS/CS
			E5b-Q	Pilot (0)	

Main frequency signals are bolded.

component, its navigation data format, and the service each signal is intended to support. The European Union has published the Galileo OS Interface Control Document, which contains significant detail on the Galileo signals in space [28]. All of the Galileo signals transmit two orthogonal signal components, where the in-phase component transmits the navigation data and the quadrature component is dataless (i.e. a pilot). These two components have a power sharing so that the dataless channel can be used to aid the receiver in the acquisition and tracking of the signal. All of the individual Galileo GNSS signal components utilize phase-shift keying modulation and RHCP for the navigation signals.

To support the various services for Galileo, three different navigation formats are being implemented: (i) a free navigation (F/NAV) format to support OS for the E5a signal on the E5a-I component, (ii) the integrity navigation (I/NAV) format to support OS and CS for the E5b signal on the E5b-I component, and the E1 signal on the E1B component, and (iii) a commercial navigation (C/NAV) format to support CS for the E6 signal on the E6-B component.

Figure 4.11 provides an overview of the Galileo signals plotted versus frequency. The Galileo E1 frequency (same as GPS L1) at 1574.42 MHz has a split-spectrum-type signal around the center frequency and interoperable with the GPS L1C signal. The Galileo E1 signal is a combined binary offset carrier (CBOC) signal that is based upon BOC signals (i.e. a BOC(1,1) and subcarrier BOC(6,1) component). The E1 in-phase E1-B component has I/NAV data encoded on it and the E1-C component (in phase quadrature to the E1-B component) has no data (i.e. pilot).

The Galileo E6 signal supports CS at a center frequency of 1278.750 MHz, with no offset carrier, and a spreading code at a rate of 5.115 MHz

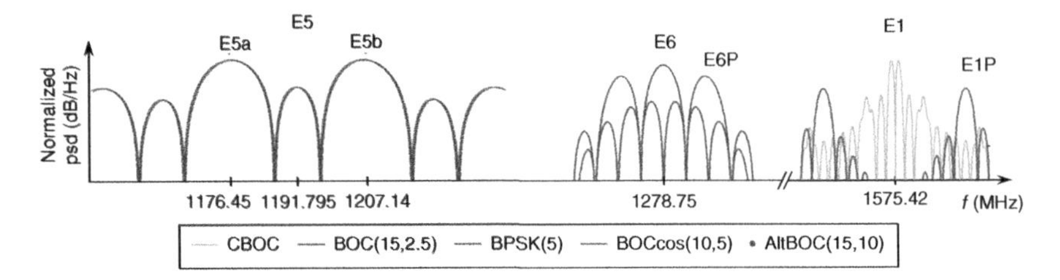

Figure 4.11 A Galileo signal spectrum. psd, power spectral density.

(5 × 1.023 MHz). The E6 signal has two signal components in quadrature, where the C/NAV message is modulated with an encrypted ranging code on the in-phase E6-B component, while there is no data on the quadrature E6-C component. The E6 signal is also planned to support the PRS service using an encrypted BOC signal to provide interoperability with the other non-BOC E6 signal components on the E6 frequency.

The E5 signal is a unique GNSS signal that has an overall center frequency of 1191.795 MHz, with two areas of maximum power at 1176.450 and 1207.140 MHz. The wideband E5 signal is generated by a modulation technique called alternative binary offset carrier (AltBOC). The generation of the E5 signal is such that it is composed of two Galileo signals that can be received and processed separately or combined by the user receiver. The first of these two signals within the composite E5 signal is the E5a, centered at 1176.450 MHz (same as the GPS L5). The E5a signal has two signal components transmitted in quadrature: the E5a-I component has the F/NAV data (in-phase) and E5a-Q quadrature component has no data on it (i.e. pilot). The second of these two signals within the composite E5 signal is the E5b, which is centered at 1207.140 MHz. The E5b signal has two signal components, transmitted in quadrature, where the in-phase E5b-I component has I/NAV data and the quadrature E5b-Q component has no data on it (i.e. pilot).

Galileo will also support a SAR service that is interoperable with the Cosmicheskaya Sistema Poiska Avariynyh Sudov-Search and Rescue Satellite Aided Tracking (COSPAS-SARSAT). Distress signals, transmitted by distress radio beacons on the 406 MHz, will be received by the Galileo SAR transponder and broadcast in the 1544–1545 MHz band to ground receiving stations to support terrestrial SAR operations. Additionally, the Galileo satellites will provide a return link message (RLM) back to the distress caller via the Galileo E1, E1-B I/NAV message.

4.5 BeiDou

BDS is the Chinese-developed GNSS. The various phases of BDS are referred to as BDS-1, BDS-2, and BDS-3. Each BDS phase has supported a mix of satellites in various orbits: MEO, geostationary Earth orbit (GEO), and inclined geosynchronous orbit (IGSO). The Chinese government performed initial experimentations on BeiDou, i.e. BDS-1 in the 2000–2003 time frame. That was followed by BDS-2 through 2012, which had an emphasis on regional development with five GEOs, five IGSO, and four MEOs. The five BDS-2 GEOs have been launched with an orbital radius of 42 164.17 km at longitude locations: 58.75 °E, 80 °E, 110.5 °E, 140 °E, and 160 °E. The BDS-3 began in 2013 to focus on a Global system. The BDS-3 space segment satellite constellation calls for 30 satellites made up of 24 MEO, 3 GEOs (80 °E, 110.5 °E, 140 °E), and 3 IGSOs. BDS uses its own datum, China Geodetic Coordinate System 2000 (CGCS2000), and time reference BeiDou time (BDT) system, relatable to UTC. Of the BDS-2 SVs, the first MEO BeiDou M-1 satellite was launched on 14 April 2007. The first BDS II IGSO SV was launched on 31 July 2010 with an inclination angle of 55° (same as GPS MEOs). The first two BDS-3 SVs were launched on 5 November 2017.

While there are multiple global and regional services planned for BeiDou, the signals transmitted by BDS have evolved. The B2I signal at 1207.14 MHz in BDS-2 will not be deployed on BDS-3, and two new signals will be transmitted by the 24 BDS-3 MEOs: the B1C (interoperable with GPS L1C and Galileo E1) and the B2a (interoperable with GPS L5 and Galileo E5a). The three BDS-3 IGSOs providing additional regional coverage over China and Southeast Asia and will transmit the BDS signals at B1C, B1I, B3I, and B2a; furthermore, the BDS-3 GEOs will transmit space-based augmentation system (SBAS)-type signals and additionally transmit an S-band service signal at 2491.751 MHz. Table 4.5 provides a summary of the key BeiDou signals and parameters; see [29, 30] for additional details on BeiDou and various signal ICDs and performance standards.

The navigation data formats for the various BDS SVs draw from techniques for GPS CNAV and CNAV-2; see [29, 30] for specific implementation details. The D2 format provides SBAS services from the BDS GEOs.

4.6 QZSS

The QZSS Navigation Service is a space-based positioning system being developed by the Japan Aerospace Exploration Agency (JAXA). The QZSS is to be interoperable with GPS and provides augmentation to the GPS over the regions covered by a particular Quasi-Zenith Satellite (QZS) or QZSS GEO. Details of the QZSS and how to interface to the various QZSS signals in space can be found in Ref. [31].

Table 4.6 illustrates the various navigation signals for the QZSS. Most of the QZSS signals are similar to the GPS signals (L1 C/A, L2C, L5, and L1C) and provide an L1 Sub-meter Level Augmentation Service (SLAS), i.e. L1S that is a

Table 4.5 Key BeiDou signals and parameters.

BeiDou phase	BeiDou signal	Frequency (MHz)	Bandwidth (MHz)	Component	Modulation	Data type (rate, sps)	Orbit
BDS-3	**B1C**	**1575.42**	32.736				MEO, IGSO
				B1C_data	BOC(1,1)	B-CNAV1 (100)	
				B1C_pilot	QMBOC(6,1,4/33)	Pilot (0)	
BDS-2 and -3	**B1I**	**1561.098**	4.092				
				B1I_data	BPSK(2)	D1 (100)	MEO, IGSO
				B1I_data		D2 (1000)	GEO
				B1I_pilot		Pilot (0)	
BDS-2 and -3	**B3I**	**1268.5**	20.46				
				B3I_data	BPSK(10)	D1 (100)	MEO, IGSO
				B3I_data		D2 (1000)	GEO
				B3I_pilot		Pilot (0)	
BDS-2	B2I	**1207.14**	20.46				
				B2I_data	BPSK(10)	D1 (100)	MEO, IGSO
				B2I_data		D2 (1000)	GEO
				B2I_pilot		Pilot (0)	
BDS-3	B2a	**1176.45**	20.46				MEO, IGSO
				B2a-I_data	BPSK(10)	B-CNAV2 (200)	
				B2a-Q (pilot)	BPSK(10)	Pilot (0)	
	B2b	**1207.14**	20.46				MEO, IGSO
				B2b-I	BPSK(10)	500	
				B2b-Q	BPSK(10)	500	

Main frequency signals are bolded.

Table 4.6 Key QZSS signals and parameters.

		Bandwidth (MHz)				Modulation			
QZSS band	Frequency (MHz)	Block I	Block II	Code type	Component	Block I	Block II	Data type (rate, sps)	Service
L1	**1575.42**	24.552	30.690						
				C/A		BPSK(1)	BPSK(1)	LNAV (50 bps)	OS (C/A)
						BOC(1,1)			L1CP
				L1C			TMBOC	Pilot (0)	L1CP
						BOC(1,1)	BOC(1,1)		L1CD
				L1S		BPSK(1)	BPSK(1)	SBAS (500)	SLAS
L2	**1227.60**	40.920	30.690	L2C	L2C	BPSK(1)	BPSK(1)	CNAV (50)	OS (L2C)
				L2CM L2CL	L2CM and L2CL combo				
L6	**1278.75**	39.000	42.000			BPSK(5)	BPSK(5)		CLAS
				Kasami	L61	BPSK(2.5)	BPSK(2.5)	L6D (250)	
				Kasami	L62	BPSK(2.5)	BPSK(2.5)	Pilot (0)/L6E (250)	
L5	**1176.45**	20.460	24.900	L5					OS (L5)
					I5	BPSK(10)	BPSK(10)	CNAV (100)	
					Q5	BPSK(10)	BPSK(10)	Pilot (0)	
				L5S		BPSK(1)	BPSK(1)	SBAS (500)	SLAS

Main frequency signals are bolded.

SBAS-type signal and on L5, i.e. L5S. Additionally, QZSS provides for a unique L6 Centimeter Level Augmentation Service (CLAS).

The first phase of the QZSS included the launch of the first Michibiki QZS on 11 September 2010 and is centered at a longitude of approximately 135 °E. (Michibiki means to guide or lead the way.) The first QZSS in operation was placed at the approximate location of right ascension of ascending node (RAAN) equal to 195° (i.e. over Japan). In 2017 three additional QZSS SVs were launched, two in QZS (QZS-2 and QZS-4) orbits and one GEO (QZS-3). The QZSs are in a highly inclined elliptical orbit (HEO), that is, geosynchronous with the Earth rotation rate. The QZSs have a semimajor axis of 42 164 km at an inclination of 43°. There are plans to place QZSs at longitude locations corresponding to RAAN of 90°, 210°, and 330°. For a user in the Southeast Asia area, the QZS will track out a "figure eight" in the sky. Most users in that area will have good visibility to the QZS vehicles 24 hours a day with slowly varying Doppler frequencies.

The QZSS GEO transmits the navigation signals listed in Table 4.6 and an additional L1Sb, and S-band signal design for QZSS safety confirmation service (Q-ANPI).

4.7 IRNSS/NAVIC

The IRNSS, operationally known as NAVIC, is a regional satellite positioning and timing system, developed by the Indian Space Research Organization (ISRO). The constellation consists of eight satellites – three GEOs (E32.5°, E83°, E131.5°) and four IGSOs (two staggered at each longitude of 55° and 11.5°) at an inclination of 27°. NAVIC is to service the areas around India, with navigation signals at L-band and S-band. In L-band, IRNSS SPS service is provided on L5, centered at 1176.45 MHz in a 24 MHz band (same as GPS L5 and Galileo E5a), and in S-band centered at 2492.028 MHz in a 14.5 MHz band. The system is to provide both SPS and restricted services (RS) for authorized users. SPS is provided via BPSK(1) modulation using Gold codes that repeat every 1 ms. The RS used BOC(5,2) with 4 ms periods. Navigation data is encoded using 600 symbols, over four subframes with interleaving FEC codes (for 50 sps) and CRC checks. See Refs. [32, 33] for additional details on IRNSS.

Problems

4.1 The relativistic effect in a GPS satellite clock, which is compensated by a deliberate clock offset, is about
(a) parts per million
(b) parts per 100 million

(c) parts per 10 billion
(d) 4.5 parts per trillion

4.2 The following component of the ephemeris error contributes the most to the range error:
(a) along-track error
(b) cross-track error
(c) both along-track and cross-track errors
(d) radial error

4.3 The differences between pseudorange and carrier phase observations are:
(a) integer ambiguity, multipath errors, and receiver noise
(b) satellite clock, integer ambiguity, multipath errors, and receiver noise
(c) integer ambiguity, ionosphere errors, multipath errors, and receiver noise
(d) satellite clock, integer ambiguity, ionosphere errors, multipath errors, and receiver noise

4.4 GPS WN started incrementing from zero at:
(a) midnight of 5–6 January 1980
(b) midnight of 5–6 January 1995
(c) midnight of 31 December 1994 to 1 January 1995
(d) midnight of 31 December 1999 to 1 January 2000

4.5 The complete set of GPS satellite ephemeris data comes once in every:
(a) 6 seconds
(b) 18 seconds
(c) 30 seconds
(d) 150 seconds

4.6 Calculate the time of the next GPS week rollover event.

4.7 How far does a Galileo satellite move during the time it takes the signal to leave the satellite and be received at Earth? Use a nominal transit time in your calculation.

4.8 Solve nonlinear algebraic equation $2\cos x - e^x = 0$ with second and first order Newton–Raphson.

4.9 Solve nonlinear algebraic equation $\cos x \cosh x = 1$ with first and second order Newton–Raphson. MATLAB programs are available in `ephemeris.m`.

4.10 An important signal parameter is the maximum Doppler shift due to satellite motion that must be accommodated by a receiver. Find its approximate value by assuming that a GPS satellite has a circular orbit with a radius of 27 000 km, an inclination angle of 55°, and a 12-hours period. Is the rotation rate of the Earth significant? At what latitude(s) would one expect to see the largest possible Doppler shift?

4.11 Another important parameter is the maximum *rate* of Doppler shift in hertz per second that a PLL must be able to track. Using the orbital parameters of the previous problem, calculate the maximum rate of Doppler shift of a GPS signal one would expect, assuming that the receiver is stationary with respect to the Earth.

4.12 Find the power spectrum of the 50-bps data stream containing the navigation message. Assume that the bit values are -1 and 1 with equal probability of occurrence, that the bits are uncorrelated random variables, and that the location of the bit boundary closest to $t = 0$ is a uniformly distributed random variable on the interval (-0.01 second, 0.01 second). (*Hint*: First find the autocorrelation function $R(\tau)$ of the bit stream and then take its Fourier transform.)

4.13 In two-dimensional positioning, the user's altitude is known, so only three satellites are needed. Thus, there are three pseudorange equations containing two position coordinates (e.g. latitude and longitude) and the receiver clock bias term B. Since the equations are nonlinear, there will generally be more than one position solution, and all solutions will be at the same altitude. Determine a procedure that isolates the correct solution.

4.14 Some civil receivers attempt to extract the L2 carrier by squaring the received waveform after it has been frequency-shifted to a lower IF. Show that the squaring process removes the P(Y)-code and the data modulation, leaving a sinusoidal signal component at twice the frequency of the original IF carrier. If the SNR in a 20-MHz IF bandwidth is -30 dB before squaring, find the SNR of the double-frequency component after squaring if it is passed through a 20-MHz bandpass filter. How narrow would the bandpass filter have to be to increase the SNR to 0 dB?

References

1 DOD (2019). Global positioning system, performance standards & specifications. GPS.gov (visited 10 March 2019).

2 GPS Directorate, Systems Engineering & Integration Interface Specification IS-GPS-200 (2018). NAVSTAR GPS space segment/navigation user segment interfaces, IS-GPS-200J, 25-APR-2018. https://www.gps.gov/technical/icwg/IS-GPS-200J.pdf (visited 17 February 2019).

3 Klobuchar, J.A. (1975). A First-Order, Worldwide, Ionospheric Time-Delay Algorithm, Air Force Cambridge Research Laboratories, Ionospheric Physics Laboratory, Project 443, AFCRL-TR-75-0502. www.ion.org/museum/item_view.cfm?cid=11&scid=9&iid=47 (accessed 01 September 2019).

4 Gold, R. (1967). Optimal binary sequences for spread spectrum multiplexing. *IEEE Transactions on Information Theory* 13 (4): 619–621.

5 Fontana, R.D., Cheung, W., Novak, P.M., and Stansell, T.A. (2001). The new L2 civil signal. *Proceedings of the 14th International Technical Meeting of the Satellite Division of The Institute of Navigation (ION GPS 2001),* Salt Lake City, UT (September 2001), pp. 617–631.

6 GPS Directorate, Systems Engineering & Integration Interface Specification IS-GPS-705 (2019). NAVSTAR GPS space segment/user segment L5 interfaces, IS-GPS-705E, 25-APR-2019. https://www.gps.gov/technical/icwg/IS-GPS-705E.pdf (visited 17 February 2019).

7 Van Dierendonck, A.J. and Spilker, J.J. Jr., (1999). Proposed civil GPS signal at 1176.45 MHz: in-phase/quadrature codes at 10.23 MHz chip rate. *Proceedings of the 55th Annual Meeting of The Institute of Navigation,* Cambridge, MA (June 1999), pp. 761–770.

8 Hegarty, C. and Van Dierendonck, A.J. (1999). Civil GPS/WAAS signal design and interference environment at 1176.45 MHz: results of RTCA SC159 WG1 activities. *Proceedings of the 12th International Technical Meeting of the Satellite Division of The Institute of Navigation (ION GPS 1999),* Nashville, TN (September 1999), p. 1727–1736.

9 Spilker, J.J. Jr., and Van Dierendonck, A.J. (2001). Proposed new L5 civil GPS codes. *Navigation, Journal of the Institute of Navigation* 48 (3): 135–144.

10 Spilker, J.J. Jr.,, Martin, E.H., and Parkinson, B.W. (1998). A family of split spectrum GPS civil signals. *Proceedings of the 11th International Technical Meeting of the Satellite Division of The Institute of Navigation (ION GPS 1998),* Nashville, TN (September 1998), pp. 1905–1914.

11 Lucia, D.J. and Anderson, J.M. (1998–1999). Analysis and recommendations for reuse of the L1 and L2 GPS spectrum. *Navigation, Journal of The Institute of Navigation* 45 (4): 251–264.

12 Barker, B.C., Betz, J.W., Clark, J.E. et al. (2000). Overview of the GPS M code signal. *Proceedings of the 2000 National Technical Meeting of The Institute of Navigation*, Anaheim, CA (January 2000), pp. 542–549.

13 Issler, J.-L., Ries, L., Bourgeade, J.-M. et al. (2004). Probabilistic approach of frequency diversity as interference mitigation means. *Proceedings of the 17th International Technical Meeting of the Satellite Division of The Institute of Navigation (ION GNSS 2004)*, Long Beach, CA (September 2004), pp. 2136–2145.

14 Betz, J., Blanco, M.A., Cahn, C.R. et al. (2006). Description of the L1C signal. *Proceedings of the 19th International Technical Meeting of the Satellite Division of The Institute of Navigation (ION GNSS 2006)*, Fort Worth, TX (September 2006), pp. 2080–2091.

15 GPS Directorate, Systems Engineering & Integration Interface Specification IS-GPD-800 (2018). NAVSTAR GPS space segment/user segment L1C interface, IS-GPS-800E, 25-APR-2018. https://www.gps.gov/technical/icwg/IS-GPS-800E.pdf (visited 17 February 2019).

16 United States Naval Observatory (USNO) (2019). GPS operational satellites (Block II/IIA/IIR/IIR-M/II-F). https://www.usno.navy.mil/USNO/time/gps/current-gps-constellation (visited 10 March 2019).

17 FAA (2008). GNSS evolutionary architecture study, Phase I—Panel Report, February 2008. https://www.gps.gov/governance/advisory/meetings/2008-03/ (accessed 01 September 2019).

18 Senior K., Coleman M. (2017). The next generation GPS time. *NAVIGATION*, 64 (4): 411–426.

19 Marquis, W., and Shaw, M. (2011) GPS III bringing new capabilities to the global community. Inside GNSS, 6 (5): 34–48.

20 Inside GNSS (2011). Septentrio, AsteRx3 receiver tracks first GLONASS CDMA signal on L3, April 12, 2011. https://insidegnss.com/septentrios-asterx3-receiver-tracks-first-glonass-cdma-signal-on-l3/ (visited 14 July 2012).

21 Information and Analysis Center For Position, Navigation, and Timing (2019). GLONASS. https://www.glonass-iac.ru/en/GLONASS/ (visited 7 March 2019).

22 Global Navigation Satellite System (GLONASS) (2008). *Interface Control Document L1, L2*, Version 5.1. Moscow: Russian Institute of Space Device Engineering. http://gauss.gge.unb.ca/GLONASS.ICD.pdf (visited 7 March 2019).

23 GLONASS (2016). *Interface Control Document, General Description of Code Division Multiple Access Signal System*, Edition 1.0. Moscow.

24 GLONASS (2016). *Interface Control Document, Code Division Multiple Access Open Service Navigation Signal*, L1 frequency band Edition 1.0. Moscow.

25 GLONASS (2016). *Interface Control Document, Code Division Multiple Access Open Service Navigation Signal*, L2 frequency band Edition 1.0. Moscow.

26 GLONASS (2016). *Interface Control Document, Code Division Multiple Access Open Service Navigation Signal*, L3 frequency band Edition 1.0. Moscow.

27 ESA, Galileo (2012). What is Galileo. http://www.esa.int/esaNA/galileo.html (visited 15 July 2012).

28 European Union, European Commission (2015). Satellite Navigation, Galileo Open Service Signal-In-Space Interface Control, OS SIS ICD, Issue 1.2, November 2015. https://galileognss.eu/wp-content/uploads/2015/12/Galileo_OS_SIS_ICD_v1.2.pdf (visited 23 February 2019).

29 BeiDou Navigation Satellite System. http://en.beidou.gov.cn/ (visited 8 March 2019).

30 China Satellite Navigation Office, BeiDou Navigation Satellite System. Official documents. http://en.beidou.gov.cn/SYSTEMS/Officialdocument/ (visited 10 March 2019).

31 Japan Aerospace Exploration Agency, Quasi-Zenith Satellites System, Quasi-Zenith Satellite System Navigation Service (2012). Interface Specification for QZSS (IS-QZSS), V1.4, February 28, 2012. http://qz-vision.jaxa.jp/USE/is-qzss/DOCS/IS-QZSS_14_E.pdf (visited 15 July 2012).

32 Indian Regional Navigation Satellite System (2017). Signal in Space ICD for Standard Positioning Service, Version 1.1, August 2017, ISRO-IRNSS-ICD-SPS-1.1.

33 Indian Regional Navigation Satellite System (2017). Signal in Space ICD for Messaging Service (IRNSS 1A), Version 1.1, August 2017, ISRO-IRNSS-ICD-MSG-1.0.

5

GNSS Antenna Design and Analysis

5.1 Applications

While there are many global navigation satellite system (GNSS) receiver and antenna systems used today, they vary significantly in size, cost, capabilities, and complexity. These variations are largely driven by the end-user application. Clearly, the characteristics of a consumer-grade GNSS antenna are very different from those used in aviation, surveying, or space applications. A performance characteristic that is very critical for one application may not be important or even desirable in another application. The performance requirements for different applications are often mapped directly into the requirements for the antenna to be used in that particular application. The following sections will discuss several key antenna performance characteristics with discussion pertaining to their application.

5.2 GNSS Antenna Performance Characteristics

The overriding requirement of any GNSS antenna is to convert the various GNSS signals intended for use from an electromagnetic wave to an electrical signal suitable for processing by the GNSS receiver. To do this effectively, there are several important performance characteristics that must be considered in the design of the GNSS antenna. Although there may be very specific performance characteristics for specific applications, the most important characteristics will be discussed here.

5.2.1 Size and Cost

In the world of antenna design, size (i.e. aperture) is arguably one of the most important factors. The eventual size of the antenna, most often constrained by the intended application, will often limit the eventual performance of the

Global Navigation Satellite Systems, Inertial Navigation, and Integration,
Fourth Edition. Mohinder S. Grewal, Angus P. Andrews, and Chris G. Bartone.

Companion website: www.wiley.com/go/grewal/gnss

antenna. For example, the size and cost constraints of consumer cell phone applications are a major factor in the performance of the GNSS antenna within the cell phone. This size constraint is based upon the real estate allocated to the GNSS antenna considering all of the other functions to be performed by the cell phone and still allow it to fit in your pocket. For other applications, the size and cost may be less constrained so that other performance requirements can be achieved. For example, in a fixed ground-based GNSS reference station application, the antenna can typically be larger to produce high-quality code and carrier phase measurements but may cost more.

5.2.2 Frequency and Bandwidth Coverage

The GNSS antenna must be sensitive to the GNSS signals' center frequency and must have sufficient bandwidth to efficiently receive these signals, but there are several other performance characteristics that are also important. Figure 5.1 graphically illustrates the various carrier frequencies and bandwidths for various GNSS bands, in the L-band (1.1–1.7 GHz).

As seen in Figure 5.1, the various GNSS frequencies for a particular GNSS typically span a considerable amount of spectrum (e.g. several hundreds of megahertz) so that ionosphere delay estimates can be done effectively for dual/multifrequency GNSS users. To have a single antenna span, the entire GNSS L-band from about 1150 to 1620 MHz would require an antenna to have about 33% bandwidth; this would likely require a broadband type of antenna design. Also evident from Figure 5.1 is that at each center frequency corresponding to a particular GNSS signal, the bandwidth requirement is not that large when considering the carrier frequency (i.e. on the order of a couple percent bandwidth in most cases); these requirements could be met with a multifrequency band antenna design, which is typically different from

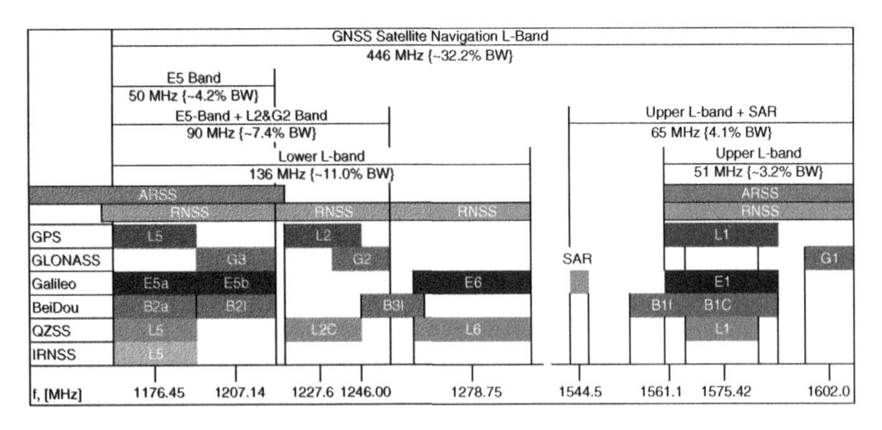

Figure 5.1 Illustration of GNSS frequency and bandwidth (BW) for various GNSSs.

a nearly continuous broadband antenna design. At first glance, the frequency and bandwidth requirements do not look very challenging, but when other performance requirements are considered for various applications, the design requirements become much more clear.

The dual/multifrequency capabilities of various GNSSs tend to find application for more high-performance users who have requirements for comprehensive ionosphere error mitigation or, to a more limited extent, frequency diversity application. Most other applications can be satisfied with the single-frequency capabilities inherent in a GNSS, such as low-cost consumer and general-purpose GNSS applications.

For example, to receive a GNSS signal at a single frequency over a relatively narrow bandwidth (e.g. L1, C/A-code), an antenna can typically be built much smaller and at lower cost with less stringent performance requirements than a multifrequency wideband GNSS antenna (e.g. L1L2L5 and C/A, P(Y), M, L5). However, even for this single-frequency narrowband GNSS antenna, other performance requirements may dominate such as size and cost. Such applications, as in the cellular phone market, demand small inexpensive antennas. As the antenna size decreases, the efficiency will be decreased and special considerations for tuning become increasingly important. Furthermore, as the size of this type of antenna decreases, the radiation characteristics typically change.

For multifrequency GNSS antennas, there are two general design approaches pursued. One approach is to design a multifrequency GNSS antenna with multiple resonances designed into the antenna structure. Such an antenna is typically a dual-frequency, dual-layer patch antenna. These types of antennas provide very good performance over a limited bandwidth but do not cover the bands between the frequencies of interest. The second general approach pursued in multifrequency GNSS antennas is a broadband design. Such a design would cover the entire band from the lowest frequency of interest to the highest frequency of interest, where the performance may be optimized and specified for the various GNSS bands of interest. While the antenna performance may vary over the entire band, the performance at the frequencies of interest can easily be verified. Examples of these types of antennas are helix and spiral types of antennas. Helix antennas have historically been used on the Global Positioning System (GPS) and Global Orbiting Navigation Satellite System (GLONASS) space vehicles (SVs) and, in a more limited extent, in the user segment. With the advent of new GNSS signals and systems, the choice of broadband spiral-based antennas is becoming more popular in recent years for advanced user equipment.

5.2.3 Radiation Pattern Characteristics

The GNSS antenna must have sufficient gain to effectively convert the GNSS electromagnetic wave into a signal voltage so that it can be processed by the GNSS receiver. In accomplishing this goal, the GNSS antenna's radiation characteristic should exhibit certain characteristics. These will be addressed with respect to the desired and undesired signals of interest.

For a terrestrial GNSS user, the GNSS antenna should provide "nearly uniform gain" with respect to the GNSS SV elevation angle in the upper-hemisphere and omnidirectional coverage with respect to the GNSS SV azimuth angle. While the description of nearly uniform gain means that a constant gain of the GNSS SV signal from zenith down to the receivers mask angle is desirable, it is difficult to achieve and not explicitly required. Some gain variation can be tolerated in the antenna coverage volume, which will produce a variation in carrier-to-noise ratio (C/N_0) for the respective SV being tracked by the receiver; however, there is a limit to the amount of gain variation that can be tolerated in the GNSS receiver. This variation is typically limited by the code cross-correlation within the GNSS receiver tracking loops. For GPS C/A-code processing, if one GPS pseudorandom noise (PRN) C/A-code is greater than another GPS C/A-code by about 20 dB, cross-correlation can be significant. A couple of dB power level variations will occur due to the GNSS signal versus elevation angle variations, which will be part of this power budget. In most cases, a gain variation of no more than about 15 dB is desirable.

A mask angle used in GNSS antennas refers to an elevation angle (e.g. 5° or 10°) above the horizon, where a GNSS signal may be tracked by the receiver but is discarded. The reason for discarding signal measurements at these low elevation angles is that these signals often have more measurement error due to atmospheric effects, signal multipath, and lower C/N_0. For some geodetic applications, the mask angle is increased to 20° to help ensure high-quality measurements but at the expense of availability of the GNSS signals. Other applications may decrease the mask angle to help increase the GNSS signal availability. Still other systems, such as indoor applications, may not implement any mask angle due to the dynamic nature of the user antenna and the challenged operational environment.

A significant number of user applications want to efficiently receive the desired GNSS SV signals of interest and at the same time minimize any undesired signals due to multipath or interference. Antenna design, ground planes, and pattern shaping can help receive the desired signals of interest and minimize the undesired signals of interest.

Most GNSS user antenna technologies involve a ground plane or planar structure so the GNSS antenna gain will naturally decrease as the elevation angles decrease. This is most often the dominant effect for decreasing C/N_0 for low-elevation GNSS SV signal reception for terrestrial users. The use of ground planes for most applications additionally helps in the mitigation of multipath.

To optimize the received power from the GNSS antenna, the polarization of the antenna should match the polarization of the incident GNSS electromagnetic wave. Most radiation characteristics are depicted graphically in 2D or 3D plots where the pattern is represented as a far-field power pattern; 2D plots provide a better quantitative illustration of the radiation characteristics of the GNSS antenna and are often called an elevation (i.e. vertical) cut or an azimuth (i.e. horizontal) cut, but this depends upon the users' frame of reference.

For certain applications, the radiation characteristics of the GNSS antenna are less important than the requirement to make the antenna a smaller size or less expensive. An example of this type of application would be the mobile, low-cost consumer market. For these applications, the antenna must be small enough to practically and economically fit within the size, cost, and real-estate constraints of the host device (e.g. a cell phone). In these types of applications, including indoor applications where multipath signals can cause large errors, they can also be used for degraded positioning if no direct signals are present. When the antenna size (i.e. aperture) is significantly decreased, the radiation pattern may have a more omnidirectional characteristic in all directions (i.e. isotropic) but will likely be affected by the other components around it.

5.2.4 Antenna Polarization and Axial Ratio

Although there is a wide variety of GNSS antennas, most are designed for right-hand circularly polarized (RHCP) to match the polarization of the incoming GNSS signal [1]. (The polarization of the GNSS electromagnetic signal is the direction that the time-harmonic electric field intensity vector travels as the wave travels away from the observation point.) For certain special cases, a GNSS antenna with polarization diversity or a linearly polarized (LP) antenna can be used. Polarization diversity can be used in some limited cases to help mitigate some known interference source, with known polarization. If an LP antenna is used to receive an RHCP GNSS signal and it is placed perpendicular to the incoming RHCP signal, 3-dB signal loss will occur due to the polarization mismatch; however, the gain of the antenna may be used to make up for this polarization mismatch loss. As will be shown later, two orthogonally LP signal components can be combined to optimally receive a circularly polarized electromagnetic wave.

In general, the polarization of an electromagnetic wave may be described as elliptical, with circular and linear polarization being special cases. The axial ratio (AR) of an antenna refers to the sensitivity of the antenna to the instantaneous electric field vector in two orthogonal polarization directions of the antenna, in particular, the magnitude of the maximum value (i.e. magnitude in the semimajor axis of the traced elliptical polarization) to the magnitude of the wave in the orthogonal direction (i.e. in the semiminor axis direction), as expressed in Eq. (5.1). Thus, for a GNSS antenna to be maximally sensitive to the incoming RHCP GNSS signal, it should have an AR of 1 (i.e. 0 dB), so that the electric field intensity vectors in the direction of maximum sensitivity are the same in the orthogonal direction. Depending upon the design type of the GNSS antenna, it is typically a good metric at boresight (i.e. in the zenith direction) for a GNSS antenna:

$$\mathrm{AR} = \frac{E_{\mathrm{major}}}{E_{\mathrm{minor}}} = \frac{E_{\mathrm{RHCP}} + E_{\mathrm{LHCP}}}{E_{\mathrm{RHCP}} - E_{\mathrm{LHCP}}} \tag{5.1}$$

where

E_{major} = magnitude of **E** in the direction of the semimajor or axis

E_{minor} = magnitude of **E** in the direction of the semiminor or axis

$E_{\mathrm{RHCP}} = \dfrac{E_\theta + jE_\phi}{\sqrt{2}}$

$E_{\mathrm{LHCP}} = \dfrac{E_\theta - jE_\phi}{\sqrt{2}}$

E_θ = complex E field in the spherical coordinate direction θ

E_ϕ = complex E field in the spherical coordinate direction ϕ

$\boldsymbol{E} = E_\theta \boldsymbol{a}_\theta + E_\phi \boldsymbol{a}_\phi$

$\boldsymbol{a}_\theta$ = unit vector in the θ direction

$\boldsymbol{a}_\phi$ = unit vector in the ϕ direction

To illustrate how RHCP (and left-hand circularly polarized [LHCP]) can be produced, consider two orthogonally placed dipole antennas as shown in Figure 5.2 from a "top-view" perspective receiving an incident RHCP signal at zenith (i.e. straight into the page). The center of the two dipoles is at the origin of the local antenna coordinate system in the x–y plane, where dipole #1 is aligned with the x-axis, dipole #2 is aligned with the y-axis, and the z-axis is pointed outward (i.e. out of the page).

Now, when the GNSS signal leaves the satellite, it can be described as propagating in a $+z_{\mathrm{SV}}$ direction (in an SV coordinate system) and represented as a normalized GNSS RHCP signal as $\mathbf{E}_{\mathrm{GNSS}}(z_{\mathrm{SV}}, t) = \cos(\omega_c t)\mathbf{a}_x + \sin(\omega_c t)\mathbf{a}_y$. This GNSS signal is incident on the GNSS antenna, where the antenna coordinate system has a $+z$ direction pointed toward zenith (i.e. in the opposite direction of the SV coordinate system $+z_{\mathrm{SV}}$). Thus, we can represent the incident normalized GNSS signal onto our local antenna coordinate systems, as shown in Figure 5.2, as $\mathbf{E}_{\mathrm{GNSS}}(z, t) = \cos(\omega_c t)\mathbf{a}_x - \sin(\omega_c t)\mathbf{a}_y$.

The normalized signal response on dipole #1 due to the incident GNSS signal $E_{\mathrm{GNSS}}(z, t)$ can be represented as $s_1(t)$, and the normalized signal on dipole #2 due to the incident GNSS signal can be represented as $s_2(t)$, where these two

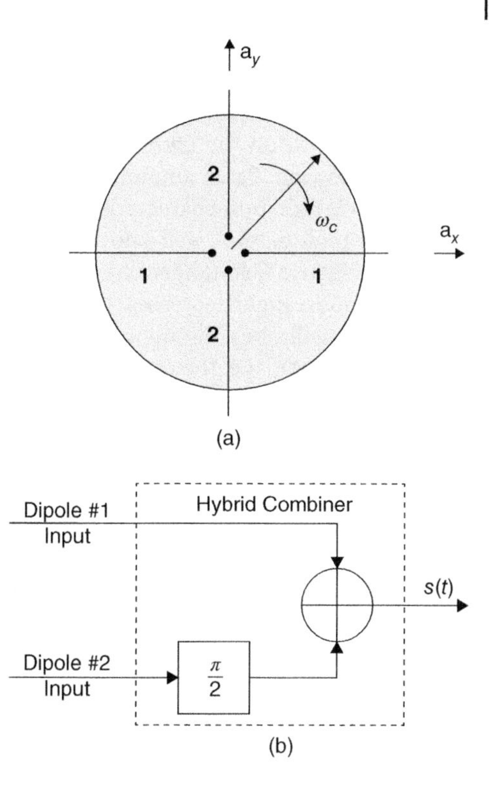

Figure 5.2 Antenna configuration for reception of a GNSS RHCP signal. (a) Top view illustrating cross dipoles and Electric Field rotation direction. (b) Hybrid power combiner to produce RHCP signal.

signals can be simply represented as shown in Eq. (5.2):

$$s_1(t) = \cos \omega_c t$$
$$s_2(t) = \sin \omega_c t \tag{5.2}$$

For the incident RHCP GNSS signal, we delay the leading E field term and then combine the signals to coherently add the signal response from each dipole. Adding the 90° delay to dipole #2, as shown in Figure 5.2, combining, then the combined signal can be represented as Eq. (5.3). When the response from the two orthogonal components is the same, then the AR will be equal to 1 (i.e. 0 dB):

$$s(t) = \cos \omega_c t + \sin \left(\omega_c t + \frac{\pi}{2} \right) = 2 \cos \omega_c t \tag{5.3}$$

If the incident signal was LHCP incident onto the antenna, as described by $\mathbf{E}(z, t) = \cos(\omega_c t)\mathbf{a}_x + \sin(\omega_c t)\mathbf{a}_y$ in the antenna coordinate system, then the signal received would naturally be 90° lag in the opposite direction and, with the 90° phase shift in the hybrid combiner, would produce an output response where the net signal from dipole #1 and dipole #2 would cancel each other, producing a zero response at the output.

Depending upon the design type of the GNSS antenna, the AR is typically a good metric at boresight (i.e. in the zenith direction) for a GNSS antenna, but the AR will typically get bigger as the elevation angle decreases. While this does depend on the antenna design, installations, and application, consider the following. Patch antennas are typically mounted on ground planes to improve their radiation characteristics, to minimize impedance variations and to mitigate multipath. A well-known boundary condition in electromagnetics is that the electric field tangent to a perfect electric conductor will go to zero. As the elevation angle decreases, the electric field component of the incoming GNSS signal will obey the boundary condition and, subsequently, the "horizontal component" (i.e. the component parallel to the ground plane) will go to zero. This will leave the vertical component (i.e. the component perpendicular to the ground plane) as the dominant component. This is one of the reasons why GNSS antennas mounted over ground planes have a gain reduction as the elevation angle decreases.

5.2.5 Directivity, Efficiency, and Gain of a GNSS Antenna

The directivity of an antenna is the ratio of the radiation intensity in a given direction, normalized by the radiation intensity averaged over all space [2]. For GNSS applications, often the shape (i.e. directivity) of the antenna radiation characteristics is very important to help maintain a more constant received signal power level for the various GNSS signals being tracked at various aspect angles from the antenna. Too much gain variation (e.g. $\gtrsim$18 dB or so) across the coverage region of the GNSS satellite signals, can cause significant cross-correlation issues in the code tracking loops between different PRNs. The amount of cross-correlation interference will depend upon the code type, frequency difference, and power level differences between the signals.

The gain of a GNSS antenna is related to the directivity, by the antenna efficiency in accordance with Eq. (5.4) [3]:

$$G(\theta,\phi) = e_{cd}D(\theta,\phi) \text{ (unitless)} \tag{5.4}$$

where

$$D(\theta,\phi) = \frac{U(\theta,\phi)}{U_{AVG}}$$

$$e_{cd} = \text{efficiency} = \frac{P_{rad}}{P_{in}}$$

P_{rad} = power radiated by the antenna (W)

P_{in} = power input to the antenna (W)

The gain for an antenna in Eq. (5.4) is often stated in units of dB (technically dBi [dB relative to an isotropic (i) radiator]), for a particular polarization. As for the efficiency of the antenna, the input power specified in Eq. (5.4) represents the power into the actual antenna terminal, and with the theory of reciprocity, this power can be viewed as the output power, at the antenna terminals for a GNSS reception antenna; likewise, power radiated can be viewed as power incident. For passive GNSS antennas, the gain is nominally 0 dBi over the upper hemisphere, but will often be slightly greater at zenith (e.g. +3 dBi), and lower (e.g. −3 dBi) at low elevation angles (e.g. 80° away from zenith). For electrically large or phased-array antennas, the gain can increase substantially above the nominal 0 dBi, and care must be taken in providing sufficient gain to the desired GNSS signals to be received while suppressing undesired signals for reception to help minimize multipath and/or interference sources.

For GNSS antennas that are integrated with other radio frequency (RF) components, including active amplifiers, often the overall gain of the antenna is specified to include these devices. This is especially true for hermetically sealed active GNSS antennas that are designed, fabricated, and sold as a single hermetically sealed package.

GNSS antennas with active components are typically supplied with a DC voltage from the GNSS receiver up the RF transmission line. Coupling between the DC and RF GNSS signals is handled on either end with a "bias-T" that separates the DC and RF signals. Typically, the voltages vary from 3 V to upward of 18 V and are often accompanied with a voltage regulation circuit to maintain constant gain in the active amplifiers and provide for some versatility when connecting various antenna and receiver combinations.

5.2.6 Antenna Impedance, Standing Wave Ratio, and Return Loss

Three additional performance parameters for a GNSS that can *help* assess the performance of a GNSS antenna are the antenna's "input" impedance, standing wave ratio (SWR), and return loss (RL). Emphasis is placed on the word *help* because these parameters should not be used in a vacuum and are not exclusive. Just because a device has good impedance, a low SWR, and high RL does not exclusively mean it is a good antenna for the application; we must still look at the radiation pattern and other performance characteristics (e.g. a 50-Ω load connected to a 50-Ω transmission line is a perfect match, has a nearly ideal SWR, and high RL, but is not a good antenna).

The theory of reciprocity in antenna theory states that, from a passive antenna perspective, the performance characteristics of the antenna will be same, whether we think of the antenna from a transmission or reception perspective. Thus, we can consider the antenna's "input" impedance (from a transmission perspective), the same as the "output" impedance (from a reception perspective), at the antenna terminals (Z_A). A desired characteristic of the GNSS antenna is to match the antenna impedance to the impedance of the transmission line that is connected to it. Most often, a transmission line with a characteristic impedance (Z_0) of 50 Ω is used. With knowledge of the transmission line characteristic impedance and the antenna impedance, the reflection coefficient (Γ_A) at the antenna terminal can be calculated as shown in Eq. (5.5). The reflection coefficient will vary from −1 to +1, where 0 is the ideal case under impedance match conditions. The reflection coefficient can then be used to calculate the SWR and match efficiency, as shown in Eq. (5.5). In the ideal case, the $\Gamma_A = 0$, SWR = 1.0, and $e_r = 1$. Some engineers prefer to use the RL when characterizing the impedance match of the antenna in dB format, which is again shown in Eq. (5.5). For example, if the RL is 20 dB, which means that the signal that gets reflected at the antenna terminal location is 20 dB down from the incident signal:

$$\Gamma_A = \frac{Z_A + Z_0}{Z_A - Z_0} = \frac{\text{SWR} - 1}{\text{SWR} + 1} \quad \text{(unitless)}$$
$$\text{SWR} = \frac{1 + |\Gamma_L|}{1 - |\Gamma_L|}, \quad \text{where } 1 \le \text{SWR} \le \infty \text{ (unitless)} \tag{5.5}$$

where

Γ_A = reflection coefficient at the antenna terminal (unitless)
Z_0 = characteristic impedance of the transmission line (Ω)
Z_A = antenna impedance (i.e. at the terminal) (Ω)
e_r = mismatch (i.e. reflection) efficiency = $(1 - |\Gamma_A|^2)$ (unitless)
RL = return loss = $-20 \log_{10}|\Gamma_A|$ (dB)

The antenna gain can also be characterized as *Realized Gain*, where the mismatch efficiency e_r is also taken into consideration, such as:

$$G_{\text{Realized}}(\theta, \phi) = e_r G(\theta, \phi) = e_r e_{\text{cd}} D(\theta, \phi) \tag{5.6}$$

5.2.7 Antenna Bandwidth

The bandwidth of a GNSS antenna refers to the range of frequencies where the performance of the antenna is satisfactory with respect to a particular performance metric [2]. For a single-frequency GNSS antenna, the range of frequencies is usually centered on the carrier frequency and must be wide enough to provide the downstream receiver system with enough signal fidelity so that the GNSS signal can effectively be processed. For example, in a low-cost C/A-code GPS application, the first null-to-null bandwidth of the signal (i.e. 2 × 1.023 MHz) is often used for a minimum bandwidth requirement. Other applications could be narrower with some signal power reductions. For other high-performance applications that use an advanced receiver design (e.g. narrow correlator), a wider bandwidth is needed to enable more effective tracking of the signal in the receiver [4]. Such applications may require up to 16 MHz in bandwidth to effectively track the C/A-code and gain the advantages provided by the advanced correlator design in the receiver.

For dual or multifrequency antennas that cover distinct bands (e.g. L1 and L2), it is appropriate to talk about the bandwidth of the antenna for each band. Certain antenna technologies lend themselves well to this type of bandwidth characterization, such as patch antennas that are typically narrowband at a given resonance frequency, but can be layered or stacked to support multiple frequency bands.

As the number of GNSSs increase, including an increase in the number of frequency bands, a more "broadband GNSS signal" design approach could be taken for the GNSS antenna. Certain antenna technologies can be selected to design a GNSS antenna that would cover the entire band from 1100 to 1700 MHz to provide sufficient antenna bandwidth to efficiently receive the GNSS signals regardless of their specific frequency in that 1100–1700 MHz band. These types of antennas are becoming increasingly popular as the number of GNSSs increase. A spiral antenna design is a good example of this type of antenna that could cover the entire GNSS band from 1100 to 1700 MHz.

As stated earlier, the bandwidth of a GNSS antenna refers to the range of frequencies where the performance of the antenna is satisfactory with respect to a particular performance metric. For various applications, the particular performance metric for satisfactory antenna performance may be specified in different ways. This satisfactory performance can refer to the gain, antenna radiation pattern characteristic, polarization, multipath performance, impedance, SWR, or some other metric. Some of these performance metrics are more complicated to measure and/or quantify versus frequency. A simple and easy to measure metric is the SWR. Often a maximum value for the SWR will be specified to help characterize the bandwidth. So, as long as the SWR is less than the specified maximum value over a range of frequencies that cover the desired signal frequency and satisfactory performance is achieved, the upper and lower limits of the maximum SWR values can be used to calculate a bandwidth. Figure 5.3 illustrates the SWR of a typical GNSS antenna plotted from 1 to 2 GHz with the 2.0 : 1.0 SWR bandwidth metric indicated.

For broadband antennas, often the SWR is not the best metric to characterize the bandwidth of the antenna. For example, a helix antenna will typically have a reasonably good SWR over a large bandwidth; however, the antenna radiation pattern characteristics will typically deteriorate before the SWR metric exceeds a specified value.

5.2.8 Antenna Noise Figure

The noise figure (NF) of a device refers to the amount of noise that is added to the output with respect to the input. The NF can be expressed as the ratio of

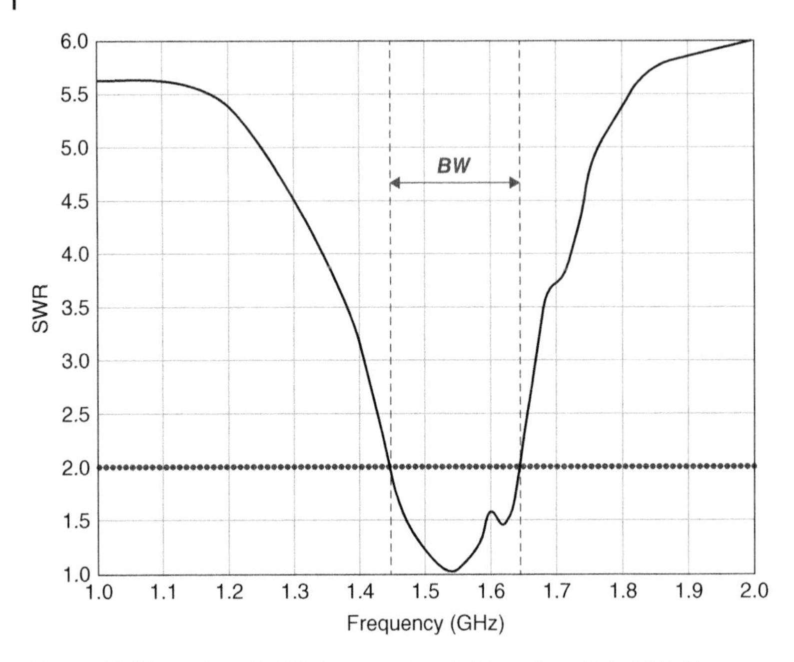

Figure 5.3 Illustration of a GNSS antenna bandwidth using a 2.0 : 1.0 SWR metric.

the input signal-to-noise ratio $(S/N)_{\text{in}}$ to output signal-to-noise ratio $(S/N)_{\text{out}}$, typically expressed in units of dB, as shown in Eq. (5.7) [5]:

$$\text{NF} = 10\log_{10}(F)\ \ (\text{dB}) \tag{5.7}$$

where

$$F = \frac{(S/N)_{\text{in}}}{(S/N)_{\text{out}}} = \text{noise factor (power ratio) (unitless)}$$

$$T_{\text{device}} = T_0(F-1)$$

$$T_0 = \text{reference temperature 290 K (US) 293 K (Japan)}$$

As shown in Eq. (5.7), the NF is often stated as noise factor, a unitless power ratio, which can be related to an actual device temperature (T_{device}).

For a GNSS antenna element, two factors that will affect the antenna NF are the antenna brightness temperature and the physical temperature of the antenna. The antenna brightness temperature is related to what the antenna is "looking at." For example, the noise at the output of the antenna terminals will be different if the GNSS antenna is operated in an outdoor environment, as opposed to inside an anechoic test chamber. The physical temperature of the antenna relates to how hot or cold, physically, the antenna is. For certain space and/or missile applications, where the antenna can see large variations in its physical temperature, the NF variations may be a consideration factor affecting the overall performance of the antenna/receiver systems.

Depending upon the application, GNSS antenna elements may be designed and operated as "passive" or "active" devices. Passive GNSS antennas are those that have GNSS antenna elements and no other active devices (i.e. amplifiers) integrated within the antenna enclosure. (On occasion, a passive bandpass filter [BPF] may be integrated within an antenna enclosure and is referred to as a passive GNSS antenna.) An active GNSS antenna refers to a GNSS antenna configuration where active amplifiers are included within the GNSS antenna enclosure. Often direct access to the antenna connections (i.e. antenna terminal) is not possible because of packaging and the desire to minimize cost, reduce size, or to hermetically seal the unit for environmental considerations. For active antennas in an integrated package, independent measurements of the gain and noise of the antenna itself are difficult to measure because direct access to the antenna terminals is not possible. An alternative approach is to measure the gain normalized by the equivalent noise temperature (G/T) of the device. This type of technique has been used in the satellite communications community for a number of years and similar techniques have been established by RTCA, Inc. for the performance characterization of active integrated GPS antennas [6].

For most general-purpose applications with active GNSS antennas, the NF is not a major factor in systems performance but should be considered for certain applications. For example, applications of GNSS antennas on fast-moving missiles/projectiles, high-sensitivity, indoor applications, and rooftop/lab installations require special attention. While a detailed NF chain analysis is beyond the scope of the material presented here, any passive loss between the passive antenna and the first active device in the receiver chain will directly add to the NF. Thus, caution should be taken when a passive antenna is used, for example, on a rooftop installation, where the first active amplifier is placed at a substantial distance from the passive antenna element. Placing a high gain, low-noise amplifier (LNA) device close to the antenna terminal output is good practice to help minimize the overall receiver system NF.

5.3 Computational Electromagnetic Models (CEMs) for GNSS Antenna Design

Historically, antenna design has been based on the analytical foundations of electromagnetic theory that has led to physical designs, fabrication, test, and often iteration to achieve satisfactory performance. In the past, the iteration step typically involved physically constructing the antenna, testing, then adjusting the physical design, and retesting. This iteration loop continued until satisfactory performance was achieved. With the advancements in recent years in computer technology and computational electromagnetic models (CEMs), the antenna design interaction loop is largely done in simulation. While there may still need to be physical design iterations, the number of physical designs can be significantly reduced with the implementation of CEMs in the antenna design process.

CEMs have become increasingly popular due to the increased efficiency, cost savings, and shorter time to market for antenna designs. For GNSS antenna design, CEMs that are based on "numerical methods" are well suited because the physical size of the GNSS antenna is not significantly larger than the wavelength of the GNSS signal. CEMs solve for the electric (E) and magnetic (H) fields across a "grid" over a region of space based on Maxwell's equations (integral form or differential form). Furthermore, the solutions of E and H can be computed in the time domain or frequency domain. Various companies and agencies continue to develop CEMs and have even combined various CEM techniques to provide software packages and products that are better suited for a wide variety of problems. In the paragraphs that follow, a few of the CEMs and methods will be discussed that are useful for GNSS antenna design.

The procedure of using CEMs typically involves many steps in defining the antenna to obtain performance predictions. For antenna design, the first step involves building a geometric model of the antenna and any surrounding objects to be simulated. The initial antenna design is based on the theoretical analytical design of the antenna. If nearby parts or objects are desired to be included in the simulation, they, too, would be geometrically built into the simulation model. Next, the electrical material properties of the components in the model need to be specified, including the conductivity (σ), the permittivity (ε), and permeability (μ). Next, the excitation on the antenna would be produced within the model. From the theory of reciprocity, this can be thought of as a simulation source, and the performance will represent the antenna as a transmission or reception antenna (assuming no active or unidirectional devices are built into the model). Most CEMs include the ability to provide a "port excitation" or a "wave port" excitation, that is, from a voltage or current on a conductor, or an electromagnetic wave propagating into the simulation region of space. CEMs solve for the E and H fields across the domain of the model in "small pieces or cells" using a geometric grid-type structure. These computations are done across the grid in the model. Some CEMs will automatically compute the grid and adapt the grid based on the frequency, span of frequencies, shape, and material properties of the components in the model. Once the E and H fields have been calculated across the grid, performance predictions of the antenna in its simulated surroundings can be computed. Performance parameters such as radiation characteristics, impedances, and SWR can readily be produced.

A popular CEM method is the "method of moments (MoM)" technique that solves for current distributions as well as E and H fields using the integral format of Maxwell's equations. MoM techniques are well suited for solving antenna problems that involve wire and wire-based array antenna designs. MoM techniques have also been called Numerical Electromagnetic Code (NEC) as originally developed to support the US Navy. Today, various companies and agencies continue to develop the core computation engine (i.e. the NEC) and have graphical user interfaces (GUIs) integrated with the NEC to provide a user-friendly interface. Popular CEMs that implement MoM techniques are FEKO [7], IE3D [8], and WIPL-D [9].

Another CEM method is the finite element method (FEM) that typically use a triangular cell structure and solves for the E and H fields using a differential equation-based implementation of the Maxwell equations. FEMs have found applications in scattering and patch antenna design applications and are often integrated with other CEM techniques to provide enhanced capability. These solution techniques (i.e. solvers) are well suited for antenna geometries the have curved or three dimensional structures. A couple of popular CEM that implements FEM techniques is the high frequency structure simulation (HFSS) [10] and solution techniques within the Computer Simulation Technology (CST) [11].

One of the earliest introduced techniques of CEMs is the finite difference time domain (FDTD) method. With the advances in computer computation capability, FDTD methods have become more efficient in recent years. FDTD methods solve for the E and H fields in a "leap frog" method across the grid in time steps. FDTD methods are well suited for broadband computations, but computation time will increase as the model becomes bigger with respect to the simulated wavelength. Most often the solvers implemented use a rectangular mesh, which may be varied. A couple of popular CEM that implements FDTD techniques is the XFdtd® [12] and CST [11].

5.4 GNSS Antenna Technologies

5.4.1 Dipole-Based GNSS Antennas

One of the most fundamental electrically sensitive antennas is the half-wave dipole antenna that operates near resonance of the desired frequency of operation. Dipole antennas are sensitive to electric fields that are colinear with the orientation of the dipole. The length is one-half of the respective wavelength and is tuned to produce a good impedance match. An ideal $\lambda/2$ dipole will have an impedance of $Z_A = 72 + j42.5\,\Omega$. Typically, the length is shortened slightly to reduce the real part of the impedance to $50\,\Omega$ and a small amount of capacitance could be added to cancel out the slight inductance of the reactive part of the antenna impedance. As antenna size decreases less than half-wavelength, the efficiency and gain of the antenna will decrease.

5.4.2 GNSS Patch Antennas

One of the most popular GNSS antenna types is the patch antenna. Patch antennas offer many advantages and a couple of disadvantages for various applications. Patch antennas are a form of microstrip antennas that are typically printed on the surface of a microwave dielectric substrate material with a ground plane on the bottom side of the dielectric substrate material. Patch antennas date back to the 1950s and today have wide spread applications in GNSS [13]. They have a low profile, which is a significant advantage for dynamic vehicles to minimize wind resistance, snow/ice buildup, and minimize breakage. The low profile is also an advantage in consumer applications such as cell phones. The low profile lends itself well to high-volume production methods and integration with RF front-end circuits such as LNAs and BPFs that can be added to the back side (bottom of the antenna ground plane) of the antenna. Patch antennas can be produced in various form factors to suit a wide variety of applications from low-cost consumer products to high-performance aviation markets. Figure 5.4 illustrates a commercially available aviation patch antenna [14], conforming to the aviation aeronautical radio, incorporated (ARINC) 743A [15] form factor.

The microstrip patch antennas start with a high-quality material that is often double-clad copper on the top and bottom, whereby some of the copper on the top is etched/removed to form the radiating element of the patch. While patch antennas can be fabricated with techniques similar to printed circuit boards, the dielectric materials that make up the substrate as well as the fabrication process need to be tightly controlled to help ensure acceptable performance over a variety of temperature and operational environments.

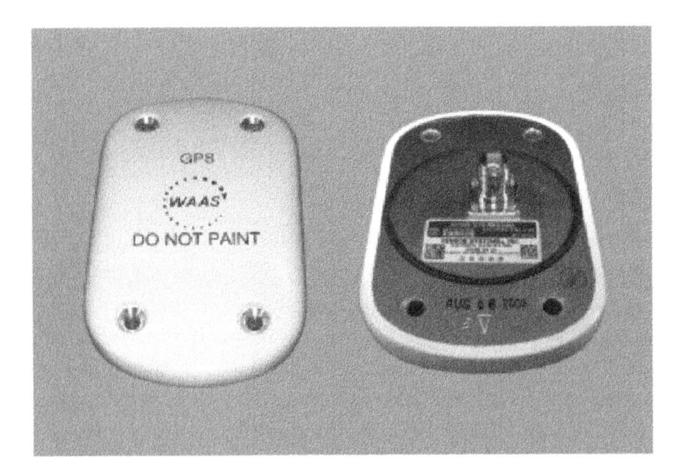

Figure 5.4 Patch antenna aviation form factors (with radome) [14]. Source: Courtesy of Sensor Systems®, Chatsworth, CA.

Higher-performance dielectric materials are made from specific dielectric compounds to help minimize variations in electrical properties during manufacturing and over various operational temperatures. In particular, the relative permittivity (ε_r) and the dielectric thermal expansion coefficient are controlled for the dielectric substrate materials. The thickness of the dielectric substrate materials is a small fraction of the wavelength (e.g. 0.5–5.0% [i.e. ~1–10 mm]) with ε_r typically in the 2–10 region. There are various manufacturers that make these types of high-frequency dielectric materials such as Rogers [16] and Taconic [17].

Microstrip patch antennas operate in a cavity resonance mode where one or two of the planner dimensions of the patch are of lengths equal to one-half of the wavelength in the dielectric substrate material (i.e. $\lambda_d/2$). With the patch element length equal to $\lambda_d/2$, the cavity will resonate at its fundamental or dominant frequency. It should be kept in mind that higher-order modes will also resonate, but these modes often produce undesirable performance characteristics for GNSS antennas (e.g. undesirable radiation pattern characteristics). There are a variety of shapes that can be used for patch antennas including square, nearly square, round, and triangular shaped. Additionally, there are a variant of feeding techniques for patch antennas including edge fed, probe fed, slot fed, and other variants. Configurations to provide a foundation for patch antenna design, and those commonly found in GNSS antennas, will be discussed next.

5.4.2.1 Edge-Fed, LP, Single-Frequency GNSS Patch Antenna

To provide a foundation for patch antenna design, consider a single-frequency, edge-fed patch antenna as shown in Figure 5.5, illustrated for two orthogonal orientations. The patch is edge and center fed and will produce LP for the individual orientations as illustrated by the magnitude of the far-field electric field

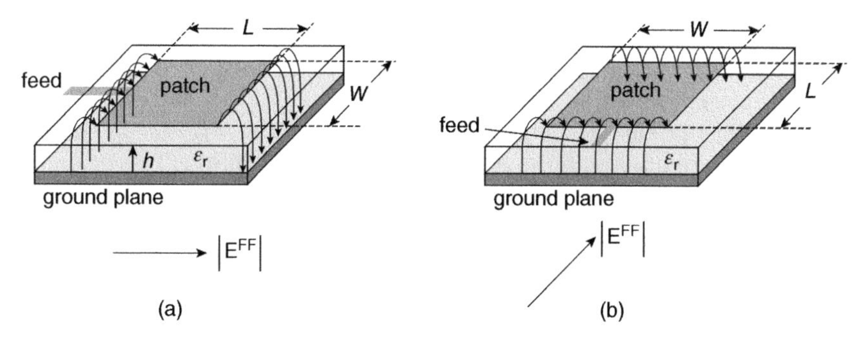

Figure 5.5 Illustration of a single-frequency edge-fed patch antenna. (a) Edge-feed patch to produce electric field aligned with L. (b) Edge-feed patch to produce electric field in direction orthogonal to (a).

vector $|\mathbf{E}^{FF}|$. For patch antennas, radiation comes from the edges of the patch between the top conductive patch and the bottom ground plane, as illustrated in Figure 5.5. The dielectric substrate is between the two conductive materials. To obtain a resonance frequency f_r, the length (L) of the patch can be set to one-half of the wavelength in the dielectric material with relative permittivity (ε_r) in accordance with Eq. (5.8). To account for the fringing of the fields at the edges of the patch, the effective length of the patch can be increased slightly; an approximation is provided in Eq. (5.8):

$$L = \frac{\lambda_d}{2} = \frac{\lambda_r}{2\sqrt{\varepsilon_r}} = \frac{c}{2f_r\sqrt{\varepsilon_r}} \text{ (m)} \tag{5.8}$$

To account for fringing:

$$L \to L_{\text{eff}} \approx\sim L + 2h$$

where

L = length of patch design for resonance (m)

L_{eff} = effective length of patch design for resonance (m)

λ_d = wavelength in dielectric (m)

λ_r = wavelength for desired resonance (m)

ε_r = relative permittivity of dielectric substrate (unitless)

c = speed of light (m/s)

The radiation from the edges of the patch antenna illustrated in Figure 5.5 act like slot radiators. When the length of the patch is adjusted "tuned" for the desired resonance frequency, at a particular instant of time, the E field will come from the ground plane to patch at radiating slot #1 (along feed edge) and from the top patch to the ground plane at radiating slot #2 (along far edge). Internal to the substrate (between the top patch and the ground plane), the E field will decrease in amplitude, moving toward the center of the patch. In the middle of the patch, the E field will be zero.

For this simple edge-fed patch, Jackson and Alexopoulos computed the approximate formulas for the input resistance and bandwidth [18] as shown in Eq. (5.9). These expressions provide good insight that an increase in bandwidth can be achieved by increasing the height of the dielectric material or increasing the width (W) of the top radiating element. Here, the bandwidth is shown using an SWR less than 2.0 metric and expressed as a fractional bandwidth (i.e. the bandwidth with respect to the resonance frequency). The impedance will be maximum at the edge of the patch and will decrease in a sinusoidal fashion to a value of 0 Ω in the middle of the patch in accordance with Eq. (5.9), where y_o represents the offset distance from the edge, along the center line:

$$Z_{\text{in}} = 90\frac{\varepsilon_r^2}{\varepsilon_r - 1}\left(\frac{L}{W}\right)^2 \ (\Omega)$$

$$\text{BW}_{\text{SWR}=2:1} = 3.77\left(\frac{\varepsilon_r - 1}{\varepsilon_r^2}\right)\left(\frac{W}{L}\right)\left(\frac{h}{\lambda}\right) \ \text{(fractional)} \qquad (5.9)$$

where

$$\frac{n}{\lambda} = 1$$

and

$$Z_{\text{in}}(y = y_o) = Z_{\text{in}}(y = 0)\cos^2\left(\frac{\pi y_o}{L}\right) \ (\Omega)$$

where

y_o = offset distance from the edge along the center line (m)

5.4.2.2 Probe-Fed, LP, Single-Frequency GNSS Patch Antenna

Probe-fed GNSS antennas are extremely popular and have the advantage of minimizing the overall size of the antenna and integrating the RF front-end amplifiers and filters on the back side of the antenna ground plane. Figure 5.6 illustrates a passive patch antenna that is probe fed with a coaxial cable or connector output. The center conductor is connected to the top patch via a hole in the dielectric material (and ground plane), and the connector ground/return is connected directly to the bottom ground plane. Once again, the probe can be connected to the edge or moved inward to help match the antenna impedance to the cable/connector impedance [19, 20]. A factor that should be considered with probe-fed patch antennas is the inductance introduced by the probe going

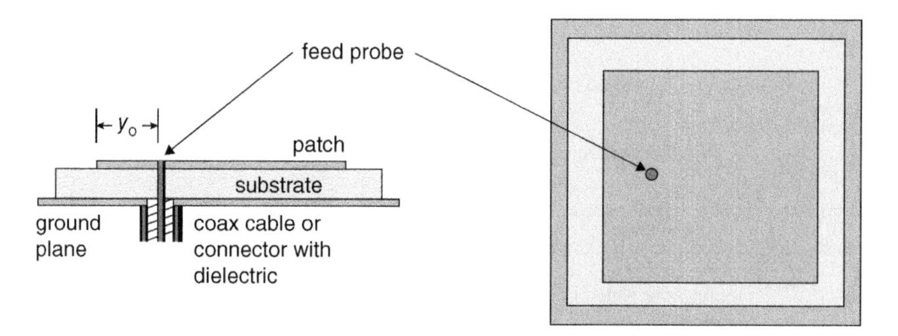

Figure 5.6 Passive probe-fed patch antenna with a coaxial/connector output.

through the dielectric substrate at the desired resonance frequency [21]. The real and imaginary parts of the input impedance for a probe-fed patch antenna can be analytically approximated shown in Eq. (5.10):

$$R_{\text{in}}(y = y_o) \cong 60\frac{\lambda}{W}\cos^2\left(\frac{\pi y_o}{L}\right) \quad (\Omega)$$

and

$$X_f \approx -\frac{\eta k h}{2\pi}\left[\ln\left(\frac{kd}{4}\right) + 0.577\right] \quad (\Omega) \tag{5.10}$$

where

η = intrinsic impedance (Ω)
k = phase constant (rad/m)
d = diameter of probe (m)

For some low-cost, low-bandwidth (e.g. C/A-code) applications, a thin substrate can be used and the added inductance can be tolerated; however, for high-quality and larger-bandwidth patch antennas, this added inductance can be compensated for. As shown in Eq. (5.10), as the height of the dielectric substrate increases, so does the inductance at the feed. There are several approaches that can be implemented to help compensate for the probe-feed inductance. One approach is to tune the patch at a slightly higher frequency (slightly above the resonance frequency), where the probe inductance will be close to zero. Another popular approach is to add a small amount of capacitance to the feed. Once again, there are several ways to add capacitance to the feed location; Figure 5.7 illustrates a common capacitive coupler ring approach. This approach simply adds a thin circular ring where the electromagnetic signal is coupled between the probe and the top patch element [22, 23]. The width of this ring is typically small (e.g. <1 mm) and will help cancel out the probe inductance.

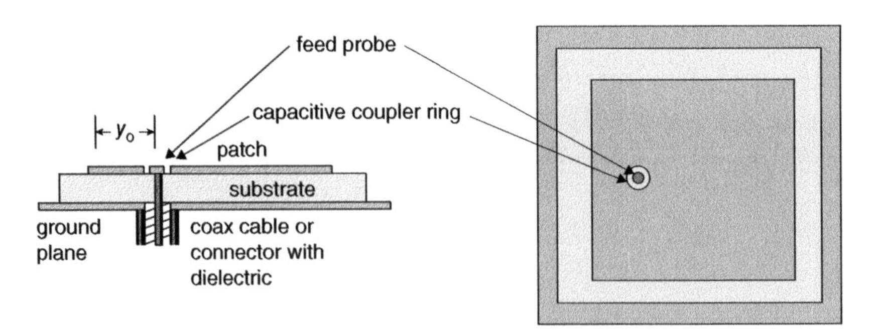

Figure 5.7 Probe-fed patch antenna with capacitive coupler ring.

5.4.2.3 Dual Probe-Fed, RHCP, Single-Frequency GNSS Patch Antenna

Each of the patch antennas presented earlier, edge-fed or probe-fed, produces (or receives) LP. In the designs presented earlier, the feed, either edge fed or probe fed, was placed in the center of the feed side (i.e. at $W/2$). To generate or receive circular polarization, each of these geometrically orthogonal signal polarization components can be combined, in a quadrature fashion, to produce either RHCP or LHCP. The combination of the two orthogonal signal components can be combined at RF using an RF hardware device such as a 90° microstrip hybrid combiner printed directly on the antenna substrate or a separate RF circuit component (i.e. similar 90° hybrid power combiner); this approach will produce an RF output that will be RHCP or LHCP, depending upon the port combination. For RHCP, the leading port from a top-view signal reception perspective will be delayed so that the two ports get combined in phase. Impedance matching of each port can be performed within the microstrip hybrid combiner for an edge-fed patch or by moving the probe feed inward from the edge. The size of each side (i.e. L) of the patch will be tuned to obtain resonance at the desired frequency; thus, the patch can be square (i.e. perfectly square). Figure 5.8 illustrates this dual probe-fed square patch for a single-frequency GNSS antenna. Here, each port can be used individually to receive each LP component or combined in quadrature for RHCP or LHCP. (Form RHCP in accordance with Eq. (5.2).)

5.4.2.4 Single Probe-Fed, RHCP, Single-Frequency GNSS Patch Antenna

In all of the previous LP cases, the antenna feed was placed in the center of the feed side. As discussed earlier, as the feed is moved inward, from the edge toward the center, the impedance will decrease; this procedure can be used to help match the input impedance of the antenna to the transmission line. If, however, the feed location is moved laterally (i.e. away for the center of the feed

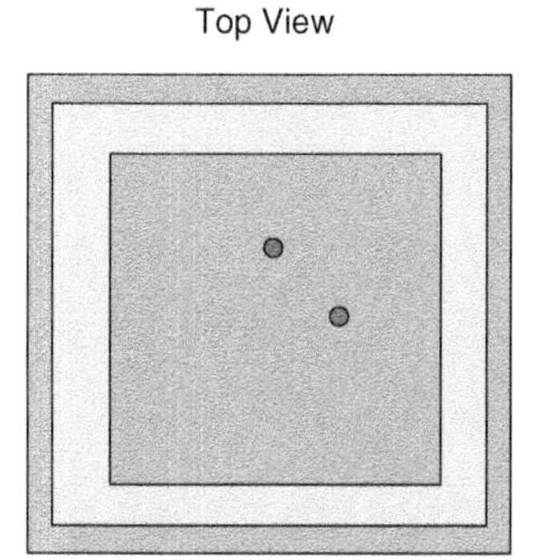

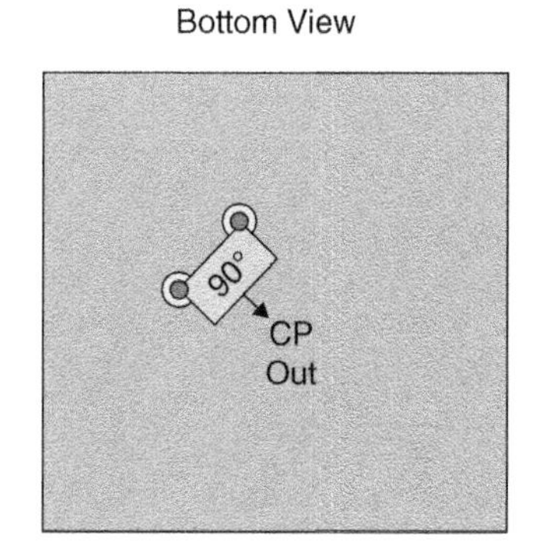

Figure 5.8 Dual probe-fed patch antenna with separate quadrature power combiner. (a) Top view illustrating dual-feed point. (b) Bottom view illustrating hybrid combiner to produce RHCP signal.

side to the left or right), then the polarization will go from being purely linear to elliptical, and possibly circular toward the diagonal of a patch antenna that is "nearly square." For an edge-fed patch, the feed can be placed directly on the corner and the impedance matched. For a probe-fed patch, the feed is most often placed on the diagonal of the nearly square patch and moved inward, away from the corner to match the impedance.

The selection of the L and W of the nearly square patch antenna is important to obtain good circular polarization (CP) performance. When the feed is placed on or near the diagonal of the patch antenna, two polarization or transmission modes will be produced internal to the patch (i.e. in the dielectric medium, that can be represented as a resonance cavity, but now with two orthogonal signals within). The key is to provide both of the two orthogonal components, at the feed location, with equal amplitude, where one signal is 45° leading a reference phase, and the other signal is 45° lagging the reference phase. Since the two signal component phases are different at the feed location, their respective resonance frequency will be different for each of the two orthogonal signals, and the reference phase should be related to the desired resonance center frequency. (Often the two internal signal components are referred to as degenerative modal signals.) The corresponding lengths of each side of the patch will be slightly different, to produce two orthogonal degenerative modal signals, with slightly different resonance frequencies. The net effect is to produce two equal amplitude signals that are 90° out of phase, at the feed location, that will effectively add to produce an RHCP (or LHCP) signal at the feed location.

The total quality factor (Q_t) for the antenna at resonance can be used as a basis for calculating the difference between the lengths of each of the sides of a nearly square patch antenna [24]. The selection of the two lengths of the nearly square patch can be designed in accordance with Eq. (5.11), where one design degenerative mode, f_{10}, is set below the desired resonance frequency (f_r), and the other design degenerative mode, f_{01}, is set to be above f_r, by a nearly equal amount. Both Q_t and the AR will go into the approximation of the BW as shown in Eq. (5.11):

$$L = W\left(1 + \frac{1}{Q_t}\right)$$

and

$$\begin{aligned} f_{10} &= \frac{f_r}{\sqrt{1+1/Q_t}} \\ f_{01} &= f_r\sqrt{1+1/Q_t} \\ \mathrm{BW} &= \frac{12\mathrm{AR}}{Q_t}\ (\%) \end{aligned} \tag{5.11}$$

where

$$L = \frac{\lambda_{dL}}{2}$$

$$W = \frac{\lambda_{dW}}{2}$$

Q_t = quality factor at resonance for the antenna

AR = axial ratio

Consider the following example. For a Q_t equal to 10, a total of 10% variation in the two degenerative signal modes (i.e. f_{10} and f_{01}) will be desired. Thus, f_{10} could be selected as 5% below f_r and f_{01} could be selected to be 5% above f_r. The lengths of each side of the nearly square patch would be different by about 10%. At boresight, if the AR is 1, then the percent bandwidth would be 1.2% or about 16 MHz at the GNSS L1 frequency.

Now, as described earlier, the corresponding lengths of each side of the patch would be slightly different to produce two orthogonal degenerative modal signals, but there are many ways to produce degenerative modal signals in a cavity. Other approaches that are seen in GNSS patch antennas are adding or subtracting tabs to a square conductive (top) patch, cutting the corner off the dielectric substrate, or any other technique to disrupt the perfectly symmetrical cavity structure of the patch to produce the two degenerative modal signals that have the same amplitude and are 90° out of phase at the feed location.

While the single-fed RHCP patch is a convenient design, the AR is usually not as good as a dual probe-fed RHCP antenna due to the asymmetry of the patch. In some implementations, the feed can be slightly adjusted off the diagonal to help match the impedance of the antenna and to help improve the AR.

Figure 5.9 is a photograph of a typical commercially available, low-cost active single-frequency RHCP GPS antenna that is probe fed. This patch

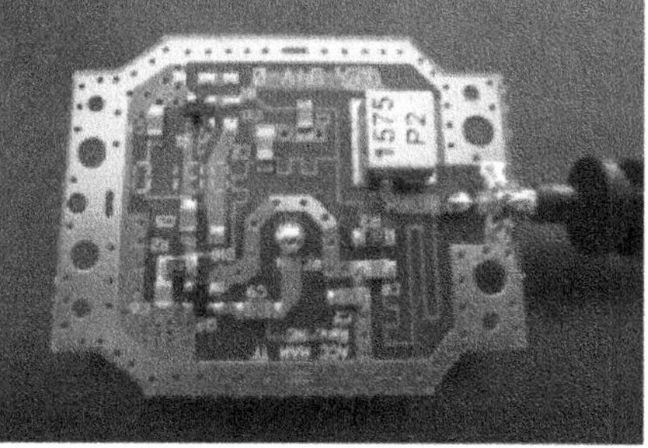

Figure 5.9 Photograph of a low-cost active single-frequency probe-fed RHCP GPS antenna (radome not shown). (a) Top patch antenna element. (b) Bottom of antenna including RF amplifier with DC bias circuit, and bandpass filter. Source: Courtesy of u-blox.

is square (not nearly square) with the corner of the dielectric substrate cut to produce the asymmetry and two degenerative modal signals within the patch. The probe feed location is slightly off of the diagonal to help provide a good impedance match and to obtain a good AR. The probe is capacitively coupled to the patch; see small thin dielectric ring around the soldered probe in Figure 5.9a. Figure 5.9b illustrates the back side of the antenna that includes the associated RF BPFs, amplifiers, and voltage bias circuit, on the bottom side of the patch ground plane. As can be seen from both photos, the bottom RF components are enclosed within an "RF can" when the bottom circuit board is screwed into the metal body base. A plastic radome covers the top of the antenna (not shown).

5.4.2.5 Dual Probe-Fed, RHCP, Multifrequency GNSS Patch Antenna

Multiple frequencies can be supported in a multifrequency patch antenna design by stacking (i.e. layering) the patch elements on top of each other vertically. Figure 5.10 illustrates a dual-frequency probe-fed LP patch design. For the dual-frequency patch illustrated in Figure 5.10, the input impedance of the first patch (i.e. lower patch) off resonance will have a very low real part (about 0 Ω) at the desired resonance frequency of the second patch (i.e. top patch). This low real part of the input impedance at the feed port (at the operating frequency of the upper patch) will effectively add in series to the input impedance, at the feed port, for the input impedance at the lower patch frequency [19]. This is generally true provided each patch operates outside the Q_t of each other's bandwidth, and that they do not operate at frequency harmonics of each other. Because each patch has a relatively narrow bandwidth, when the frequency bands to be supported are fairly far apart (e.g. greater than at least 10%) [19], then there is no prohibitive mutual coupling between the two patch elements. Now, this isolation is not exclusive, and in

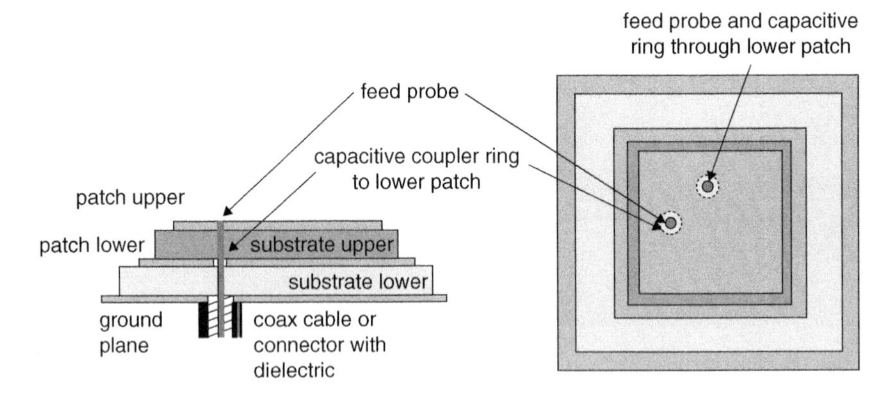

Figure 5.10 Dual probe-fed, RHCP, multifrequency GNSS patch antenna.

reality, each patch does affect each other, but multiple resonance frequencies can be supported with proper tuning.

In Figure 5.10, the feed comes up from the bottom, through the ground plane on the bottom of the lower substrate, and is typically capacitively coupled to the lower patch, and then either physically connected to the upper patch element or, again, capacitively coupled to the top patch [22, 23]. The return/ground of the feed is attached to the bottom of the ground plane.

The various design techniques for probe-fed LP, single-frequency patch antennas can be applied to multifrequency patch antennas, with the capacitive coupling technique illustrated in Figure 5.10. As presented earlier for the probe-fed LP, single-frequency patch antennas, the impedance of the antenna can be matched to the transmission line/connector by moving the feed location from the edge toward the center of the patch. Keeping in mind that the length of each patch is determined by the desired resonance frequency, careful attention must be given to the final feed location to obtain good impedance matches at all of the desired resonance frequencies. Often this is a compromise between the impedance match obtained at the multiple frequencies to be supported by the antenna. With the dual probe-fed RHCP approach to support multiple frequencies, each of the LP probe ports can then be combined with a hybrid (i.e. 90°) power combiner to produce an RHCP output signal.

5.4.3 Survey-Grade/Reference GNSS Antennas

This section is devoted to what can be categorized as high-quality, survey-grade, geodetic, and/or reference station GNSS antennas. There are a significant number of these types of antennas used for various high-performance GNSS applications, including geodetic survey, high-precision farming/construction/machine control, and fixed GNSS reference stations. Almost all are dual or multifrequency to support removal of the error introduced due to the ionosphere. These types of applications also place a premium on performance including multipath mitigation to enable high precision. Most often, these types of antennas are larger than simple patch antennas and are more costly, even though they may incorporate patch antenna technologies into their design. Once again, there are a wide variety of designs in the marketplace, and this section will present a few of the more common configurations.

For high accuracy and real time kinematic (RTK) type applications, calibration and code and/or phase delay compensation may be on the GNSS measurement prior to user solution computations.

5.4.3.1 Choke Ring-Based GNSS Antennas

The choke ring-based antenna was originally introduced to the GPS community by the Jet Proposal Laboratory (JPL) and was used with a Dorne &

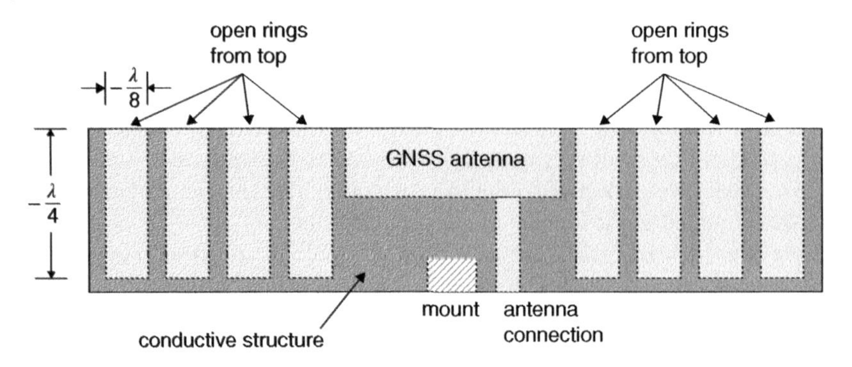

Figure 5.11 Typical choke ring-based GNSS antenna configuration.

Margolin dipole-based GPS antenna element. The choke ring is essentially a shaped ground plane made of conductive material forming a series of concentric circular troughs for the purpose of mitigating multipath. Figure 5.11 illustrates a common configuration of a typical (i.e. 2D) choke ring utilized in GNSS applications.

Typical configurations include three to four choke rings, with ring depth slightly greater than a quarter wavelength and a width between rings slightly greater than an eight wavelength [25]. The geometric variation of the ground plane provides a reduction in gain for lower elevation angles to the GNSS SVs. Multipath and interference signals below the horizon (i.e. 0° elevation angle) are diffracted by the choke rings. The diffractions will induce secondary currents on the choke ring structure, and some current will find its way to the antenna elements but will be substantially reduced had the choke ring structure not been present.

The physical dimensions of the choke ring ground plane structure can be optimized for a single GNSS frequency, or designed to be a compromise in performance for multiple GNSS frequencies that are several hundreds of megahertz apart.

Various antenna designs can be used at the center of the choke ring as the radiating element. Initially, a domed (i.e. bubblelike) radome housing was commonly used to enclose a dipole-based GPS antenna. Today, multifrequency patch antennas are also commonly placed at the center of GNSS choke ring antennas.

Variations in the traditional choke ring structure have also been used to help mitigate multipath error, including a variation of depth [26, 27], use of a single choke ring [28], and use of a 3D choke ring structure [29]. Figure 5.12 illustrates a 3D choke ring structure [30]. The latest generation of 3D choke ring structures provide for a more uniform gain in the upper hemisphere and better suppression of the gain at negative elevation angles for multipath and interference mitigation.

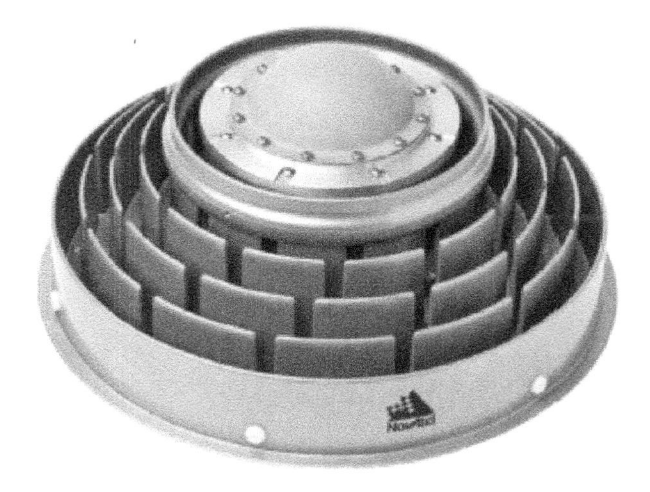

Figure 5.12 Photograph of a 3D choke ring [30]. Source: Courtesy of NovAtel.

5.4.3.2 Advanced Planner-Based GNSS Antennas

Aside from choke ring-based antennas, there is another class of high-performance GNSS antennas that are mostly planar (i.e. microwave circuit board based) in design used in geodesy and reference station applications. While there are several antennas that can be categorized as such, information on three antennas, the Roke geodetic, NovAtel Pinwheel, and Trimble Zephyr antenna, will be presented here.

The Roke geodetic-grade antenna is a broadband spiral-type antenna designed to support triple-frequency GNSS bands [31]. The antenna is a cavity backed spiral with a unique ground plane structure to help minimize multipath [32]. Figure 5.13 is a photograph of the spiral radiating element above the cavity backed structure (i.e. metal can). A cavity backed structure attempts to minimize the reflection off the bottom of the can with special absorbing material within, and what is left reflects in phase with the upward signal in the upper hemisphere. The radiation spiral elements are typically placed $\lambda/4$ above the inside bottom of the can structure. (The total travel distance of the signal inside the can will be 180° and, with a 180 phase shift caused by the normal incident reflection from the metallic can structure, will cause the reflected signal [at zenith] to be in-phase with the upward signal.)

One of the latest GNSS antenna that receives all of the GNSS constellation signals in L-band with low and known phase center variations is the NovAtel VEXXIS® GNSS-850 antenna [33, 34]. This type of planar antenna design, illustrated in Figure 5.14 utilizes a circular patch type element combined with loaded parasitic elements that provides increased bandwidth and gain suppression at low and negative elevation angles [33]. The typical normalized

Figure 5.13 Spiral GNSS antenna (spiral and can only) [31]. Source: Courtesy of Roke.

Figure 5.14 GNSS-850 antenna (radome removed) plane [34]. Source: Courtesy of NovAtel.

radiation characteristics for this wideband GNSS antenna are illustrated in Figure 5.15 in the L1/E1/B1C, L2, L5/E5a, and E6/L6 frequency bands, for both RHCP (outer trace) and LHCP (inner trace). This GNSS antenna provides a reduction of RHCP gain at low elevation angles and significant gain suppression below the horizon to mitigate multipath. Additionally, the LHCP gain is surprised by typically less than 20 dB from the RHCP. The antenna also exhibits good gain and gain flatness across the GNSS L-band (i.e. <2 dB and 3–5 dBic) and consisting gain.

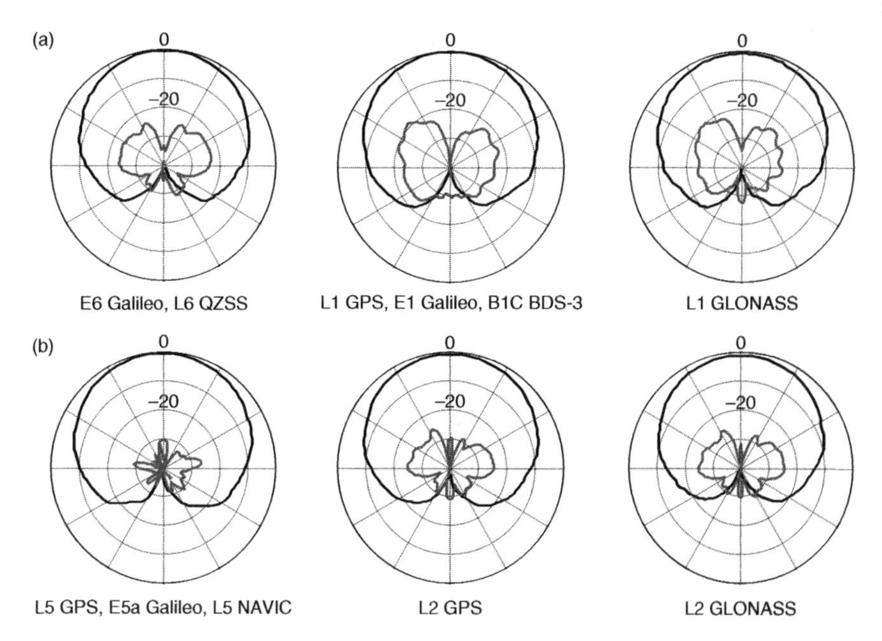

Figure 5.15 Typical radiation characteristics of the wideband GNSS-850 antenna in elevation plane. RHCP is outer trace, LHCP inner trace [34]. (a) E6 Galileo, L6 QZSS; (b) L1 GPS, E1 Galileo, B1C BDS-3; (c) L1 GLONASS; (d) L5 GPS, E5a Galileo, L5 NAVIC; (e) L2 GPS; (f) L2 GLONASS. Source: Courtesy of NovAtel.

The Trimble Zephyr GNSS antenna is another planar-type GNSS antenna that implements advanced technologies to help mitigate multipath for mainly geodetic and reference ground station applications. The Zephyr antenna utilizes a six-feed patch antenna design with a resistively tapered ground plane, where the surface resistance increases as the distance from the center of antenna increases to provide for enhanced performance and multipath mitigation [35].

5.5 Principles of Adaptable Phased-Array Antennas

Phased-array antennas implement multiple antenna elements where the physical orientation, phase, and/or amplitude can be controlled to obtain superior performance above that of a single antenna (i.e. a fixed reception pattern antenna [FRPA]). Adaptable phased-array antennas, often called smart antennas or controlled reception pattern antenna (CRPA) in the GNSS community, have the capability to control and change the net antenna radiation pattern characteristics as a result of sensing the environmental conditions that are presented to the GNSS antenna and/or receiver system. The sensing of the environmental conditions presented to the GNSS antenna and/or receiver system can include the following: sensing the host platform attitude with a host inertial navigation system (INS) or integrated inertial measurement unit (IMU); decoding or receiving information pertaining to the location of the GNSS SVs to be received; measuring/estimating the composite signal power level within the entire band; and/or estimating the signal power levels of individual signals within the band. When the sensing and adjusting is done iteratively over time, the process is adaptive to account for the changing environmental conditions to optimize the GNSS signal measurements.

Various signal processing techniques are implemented in GNSS adaptable phased-array antennas. Space adaptive processing (SAP) generally refers to a technique where the spatial relationship between the antenna elements in the array is used to optimize the antenna/receiver performance. SAP can be implemented with a power minimization, maximizing the signal-to-noise ratio, or a digital beamforming approach. Space–time adaptive processing (STAP), also known as spatial temporal adaptive processing techniques add filtering, typically implemented digitally (e.g. finite impulse response [FIR] filter) to each antenna reception path to help increase the degrees of freedom and bandwidth response of array. Space–frequency adaptive processing (SFAP) techniques perform signal processing in the frequency domain using digital filtering for interference/jamming signal mitigation. Digital beamforming techniques explicitly perform pattern shaping based on the desired signal and interference signals (or just the desired signal) locations and power levels.

A general block diagram of an antenna array that is used in an adaptive method for GNSS applications is shown in Figure 5.16, including a description of the block functions and notation listed below the block diagram. For each of the N antenna elements in the array, the individual antenna elements will have a transfer function that will vary based on frequency and aspect angle, $T_n(f, \theta, \phi)$, which will feed front-end RF components such as RF amplifiers, filters, and other components that can be represented by the transfer function $F_n(f, \theta, \phi)$. At the output of each of these antenna paths, complex weights (i.e. amplitude and phase modification) are applied to each signal path. (When STAP is not performed, the delay unit [DU] has no delay [i.e. $j = 0$].) After the complex antenna weights have been applied, the signals from each path can be combined, as represented in Eq. (5.12) for non-STAP processing, where the desired signals (s_d), interference (i), and noise (n) are received as the total signal (x), assuming the effects of $T_n(f, \theta, \phi)$ and $F_n(f, \theta, \phi)$ have been compensated for in an antenna calibration process. The output of the antenna array in Eq. (5.12) is shown as a channel output $y_h(k)$; this can be a single RF or intermediate frequency (IF) port, or on a per SV basis depending upon the algorithm used to calculate the antenna weights, and the receiver configuration. The effects from the geometric position of the antenna elements are included in x, in the form of the antenna steering vector (i.e. geometric

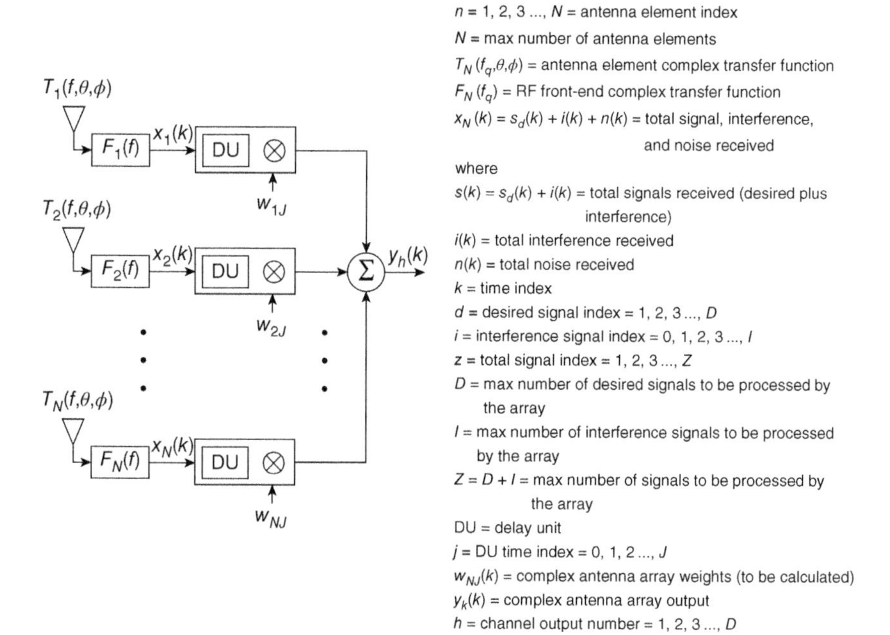

Figure 5.16 General block diagram of a GNSS adaptive antenna array.

antenna array factor):

$$y_h(k) = \sum_{n=1}^{N} w_n(k)x_n(k), \quad \text{non-STAP (neglecting } J\text{)} \tag{5.12}$$

Without loss of generality, the signals in Eq. (5.12) are represented as discrete signals with time index k. Additionally, the details of where the functions of Figure 5.16 are performed, and what methods are used to calculate the complex antenna weights will depend upon the performance requirements for the GNSS antenna and receiver systems, technologies employed, and user configuration constraints.

Historically, CRPA used with GPS receivers has been used by military users to help mitigate intentional jamming and unintentional interference [36]. These configurations typically involved the CRPA antenna, associated antenna electronics (AE), and a single-RF (or IF) input port on a GPS receiver [37]. Initial CRPA configuration performed an adaptive nulling technique to place nulls at the spatial location of jamming/interference sources to provide a single output to a GPS receiver. The objective of the adaptive nulling algorithm is to adjust the weighting of the antenna elements so that the power in the sum of the weighted signal output is minimized, subject to a constraint that prevents the minimized power from being below a certain value. Typically, the constraint is provided by fixing the weight of one of the antenna elements (i.e. a reference antenna element) and by allowing the other weights to be adjusted. In mathematical terms, the weights are adjusted to minimize the average power of the antenna output with respect to the reference element. Without loss of generality, it can be assumed that the constraint is applied by fixing the weight for the reference element path to unity. Thus, there are $N-1$ "degrees of freedom" in adjusting the weights; as many as $N-1$ nulls in the antenna spatial pattern can be generated.

The adaptive nulling power minimization technique described earlier works well to minimize the effects from the interference source but can have a negative effect on the desired GNSS signals to be tracked because the location of the desired signal locations is not considered in the antenna weight calculations. The most severe case is when the interference source is in the same direction as a desired signal direction. For certain high-performance applications, the distortion to the code and carrier phase measurements is significant, and more advanced techniques can be applied.

5.5.1 Digital Beamforming Adaptive Antenna Array Formulations

In addition to placing nulls in the directions of interference sources, directional beams can be pointed in the direction of the desired GNSS signals to help minimize the effects from the interference source on the GNSS code and carrier measurements. This type of technique is typically referred to as digital beamforming, and one popular algorithm is referred to as minimum variance distortionless response (MVDR) whereby the desired signal is passed undistorted, after application of the complex antenna array weights, while minimizing the output noise variance [38–40]. Theoretically, this type of technique will not distort the code and carrier measurement from the SV signal being processed, but practical limitation with the degrees of freedom in the antenna array, locations of interference sources with respect to the location of the desired SV signal, interference power level, and waveform type will limit the effectiveness of this distortionless processing of the desired signal. These types of digital beamforming algorithms are well suited for digitally based GNSS receiver signal processing (i.e. software-defined radio)-based architectures, where each GNSS signal can be processed, individually, in a digital receiver channel. Each of the N antenna path signal data will have the antenna weights applied, different for each SV to be processed, and will produce D digital channel outputs for code and carrier tracking. A separate set of weights is required for each beam pointed in the direction of each desired GNSS signal to be tracked. The complex antenna weights are typically computed on the basis of antenna attitude information using either the host INS or integrated IMU, and GNSS satellite position data (e.g. ephemeris data), so that each beam points toward a satellite.

For example, the output of channel 1 of the antenna array output to be applied to the digital channel 1 of the digital GNSS receiver can be represented as in Eq. (5.13) [38–40]. For distortionless processing of the desired signal s_{d1}, the product of the geometric antenna steering vector (i.e. antenna array factor, pointed in the estimated direction of the desired signal) and the Hermitian (H) transpose of the complex antenna weights (yet to be determined) should be 1:

$$\mathbf{y}_1(k) = \underbrace{\mathbf{w}^H \mathbf{a}_1(\theta_1, \phi_1) s_{d1} + \mathbf{w}^H \mathbf{u}}_{\text{want this to be 1}} \quad \text{for } h = 1 \tag{5.13}$$

where

$\mathbf{a}_1(\theta_1, \phi_1)$ = the N-element geometric antenna array steering vector in the direction of (θ_1, ϕ_1) for the desired signal s_{d1}

$\mathbf{w}$ = is the complex $[N \times 1]$ antenna array weights (yet to be determined)

$[\,]^H$ = Hermitian transpose (i.e. complex conjugate transpose)

$\mathbf{u}$ = undesired signal vector (interference + noise)

The expected value of Eq. (5.13) will be s_{d1}, and the variance can be calculated as $\text{var}[\mathbf{y}_1] = \mathbf{w}^H \mathbf{R}_{uu} \mathbf{w}$, where the undesired signal correlation matrix is expressed as $\mathbf{R}_{uu} = E[\mathbf{u}\mathbf{u}^H]$. To optimally solve for the complex antenna weights for the MVDR process, the method of Lagrange can be used to define a cost function that is a linear combination of the variance of the output and the constraint that $\mathbf{w}^H \mathbf{a}_1(\theta_1, \phi_1) = 1$. Minimizing this cost function leads to the solution for the antenna weight shown in Eq. (5.14), for channel 1 [39–41]:

$$\mathbf{w}_{\text{mv1}} = \frac{\mathbf{R}_{uu}^{-1} \mathbf{a}_1(\theta_1, \phi_1)}{\mathbf{a}_1^H(\theta_1, \phi_1) \mathbf{R}_{uu}^{-1} \mathbf{a}_1(\theta_1, \phi_1)} \quad \text{(for a single signal per receiver)} \tag{5.14}$$

where

$\mathbf{R}_{uu}$ = $E[\mathbf{u}\mathbf{u}^H]$ = undesired signal correlation matrix

$E[\,]$ = expected value function

The complex antenna weights expressed in Eq. (5.14) would be used to process the desired signal 1, that is, s_{d1}, and applied to digital receiver channel 1. This process places a beam in the direction of s_{d1}. A similar process would be completed for each of the other desired GNSS signals to be processed into each of the digital GNSS receiver channels.

Figure 5.17 illustrates a typical theoretical performance of a seven-element CRPA-type configuration array factor for a desired signal at an elevation angle of 80° and azimuth angle of 90°, where the view angle is at an elevation angle of 30° and azimuth angle of 10°. The azimuth scale represents the horizon and

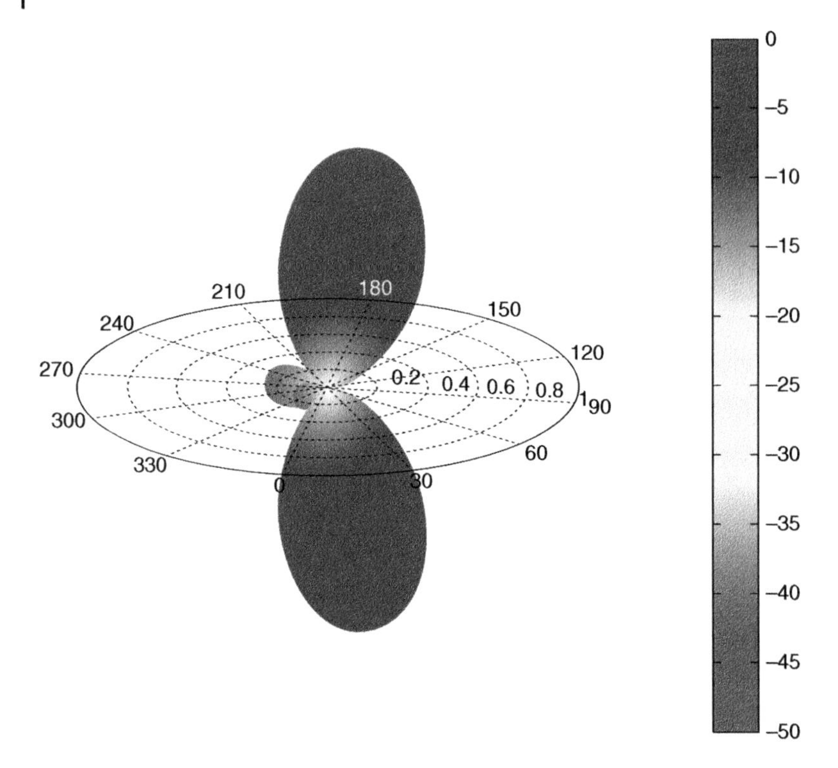

Figure 5.17 Illustration of theoretical seven-element CRPA array factor (with no ground plane). Source: From Ref. [41].

the upper hemisphere would represent the radiation characteristics pointed toward the SV (the lower half of the theoretical array factor is shown but in reality would be suppressed by the ground plane) [41].

In addition to digital beamforming to place a directional beam in the direction of the desired signal to be digitally processed, nulls can be placed in the direction of interference sources while minimizing the output variance of the desired signal. This can be accomplished by including an estimate of the interference signal direction in the antenna steering vector and including this in the undesired signal correlation matrix shown in Eq. (5.14). A similar process would be completed for each of the other desired GNSS signals to be processed into each of the digital GNSS receiver channels.

Using a reference element in the antenna array, the degrees of freedom available for interference mitigation will be $N-1$, but the effectiveness for interference mitigation will include: a function of the number of elements in the array; locations of interference sources with respect to the location of the desired SV signal; interference power levels; and waveform type. It has been shown that strong and broadband interference signals consume more than just one degree of freedom in the array processing [42]. Methods to increase the degrees of freedom are to increase the number of geometric elements [41] or to implement advanced digital signal processing with the same number of elements.

5.5.2 STAP

The STAP technique provides for better interference mitigation above the basic SAP processing technique when the geometric degrees of freedom in the antenna structure are consumed by increasing interference power and/or waveforms [42]. To help increase the interference mitigation performance, digital filters (i.e. FIR) can be added to each antenna path to help increase the bandwidth response of the array. In Figure 5.16, the DU function is expanded from 0 to some integer number (i.e. J), and each of the J weights (i.e. $w_{nj}(j)$) are applied to each delayed version of the signal (i.e. $x_{nj}(j)$). The weighted outputs of each channel, $x_{nJ}(k)$, are then summed to produce the antenna array output in accordance with Eq. (5.15):

$$x_{nJ}(k) = \sum_{j=1}^{J} w_{nj}(j)x_{nj}(j)$$

$$y_h(k) = \sum_{j=1}^{J} w_{nJ}(k), \quad \text{with STAP (finite } J) \tag{5.15}$$

5.5.3 SFAP

The interference reduction capability of SAP-only processing is best when the interfering signals are narrowband. Broadband jammers present a problem, because for a given set of antenna weights, the antenna spatial gain pattern varies with frequency. Thus, a set of weights optimal at one frequency will not be optimal at another frequency. SFAP solves this problem by dividing the frequency band into multiple narrow sub-bands, typically by using a fast Fourier transform. For each sub-band, a set of optimal weights is used to obtain a corresponding desired antenna spatial pattern for that sub-band. By combining the nullformed or beamformed signals from all the subbands, an optimal wideband antenna spatial pattern is obtained. Similar results can be obtained by using a bank of narrowband filters in the time domain, as presented previously [43].

5.5.4 Configurations of Adaptable Phased-Array Antennas

Various configurations of adaptable phased-array antenna have been used, mostly by military users for air and sea applications. The seven-element CRPA (e.g. CRPA and CRPA-2) with AE (e.g. AE-1) has been integrated with single RF and IF input GPS receivers to form nulls in the location of interference sources. The AE does a significant amount of the processing to adaptively process the signals received from the CRPA antenna for null forming. A modernized version of the CRPA-2/AE-1 configuration is the GPS Antenna System-1 (GAS-1). The follow-on to the GAS-1 is the Advanced Digital Antenna Production (ADAP) system, which will provide for enhanced interference mitigation and beamforming capabilities [44].

One of the newest generations of GPS antenna/receiver systems to implement digital beamforming and nulling technology, with STAP, is the GPS Spatial Temporal Anti-Jam Receiver (G-STAR). The G-STAR has the ability to handle narrowband and wideband interference sources with its STAP and to place nulls toward the interference sources while directing high gain beams in the directions of the desired SV signals [45].

Due to the physical size of the seven-element CRPA, efforts have gone into developing smaller adaptable phased-array antennas for various platforms, especially smaller vehicles. Small or compact CRPAs have emerged and several four-element CRPA-type configurations with a smaller footprint are available [46].

5.5.5 Relative Merits of Adaptable Phased-Array Antennas

- Adaptive nulling is much simpler and cheaper than beamforming since only one output emerges from the process, enabling use with a more traditional single-input GNSS receiver.
- A beamforming antenna produces one output for each beam, so a considerably more complex receiver is required to process each output independently; typically implemented in a digital receiver, capable of processing each channel independently.
- Beamforming can produce significant spatial gain in the direction of the GNSS satellites, while adaptive nulling makes no attempt to maximize gain in the desired directions.
- Jamming reduction with adaptive nulling antenna; techniques can be limited for strong or broadband interference waveforms; STAP- and SFAP-type techniques can be added to increase the array bandwidth response and to improve interference mitigation robustness.
- Beamformers tend to have high spatial gain in the direction of desired signals and lower gain in other directions. Multipath arriving from a low-gain direction is therefore attenuated relative to the desired signal.
- Because of the large physical extent of the antenna array, beamformers and nullers have the common problem of causing biases in signal delay caused by movement of the antenna phase center as a function of the weight values. In some cases, biases of 100° in carrier phase and 1 m in code phase can occur. For high-precision systems, these errors can be significant and require compensation/calibration [47].

5.6 Application Calibration/Compensation Considerations

This section addresses issues regarding calibration of GNSS antennas, compensation for group and phase delays due to the antenna design, surroundings, or operational effects on the code and carrier phase measurements. Depending upon the end-user environmental and performance requirements, the level of attention to code and carrier phase measurement compensation may vary. While the GNSS antenna design requirement is to provide the GNSS signals to the GNSS receiver so that they can effectively be processed, the quality and fidelity will depend on the application, cost, size, and so on.

Generally, the larger an antenna is, and the closer things get to the radiating element of an antenna, the more variation there will be in antenna performance. At the antenna design level, items (ground planes, mounts, radomes, lightning rods, filters, power supply, etc.) in close proximity can have a significant effect on the performance. This is especially true for small consumer products that must conform to a compact form factor. Typically, other device components close to the antenna are modeled in a CEM along with the antenna so that performance predictions for the antenna "as installed in its operational environment" can be accomplished. This is also true for radome structures that will shift the operating frequency of the antenna downward. For example, a patch antenna with a plastic-type radome a couple of millimeters thick, placed a couple of millimeters above the radiating patch, will shift the operating frequency downward on the order of 4–6 MHz, which, of course, changes the amplitude and phase response of the antenna. This type of compensation must be done at the design stage of the antenna, including its structure.

After the antenna has been produced, for high-performance and/or high-integrity applications, calibration/compensation of the code and/or carrier phase measurements may be needed with respect to an antenna reference point (ARP). The ARP is a physical point on the antenna that code and/or carrier phase measurements are referenced to. Additionally, depending upon application, it is also useful to establish an azimuth reference marker on the antenna to calibrate/compensate for possible azimuth variation of the code and/or carrier phase measurements. Most often, this is accomplished with a "north marker" tab for stationary applications, or aligning the antenna in a particular fixed direction with respect to the body frame of a dynamically moving vehicle.

For survey-grade/reference GNSS antennas used in geodetic and general-purpose reference station-type applications, the carrier phase and/or code phase compensation is often a very important item that must be addressed to enable high-accuracy kinematic processing with the carrier phase or for high-integrity applications. For multifrequency antennas, the US National Oceanic and Atmospheric Administration, National Geodetic

Service (NGS) provides individual absolute calibration for a wide variety of multifrequency GNSS antennas [48]. Additionally, commercial companies have provided GNSS user equipment and GNSS SV antenna calibration services [49].

For high-performance GNSS antennas used in GNSS reference station-type applications, the code phase compensation is often a very important item that must be addressed to enable high accuracy and high integrity. Reference station applications used to support precision landing systems must limit the multipath caused by the ground and are often large antennas. Larger antennas will typically have more code and carrier phase variation versus the spatial coordinates to the satellite that are being tracked. For these types of applications, antenna modeling [50], anechoic chamber testing [51], and field testing [52] can be used to build confidence in the code and/or carrier phase compensation/calibration process employed.

The adaptable phased-array antennas presented earlier have very unique antenna calibration requirements. First, by their nature of being in an antenna array, the amplitude and phase response of each individual antenna element, $T_n(f, \theta, \phi)$ in Figure 5.16, will be different, based on their location in the array. Additionally, these transfer functions may, and will likely change, based on the "as installed" (i.e. *in situ*) configuration of the antenna array. Furthermore, when interference is present, the complex amplitude weight algorithm will compute and apply the antenna weights to each signal path in accordance with the technique used. These techniques, even for an MVDR algorithm, can produce significant amplitude and phase distortion depending upon the degrees of freedom in the antenna array, locations of interference sources with respect to the location of the desired SV signal, interference power level, and waveform type. Significant effort has gone into providing pre-mission and online calibration for these types of applications [47].

For other applications, such as wraparound microstrip GNSS antennas on spinning missiles, projectiles, or launch vehicles, measurement compensation is important. Since the code phase is unambiguous, provided the signal antenna gain pattern presented to the SV is fairly uniform, and the code tracking loop is able to maintain reasonable C/N_0, then the code measurements should not see significant error. However, significant attention must be given to the carrier phase measurement if it is continuously tracked and used. Since the carrier phase measurement is ambiguous, with a continuously tracked carrier phase measurement on-board a spinning vehicle, the carrier phase measurement will "wind up" depending upon the vehicle spin rate and aspect angle to the SV. An IMU or INS can be used to compensate for this carrier phase windup to "unwind" the carrier phase measurement, based upon the spin rate and attitude with respect to the aspect angle to the SV signal being tracked.

As a final note on antenna arrays for GNSS applications, their size, cost, complexity, and integration into certain platforms may be prohibitive for certain applications (e.g. unmanned aerial system (UAS), manpack, vehicular). Additionally Export Administration Regulations (EAR) or International Traffic in Arms Regulations (ITAR) may prohibit certain types of configurations and/or capabilities. For these types of applications, smaller, less robust, less complex solutions may be well suited for the applications. Single antenna dual polarization techniques have been implemented to adjust the polarization components to cancel interference [53]. Other unique single-aperture, multi-feed approaches have implemented quadrant-by-quadrant beam forming (and null steering) to provide gain reduction in commanded regions [54, 55].

Problems

1 A passive GNSS antenna with a measured input impedance of $Z_A = 45 - j15\,\Omega$ is connected to a small transmission line (RF microstrip line) with a characteristic impedance of $Z_o = 52 + j10\,\Omega$. At this interface connection, what is the following?
(a) Reflection coefficient
(b) Standing wave ratio
(c) Mismatch efficiency
(d) Return loss

2 Consider an antenna coordinate system and crossed-dipole antenna configuration presented in Figure 5.2, with an incident GNSS signal for the form $\boldsymbol{E}_{\text{GNSS}}(z,t) = \cos(\omega_c t)\mathbf{a}_x - \sin(\omega_c t)\mathbf{a}_y$. If the input ports of the 90° power hybrid combiner are reversed (i.e. dipole port 1 is delayed by 90°), calculate the output signal $s(t)$.

3 Design an edge probe-fed square patch antenna to operate at the GPS L1 frequency with a Rogers RT/duroid® 6002 dielectric substrate that has a relative permittivity of 2.9 and a height of 3.0 mm. The probe feed has a diameter of 0.7 mm and is to be placed at the center of the feed side. Perform the following:
(a) Sketch the patch antenna from the side and top; label all important items.
(b) What is the physical size (i.e. length and width) of the top radiation patch elements?
(c) What is the input impedance for the antenna?
(d) What polarization will the patch antenna be sensitive to?
(e) Plot the real part of the input impedance if the feed is moved from the edge to the center, and then to the far edge, along the center line of the patch.

(f) At what distance from the edge of the patch would the real part of the input impedance be 50 Ω?

4 Design a single, probe-fed nearly square patch antenna to receive the RHCP GPS L1. The probe feed is to be placed along the diagonal of the patch. The performance requirement is to have an AR no more than 2 (at zenith). The dielectric substrate has a relative permittivity of 9.8 and a height (i.e. vertical thickness) of 6 mm. The antenna is expected to have a total Q of 10. Perform the following:

(a) Calculate the physical size (i.e. length and width) of the top radiation patch elements.
(b) What are the two degenerative modal signal frequencies for your design?
(c) Sketch the patch antenna from the top view; label all important items.

References

1 IEEE Std 211-1997(R2003) (1997). *IEEE Standard Definitions of Terms for Radio Wave Propagation*. New York: IEEE, Inc.
2 IEEE Std 145-1993(R2004) (1983). *IEEE Standard Definitions of Terms for Antennas*. New York: IEEE, Inc.
3 Balanis, C.A. *Antenna Theory, Analysis and Design*, 4e. Hoboken, NJ: Wiley.
4 Van Dierendonck, A.J., Fenton, P., and Ford, T. (1992). Theory and performance of narrow correlator spacing in a GPS receiver. *ION Navigation Journal* 39 (3): 265–284.
5 Pratt, T., Bostian, C., and Allnutt, J. (2003). *Satellite Communications*, 2e. New York: Wiley.
6 RTCA, Inc. (2006). MOPS for GNSS Active Antenna Equipment in L1 Frequency Band, SC-159, DO-301, 13 December 2006.
7 Altair (2019). Altair Feko™ overview. https://altairhyperworks.com/product/FEKO (visited 20 March 2019).
8 Mentor® (2019). IE3D, HyperLynx Full-Wave Solver. https://www.mentor.com/pcb/hyperlynx/full-wave-solver/ (visited 26 May 2012).
9 WIPL-D (2019). https://wipl-d.com// (visited 20 March 2019).
10 HFSS (2019). ANSYS electromagnetics. https://www.ansys.com/products/electronics/ansys-hfss (visited 20 March 2019).
11 Dassault Systemes (2019). CST-computer simulation technology. https://www.cst.com/ (visited 20 March 2019).
12 REMCOM (2019). Electromagnetic simulation software. https://www.remcom.com/ (visited 20 March 2019).

13 Deschmps, G. and Sichak, W. (1953). Microstrip microwave antennas. In: *Proceedings of the 3rd Symposium on USAF Antenna Research and Development Program (18–22 October 1953).*

14 Sensor Systems. GPS S67-1575-135 GPS Antenna. http://www.sensorantennas.com/product/gps-waas-arinc-antenna/ (visited 17 November 2012).

15 ARINC (2001). GNSS Sensor Arinc Characteristic 743A-4, published 27 December 2001.

16 Rogers Corporation (2012). Advanced circuit materials. http://www.rogerscorp.com/acm/literature.aspx (visited 24 May 2012).

17 Taconic (2012). RF & Microwave Laminates. https://www.4taconic.com/page/microwave--rf-laminates-66.html (visited 24 May 2012).

18 Jackson, D.R. and Alexopoulos, N.G. (1991). Simple approximate formulas for input resistance, bandwidth, and efficiency of a resonant rectangular patch. *IEEE Transactions on Antennas and Propagation* 3: 407–410.

19 Johnson, R.C. (1993). *Antenna Engineering Handbook*, Chapter 7 (ed. R.E. Munson). New York: McGraw Hill.

20 Richards, W.F. (1998). Microstrip antennas. In: *Antenna Handbook* (eds. Y.T. Lo and S.W. Lee). New York: Van Nostrand Reinhold Co.

21 Richardson, W.F., Zinecker, J.R., Clark, R.D., and Long, S.A. (1983). Experimental and theoretical investigation of the inductance associated with a microstrip antenna feed. *Electromagnetics* 3 (3–4): 327–346.

22 Garg, R., Bhartia, P., Bahl, I., and Ittipiboon, A. (2001). *Microstrip Antenna Design Handbook*. Boston, MA: Artech House.

23 Hall, P.S. (1987). Probe compensation in thick microstrip patches. *Electronics Letters* 23: 606–607.

24 Richardson, W.F., Lo, Y.T., and Harrison, D.D. (1981). An improved theory of microstrip antennas with applications. *IEEE Transactions on Antennas and Propagation* 29 (1): 38–46.

25 UNAVCO (2012). Choke ring antenna calibrations. http://facility.unavco.org/kb/questions/311/Choke+Ring+Antenna+Calibrations (visited 28 May 2012).

26 Filippov, V., Tatarnicov, D., Ashjaee, J. et al. (1998). The first dual-depth dual-frequency choke ring. In: *Proceedings of the 11th International Technical Meeting of the Satellite Division of The Institute of Navigation (ION GPS 1998)*, 1035–1040. TN: Nashville.

27 Javad (2012). Choke ring theory. http://www.javad.com/jns/index.html?/jns/technology/Choke%20Ring%20Theory.html (visited 28 May 2012).

28 Thornberg, D.B., Thornberg, D.S., DiBenedetto, M.F. et al. (2003). LAAS integrated multipath-limiting antenna. *ION Navigation Journal* 50 (2): 117–130.

29 Kunysz, W. (2003). A three dimensional choke ring ground plane antenna. In: *ION GPS/GNSS 2003, Portland, OR (9–12 September 2003)*, 1883–1888.

30 NovAtel (2012). Antennas GNSS-705. http://www.novatel.com/assets/Documents/Papers/GNSS-750.pdf (visited 4 June 2012).
31 Roke (2012). Roke triple GNSS geodetic-grade antenna. https://www.roke.co.uk/~/media/Files/C/Chemring-Roke/documents/042-GNSS-Antenna.pdf (visited 7 June 2012).
32 Granger, R. and Simpson, S. (2008). An analysis of multipath mitigation techniques suitable for geodetic antennas. In: *Proceedings of the 21st International Technical Meeting of the Satellite Division of The Institute of Navigation (ION GNSS 2008), Savannah, GA (16–19 September 2008)*, 2755–2765.
33 Yang, N. and Freestone, J. (2016). High-performance GNSS antennas with phase-reversal quadrature feeding network and parasitic circular array. In: *Proceedings of the 29th International Technical Meeting of the Satellite Division of The Institute of Navigation (ION GNSS+ 2016), Portland, Oregon (September 2016)*, 364–372. https://doi.org/10.33012/2016.14833.
34 Gerein, N. and Freestone, J. (2019). NovAtel GNSS-800 Antenna Information files.
35 Krantz, E., Riley, S., and Large, P. (2001). The design and performance of the zephyr geodetic antenna. In: *Proceedings of the 14th International Technical Meeting of the Satellite Division of The Institute of Navigation (ION GPS 2001), Salt Lake City, UT (11–14 September 2001)*, 1942–1951.
36 NAVSTAR (1996). Navstar GPS User Equipment Introduction. http://www.navcen.uscg.gov/pubs/gps/gpsuser/gpsuser.pdf (visited 12 September 2010).
37 NSSRM (2012). National security space road map. http://www.fas.org/spp/military/program/nssrm/initiatives/crpa.htm (visited 28 May 2012).
38 Compton, R.T. (1988). *Adaptive Antennas*. Englewood Cliffs, NJ: Prentice Hall.
39 Godara, L.C. (2004). *Smart Antennas*. Boca Raton, FL: CRC Press.
40 Gross, F. (2005). *Smart Antennas for Wireless Communications*. New York: McGraw Hill.
41 Bartone, C.G. and Stansell, T. (2011). A multi-circular ring CRPA for robust GNSS performance in an interference and multipath environment. In: *Proceedings of the 24th International Technical Meeting of the Satellite Division of The Institute of Navigation (ION GNSS 2011), Portland, OR (20–23 September 2011)*, 1129–1139.
42 Moore, T.D. and Gupta, I.J. (2003). The effect of interference power and bandwidth in space-time adaptive processing. In: *Institute of Navigation, 59th Annual Meeting (23–25 June 2003)*.
43 Gupta, I.J. and Moore, T.D. (2001). Space-frequency adaptive processing (SFAP) for interference suppression in GPS receivers. In: *Proceedings of the 2001 National Technical Meeting of The Institute of Navigation, Long Beach, CA, January 2001*, 377–385.

44 USAF (2012). *Advanced Digital Antenna Production System*. Los Angeles, CA: U.S. Air Force. https://govtribe.com/opportunity/federal-contract-opportunity/advanced-digital-antenna-production-adap-system-1255 (visited 31 May 2012).
45 International On-line Defense Magazine (2012). Defense update, GPS anti jamming techniques. http://defense-update.com/products/g/GPS-STAP.htm (visited 31 May 2012).
46 ITT (2012). DM N100-3 Series GPS Antenna, ITT Exelis, Antenna Products and Technologies. http://www.exelisinc.com/capabilities/Antennas/Documents/N100-3_Series.pdf (visited 31 May 2012).
47 O'Brien, A. (2009). Adaptive antenna arrays for precision GNSS receivers. PhD dissertation. Ohio State University. https://etd.ohiolink.edu/ap/10?0::NO:10:P10_ACCESSION_NUM:osu1259170076 (visited 30 May 2012).
48 NGS (2012). National Oceanic and Atmospheric Administration, National Geodetic Service, Antenna Calibrations. https://www.ngs.noaa.gov/ANTCAL/index.xhtml (visited 20 March 2019).
49 Geo++ (2012). SMART in positioning. http://www.geopp.de/ (visited 20 March 2019).
50 Aloi, D. (2004). Analysis of LAAS integrated multipath limiting antennas using high-fidelity electromagnetic models. In: *Proceedings of the 17th International Technical Meeting of the Satellite Division of The Institute of Navigation (ION GNSS 2004), Long Beach, CA, September 2004*, 2303–2315.
51 van Graas, F., Bartone, C., and Arthur, T. (2004). GPS antenna phase and group delay corrections. In: *Proceedings of the 2004 National Technical Meeting of The Institute of Navigation, San Diego, CA, January 2004*, 399–408.
52 Lopez, A.R. (2008). LAAS/GBAS ground reference antenna with enhanced mitigation of ground multipath. In: *Proceedings of the 2008 National Technical Meeting of The Institute of Navigation, San Diego, CA, January 2008*, 389–393.
53 Mario M. Casabona, Murray W. Rosen, George A. Silverman (1998). Interference cancellation system for global positioning satellite receivers. US Patent 5,712,641, dated 27 January 1998.
54 Bartone, C. and Schopis, J. (2015). Patch antenna asymmetry performance considerations & mitigation. In: *Proceedings of the 28th International Technical Meeting of The Satellite Division of the Institute of Navigation (ION GNSS+ 2015), Tampa, FL, (September 2015)*, 907–914.
55 Bartone, C. (2016). Single-aperture patch antenna with pattern control. In: *IEEE Proceedings of the IEEE/ION Position, Location, and Navigation Symposium (PLANS), Savanna, GA (12–14 April 2016)*, Peer reviewed session.

6

GNSS Receiver Design and Analysis

6.1 Receiver Design Choices

6.1.1 Global Navigation Satellite System (GNSS) Application to Be Supported

One of the most important factors in global navigation satellite system (GNSS) receiver design is determining what application the receiver is going to be used in. An aviation receiver used for safety of live applications has very different requirements compared with a geodetic surveying receiver or a low-cost consumer-grade receiver used in a mobile cell phone. Many of the receiver design characteristics will be very different, but some characteristics will functionally be the same.

The intended application often drives the requirement to use a single GNSS (e.g. global positioning system [GPS] only) or a multiconstellation GNSS. For example, in low-cost consumer applications, a GPS L1-only receiver may be favored. For a more robust consumer application, a single-frequency but multiconstellation approach may provide better performance, for example, L1-only GPS and global orbiting navigation satellite system (GLONASS) and/or Galileo and/or Compass/BeiDou. For a high-quality geodetic application where accuracy is paramount, a multifrequency approach will be favored for high-quality code and carrier phase measurements (e.g. L1 and L2 GPS), and multiconstellation support may be of secondary consideration. Aviation applications warrant either augmented single-frequency GNSS services or dual-frequency GNSS support such as L1 GPS with wide-area augmentation system (WAAS), GPS L1 and L5, or Galileo E1 and E5 support, where accuracy, continuity, integrity, and availability are key performance requirements.

6.1.2 Single or Multifrequency Support

Single-frequency GNSS support is often favored when the design constraints for multifrequency support in the GNSS receiver become too prohibitive for

Global Navigation Satellite Systems, Inertial Navigation, and Integration,
Fourth Edition. Mohinder S. Grewal, Angus P. Andrews, and Chris G. Bartone.

Companion website: www.wiley.com/go/grewal/gnss

the application. This can include the size, cost, or power consumption needed to support a multifrequency GNSS receiver. If the performance requirements placed on the GNSS receiver can be met with a single-frequency GNSS system, then that is almost always favored. Some aviation applications utilize the single-frequency GNSS with augmentation, such as GPS with satellite based augmentation system (SBAS) to satisfy ionosphere corrections and integrity requirements.

Multifrequency support in a GNSS receiver is most often driven by the requirement for enhanced accuracy and/or integrity for the intended applications. While there are some advantages for triple-frequency utilization in carrier phase applications, most pseudorange-based GNSS receiver applications benefit from dual-frequency support for ionosphere error measurement correction and/or frequency band protection utilization. Dual-frequency support in the GNSS receiver can also provide increased integrity either through removal of the dominate ionosphere bias error and/or provided integrity data (e.g. Galileo E5b).

6.1.2.1 Dual-Frequency Ionosphere Correction

Because the error caused by the ionosphere is largely inversely proportional to the square of frequency, it can be calculated in a dual-frequency GNSS receiver by comparing the pseudorange measurements obtained on two frequencies. Using two GNSS pseudorange measurements, generally the further away these two frequencies are, the better the ionosphere error prediction will be. As will be presented in Chapter 7, the ionosphere error can be predicted by scaling and subtracting two pseudorange measurements. The predicted ionosphere error can then be subtracted from the measured pseudoranges. This process removes the dominate ionosphere bias error but does increase the noise. This increased noise can then be averaged/smoothed. With the dominate bias removed, the smoothing time can increase substantially over any single-frequency smoothing approach [1].

6.1.2.2 Improved Carrier Phase Ambiguity Resolution in High-Accuracy Differential Positioning

High-precision receivers, such as those used in surveying, use carrier phase measurements to obtain very precise range estimations. However, the periodic nature of the carrier makes the measurements highly ambiguous. Therefore, the solution of the positioning equations yields a grid of possible positions separated by a finite number of distances depending on code and carrier combinations, carrier phase ambiguity initialization, geometry, and measurement error. Removal of the ambiguity is accomplished by using additional information in the form of code pseudorange measurements, changes in satellite geometry, or the use of more satellites. In general, ambiguity resolution becomes less difficult as the effective frequency of the carrier decreases. For example, by using both the GPS L1 and L2 carriers, a virtual carrier frequency of L1 − L2 = 1575.42 − 1227.60 = 347.82 MHz can be obtained, which has a wavelength of about 86 cm as compared with the 19 cm wavelength of the L1 carrier; this particular combination is often referred to as the *wide-lane* combination. Ambiguity resolution can therefore be made faster and more reliable by using this difference frequency, albeit with slightly less accuracy than with the L1-only solution. Various dual-frequency code and carrier combinations are possible [2] as well as triple-frequency techniques [3].

6.1.3 Number of Channels

GPS receivers must observe and measure GNSS navigation signals from at least four satellites to obtain three-dimensional position, velocity, and user clock error estimates. If the user altitude is known with sufficient accuracy, three satellites will suffice. There are several choices as to how the signal observations from a multiplicity of satellites can be implemented. Today, almost all GNSS receivers are considered *all-in-view receivers* that have enough channels to receive all desired satellites that are visible for a particular GNSS. For a single-frequency, single GNSS application, in most cases, 12 or fewer useful satellites are visible at any given time; for this reason, modern receivers typically have approximately 12 channels, with perhaps several channels being used for acquisition of new satellites or noise calculations whereby the remainder are for tracking. Use of more than the minimum of four satellites will materially improve the accuracy of the user solution by using an overdetermined solution.

For dual-frequency single GNSS applications, a 24-channel receiver would be considered an all-in-view receiver (e.g. GPS L1 and L2).

Additional receiver channels, on the order of 50 or so, provide added benefit to support either advanced measurement processing, such as multipath mitigation [4], or for multiconstellation GNSS support.

As semiconductor technology has advanced, so has the ability to place thousands of digital correlator channels on a single semiconductor device. Receiver architectures have been developed that advertise tens of thousands of digital correlator channels within a single device to rapidly search a multitude of carrier and code phase offsets simultaneously [5]. These types of GNSS correlator engines can be combined with microprocessors to provide a host-based GNSS user solution.

6.1.4 Code Selections

The signal code formats needed to support the planned GNSS service need to be supported by the GNSS receiver. The code rates and format will have an impact on the signal bandwidths, processor speed, memory, and eventual power consumption needed. Single lower-rate codes, such as the GPS

C/A-code, can most easily be supported, whereas the Galileo E5a/b signal will demand the most bandwidth and processing. Codes that have error correction, authentication, and/or encryption codes on them will require additional functionality to be processed within the GNSS receiver.

Commercial receivers can recover the L2 carrier without knowledge of the code modulation simply by squaring the received signal waveform or by taking its absolute value. More advanced squaring techniques take advantage of the underlying P(Y) code periodicity to obtain *semi-codeless* tracking to produce pseudorange measurements on the GPS L2 frequency (without use of the L2C code) [6]. Because the a priori signal-to-noise ratio (SNR) is so small, the SNR of the recovered carrier will be reduced by as much as 33 dB because the squaring of the signal greatly increases the noise power relative to that of the signal. However, the squared signal has an extremely small bandwidth so that narrowband filtering can make up the difference to produce a pseudorange estimate on the L2 frequency [6].

6.1.5 Differential Capability

Differential global navigation satellite system (DGNSS) is a powerful technique for improving the performance of a GNSS user solution. The performance increase can be realized in terms of accuracy, integrity, or availability for the particular application. This concept involves the use of not only the user's receiver (sometimes called *the remote or roving* unit) but also typically a *reference or monitor receiver* and a *supporting data delivery method*. The complete treatment of DGNSS will be presented in Chapter 8. There are many ways to implement DGNSS and the implementation methods within the GNSS receiver can be just as varied.

DGNSS removes common systematic errors common to the user and monitor receiver. In a network-based DGNSS, there may be many sources of measurements or corrections. This chapter will focus on the corrections provided by or applied to a single GNSS receiver. The major sources of errors common to the reference and remote receivers, which can be removed (or mostly removed) by differential operation, are the following:

1. *Ionosphere delays*. Ionosphere signal propagation group delay, which is discussed further in Chapter 8, can be up to about 80 m during the day to 3–6 m at night. Receivers that utilize dual frequencies can largely remove these bias errors by applying the inverse square law dependence of delay on frequency. DGNSS will significantly remove this error contribution for single-frequency GNSS users.
2. *Troposphere delays*. These delays, which occur in the lower atmosphere, are usually smaller than ionosphere errors and typically are in the 1–3 m range for higher-elevation satellites but can be significantly larger at low satellite

elevation angles (e.g. up to about 40 m). The troposphere errors are difficult to measure directly with GNSS receivers and are often mitigated through a model for non-DGNSS users. DGNSS will significantly remove the troposphere error contribution for all GNSS users.
3. *Ephemeris errors.* Ephemeris orbit errors, which are the difference between the actual satellite location and the location predicted by satellite ephemeris orbital data, are typically less than 2 m and will undoubtedly become smaller as satellite tracking technology improves. DGNSS will significantly remove the orbit error contribution for all GNSS users.
4. *Satellite clock errors.* These errors are the difference between the actual satellite clock time and the predicted satellite clock time, after applying the satellite clock error predictions from the satellite data. DGNSS will significantly remove the satellite transmitter clock error contribution for all GNSS users.

Differential operation can almost completely remove satellite clock and orbit ephemeris errors. For these quantities, the quality of correction has little dependence on the separation of the reference and roving receivers. The degree of correction that can be achieved for ionosphere and troposphere delays is excellent when the two receivers are in close proximity, such that the error terms are the same (i.e. do not decorrelate), for example, up to 20 km or so. At larger separations, the ionosphere and troposphere propagation delays to the receivers become less correlated, and residual errors after correction are correspondingly larger. Nonetheless, substantial corrections can often be made with receiver separations as large as 100–200 km. The amount of error mitigation needs to be compared with the performance requirements for the DGNSS applications. DGNSS is ineffective against noncommon errors such as multipath and receiver noise because these errors are strictly local to each of the receivers.

6.1.5.1 Corrections Formats

In the broadest sense, there are several ways that differential corrections can be made and formatted for use. In a *solution-domain* approach, the reference station computes the position error that results from pseudorange measurements to a set of satellites, and this is applied as a correction to the user's computed position; a significant drawback to the solution-domain approach is that the user and reference station must use exactly the same set of satellites if the position correction is to be valid. Thus, the position domain DGNSS approach is not popular.

In the *measurement domain*, corrections are determined for pseudorange measurements to each satellite in view of the reference receiver, and the user simply applies the corrections corresponding to the satellites the roving receiver is tracking. Reference to a "lumped correction" means all of the error corrections are together in a single correction. Another method, which is still measurement based, decomposes error terms into individual terms (i.e. one for the ionosphere and one for orbit errors). Ground-based augmentation system (GBAS) and precise point positioning (PPP) DGNSS solution approaches tend to favor decomposition of error source terms.

In the majority of cases, it is important that corrections be applied as soon as the user has enough measurements to obtain a user solution, and for the corresponding ephemeris data set. Issuance of data (IOD) parameters can be used to ensure the correction data "matches" up with the correct basis in which the corrections were formed. When the user needs to know its corrected position in real time, current corrections can be transmitted from the reference receiver to the user via a data delivery method that may include a terrestrial or satellite link. The users can then receive, verify, and use the corrections in the user solution calculations. This capability requires a user receiver input port for receiving and using differential correction messages. While a user unique format could be used, standardized formats of these messages have been established by the radio technical commission for maritime service (RTCM) Special Committee 104 (SC-104). Various versions of the RTCM SC-104 standard have been used over the years. Earlier versions tended to concentrate on robust pseudorange correction formats (e.g. Version 2.3), while later versions have emphasized carrier phase corrections (e.g. Version 3.0) and real-time delivery methods such as Networked Transport of RTCM via Internet Protocol (Ntrip) [7].

In some applications, such as surveying or non-real-time research truth reference systems, it may not be necessary to obtain differentially corrected position solutions in real time. In these applications, it is common practice to obtain corrected positions at a later time by bringing together recorded data from both receivers.

6.1.6 Aiding Inputs

Although various GNSS receivers can operate in a stand-alone system, navigation accuracy, coverage, and/or system availability can be materially improved if additional information supplements the GNSS receiver to aid in acquiring and/or tracking the received GNSS signals. Basic GNSS receiver aiding sources include the following:

1. *Inertial navigation system (INS) aiding*. Although GNSS navigation is potentially very accurate, periods of poor signal availability, jamming, and high-dynamics platform operations often limit its capability. INSs are relatively immune to these situations and thus offer powerful leverage in performance under these conditions. On the other hand, the fundamental limitation of INS long-term drift is overcome by the inherent calibration capability provided by a GNSS. Incorporation of INS measurements is readily achieved through Kalman filtering.

2. *Aiding with additional navigation inputs.* Kalman filtering can also use additional measurement data from navigation systems, such as vehicular wheel sensors and magnetic compasses, to improve navigation accuracy and reliability.
3. *Altimeter aiding.* A fundamental property of GNSS satellite geometry typically causes the greatest error in the user solution to be in the vertical direction. Vertical error can be reduced by inputs from an absolute barometric pressure altitude sensor; however, a more common integration method is to relate the vertical solution height to a height reference, such as the ground using a barometric, radar, or laser altimeter sensor.
4. *User clock aiding.* An external clock with high stability and accuracy can be used by the user equipment to improve the user solution performance, but often only practical for stationary reference or time reference receiver applications. It can be continuously calibrated when enough satellite signals are available to obtain precise GPS time. During periods of poor satellite visibility, it can be used to reduce the number of satellites needed for user solution determination.
5. *Assisted GPS (A-GPS).* An assistance technique that has been applied to the indoor cellular market is A-GPS that provides a mobile station (MS), that is, a handset with assistance data from the cellular base station (BS). A-GPS data broadcast from a BS to an MS via the cellular network can provide the MS with the GPS broadcast navigation data and GPS system time to aid the receiver in initial GPS signal acquisition. These data allow the receiver to remove the space vehicle (SV) position, Doppler, 50 bps navigation data, and GPS system time uncertainly to expedite the initial signal acquisition. This is very useful in low SNR indoor environments [8].

6.2 Receiver Architecture

Although there are many variations in GNSS receiver design, all receivers must perform certain basic functions. Figure 6.1 illustrates a generic block diagram illustrating these basic functions that are performed by GNSS receivers. The GNSS antenna was discussed in detail in the previous Chapter 5. As depicted in Figure 6.1, the GNSS antenna is illustrated as a passive device. We will now discuss the main GNSS receiver functions in further detail.

6.2.1 Radio Frequency (RF) Front End

The purpose of the receiver radio frequency (RF) front end is to filter, amplify, and typically down-convert the incoming GNSS signal to an intermediate frequency (IF) signal that can be processed. Figure 6.2 illustrates a more detailed depiction of the RF front end and IF signal conditioning circuit. For

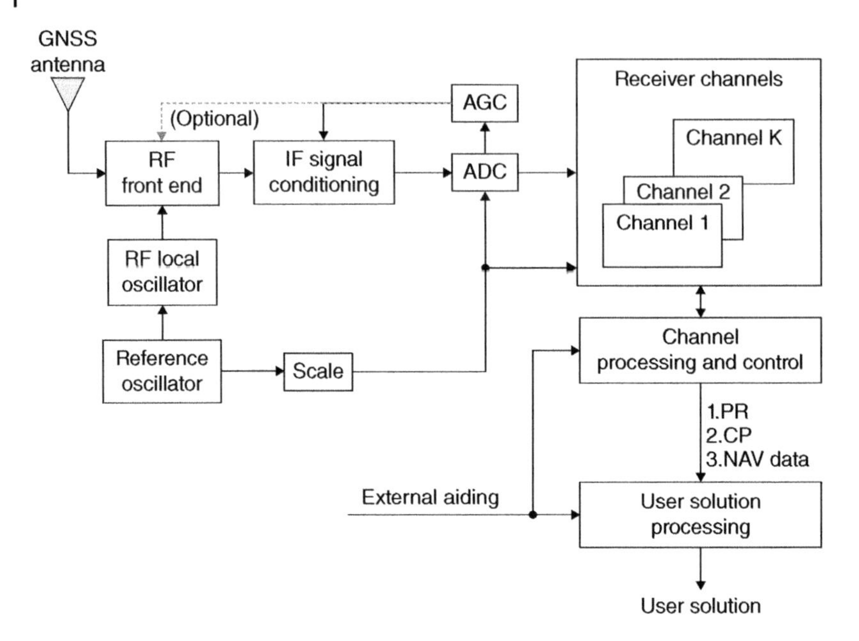

Figure 6.1 Generic block diagram of GNSS receiver. PR: pseudorange; CP: carrier phase; NAV: navigation.

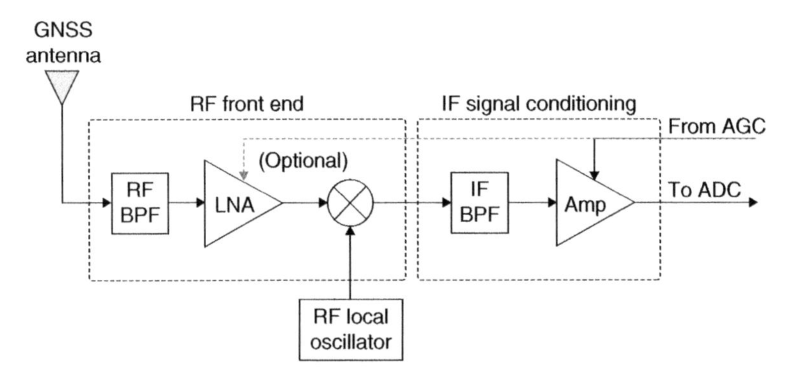

Figure 6.2 Generic GNSS receiver RF front end and IF signal conditioning circuit.

high-quality RF front-end circuits, an RF bandpass filter (BPF) is often placed directly after the passive antenna terminals. This RF passive BPF can be used to reduce out-of-band interference without degradation of the GPS signal waveform. Generally, the bandwidth of a BPF should be sufficient to pass, largely undistorted, the desired signal and have a sharp-cutoff out-of-band for signal rejection. However, the small ratio of passband width to carrier frequency makes the design of such filters unpractical for most GNSS receivers (and even undesirable at this stage of the receiver). Consequently, filters with wider skirts are commonly used as a first stage of filtering, which helps prevent front-end overloading by strong out-of-band interference sources, and the sharp-cutoff filters are used later after down-conversion to an IF. Furthermore, this RF BPF should be low loss (e.g. less than 1 dB or so) to keep the overall noise figure down and functions to provide two benefits: First, the RF BPF prior to the amplifier will attenuate the image frequency that can be directly down-converted to the IF frequency [9]. Second, the RF BPF will help attenuate any strong out-of-band interference from entering the low-noise amplifier (LNA) and will help prevent it from saturation.

As was pointed out earlier, the GNSS signal power available at the receiver antenna output terminals is extremely small and can easily be masked by interference from more powerful signals adjacent to the GNSS passband. To make the signal usable for digital processing at a later stage, RF amplification in the receiver front end typically provides as much as 35–55 dB of gain. The first-stage active amplifier (i.e. LNA) should have the highest gain and lowest noise to help minimize the overall receiver chain noise figure; however, not all the gain needs to be provided at the RF stage [10].

6.2.2 Frequency Down-Conversion and IF Amplification

After amplification in the GNSS receiver front end, the GNSS signal is typically converted to a lower IF, for further amplification and filtering. Down-conversion accomplishes several objectives:

1. The total amount of signal amplification needed by the receiver exceeds the amount that can be performed efficiently in the receiver RF front end at the GNSS carrier frequency. Excessive amplification at a particular stage in the receiver can result in parasitic feedback oscillation, which is difficult to control. In addition, since sharp-cutoff filters with a GPS signal bandwidth are not practical at the L-band, excessive front-end gain makes the end-stage amplifiers vulnerable to overloading by strong nearby out-of-band signals. By providing additional amplification at an IF different from the received signal frequency, a large amount of gain can be realized without the tendency toward oscillation.
2. The amplifiers used at either a first or second IF frequency are typically lower cost than an RF amplifier, which is especially true for the first RF LNA.
3. By converting the signal to a lower frequency, the signal bandwidth is unaffected, and the increased ratio of bandwidth to center frequency permits the design of sharp-cutoff BPFs. These filters can be placed ahead of the IF amplifiers to prevent saturation by strong out-of-band signals. The filtering is often by means of a high-order filter or surface acoustic wave (SAW) devices.

4. Conversion of the signal to a lower frequency makes the sampling of the signal required for digital processing much more feasible for small low-cost and low-power consumption applications.

Down-conversion is accomplished by multiplying the GNSS signal by a sinusoid called the *local oscillator signal* in a device called a *mixer*. The local oscillator frequency is either larger or smaller than the GNSS carrier frequency by an amount equal to the IF stage. Typically, the IF signal is selected to be the difference between the signal and local oscillator frequencies. The sum frequency components are also produced, but this is eliminated by a simple BPF following the mixer. An incoming signal either above or below the local oscillator frequency by an amount equal to the IF will produce an IF signal, but only one of the two signals is desired. The other signal, called the image, can be eliminated by bandpass filtering of the desired signal prior to down-conversion. However, since the frequency separation of the desired and image signals is twice the IF, the filtering becomes difficult if a single down-conversion to a low IF is attempted. For this reason, down-conversion is often accomplished in more than one stage, with a relatively high first IF (30–100 MHz) to permit image rejection. Either a single stage of mixing (illustrated in Figure 6.2) or double stage of mixing (i.e. super heterodyne) design can be utilized.

Whether it is single stage or multistage, down-conversion typically provides a final IF that is low enough to be digitally sampled at feasible sampling rates. In low-cost receivers, typical final IFs range from 4 to 20 MHz with bandwidths that have been filtered down to pass a substantial part of the GNSS code bandwidth (e.g. 2 MHz or so for a GPS C/A-code). This permits a relatively low digital sampling rate and at the same time keeps the lower edge of the signal spectrum well above 0 Hz to prevent spectral foldover. However, for adequate image rejection, either multistage down-conversion or a special single-stage image rejection mixer is required. In more advanced receivers there is a trend toward single conversion to a signal at a relatively high IF (30–100 MHz) because advances in technology permit sampling and digitizing even at these high frequencies.

Additionally, the theory of bandpass sampling has been applied to GNSS receiver designs. This theory allows for sampling frequency not to be a minimum of two times the carrier (or IF) frequency, but selected to be at least two times the bandwidth of the bandpass signal, with constraints based on the carrier frequency (or IF) and actual sampling frequency selected [11]. With regard to GNSS signal processing, the GNSS signal can be converted to an IF and then the signal is bandpass sampled to alias down the desired baseband signal band [12].

6.2.2.1 SNR

An important aspect of receiver design is the calculation of signal quality as measured by the SNR in the receiver IF bandwidth. Typical IF bandwidths range from about 2 MHz in low-cost receivers to the full GPS signal bandwidth on the order of 20 MHz in high-end units, and the dominant type of noise is the thermal noise in the first RF amplifier stage of the receiver front end (or the antenna preamplifier if it is used). The noise power in this bandwidth is given by Eq. (6.1):

$$N = kT_eB \tag{6.1}$$

where $k = 1.3806 \times 10^{-23}$ J/K, B is the bandwidth in hertz, and T_e is the effective noise temperature in Kelvin. The effective noise temperature is a function of sky noise, antenna noise temperature, line losses, receiver noise temperature, and ambient temperature. A typical effective noise temperature for a GNSS receiver is 513 K, resulting in a noise power of about −138.5 dBW in a 2 MHz bandwidth and −128.5 dBW in a 20 MHz bandwidth. The SNR is defined as the ratio of signal power to noise power in the IF bandwidth or the difference of these powers when expressed in decibels. Assuming a nominal GNSS signal power level of −160.0 dBW, the SNR in a 20 MHz bandwidth is seen to be −160.0 − (−128.5) = −31.5 dB. About 90% of the C/A-code power lies in a 2 MHz bandwidth, so there is only about 0.5 dB loss in signal power. Consequently the SNR in a 2 MHz bandwidth is (−160.0 − 0.5) − (−138.5) = −21.5 dB. In either case, it is evident that the signal is completely masked by noise. Further processing to elevate the signal above the noise will be discussed subsequently.

6.2.3 Analog-to-Digital Conversion and Automatic Gain Control

In modern GNSS receivers, digital signal processing is used to track the GNSS signal; make pseudorange, Doppler, and carrier phase measurements; and demodulate the navigation data bit stream. For this purpose, the signal is sampled and digitized by an analog-to-digital converter (ADC) to be processed digitally. In most receivers, the final IF signal is sampled, but in some, the final IF signal is converted down to an analog baseband signal prior to sampling. The sampling rate must be chosen so that there is no spectral aliasing of the sampled signal; this generally will be several times the final IF bandwidth (2–20 MHz).

Most low-cost receivers use 1-bit quantization of the digitized samples, which not only is a very-low-cost method of analog-to-digital conversion but also has the additional advantage that its performance is insensitive to changes in voltage levels. Thus, the receiver needs no automatic gain control (AGC). At first glance, it would appear that 1-bit quantization would introduce severe signal distortion. However, the noise, which is Gaussian and typically much greater than the signal at this stage, introduces a dithering effect that, when statistically averaged, results in an essentially linear signal component. One-bit quantization does introduce some loss in SNR, typically about 2 dB, but in low-cost receivers, this is an acceptable tradeoff. A major disadvantage of

1-bit quantization is that it exhibits a capture effect in the presence of strong interfering signals and is therefore quite susceptible to jamming.

Typical high-end receivers use anywhere from 1.5-bit (three-level) to 3-bit (eight-level) sample quantization. Three-bit quantization essentially eliminates the SNR degradation found in 1-bit quantization and materially improves performance in the presence of jamming signals. However, to gain the advantages of multibit quantization, the ADC input signal level must exactly match the ADC dynamic range. Thus, the receiver must have AGC to keep the ADC input level constant. Some military receivers use even more than 3-bit quantization to extend the dynamic range so that jamming signals are less likely to saturate the ADC. Additionally, there are some instrumentation-grade software-defined GNSS receivers that use a high number of sampling bits (i.e. 14) to capture and analyze anomalous GNSS signal events and interference [13].

6.2.4 Baseband Signal Processing

Baseband signal processing refers to a collection of high-speed algorithms implemented in dedicated hardware and controlled by software that acquire and track the GNSS signal, provide measurements of code phase and carrier phase for pseudorange and carrier phase measurements, and extract the navigation data. These baseband signal processing functions are performed by the receiver channels and associated channel processing and control functions depicted in Figure 6.1. The functionality of the code and carrier signal acquisition and tracking functions as well as the measurements that they produce will be presented.

6.3 Signal Acquisition and Tracking

When a GNSS receiver is turned on, a sequence of operations must ensue before information in a GNSS signal can be accessed and used to provide a user solution. These operations are typically as follows:

1. Hypothesize about the user location.
2. Hypothesize about which GNSS satellites are visible to the antenna.
3. Estimating the approximate Doppler frequency of each visible satellite.
4. Searching for the signal both in frequency and code phase.
5. Detecting the presence of a signal code and carrier and confirming detection.
6. Tracking the code phase.
7. Tracking the carrier phase.
8. Performing data bit synchronization.
9. Decoding the navigation data.

In many GNSS receiver applications, it is desirable to minimize the time from when the receiver is first turned on until the first user solution is obtained. This time interval is commonly called *time to first fix* (TTFF). Depending on receiver characteristics, the TTFF might range from several seconds to several minutes, depending upon the amount and quality of the information the GNSS receivers has when it begins its searching process. Often the initial conditions of a GNSS receiver are referred to as a "cold start," "warm start," or "hot start." While these terms are subjective and vary, generally, a cold start means that the GNSS receiver has no information pertaining to its location (or previous location), no time, and no almanac data to aid in initial acquisition. A warm start refers to when the receiver may have some of the data items previously mentioned; for example, it may have a reasonably accurate estimate of location (where it was previously turned off and was not moved substantially) and recent GNSS satellite almanac (i.e. from the previous day). A hot start is when the receiver has an accurate initial estimate of its location, an accurate estimate of GNSS system time, and accurate GNSS almanac (or even good GNSS ephemeris); this can occur if the receiver had a momentary outage, received assistance data via A-GPS, or other network providers.

6.3.1 Hypothesize About the User Location

Most GNSS receivers, when turned off, will store their location, so that upon power up, they can use the last location, along with other data to determine parameters to help it acquire the code and carrier phase of the various GNSS satellites. This location is typically stored in memory, often supported with a battery. Receivers that have not stored their last location or have lost their last location in memory will often begin to search for satellites at some fixed location, for example, the receiver manufacture location.

In GNSS receivers that have been moved a substantial distance (e.g. to a different continent) and begin the search based on their last stored location, searching can be substantially long as the receiver is searching for satellites that may be on the other side of the Earth. In these cases, it is often best to load the user location into the receiver or to clear the almanac and allow the receiver to begin a new acquisition process.

6.3.2 Hypothesize About Which GNSS Satellites Are Visible

An important consideration in minimizing the TTFF is to avoid a fruitless search for those satellite signals that may be blocked by the Earth, are unhealthy, or are unavailable. A receiver can restrict its search to only those satellites that are visible if it knows its approximate location (even within several hundred miles) and approximate time (within approximately 10 minutes) and has satellite almanac data obtained within the last several months. The approximate location can be manually entered by the user or it can be the position obtained by the receiver when it was last in operation. The approximate time can also be entered manually, but most receivers have a sufficiently accurate real-time clock that operates continuously, even when the receiver is off.

Using the approximate time, approximate position, and almanac data, the receiver calculates the elevation angle of each satellite and identifies the visible satellites as those whose elevation angle is greater than a specified value, called the *mask angle*, which has typical values of 5–20°. At elevation angles below the mask angle, atmospheric and multipath delays tend to make the signals unreliable.

Most receivers automatically update the almanac data when in use, but if the receiver is just "out of the box," or has not been used for many months, it will need to search "blind" (i.e. cold start) for a satellite signal to collect the needed almanac. In this case, the receiver will not know which satellites are visible, so it simply must work its way down a predetermined list of satellites until a signal is found. Although such a "blind" search may take an appreciable length of time, it is infrequently needed.

6.3.3 Signal Doppler Estimation

The TTFF can be further reduced if the approximate Doppler shifts of the visible satellite signals are known. This permits the receiver to establish a frequency search pattern in which the most likely frequencies of reception are searched first. The expected Doppler shifts can be calculated from knowledge of approximate user position, approximate time, and position estimates of the satellites using valid almanac data. The greatest benefit is obtained if the receiver has a reasonably accurate clock reference oscillator.

However, once the first satellite signal is found, a fairly good estimate of receiver clock frequency error can be determined by comparing the predicted Doppler shift with the measured Doppler shift. This error can then be subtracted out while searching in frequency for the remaining satellites, thus significantly reducing the range of frequencies that need to be searched.

6.3.4 Search for Signal in Frequency and Code Phase

There are various techniques available and used in GNSS receivers to acquire and track the GNSS signal carrier phase and code phase. Traditional techniques involve sequentially searching for the desired GNSS signal in frequency and code delay. While some techniques are more popular than others, some amount of searching and confirmation is required. The traditional searching techniques will be emphasized here. Since GNSS signals are radio signals, one might assume that they could be received simply by setting a dial to a particular frequency, as is done with AM and FM broadcast band receivers. Unfortunately, this is not the case.

GNSS signals are *spread-spectrum* signals in which the codes spread the total signal power over a wide bandwidth. The signals are therefore virtually undetectable unless they are *despread or correlated* with a replica code in the receiver that is precisely aligned with the received code. Since the signal cannot be detected until alignment has been achieved, a search over the possible alignment positions (code phase search) is required.

Almost all current GNSS receivers are multichannel units in which each channel is assigned a satellite pseudorandom noise (PRN) code and carrier frequency, and processing in the channels is carried out simultaneously. Thus, simultaneous searches can be made for all usable satellites when the receiver is turned on. Because the search in each channel consists of sequencing through the possible frequency and code delay steps in time, it is called a *sequential search.* In this case, the expected time required to acquire as many as eight satellites is typically 30–100 seconds, depending on the specific search parameters used.

A relatively narrow postdespreading bandwidth (perhaps 100–1000 Hz) is required to raise the SNR to detectable and/or usable levels. However, because of the high carrier frequencies and large satellite velocities used by various GNSSs, for terrestrial users, the received signals can have large Doppler shifts (as much as ±5 kHz) with the SVs in a medium Earth orbit (MEO), which may vary rapidly (by as much as 1 Hz/s). The observed Doppler shift also varies with location on Earth, so that the received frequency will generally be unknown a priori. Furthermore, the frequency error in typical receiver reference oscillators will typically cause several kilohertz or more of frequency uncertainty at the L-band. Thus, in addition to the code search, there is also the need for a search in frequency. At a given estimated Doppler and user clock error, a *frequency bin* on the order of 500 Hz can be used in the acquisition process.

Therefore, a GPS receiver must conduct a two-dimensional search in order to find each satellite signal, where the dimensions are code delay and carrier frequency uncertainty. A search must be conducted across the delay range of the code for each frequency bin searched. Depending upon the accuracy of the estimated satellite locations, user location, and time, the full code phase, or a limited code phase search, may be performed. A generic method for conducting the search is illustrated in Figure 6.3, where the digital IF signal from the AGC is split and then multiplied by a locally generated version of the carrier (really the IF signal) in quadrature (i.e. where the signal from the ADC is multiplied by a sin function signal to produce an in-phase (I) component, and the other split ADC is multiplied by a cosine function signal to produce a quadrature (Q) component). These I and Q components are then multiplied by delayed replicas of the code and then passed to a signal integrator (i.e. integrate and dump) circuit

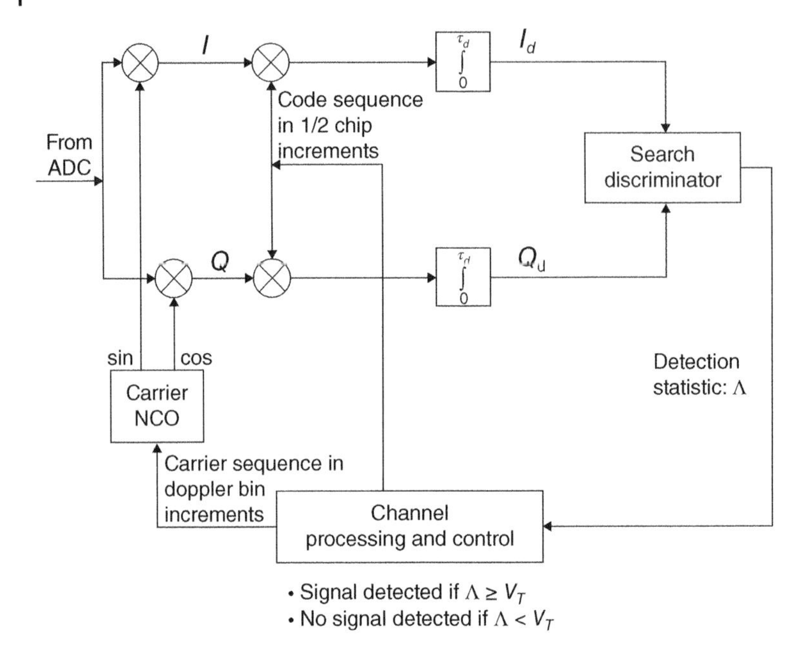

Figure 6.3 Signal search method.

that has a relatively small bandwidth (e.g. 100–1000 Hz where the inverse represents T_d [the detection integration time shown in Figure 6.3]). These I and Q correlation outputs are then fed into a search discriminator. While the search discriminator function may be implemented in various ways, one approach is to provide an output from the search discriminator that is proportional to $I^2 + Q^2$. The output energy of the search discriminator (Λ) serves as a signal detection statistic that can be compared with a threshold (V_T). This output will be significant only if both the selected code delay and frequency translation match that of the received signal. When the energy exceeds a predetermined threshold V_T, a tentative decision is made that a signal is being received synchronously in code phase and frequency, subject to later confirmation. The value chosen for the threshold V_T is a compromise between the conflicting goals of maximizing the probability of detecting (P_D) the signal when it is actually present at a given Doppler and code delay and minimizing the probability of false alarm (P_{FA}) when it is not.

For GNSS codes that have a relatively short period (e.g. GPS C/A-code), the unknown code delay of the signal can be considered to be uniformly distributed over its range so that each delay value is equally likely. Thus, the delays used in the code search can simply sequence from 0 to 1023.5 chips in 0.5-chip increments. For GNSS codes that have a long period, knowledge of GNSS system time, satellite location, estimated user location, and clock uncertainties (satellite and user) can help the receiver reduce the search for the correct code delay.

6.3.4.1 Sequential Searching in Code Delay

For each frequency searched, the receiver generates the PRN code corresponding to the hypothesized satellite signal and moves the delay of this code in discrete steps (typically 0.5 chip) until approximate alignment with the received code (and also a match in Doppler) is indicated when the correlator output energy exceeds the threshold V_T. A step size of 0.5 code chip, which is used by many GPS receivers, is an acceptable compromise between the conflicting requirements of search speed (enhanced by a larger step size) and guaranteeing a code delay that will be located near the peak value of the code correlation function (enhanced by a smaller step size).

An important parameter in the code search is the dwell time used for each code delay position since it influences both the search speed and the detection/false-alarm performance. The dwell time is typically an integral multiple of 1 ms to assure that the correct correlation function, using the full range of code states, is obtained. Satisfactory performance is obtained with dwell times from 1 to 4 ms in most GNSS receivers that have navigation data on the respective channel, but longer dwell times are sometimes used to increase detection capability in weak-signal environments. However, if the dwell time for the search is a substantial fraction of 20 ms (the duration of one typical navigation data bit), it becomes increasingly probable that a bit transition of the 50 Hz data modulation will destroy the coherent processing of the correlator during the search and lead to a missed detection. This imposes a practical limit for a search using coherent detection on a GNSS signal with unknown data. For signal acquisition of a dataless GNSS signal (i.e. pilot signal), no data transmission is present, so integration time is merely limited by the time spent deciding if the signal is present.

The simplest type of code search uses a fixed dwell time, a single detection threshold value V_T, and a simple yes/no binary decision as to the presence of a signal. Many receivers achieve considerable improvement in search speed by using a sequential detection technique in which the overall dwell time is conditioned on a ternary decision involving an upper and a lower detection threshold. Details on this approach can be found in the treatise by Wald [14].

6.3.4.2 Sequential Searching in Frequency

The range of frequency uncertainty that must be searched is a function of the accuracy of the receiver reference oscillator; how well the approximate user position is known, the velocity of the user relative to the satellite. The first step in the search is to use stored almanac data to obtain an estimate of the Doppler shift of the satellite signal. An interval $(f_{\text{lower}}, f_{\text{upper}})$ of frequencies to be searched is then established. The center of the interval is located at $f_c + f_d$, where f_c is the GNSS carrier frequency to be searched and f_d is the estimated carrier Doppler shift. The width of the search interval is made large enough to account for worst-case errors in the receiver reference oscillator, in the estimate of user position, and in the user clock. Without any estimate of the Doppler shift, a typical range for the frequency search interval is $f_c \pm 5$ kHz for a terrestrial user without substantial velocity. If the user has substantial velocity, then this uncertainty should also go into the frequency uncertainty. For space-based receivers, the frequency uncertainty will likely have to be extended depending upon the user satellite orbit and relative velocity to the GNSS satellite. Extending the search window from ±5 to ±20 kHz is often required.

The frequency search is conducted in N discrete frequency steps that cover the entire search interval. The value of N is $(f_{\text{upper}} - f_{\text{lower}})/\Delta f$, where Δf is the spacing between adjacent frequencies (i.e. frequency bin width). The bin width is determined by the effective bandwidth of the correlator. For the coherent processing used in many GPS receivers, the frequency bin width is approximately the reciprocal of the search dwell time. Thus, typical values of Δf are 250–1000 Hz. Assuming a ±5 kHz frequency search range, the N number of frequency steps to cover the entire search interval with a 500 Hz frequency bin would thus be 20 discrete frequency steps.

6.3.4.3 Frequency Search Strategy

Because the received signal frequency is more likely to be near to, rather than far from, the Doppler estimate, the expected time to detect the signal can be minimized by starting the search at the estimated frequency and expanding in an outward direction by alternately selecting frequencies above and below the estimate. Figure 6.4 illustrates a frequency search strategy where the initial frequency offset is estimated to be 1500 Hz and a Doppler bin of 500 Hz is used over 10 steps.

6.3.4.4 Parallel and Hybrid Search Methods

Certain applications demand that the satellites be acquired much more rapidly (perhaps within a few seconds). This can be accomplished by using a *parallel search* technique in which extra hardware permits many frequencies and code delays to be searched at the same time. Still other techniques that take the advantages in semiconductor technology use a substantially large amount of correlation operations to process nearly all possible frequency and code delays simultaneously [5].

Other techniques involve estimating the frequency and code delay using a block of data and Fourier transform methods [15]. These types of techniques take a sample block of digital data and perform fast Fourier transform to estimate the frequency and code phase uncertainty. These uncertainties can then be used to hand over to refined time-domain tracking operations to provide a hybrid acquisition and tracking approach in a GNSS receiver [16].

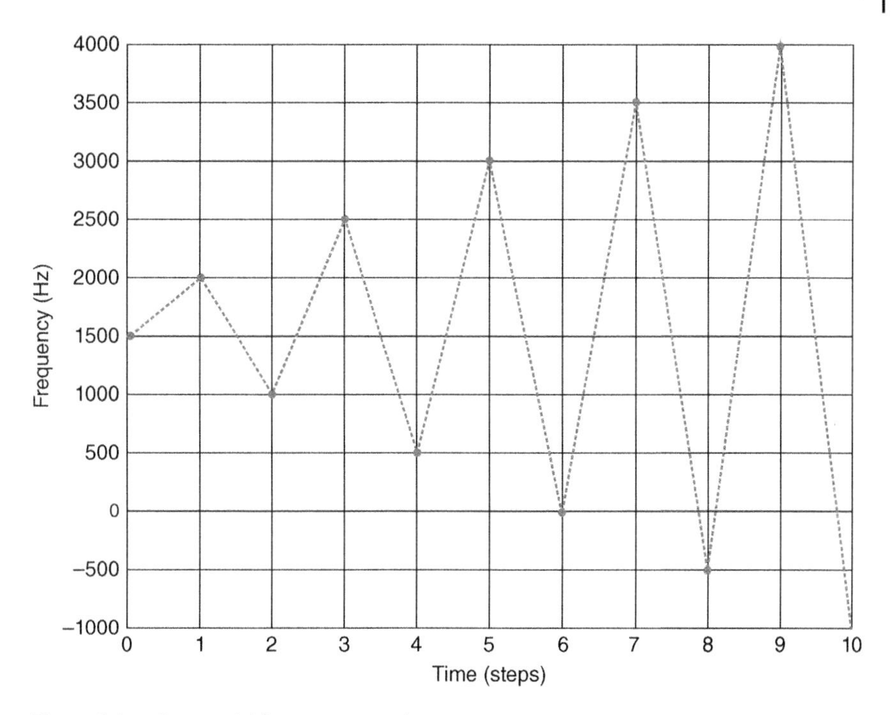

Figure 6.4 Sequential frequency search strategy.

6.3.5 Signal Detection and Confirmation

As previously mentioned, there is a tradeoff between the probability of detection P_D and the probability of false alarm P_{FA}. As the detection threshold V_T is decreased, P_D increases, but P_{FA} also increases, as illustrated in Figure 6.5. Thus, the challenge in receiver design is to achieve a sufficiently large P_D so that a signal will not be missed but at the same time to keep P_{FA} small enough to avoid difficulties with false detections. When a false detection occurs, the receiver will try to lock onto and track a nonexistent signal. By the time the failure to track becomes evident, the receiver will have to initiate a completely new search for the signal. On the other hand, when a detection failure occurs (i.e. a missed detection), the receiver will waste time continuing to search remaining search cells that contain no signal, after which a new search must be initiated.

6.3.5.1 Detection Confirmation

One way to achieve both a large P_D and a small P_{FA} is to increase the dwell time so that the relative noise component of the detection statistic is reduced. However, to reliably acquire weak GNSS signals, the required dwell time may result in unacceptably slow search speed. An effective way around this problem is to use some form of *detection confirmation*.

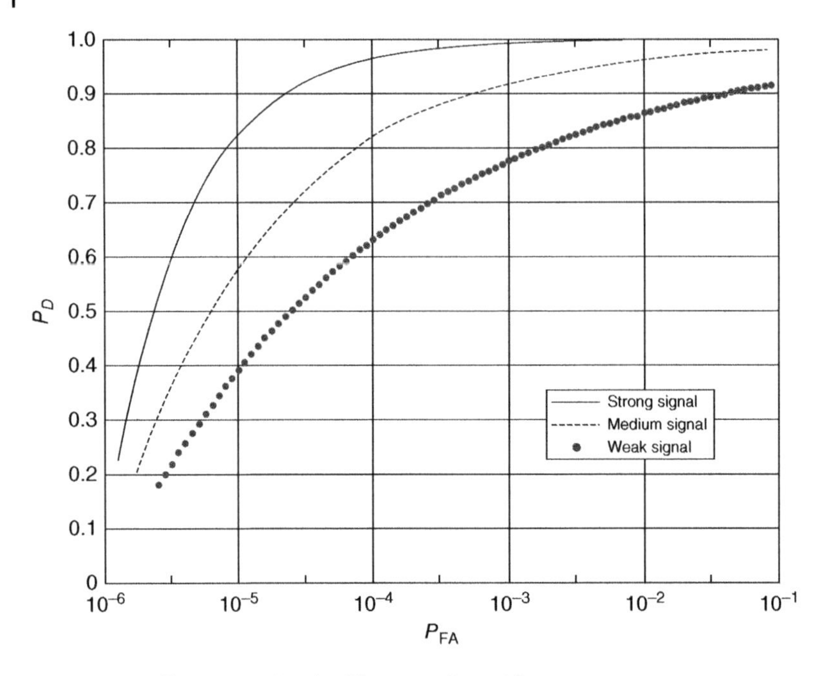

Figure 6.5 Illustration of tradeoff between P_D and P_{FA}.

To illustrate a detection confirmation process, suppose that to obtain the detection probability $P_D = 0.95$ with a typical medium-strength GPS signal, we obtain the false-alarm probability $P_{FA} = 10^{-3}$. (These are typical values for a fixed search dwell time of 3 ms.) This means that on the average, there will be one false detection in every 1000 frequency/code cells searched. Consider the following example for a typical two-dimensional GPS search region that may contain as many as 40 frequency bins and 2046 code delay positions for a GPS C/A-code for a total of $40 \times 2046 = 81\,840$ such cells. Thus, we could expect about 82 false detections in the full search region. Given the implications of a false detection discussed previously, this is clearly unacceptable.

However, suppose that we change the rules for what happens when a detection (false or otherwise) occurs by performing a confirmation of detection before turning the signal over to the tracking loops. Because a false detection takes place only once in 1000 search cells, it is possible to use a much longer dwell (or a sequence of repeated dwell) for purposes of confirmation without markedly increasing the overall search speed, yet the confirmation process will have an extremely high probability of being correct. In the event that confirmation indicates no signal, the search can continue without interruption by the large time delay inherent in detecting the failure to track. In addition to using longer dwell times, the confirmation process can also perform a *local*

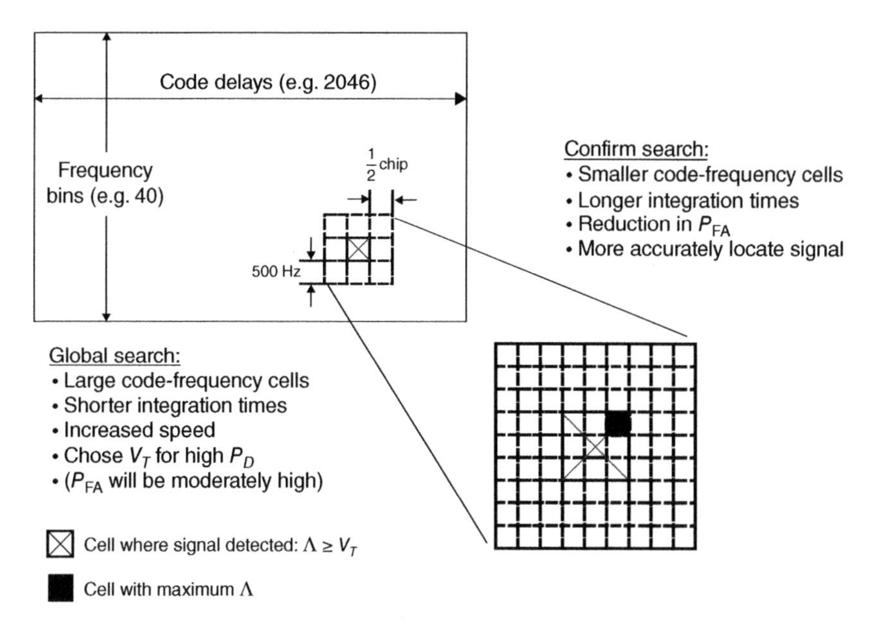

Figure 6.6 Global and confirmation search regions.

search in which the frequency/code cell size is smaller than that of the main, or *global*, search, thus providing a more accurate estimate of signal frequency and code phase when detection is confirmed. Figure 6.6 depicts this process. The global search uses a detection threshold, V_T, which provides a high P_D and a moderate value of P_{FA}. Whenever the detection statistic Λ exceeds V_T at a frequency/delay cell, a confirmation search is performed in a local region surrounding that cell. The local region is subdivided into smaller cells to obtain better frequency delay resolution, and a longer dwell time is used in forming the detection statistic Λ. The longer dwell time makes it possible to use a value of V_T that provides both a high P_D and a low P_{FA}.

Some GNSS receivers use a simple adaptive search in which shorter dwell times are first used to permit rapid acquisition of moderate to strong signals (e.g. high elevation satellites). Whenever a search for a particular satellite is unsuccessful, or it is known that the satellite is at a low elevation angle, it is likely that the signal from that satellite is relatively weak, so the receiver can increase the dwell time and start a new search that is slower but has better performance in acquiring weak signals.

6.3.5.2 Coordination of Frequency Tuning and Code Chipping Rate

As the receiver is tuned in frequency during searching, it is advantageous to advance or retard the code chipping rate of the receiver-generated code so that it is in accordance with the carrier Doppler shift under consideration. The relationship between carrier Doppler shift and the advance or retard of code chipping rate (i.e. precession rate) is given by $p(t) = f_d/N_{cyc}$, where $p(t)$ is the code precession rate in chips per second, f_d is the carrier Doppler shift (Hz), and N_{cyc} represents the number of carrier cycles in one chip of the respective code (e.g. 1540 for GPS L1 C/A-code). Here, a positive precession rate is interpreted as an increase in the chipping rate. Precession is not required while searching because the dwell times are so short. However, when detection of the signal occurs, it is important to match the incoming and reference code rates during the longer time required for detection confirmation and/or initiation of code tracking to take place.

6.3.6 Code Tracking Loop

At the time of detection confirmation, the receiver-generated local reference code will be in approximate alignment with that of the received signal (usually within 0.5 chip), and the reference code chipping rate will be approximately that of the received signal. Additionally, the frequency of the received signal will be known to within the frequency bin width Δf. However, unless further measures are taken, the residual Doppler on the received signal will eventually cause the received and reference codes to drift out of alignment and the signal frequency to drift outside the frequency bin at which detection occurred. If the code alignment error exceeds one chip in magnitude, the incoming signal will no longer despread and will disappear below the noise level. (Additionally, the correlated signal will also decrease in value as the signal frequency rate and locally generated frequency rate become less similar, and will eventually disappear if it drifts outside the detection frequency bin.) Thus, there is the need to continually adjust the timing of the locally generated reference code so that it maintains accurate alignment with the received code, a process called *code tracking*. The process of maintaining accurate frequency tuning to the received signal carrier, called *carrier tracking*, is also necessary and will be discussed in following sections.

Code tracking is initiated as soon as signal detection is confirmed, and the goal is to make the receiver's locally generated code line up with incoming code as precisely as possible. There are two objectives in maintaining alignment:

1. *Code correlation (i.e. signal despreading).* The first objective is to fully despread the received signal so that it is no longer below the noise and so that information contained in the carrier and the underlying navigation data can be recovered.
2. *Pseudorange measurements.* The second objective is to enable precise measurement of the time of arrival (TOA) of the received code for purposes of producing a pseudorange measurement. Such measurements cannot be made directly from the received signal since it is below the noise level. Therefore, a code tracking loop, which has a large processing gain, is employed

to generate a reference code precisely aligned with that of the received signal. This enables pseudorange measurements to be made using the reference code instead of the much noisier received signal code waveform.

Figure 6.7 illustrates a generic code tracking loop within the receiver channels shown in Figure 6.1. The digitized IF signal, which has a wide bandwidth due to the spreading code modulation, is completely obscured by noise. The signal power is raised above the noise power by *despreading (or code correlation)*, where the digitized IF signal is multiplied by a receiver-generated replica of the code precisely time-aligned with the code of the received signal. The code tracking loop shown in Figure 6.7 works together with the carrier tracking loop to iteratively acquire and track the received GNSS signal. The carrier numerically controlled oscillator (NCO) is typically multiplied by the split received digital IF signal, as is done in the carrier tracking loop shown in Figure 6.3, to produce individual baseband I and Q signals. These I and Q signals are then typically multiplied by a delayed version of the locally generated spreading code that is controlled by the code NCO. The code tracking loop in Figure 6.7 is known as a delay-lock loop (DLL) whereby early (E), punctual (P), and late (L) versions of the locally generated spreading code are produced to be used in the correlation process. In typical GNSS receivers, the early and late codes, respectively,

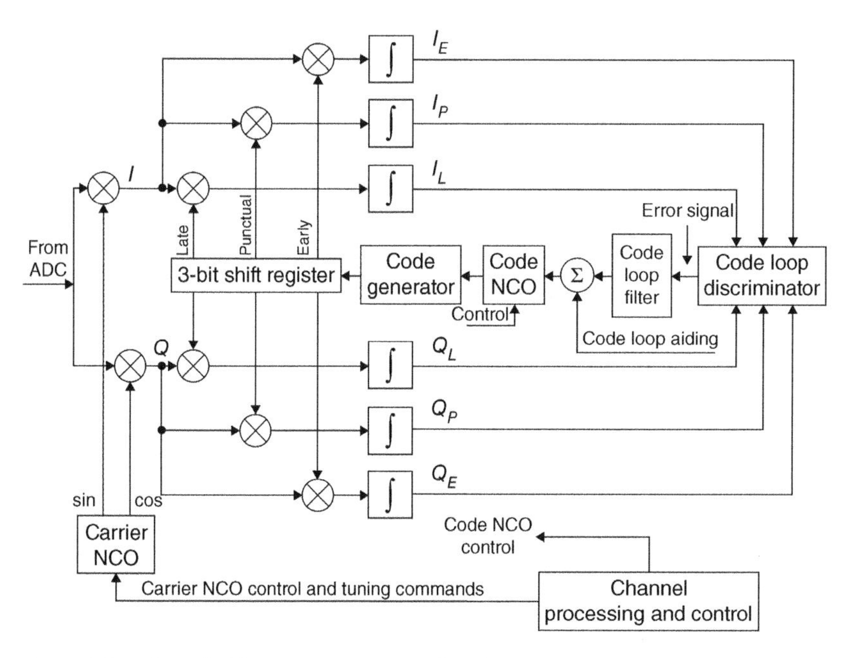

Figure 6.7 Generic GNSS receiver code tracking loop.

lead and lag the punctual code by 0.05–1.0 code chips and maintain these relative positions during the code tracking process. These versions are used to produce delayed versions of the correlated signal for the purpose of code tracking. (Strictly speaking, the code tracking loop only tracks the code phase, and the carrier tracking loop tracks the carrier phase and performs data recovery, which will be discussed later in Sections 6.3.6 and 6.3.7, but there are significant functional components that are the same.)

The mixer and integrator in each correlator channel are performed in parallel. The output magnitude of each correlator is proportional to the cross correlation of its received and reference codes, where the cross-correlation function has the triangular-shaped function, with its peak occurring when the two codes are aligned. The integration process that is performed prior to the code loop discriminator is controlled by a predetection integration interval (PDI) that determines the integration time (i.e. the number of samples to integrate based on the sample rate). The amount of time the samples can be integrated will depend upon the duration of a navigation data bit, if navigation data bit synchronization has occurred, if the navigation data has been removed, or if the navigation data is not present for a dataless GNSS signal (i.e. pilot signal). If the navigation data has been synchronized and removed, often referred to as "data wipe-off," then the predetection integration times can exceed the data bit boundaries and enable better acquisition and tracking performance in low SNR environments. If the data transitions have not been removed, or are still unknown, data removal can be done within the code loop discriminator functional block by squaring, taking the absolute value or extracting the envelope of each of the correlation functions (i.e. output from each of the integrators).

When attempting to acquire a GNSS signal that is encoded with navigation date and the code tracking loop is first turned on, the integration time T for the correlators is usually no more than a few milliseconds, in order to minimize corruption of the correlation process by data bit transitions of the navigation data bit stream whose locations in time are not yet known. However, after bit synchronization has located the data bit boundaries, the integration interval can span a full data bit (e.g. 20 ms for the GPS C/A-code) in order to achieve a maximum contribution to processing gain. When first attempting to acquire a GNSS signal that is not encoded with navigation, the PDI time T for the correlators can be made very long, depending upon the dynamics of the platform.

Various discriminator functions can be used in the code tracking process. Figure 6.7 depicts a generic code tracking loop that can be operated in a noncoherent or coherent fashion. The noncoherent DLL utilizes the early and late correlator channels to form a discriminator function that produces an error signal, $e_c(\tau)$, which can be used as a basis to "speed up" or "slow down" the locally generated codes so that the locally generated punctual code is driven to align with the incoming signal. A common early minus late noncoherent discriminator function is shown in Eq. (6.2) to produce a code delay error signal:

$$e_c(\tau) = \frac{(I_E^2 + Q_E^2) - (I_L^2 + Q_L^2)}{(I_E^2 + Q_E^2) + (I_L^2 + Q_L^2)} \tag{6.2}$$

The discriminator function in Eq. (6.2) has been normalized by the power in the early and late channels. Normalizing the discriminator function is useful to help minimize the error signal variations that may result in SNR variations and if the receiver is to implement dynamic discriminator functions.

Figure 6.8 illustrates a typical open-loop error signal produced by a one-chip spaced early minus late DLL verse delay (τ) between the locally generated code and the incoming received code. The error signal is slightly rounded due to filtering with its stable operating point along the linear portion of the function close to 0 delay. When the code loop eventually locks onto the received code, the stable operating point will be between the two peaks on the nearly linear slope of the discriminator function output (i.e. this error signal).

Alignment of the locally generated punctual code with the received code is maintained by using the error signal to advance or delay the reference code generator rate using the discriminator error output. For example, depending upon the polarity of the slope (i.e. sensitivity) of the code NCO, when the error signal is positive, the reference code rate will speed up, and when the error signal is negative, the reference code rate will slow down. Since $e_c(\tau)$ is generally quite noisy, it is sent through a low-pass *loop filter* before it controls the clock rate of the code NCO, which drives the local reference code generator, as indicated in Figure 6.7. The bandwidth of this filter is usually quite small, resulting in a closed-loop bandwidth typically less than 1 Hz. The bandwidth of this code

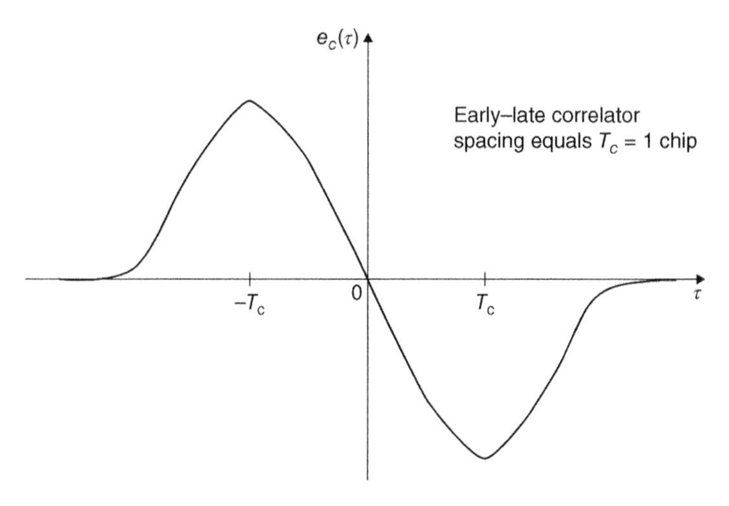

Figure 6.8 Code tracking loop error signal (open loop).

loop filter can be varied based on the expected dynamics of the user platform to ensure code tracking is not lost under high dynamics while minimizing noise in the code tracking process.

6.3.6.1 Code Loop Bandwidth Considerations

The bandwidth of the code tracking loop is determined primarily by the loop filter and needs to be narrow for best ranging accuracy but wide enough to avoid loss of lock if the receiver is subject to large accelerations that can suddenly change the apparent chipping rate of the received code. Excessive accelerations cause loss of lock by moving the received and reference codes too far out of alignment before the loop can adequately respond. Once the alignment error exceeds approximately one code chip, the loop loses lock because it no longer has the ability to form the proper error signal.

In low-dynamics applications with lower-cost receivers, code tracking loop bandwidths on the order of 1 Hz permit acceptable performance in handheld units and in receivers with moderate dynamics (e.g. in automobiles). For high-dynamics applications, such as missile platforms, loop bandwidths might be on the order of 10 Hz or larger. In surveying applications, which have no appreciable dynamics, loop bandwidths can be as small as 0.01 Hz to obtain the required ranging accuracy. Both tracking accuracy and the ability to handle dynamics are greatly enhanced by means of *carrier aiding* from the receiver's carrier phase tracking loop, which will be discussed subsequently.

6.3.6.2 Coherent Versus Noncoherent Code Tracking

In general, the outputs of each of the correlator channels shown in Figure 6.7 are complex (i.e. have amplitude and phase). A noncoherent code track look does not explicitly need to know the incoming carrier phase since the energies in I and Q channels are combined. Noncoherent code tracking loops will often remove the phase transitions by squaring, taking the absolute value, or rectifying the correlator outputs. Phase transitions can come about because of navigation data that are modulated within the received GNSS signal, and the data bit boundaries have not yet been determined. A distinguishing feature of a noncoherent code tracking loop is its insensitivity to the phase of the received signal. Insensitivity to phase is desirable when the loop is first turned on since, at that time, the signal phase is random and not yet under any control.

However, once the phase of the signal is being tracked, a *coherent* code tracker can be employed, in which the outputs of the early and late correlators are purely real, and all of the signal power will be concentrated, or synchronized, with the I channel. In this situation, the loop error signal can be formed directly from the difference of the early and late squared magnitudes from only the I correlators. By avoiding the noise in the Q correlator outputs, a 3 dB SNR advantage is thereby gained in tracking the code.

With the code loop tracking the incoming signal, the locally generated punctual code is synchronized to the incoming coded signal. Once this occurs, not only can the Q channel not be used, but the energy in the P cannel can be used to help improve the performance of the code tracking loop. A discriminator function of this kind is called the dot product discriminator function [17, 18].

However, a price is paid in that the code loop error signal becomes sensitive to phase error in tracking the carrier. If phase tracking is ever lost, failure of the code tracking loop could occur (without additional aiding). This is a major disadvantage, especially in mobile applications where the signal can vary rapidly in magnitude and phase. Since noncoherent operation is much more robust in this regard and is still needed when code tracking is initiated, most GNSS receivers will use a noncoherent code tracking loop; however, some will use a hybrid approach and switch from noncoherent to coherent code tracking once synchronization has occurred.

6.3.7 Carrier Phase Tracking Loops

The purposes of tracking the carrier phase of the received GNSS signal are to:

1. Obtain a phase reference for coherent detection of the GNSS phase modulated data.
2. Provide precise velocity measurements (via phase rate).
3. Obtain integrated Doppler for rate aiding of the code tracking loop.
4. Obtain precise carrier phase measurements for use in high-accuracy receivers.

Tracking of carrier phase is usually accomplished by a phase-lock loop (PLL). A Costas-type PLL or its equivalent can be used to prevent loss of phase coherence with the received GNSS signal that has phase modulated navigation data on the GNSS carrier signal. The origin of the Costas PLL is described in Ref. [19]. Figure 6.9 is a block diagram of a generic carrier tracking loop that is contained within each of the receiver channels shown in Figure 6.1.

In Figure 6.9, the output of the receiver IF is converted to a complex baseband signal by multiplying the signal by both the in-phase and quadrature-phase outputs of a carrier NCO. The carrier NCO produces the in-phase (i.e. sin function) and quadrature-phase (i.e. cosine function). The code is removed (assume the code tracking loop is synchronized) with the punctual code from the code tracking loop, and the resulting signal is integrated. For a GNSS signal that has navigation data encoded, the signal is then integrated over each navigation data bit interval to form a sequence of phasors. The phase angle of each phasor is the phase difference between the received signal carrier and the carrier NCO output during the navigation data bit integration interval. For a classical Costas loop, the discriminator function illustrated in Figure 6.9 is a multiplier to produce a carrier loop phase error signal by multiplying together the I_p and Q_p

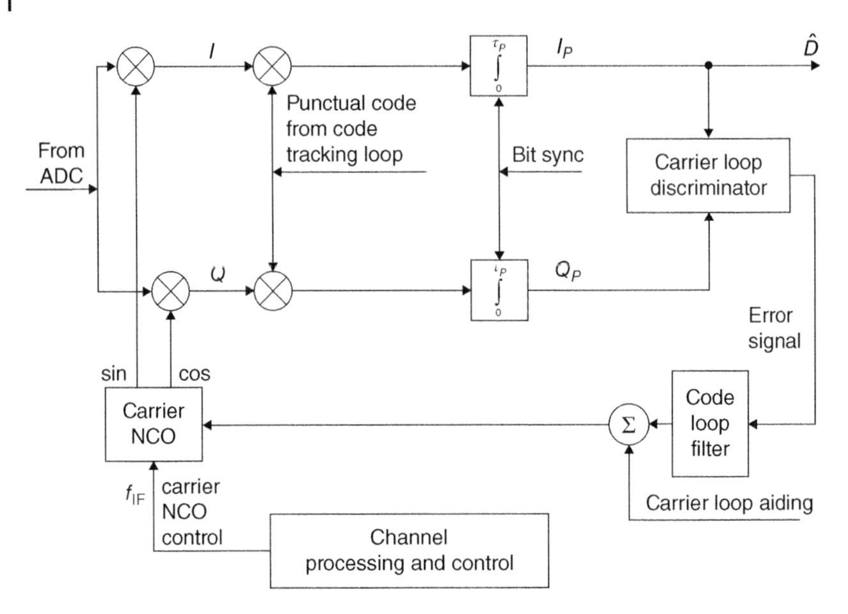

Figure 6.9 Generic GNSS receiver carrier tracking loop.

components of each phasor. This error signal is unaffected by the phase modulated navigation data because the modulation appears on both I_p and Q_p and is removed in forming the $I_p \times Q_p$ product. After passing through a carrier low-pass loop filter, the error signal controls the carrier NCO phase to drive the loop error signal $I_p \times Q_p$ to zero (the phase-locked condition).

Because the Costas loop is unaffected by the data modulation, it will achieve phase lock at two stable points where the NCO output phase differs from that of the signal carrier by either 0° or 180°, respectively. This can be seen by considering $I = A \cos\theta$ and $Q = A \sin\theta$, where A is the phasor amplitude and θ is its phase. Then, the product of the I_p and Q_p channels is shown in Eq. (6.3):

$$\phi_e = I_p \times Q_p = A^2 \cos\theta \sin\theta = \frac{1}{2}A^2 \sin 2\theta \tag{6.3}$$

There are ambiguous values of θ in $(0, 2\pi)$, where the error signal $I_p \times Q_p = 0$. Two of these are the stable points, namely, $\theta = 0°$ and $\theta = 180°$, toward which the loop tends to return if disturbed. Since $\sin 2\theta$ is unchanged by 180° changes in θ caused by the data bits, the data modulation will have no effect. At either of the two stable points, the Q_p integrator output is nominally zero and the I_p integrator output contains the demodulated data stream, but with a polarity ambiguity that can be removed by observing the navigation data frame preamble. Thus, the Costas loop has the additional feature of serving as a data demodulator.

In the Costas loop design shown, the phase of the received signal is measured by comparing the phase of the carrier NCO output with a reference signal. Normally, the reference signal frequency is a rational multiple of the same crystal-controlled oscillator that is used in frequency shifting the GNSS signal down to the last IF. When the carrier NCO is locked to the phase of the incoming signal, the measured phase rate will typically be in the range of ± 5 kHz due to signal Doppler shift. Two types of phase measurements are usually performed on a periodic basis (the period might be every navigation data interval). The first is an accurate measurement of the phase modulo 2π, which is used in precision carrier phase ranging. The second is the number of cycles (including the fractional part) of phase change that have occurred from a defined point in time up to the present time. The latter measurement is often called integrated Doppler and is used for aiding the code tracking loop. By subtracting consecutive integrated Doppler measurements, extremely accurate average frequency measurements can be made, which can be used by the navigation filter to accurately determine user velocity.

Although the Costas loop is not disturbed by the presence of data modulation, at low SNR, its performance degrades considerably from that of a loop designed for a pure carrier. The degradation is due to the noise × noise component of the $I_p \times Q_p$ error signal. Furthermore, the data bit duration of the I_p and Q_p integrations represents a limit to the amount of coherent processing that can be achieved. If it is assumed that the maximum acceptable bit error rate (BER) for the navigation data demodulation is 10^{-5}, GPS signals typically become unusable when C/N_0 falls below about 25 dBHz.

There are various carrier phase discriminator functions that can be implemented to successfully track the carrier phase. While the classical Costas loop discriminator function merely multiplies the punctual I_P and Q_P channel together and applies the error signal to the carrier loop filter, an alternative optimum PLL discriminator function takes the punctual I_P and Q_P channels and implements a full four quadrant arctan function (i.e. atan2), which can be implemented with a memory lookup table; for this phase discriminator, the baseband phase is defined by Eq. (6.4):

$$\phi_e = \text{atan}\,2\,(Q_P, I_P) \tag{6.4}$$

The design bandwidth of the PLL is determined by the SNR, desired tracking accuracy, signal dynamics, and ability to "pull in" when acquiring the signal or when the lock is momentarily lost.

6.3.7.1 PLL Capture Range

An important characteristic of the PLL is the ability to pull in to the frequency of a received signal. When the PLL is first turned on, assuming code acquisition, the difference between the signal carrier frequency and the carrier NCO frequency must be sufficiently small or the PLL will not lock. In typical GNSS applications, the PLL must have a relatively small bandwidth (1–10 Hz) to prevent loss of lock due to noise. However, this results in a small pull-in (or capture) range (perhaps only 3–30 Hz), which would require small (hence many) frequency bins in the signal acquisition search algorithm. Advanced PLL designs or use of a frequency lock loop (FLL) can enhance the PLL capture range performance.

6.3.7.2 PLL Order

The *order* of a PLL refers to the number of integrators within the tracking loop and thereby its capability to track different types of signal dynamics. Most GPS receivers use second- or third-order PLLs. A second-order loop can track a constant rate of phase change (i.e. constant frequency) with zero average phase error and a constant rate of frequency change with a nonzero but constant phase error. A third-order loop can track a constant rate of frequency change with zero average phase error and a constant acceleration of frequency with nonzero but constant phase error. Most receivers typically use a second-order PLL with fairly low bandwidth because the user dynamics are minimal and the rate of change of the signal frequency due to satellite motion is sufficiently low (<1 Hz/s) such that phase tracking error is negligible. On the other hand, receivers designed for high dynamics (i.e. missiles) will sometimes use third-order or even higher-order PLLs to avoid loss of lock due to the large accelerations encountered.

The price paid for using higher-order PLLs is a somewhat lower robust performance in the presence of noise. If independent measurements of platform dynamics are available (such as accelerometer or INS outputs), they can be used to aid the PLL by reducing stress on the loop. This can be advantageous because it often renders the use of higher-order loops unnecessary.

6.3.7.3 Use of Frequency-Lock Loops (FLLs) for Carrier Capture

Some receivers avoid the conflicting demands of the need for a small bandwidth and a large capture range in the PLL by using an FLL to aid the PLL. The capture range of an FLL is typically much larger than that of a PLL, but the FLL cannot lock onto nor track the actual phase of the received signal. Therefore, an FLL is often used to pull the carrier NCO frequency into the capture range of the PLL. A typical FLL design is similar to the PLL shown in Figure 6.9, except a different discriminator function is used, and there is no navigation data output. The FLL generates a loop error signal that is approximately proportional to the rotation rate of the baseband signal phasor and is derived from the vector cross product of successive baseband phasors where a fixed delay is used, typically, 1–5 ms. If we let t_1 correspond to the outputs at a given time, and t_2 correspond to the corresponding outputs 1–5 ms later, then a maximum likelihood (ML) estimator for the carrier FLL is shown in Eq. (6.5), where the error signal here is frequency error [20]:

$$f_e = \frac{\text{atan2}\,(\text{cross, dot})}{t_2 - t_1} \tag{6.5}$$

where

$$\text{cross} = I_{\text{P1}} \times Q_{\text{P2}} - I_{\text{P2}} \times Q_{\text{P1}}$$
$$\text{dot} = I_{\text{P1}} \times I_{\text{P2}} + Q_{\text{P1}} \times Q_{\text{P2}}$$

6.3.8 Bit Synchronization

For GNSS signals that have navigation data encoded, synchronizing to the data bit edges is necessary in order to coherently integrate the punctual channel and to optimally recover the navigation data bits. Before bit synchronization can occur, the PLL must be locked to the GNSS signal. This can be accomplished by initially running a Costas loop with a relatively short integration time (e.g. 1 ms) where each interval of integration is over one period of the code (e.g. GPS C/A-code), starting and ending at the code epoch. Since the navigation data bit transitions can occur at code epochs using phase modulation, there can be no bit transitions while integration is taking place. When the PLL achieves lock, the output of the I_p integrator will change as a function of the navigation data bits. Once the navigation data bit boundaries have been determined, then the output of the punctual integrator can be integrated over the entire number of codes used to encode a single data bit (e.g. 20 for GPS C/A-code).

A simple method of bit synchronization is to clock a modulo counter with the epochs of the receiver-generated reference code and to record the count each time the polarity of the I_p integrator output changes. The modulo of the counter will be the number of code epochs that are used to encode the navigation data. For example, with the GPS C/A-code, 20 C/A-codes are transmitted for each navigation data bit; thus, a modulo 20 counter should be used in the clock recovery process. A histogram of the frequency of each count is constructed, and the count having the highest frequency identifies the epochs that mark the data bit boundaries. (Counts corresponding to lower frequencies will correspond to multiple data bits of the same polarity and are of limited value in the navigation data clock recovery process.)

6.3.9 Data Bit Demodulation

Once navigation data bit synchronization has been achieved, demodulation of the navigation data bits can occur. As previously described, many GNSS receivers demodulate the data by integrating the I_p component of the baseband phasor generated by a Costas loop, which tracks the carrier phase. Each data bit is generated by integrating the I_p component over a number of code epochs used to encode at single navigation data bit (e.g. 20 ms interval for the GPS C/A-code) from one data bit boundary to the next. The Costas loop causes a polarity ambiguity of the data bits that can be resolved by observation of the subframe preamble in the navigation message data.

6.4 Extraction of Information for User Solution

After carrier, code, and data clock synchronization, the navigation data decoding can be performed to extract the navigation information encoded onto the GNSS broadcast. This navigation information will be used by the GNSS receiver to calculate several important parameters for the user solution including the determination of the following:

1. Signal transmission time.
2. Position and velocity of each satellite.
3. Pseudorange measurements.
4. Delta pseudorange measurements.
5. Carrier Doppler measurements.
6. Integrated Doppler measurements.

6.4.1 Signal Transmission Time Information

The GNSS receiver can calculate the time of transmission of the GNSS signal (i.e. the GNSS time that the signal left the satellite) by decoding the GNSS system time encoded onto the broadcast and counting the number of code cycles and chips that have lapsed until the receiver has correlated to the spreading code. In particular, for GPS, the Z-count represents the number of 1.5 s X1 epochs that have elapsed since the beginning of the week. The time of week (TOW), part of the Z-count, is encoded at the beginning of every subframe and represents a GPS system time reference to the beginning of the next subframe. (The receiver will still need to apply the SV transmitter clock error correction to correct the actual SV clock used to this reference time encoded as the TOW count.) After the code and carrier tracking loops have synchronized to the received signal, the receiver will then count the whole number of code epochs, the whole number of code chips, and the fractional number of code chips that have lapsed since that Z-count, from the code NCO (with respect to the punctual code). This count is performed within the GNSS receiver by a function that is commonly referred to as the *code accumulator*. The information that goes into the code accumulator count is obtained from the navigation data and the code NCO. (The part of the code accumulator that counts the whole and fractional code chips since the last code epoch is often referred to as the *code state*, which can be obtained from the code NCO.) Thus, the total number (whole and fractional) of code chips that have lapsed since the Z-count is used to calculate the time of transmission (t_T).

6.4.2 Ephemeris Data for Satellite Position and Velocity

The ephemeris data permit the position and velocity of each satellite to be computed at the signal transmission time. The calculations are outlined in Table 4.2.

6.4.3 Pseudorange Measurements Formulation Using Code Phase

In an ideal system, with no clock, atmospheric, or measurement errors, finding the three-dimensional position of a user would consist of determining the *true range,* that is, the distance of the user from each of three or more satellites having known positions in space, and mathematically solving for a point in space where that set of ranges would occur. The range to each satellite can be determined by measuring how long it takes for the signal to propagate from the satellite to the receiver and multiplying the propagation time by the speed of light.

Unfortunately, this method of computing range would require very accurate synchronization of the satellite and receiver clocks used for the time measurements. GNSS satellites use very accurate and stable atomic clocks that are corrected from the supporting ground control segment, but it is often impractical to provide a comparable clock in a receiver. The problem of user clock synchronization is circumvented in GPS by treating the receiver clock error as an additional unknown in the navigation equations and using measurements from an additional satellite to provide enough equations for a user solution for time as well as for position. Thus, the receiver can use an inexpensive clock to make its measurements. Since the user clock error is common to all of the measurements, it will eventually cancel in the user solution. Such an approach leads to perhaps the most fundamental measurement made by a GNSS receiver, the *pseudorange* measurement from SV i, computed as shown in Eq. (6.6):

$$\rho_i = c(t_R - t_{Ti}) \tag{6.6}$$

where

c = GNSS propagation constant (i.e. speed of light) (m/s)

t_R = time of reception (s)

t_{Ti} = time of transmission from SVi (s)

In Eq. (6.6), t_R is the time at which a specific, identifiable portion of the signal is received; this is often derived from the reference clock at a particular sample time. The t_{Ti} is calculated from the code accumulator, with reference to the GNSS system time encoded in the broadcast (i.e. Z-count for GPS), which is the time that same portion of the signal was transmitted, that is, currently being correlated in the code tracking loop for SV i. The GNSS propagation constant is essentially the speed of light ($2.997\,924\,58 \times 10^8$ m/s), as defined in the GNSS interface specification [21]. It is important to note that t_R is measured according to the receiver clock, which may have a large time error, which is one of the reasons why the code phase measurement produced by the receiver is called a pseudorange rather than a range measurement. Additionally, the raw pseudorange measurement will also contain the transmitter clock error, but this can largely be removed by applying the SV transmitter clock error corrections encoded in the broadcast.

Figure 6.10 shows the pseudorange measurement concept with four GNSS satellites. The raw pseudorange measurements are simultaneous snapshots at time t_R of the states of the received codes from the illustrated four satellites. This is accomplished indirectly by observation of the receiver's locally generated code state from each code tracking loop. The code state is a real number (whole and fractional chips since the last code epoch interval or the GNSS reference time for a long PRN code that has not repeated since the last GNSS reference time). For example, for the GPS C/A-code, the code state will be in the interval (0, 1023), and this will be added to the C/A-code epochs (i.e. multiples of 1 ms C/A-code epochs) within a subframe to get back to the edge of the GPS subframe, where the GPS subframe reference time is encoded as the Z-count (1.5 s X1 epochs). For a long PRN code that does not repeat since the GNSS reference time (e.g. GPS Z-count), the code accumulator (and hence code state) will represent the whole and fractional chips since the GPS reference time. In Figure 6.10, the time of epoch edge to each SV i, (t_{ei}) represents the time at the last code epoch edge (i.e. the beginning of the code state count [integer and fraction]). Thus, the time of transmission can be calculated as shown in Eq. (6.7):

$$t_{Ti} = t_{ei} - \frac{X_{ei}}{R_c} \tag{6.7}$$

where

t_{ei} = time of epoch edge (s)

X_{ei} = code state count (real number)

R_c = rate of code (1/s)

6.4.3.1 Pseudorange Positioning Equations

If pseudorange measurements can be made from at least four GNSS satellites, enough information exists to solve for the unknown GNSS user position (X, Y, Z) and for the receiver clock error C_b (often called the user receiver *clock*

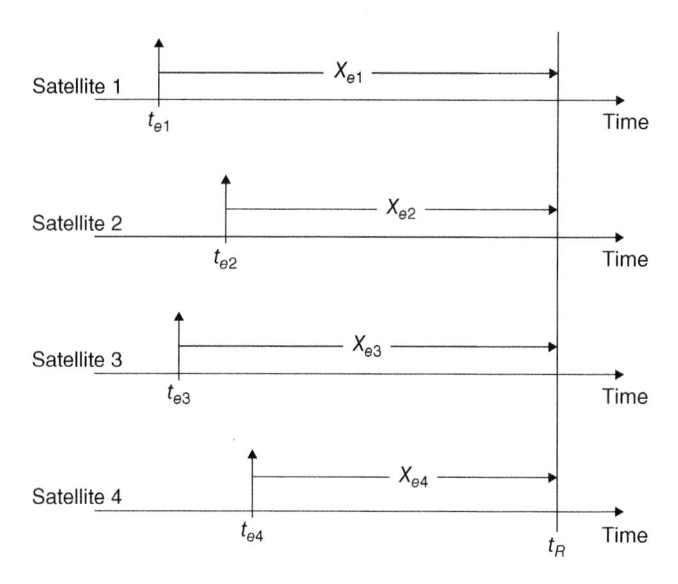

Figure 6.10 Pseudorange measurement concept.

bias), expressed in units of meters. The equations are set up by equating the measured pseudorange (ρ_i) to each satellite i with the corresponding unknown user-to-satellite distance, plus the distance error due to receiver clock bias as shown in Eq. (6.8):

$$\rho_i = \sqrt{(x_i - X)^2 + (y_i - Y)^2 + (z_i - Z)^2} + C_b \ \ (\mathrm{m}) \tag{6.8}$$

where ρ_i denotes the measured pseudorange of the ith satellite whose position in Earth-centered, Earth-fixed (ECEF) coordinates at t_{Ti} is (x_i, y_i, z_i) and $i = 1$, 2, 3 up to n, where $n \geq 4$ is the number of satellites observed. The unknowns in this nonlinear system of equations are the user position (X, Y, Z) in ECEF coordinates and the receiver clock bias C_b.

6.4.4 Measurements Using Carrier Phase

Although pseudorange measurements using the code are the most commonly employed because they provide an unambiguous range measurement, a much higher level of measurement precision can be obtained by measuring the received carrier phase of the GNSS signal. Because the carrier waveform has a very short period (6.35×10^{-10} seconds [or 19.0 cm] at the L1 frequency), the noise-induced error in measuring signal delay by means of phase measurements is typically 10–100 times smaller than that encountered in code delay measurements.

However, carrier phase measurements are highly ambiguous because phase measurements are simply modulo 2π numbers. Without further information, such measurements determine only the fractional part of the pseudorange when measured in carrier wavelengths. Additional processing is required to effectively remove the carrier cycle ambiguity in the carrier phase measurements. This process is often referred to as carrier cycle *ambiguity resolution,* which determines and/or removes ambiguous integer number of wavelengths in the carrier phase measurement. The relation between the ambiguous carrier phase and the unambiguous pseudoranges can be expressed as Eq. (6.9):

$$\rho_i = (\phi_i + N_i)\,\lambda \tag{6.9}$$

where

ρ_i = code phase measurement, that is, pseudorange (m)

ϕ_i = carrier phase measurement (cycles)

N_i = carrier cycle ambiguity (integer)

λ = carrier wavelength (m)

The measurements required for the determination of the N_i carrier cycle ambiguities are most often differential in nature and may be from single or multifrequency GNSS measurements. Since the code measurements are unambiguous, they significantly narrow the range of admissible integer values for the N_i. Differential techniques significantly help mitigate common systematic errors. Additionally, measurements made on two or more GNSS frequencies can help improve the speed and robustness of the carrier cycle ambiguity process. For example, for GPS, the L1 and L2 signals can be used to obtain a virtual carrier frequency equal to the difference of the two carrier frequencies (1575.42 − 1227.60 = 347.82 MHz). This combination is often referred to as the GPS *wide-lane solution.* The 86.3 cm wavelength of this virtual carrier thins out the density of carrier cycle ambiguities by a factor of about 4.5, making the ambiguity resolution process much more robust. Single-frequency, various code and carrier combinations, and triple-frequency techniques can be implemented with various performance advantages in terms of accuracy, speed, integrity, and robustness of the ambiguity solution [2, 22]. A popular technique is the least-squares ambiguity decorrelation adjustment (LAMBDA) method [23]. The LAMBDA method decorrelates the double-difference measurements to enable a more efficient search for the carrier cycle ambiguities.

In GNSS receivers, the carrier phase is usually measured by sampling the phase of the locally generated carrier signal in the carrier tracking loop. In most receivers, this locally generated signal is the carrier NCO that tracks the phase of the incoming signal at an IF. The signal phase is preserved when the incoming signal is frequency down-converted. The carrier NCO is designed to provide a digital output of its instantaneous phase (or frequency) in response to a sampling signal. Phase-based measurements are made by simultaneously sampling at time t_R the phases of the carrier NCOs tracking the various satellites signals. As with all receiver measurements, the reference for the phase measurements is the receiver's clock reference oscillator.

6.4.5 Carrier Doppler Measurement

Measurement of the received carrier frequency provides information that can be used to determine the velocity vector of the user. Although this could be done by forming differences of code-based position estimates, frequency measurement is inherently much more accurate and has faster response time in the presence of user dynamics. The equations relating the measurements of Doppler shift to the user velocity are shown in Eq. (6.10):

$$f_{di} = (\mathbf{v} \cdot \mathbf{u}_i - \mathbf{v}_i \cdot \mathbf{u}_i)\frac{1}{\lambda} + f_b \text{ (Hz)} \tag{6.10}$$

where

$\mathbf{v}$ = user velocity vector (m/s)

$\mathbf{u}_i$ = unit vector pointing from user to SV i (unitless)

$\mathbf{v}_i$ = SV i velocity vector (m/s)

λ = carrier wavelength (m)

f_b = user receiver clock frequency error (Hz)

In Eq. (6.10), the unknowns are the user velocity vector ($\mathbf{v}$) and the receiver reference clock frequency error (f_b), and the known quantities are the carrier wavelength, the measured Doppler shifts f_{di}, satellite velocity vectors $\mathbf{v}_i$, and unit satellite direction vectors $\mathbf{u}_i$ for each satellite index i. The unit vectors $\mathbf{u}_i$ are determined by computing the user-to-ith satellite displacement vectors $\mathbf{r}_i$ and normalized to unit length as shown in Eq. (6.11):

$$\begin{aligned} \mathbf{r}_i &= \sqrt{(x_i - X)^2 + (y_i - Y)^2 + (z_i - Z)^2} \text{ (m)} \\ u_i &= \frac{\mathbf{r}_i}{|\mathbf{r}_i|} \text{ (unitless)} \end{aligned} \tag{6.11}$$

In the expressions of Eq. (6.11), the ith satellite position (x_i, y_i, z_i) at time t_{Ti} is computed from the ephemeris data, and the user position (X, Y, Z) can be determined from the user position solution of the basic positioning equations using the unambiguous code.

In GNSS receivers, the Doppler measurements f_{di} are usually derived by sampling the frequency setting of the carrier NCO that tracks the phase of the incoming signal. An alternate method is to count the output cycles of the carrier NCO over a relatively short time period, perhaps one second or less. However, in either case, the measured Doppler shift is not the raw measurement itself but the deviation from what the nominally carrier NCO measurement would be without any signal Doppler shift, assuming that the receiver reference clock oscillator had no error.

6.4.6 Integrated Doppler Measurements

Integrated Doppler can be defined as the number of carrier cycles of Doppler shift that have occurred in a given interval (t_0, t). For the *i*th satellite, the relation between integrated Doppler F_{di} and Doppler shift f_{di} is given by Eq. (6.12):

$$F_{di}(t) = \int_{t_0}^{t} f_{di} dt \tag{6.12}$$

However, accurate calculation of integrated Doppler according to this relation would require that the Doppler measurement be a continuous function of time. Instead, GNSS receivers use the output of the carrier NCO in the carrier tracking loop shown previously in Figure 6.9. As shown in the carrier tracking loop of Figure 6.9, the carrier loop error signal is filtered by the carrier loop filter (with optional aiding) and then applied to the carrier NCO. (The nominal carrier rate [scaled to IF] is commanded by the channel processing and command function.) The incremental carrier NCO increments are accumulated, and the number of whole carrier cycles (N_{cc}) that have occurred since initial time can then be counted directly, and the fractional cycles (ϕ_{cc}) are determined. Thus, the integrated Doppler measurement in a GNSS receiver is the addition of the whole carrier cycles plus the fractional cycle since initialization. (The carrier phase integration [a.k.a. accumulation] is similar to the code accumulation performed to determine the time of transmission.) Thus, the raw integrated Doppler measurement (a.k.a., carrier phase measurements) can be represented as shown in Eq. (6.13):

$$\phi_i = (N_{cc,i} + \phi_{cc,i}) \text{ (cycles)}$$

where

$N_{cc,i}$ = whole carrier cycle (cycles)

$\phi_{cc,i}$ = fractional carrier phase cycles (cycles)

and

$$\Phi_i = \phi_i \lambda \text{ (m)} \tag{6.13}$$

where

λ = carrier wavelength (m)

Since the integrated Doppler measurements are ambiguous, they can be produced as arbitrary numbers or scaled to "look like" measurements on the order of the unambiguous pseudorange measurements. Additionally, the integrated Doppler measurements can be reported in units of (cycles) or multiplied by the respective wavelength and reported in units of meters, as shown in the lower half of Eq. (6.13). (Note that the carrier wavelength is the nominal GNSS wavelength and does not include any Doppler for the respective SV i.)

Integrated Doppler measurements have several uses:

1. *Accurate measurement of receiver displacement over time.* The motion of the receiver causes a change in the Doppler shift of the incoming signal. Thus, by counting carrier cycles to obtain integrated Doppler, precise estimates of the *change* in position (*delta position*) of the user over a given time interval can be obtained. The error in these estimates is much smaller than the error in establishing the absolute position using the code measurements. Often these types of measurements are referred to as delta pseudorange measurements and are shown in Eq. (6.14), based on code phase and carrier phase, integrated Doppler measurements:

   $$\Delta\rho_i(k) = \rho_i(k) - \rho_i(k - t_k) \ \text{(m)}$$

   or

   $$\Delta\Phi_i(k) = \Phi_i(k) - \Phi_i(k - t_k) \ \text{(m)} \tag{6.14}$$

   The capability of accurately measuring changes in position is used extensively in *real-time kinematic* application with DGNSS. In real time kinematic (RTK)-type surveying applications, the user needs to determine the locations of many points in a given area with great accuracy (perhaps to within a few centimeters). When the receiver is first turned on, it may take a relatively long time to acquire the satellites, to make both code phase and carrier phase measurements, and to resolve carrier phase ambiguities so that the location of the first surveyed point can be determined. However, once this is done, the relative displacements of the remaining points can be found very rapidly and accurately by transporting the receiver from point to point while it continues to make integrated Doppler measurements. Most often these are done in a differential sense, and the carrier cycle ambiguities that are indeed solved for are the double-difference (or differenced) ambiguities.
2. *Positioning based on received signal phase trajectories.* In another form of differential GPS, a fixed receiver is used to measure the integrated Doppler function, or *phase trajectory curve,* from each satellite over relatively long periods of time (perhaps 5–20 minutes). The position of the receiver can be determined by solving a system of equations relating the shape of the trajectories to the receiver location. The accuracy of this positioning technique, typically within a few decimeters, is not as good as that obtained by resolving the carrier cycle ambiguities but has the advantage that there is no phase

ambiguity. Some handheld GPS receivers employ this technique to obtain relatively good positioning accuracy at low cost.

3. *Carrier rate aiding for the code tracking loop.* In the code tracking loop, proper code alignment is achieved by using observations of the loop error signal to determine whether to advance or retard the state of the otherwise free-running receiver-generated code replica. Because the error signal is relatively noisy, a narrow loop bandwidth is desirable to maintain good pseudorange accuracy. However, this degrades the ability of the loop to maintain accurate tracking in applications where the receiver is subject to substantial accelerations. The difficulty can be substantially mitigated with *carrier rate aiding*, in which the primary code advance/retard commands are not explicitly derived from the code discriminator (early–late correlator) error signal but instead are primarily derived from the Doppler-induced accumulation of carrier cycles in the integrated Doppler function. For example, with the GPS C/A-code, there are 1540 carrier cycles per C/A-code chip, so the code will therefore be advanced by precisely one chip for every 1540 cycles of accumulated count of integrated Doppler. The advantage of this approach is that, even in the presence of dynamics, the integrated Doppler can track the received code *rate* very accurately. As a consequence, the error signal from the code discriminator is largely "decoupled" from the dynamics and can be used for very small and infrequent adjustments to the code generator.
4. *Postcorrelation carrier smoothing of the code measurement.* After the pseudorange and carrier phase measurements have been produced, the code measurements can be smoothed by the carrier phase measurements [24]. Smoothing time will be limited (e.g. <100 seconds) for single-frequency GNSS users but can be extended substantially for multifrequency users [3, 25].

6.5 Theoretical Considerations in Pseudorange, Carrier Phase, and Frequency Estimations

In a GNSS receiver, the measurement error will be limited by thermal noise so it is useful to know the best performance that is theoretically possible in its presence (without additional error sources). Theoretical bounds on errors in estimating code-based and carrier-based measurements, as well as in Doppler frequency estimates, have been developed within a branch of mathematical statistics called *estimation theory*. Using estimation theory, an estimation approach called the *method of ML* can often approach theoretically optimum performance. ML estimates of pseudorange, carrier phase, and frequency are *unbiased*, which means that the expected value of the error due to random noise is zero.

An important lower bound on the error variance of any unbiased estimator is provided by the *Cramer–Rao bound*, and any estimator that reaches this lower limit is called a *minimum-variance unbiased estimator* (MVUE). It can be shown that at the typical SNRs encountered in GNSS, ML estimates of code-based pseudorange, carrier-based measurements, and carrier frequency are all MVUEs. Thus, these estimators are optimal in the sense that no unbiased estimator has a smaller error variance [26].

6.5.1 Theoretical Error Bounds for Code Phase Measurement

As shown in Eq. (6.15), the ML estimate τ_{ML} of signal delay based on code measurements is obtained by maximizing the cross correlation of the received code $c_{\mathrm{rec}}(t)$ with a reference code $c_{\mathrm{ref}}(t)$ that is an identical replica (including bandlimiting) of the received code, where $(0, T)$ is the signal observation interval:

$$\tau_{\mathrm{ML}} = \max_{\tau} \int_0^T c_{\mathrm{rec}}(t) c_{\mathrm{ref}}(t-\tau)\, dt \tag{6.15}$$

where

$$\begin{aligned} c_{\mathrm{rec}}(t) &= \text{received code} \\ c_{\mathrm{ref}}(t-\tau) &= \text{reference code} \\ c'_{\mathrm{Rec}}(t) &= \text{derivative} \end{aligned}$$

Here we assume coherent processing for purposes of simplicity. This estimator is an MVUE, and it can be shown that the error variance of τ_{ML} (which equals the Cramer–Rao bound) is expressed in Eq. (6.16):

$$\sigma^2_{\tau\mathrm{ML}} = \frac{N_0}{2\int_0^T [c'_{\mathrm{rec}}(t)]^2 dt} \tag{6.16}$$

This is a fundamental relation that in temporal terms states that the error variance is proportional to the power spectral density N_0 of the noise and is inversely proportional to the integrated square of the derivative of the received code waveform. It is generally more convenient to use an expression for the standard deviation, rather than the variance, of delay error, in terms of the bandwidth of the code. For the GPS C/A-code, this standard deviation can be calculated as a function of the C/N_0, bandwidth, and integration time as shown in Eq. (6.17) [27]:

$$\sigma_{\tau_{\mathrm{ML}}} = \frac{3.444 \times 10^{-4}}{\sqrt{(C/N_0)WT}} \tag{6.17}$$

The expression in Eq. (6.17) assumes that the received code waveform has been bandlimited by an ideal low-pass filter with one-sided bandwidth W. The

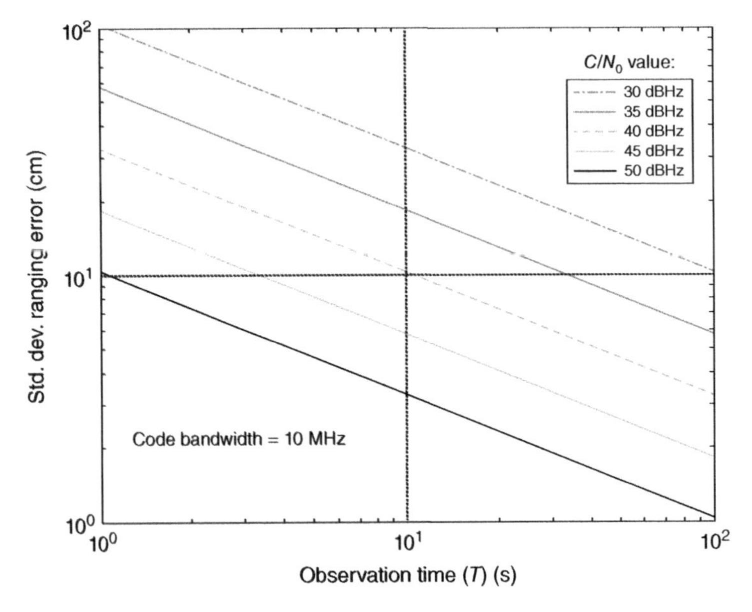

Figure 6.11 Theoretically achievable C/A-code pseudorange error.

signal observation time is denoted by T, and C/N_0 is the ratio of power in the code waveform to the one-sided power spectral density of the noise. A similar expression is obtained for the error variance using the GPS P(Y)-code, except the numerator is $\sqrt{10}$ times smaller.

Figure 6.11 shows the theoretically achievable pseudoranging standard deviation error using the GPS C/A-code as a function of signal observation time for various C/N_0 values. The error is surprisingly small if the code bandwidth is sufficiently large. As an example, for a moderately strong signal with $C/N_0 = 31{,}623$ (45 dBHz), a bandwidth $W = 10\,\text{MHz}$, and a signal observation time of 1 second, the standard deviation of the ML delay estimate obtained from Eq. (6.17) is about 6.2×10^{-10} s, corresponding to 18.6 cm after multiplying by the speed of light.

6.5.2 Theoretical Error Bounds for Carrier Phase Measurements

At typical GNSS SNRs, the ML estimate τ_{ML} of signal delay using the carrier phase is an MVUE, and the error standard deviation can be expressed as Eq. (6.18):

$$\sigma_{\tau_{\text{ML}}} = \frac{1}{2\pi f_c \sqrt{2(C/N_0)T}} \tag{6.18}$$

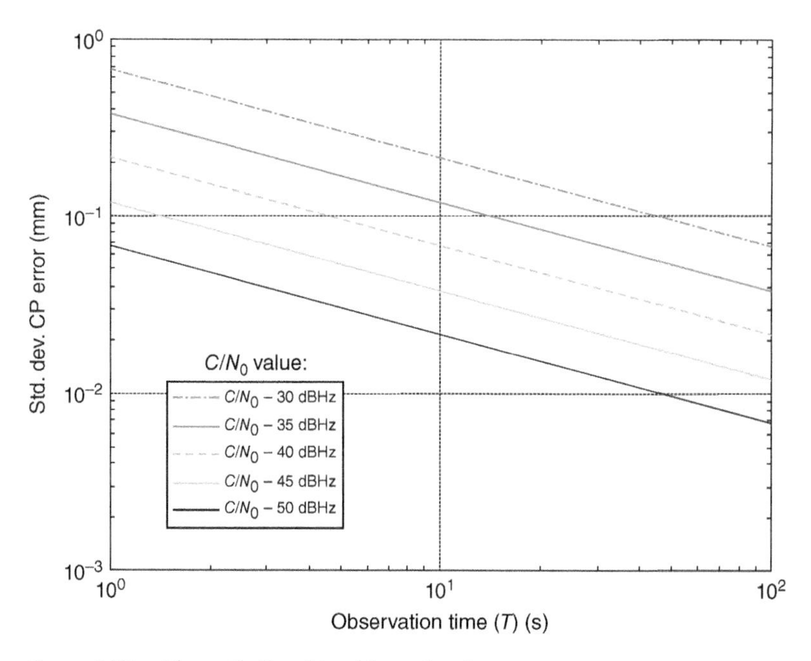

Figure 6.12 Theoretically achievable carrier phase measurement error.

In Eq. (6.18) the GNSS carrier frequency is represented as f_c, and C/N_0 and T are the same as described for Eq. (6.17). Figure 6.12 shows the theoretically achievable carrier phase standard deviation error at the L1 frequency as a function of signal observation time for various C/N_0 values. This result is also reasonably accurate for a carrier tracking loop if T is set equal to the reciprocal of the loop bandwidth. As an example of the much greater accuracy of carrier phase measurement compared with code pseudorange measurement, a signal at $C/N_0 = 45$ dBHz observed for 1 second can theoretically yield an error standard deviation of 4×10^{-13} seconds, which corresponds to only 0.12 mm. However, typical errors of 1–3 mm are experienced in most receivers as a result of random phase jitter in the reference oscillator.

6.5.3 Theoretical Error Bounds for Frequency Measurement

The ML estimate τ_{ML} of the carrier frequency is also an MVUE, and its error standard deviation can be expressed as Eq. (6.19):

$$\sigma_{\tau_{\mathrm{ML}}} = \frac{3}{\sqrt{2\pi^2 (C/N_0) T^3}} \tag{6.19}$$

Figure 6.13 shows the theoretically achievable frequency estimation error as a function of signal observation time for various C/N_0 values. A one-second

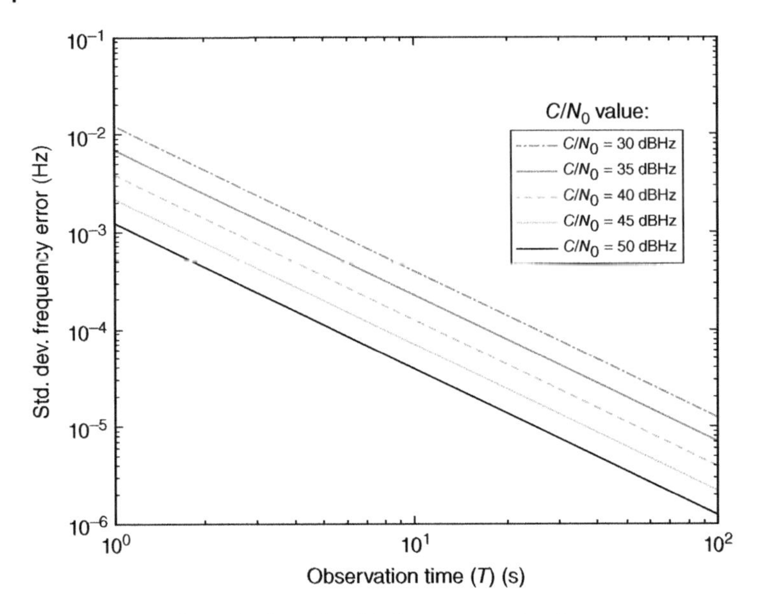

Figure 6.13 Theoretically achievable frequency estimation error.

observation of a despread GNSS carrier with $C/N_0 = 45$ dBHz yields a theoretical error standard deviation of about 0.002 Hz, which could also be obtained with a phase tracking loop having a bandwidth of 1 Hz. As in the case of phase estimation, however, phase jitter in the receiver reference oscillator yields frequency error standard deviations from 0.05 to 0.1 Hz.

6.6 High-Sensitivity A-GPS Systems

Over the last decade, a significant amount of emphasis and development has been made in the area of high-sensitivity GPS receivers for use in poor signal environments. More generally, such receivers can be designed and used with any GNSS, such as the GLONASS and Galileo. A major application is incorporation of such receivers in cell phones, thus enabling a user to automatically transmit its location to rescue authorities in emergency 911 (E-911) calls. The utility of using GPS for such E-911 applications has been facilitated by the removal of selective availability (SA) to GPS on 1 May 2000 and the US mandated accuracies put into law by the US Congress to support E-911 [28, 29]. Such a receiver must be able to reliably operate deep within buildings or heavy vegetation, which severely attenuates the GPS signals.

In order to achieve the reliable and rapid positioning for such applications, assisting data from a BS receiver (the server) at a location having good signal reception are sent to the user's MS receiver (the client). The assisting data can include BS location, satellite ephemeris data, the demodulated navigation data bit stream, frequency calibration data, and timing information. In addition, the BS can provide pseudorange and/or carrier phase measurement information that enable differential operation. The assisting data can be transmitted via a cell phone or other radio links. In some cases, the assisting information can be transmitted over the Internet and relayed to a cell phone via a local-area wireless link, or the cellular network directly. Significant effort has gone into developing standards and formats to support A-GPS for cellular network and handset providers. These efforts have been coordinated by the 3rd Generation Partnership Project (3GPP) [8].

Assistance data can not only increase the sensitivity of the client receiver but can also significantly reduce the time required to obtain a position solution. A typical stand-alone GPS receiver can acquire signals down to about −145 dBm and might require a minute or more to obtain a position from a cold start. On the other hand, high-sensitivity A-GPS receivers are currently being produced by a number of major manufacturers who are claiming sensitivities in the −155 to −165 dBm range and a cold start TTFF as low as one second [30]. To gain the required sensitivity and processing speed, A-GPS receivers usually capture several seconds of received signals in a memory that can be accessed at high speed to facilitate the signal processing operations.

6.6.1 How Assisting Data Improves Receiver Performance

6.6.1.1 Reduction of Frequency Uncertainty

To achieve rapid positioning, the range of frequency uncertainty in acquiring the satellites at the client receiver must be reduced as much as practicable in order to reduce the search time. Reducing the number of searched frequency bins also increases receiver sensitivity because the acquisition false-alarm rate is reduced. Two ways that frequency uncertainty can be reduced are as follows:

1. *Transmission of Doppler information.* The server *can* accurately calculate signal Doppler shifts at its location and transmit them to the user. For best results, the user must either be reasonably close to the server's receiver or must know its approximate position to avoid excessive uncompensated differential Doppler shift between server and client. For cellular applications, the BS to MS distance is usually within a couple of miles and this Doppler estimate is sufficiently accurate.
2. *Transmission of a frequency reference.* If only Doppler information is transmitted to the user, the frequency uncertainty of the client receiver local oscillator still remains an obstacle to rapid acquisition. Today's technology can produce oscillators that have a frequency uncertainty on the order of 1 ppm at a cost low enough to permit incorporation into a consumer

product such as a cell phone. Even so, at the GNSS L1 frequency, 1 ppm translates into about ±1575 Hz of frequency uncertainty. Thus, even with an accurate Doppler estimate from the BS, the MS must search over this frequency uncertainty induced by the receiver's reference oscillator. Once the first satellite is acquired, the local oscillator offset can be determined (using the accurate Doppler estimate from the BS), and the frequency uncertainty in searching for the remaining satellites can thereby be reduced to a small value. To remedy the problem of acquiring the first satellite in a sufficiently short time, some A-GPS implementations use an accurate frequency reference transmitted from the server to the client, in addition to satellite Doppler measurements. However, this requirement complicates the design of the server-to-client communication system and is undesirable when trying to use an existing communication system for assisting purposes. If the communication system is a cell phone network, every cell tower would need to transmit a precise frequency reference or encode this timing into a message.

6.6.1.2 Determination of Accurate Time

In order to obtain accurate pseudoranges, a conventional GPS receiver obtains time information from the navigation data message that permits the precise GPS time of transmission of any part of the received signal to be determined at the receiver. When a group of pseudorange measurements is made, the time of transmission from each satellite is used for two purposes: (i) to obtain an accurate position of each satellite at the time of transmission and (ii) to compute the pseudorange by computing the difference between signal reception time (according to the receiver clock) and transmission time, as shown previously in Eq. (6.6).

In order to obtain time information from the received GPS signal, a conventional receiver must go through the steps of acquiring the satellite signal, tracking it in code phase and carrier phase with a PLL to form a coherent reference for data demodulation, achieve bit synchronization, demodulate the data, achieve frame synchronization, locate the portion of the navigation message that contains the GPS system time information (i.e. Z-count), and, finally, continue to keep track of time (usually by counting code epochs as they are received).

However, it is desirable to avoid these numerous and time-consuming steps in a positioning system that must reliably obtain a position within several seconds of startup in a weak-signal environment. Because the legacy navigation data message encodes GPS time information only once per six seconds subframe, the receiver may have to wait a minimum of six seconds to obtain it (additionally, more time is needed to phase-lock to the signal and to achieve bit and frame synchronization). Furthermore, if the signals are below about

−154 dBm, demodulation of the navigation data message has an error rate that may preclude the reading of time from the signal.

If the position of the client is known with sufficient accuracy (even within 100 km), it is possible to resolve the difference in times of transmission. This is possible because the times of transmission of the C/A-code epochs are known to be integer multiples of 1 ms according to SV time (which can be corrected to GPS time using slowly changing time correction data sent from the server). This integer ambiguity in differences of time transmission is resolved by using approximate ranges to the satellites, which are calculated from the approximate position of the client and insertion of approximate time into satellite ephemeris data sent by the server to the client. For this purpose, the accuracy of the approximated time needs to be sufficiently small to avoid excessive uncertainty in the satellite positions. Generally, a time accuracy of better than 10 seconds will suffice for this purpose.

Once the ambiguity of the differences in transmission times has been resolved, accurate positioning is possible if the positions of the satellites at transmission time are known with an accuracy comparable to the positioning accuracy desired.

However, since the satellites are moving at a tangential orbital velocity of approximately 3800 m/s, the accuracy in knowledge of signal transmission time for the purpose of locating the satellites must be significantly more accurate than that required for the ambiguity resolution previously described.

Most weak-signal A-GPS rapid positioning systems obtain the necessary time accuracy for locating the satellites by using time information transmitted from the server. It is important to recognize that such time information must be in "real time"; that is, it must have a sufficiently small uncertainty in latency as it arrives at the client receiver. For example, a latency uncertainty of 0.1 second could result in a satellite position error of 380 m along its orbital path, causing a positioning error of the same order of magnitude. Transmission of time from the server with small latency uncertainty has an impact on the design of the server-to-client communication system such as cellular networks, to provide A-GPS positioning services.

6.6.1.3 Transmission of Satellite Ephemeris Data

Due to the structure of the GPS legacy navigation message, up to 36 seconds (i.e. 2× the first three subframes, worst case) is required for a stand-alone GPS receiver to obtain the ephemeris data necessary to determine the position of a satellite. This delay is undesirable in emergency applications. Furthermore, in indoor operation, the signal is likely to be too weak to demodulate the ephemeris data. The problem is solved if the server transmits the data to the client via a high-speed communication link. The Internet can even be used for this purpose if the client receiver has access to a high-speed Internet connection or the network provider can provide this information directly using the A-GPS standard messaging [8].

6.6.1.4 Provision of Approximate Client Location

Some servers (e.g. a cell phone network) can transmit the approximate position of the client receiver to the user. As mentioned previously, this information can be used to resolve the ambiguity in times of signal transmission from the satellites.

6.6.1.5 Transmission of the Demodulated Navigation Bit Stream

The ultimate achievable receiver sensitivity is affected by the length of the signal capture interval and the presence of navigation data modulation on the GPS signal.

Fully coherent processing. If the GPS signal were modulated only by the C/A-code and contained no navigation data modulation, maximum theoretically possible acquisition sensitivity would result from fully coherent delay and Doppler processing. In this form of processing, the baseband signal in the receiver is frequency-shifted and precession-compensated in steps (frequency bins), and for each step, the signal is cross correlated with a replica of the C/A-code spanning the entire signal observation interval. Alternatively, the 1 ms periods of the C/A-code could be synchronously summed prior to cross correlation.However, the presence of the 50 bps navigation data modulation precludes the use of fully coherent processing over signal capture intervals exceeding 20 ms unless some means is available to reliably strip the data modulation from the signal. If the server can send the demodulated data bit stream to the client, the data modulation can be stripped from the user's received signal, thus enabling fully coherent processing. However, the timing of the bit stream must be known with reasonable accuracy (within approximately 1 ms); otherwise, a search for time alignment must be made.

Partially coherent processing. In the absence of a demodulated data bit stream from the server, a common method of dealing with the presence of data modulation is to coherently process the signal within each data bit interval, followed by noncoherent summation of the results. Assuming that the timing of the data bit boundaries is available from the server, the usual implementation of this technique is to first coherently sum the 20 periods of the complex baseband C/A-coded signal within each data bit. For each data bit, a waveform is produced that contains one, 1 ms period of the C/A-code, with a processing gain of $10 \log_{10}(20) = 13\,\text{dB}$. Each waveform is then cross correlated with a replica of the C/A-code to produce a complex-valued cross-correlation function. The squared magnitudes of the cross-correlation functions are computed and summed to produce a single function spanning 1 ms, and the location of the peak value of the function is the signal delay estimate. We shall call this form of processing *partially coherent*.When one second of the received signal is observed by the user, fully coherent processing provides approximately a 3–4 dB improvement over partially coherent processing. It is important to note that fully coherent processing has a major drawback without assisted navigation data in that many more code delay and frequency bins must be processed, which either dramatically slows down processing speed or requires a large amount of parallel processing to maintain that speed.

Data detection and removal by the client receiver. An alternate method of achieving fully coherent processing is to have the client receiver detect the data bits and use them to homogenize the polarity of the signal, thus permitting coherent processing over the full signal capture interval. In order for this method to be effective, the signal must be strong enough to ensure reliable data bit detection. Furthermore, a phase reference is needed, and it should be estimated using the entire signal observation. A practical technique for estimating phase, which approaches theoretically optimum results, is the method of ML. We shall call this methodology *coherent processing with data stripping*, or simply *data stripping* for short. (Some refer to this as *data wipe-off*.) At low signal power levels, its performance approaches that of partially coherent processing, and at high signal levels, its performance approaches that of fully coherent processing. At first glance, it seems that data stripping might give a worthwhile advantage over partially coherent processing. However, it shares a common disadvantage with fully coherent processing in that a larger number of code delay and frequency bins must be processed. Additionally, when the navigation data changes, resulting from an ephemeris update, the data change must be detected and new navigation should be used.

6.6.1.6 Server-Provided Location

Some servers (e.g. a cell phone network) support a user solution for the MS. In these types of networks, raw measurement data are collected at the MS and then these raw measurement data are sent to the BS. The MS still needs to synchronize the carrier and code phase so that valid (i.e. correlated) measurement data are available. At the BS, the GPS ephemeris and error mitigation data are available, and the MS user solution is calculated. The BS then transmits back to the MS its position solution for use. This type of approach takes the advantage of utilizing the full information that may be available at the BS and supporting network and reduce computational burden on the MS.

6.6.2 Factors Affecting High-Sensitivity Receivers

In a good signal environment, a certain amount of SNR implementation loss is tolerable. Typical stand-alone GPS receivers for outdoor use may have total losses as great as 3–6 dB. However, in a high-sensitivity receiver, the maintenance of every decibel of SNR is important, thus requiring attention to minimizing losses that would otherwise not be of concern. The following are some of the more important issues that arise in high-sensitivity receiver design.

6.6.2.1 Antenna and Low-Noise RF Design

A good antenna and a low-noise receiver RF front end are very important elements of a high-sensitivity receiver, the details of which were discussed in Chapter 5.

6.6.2.2 Degradation due to Signal Phase Variations

With fully coherent processing over long time intervals, performance is adversely affected by signal phase variations from sources including Doppler curvature due to satellite motion, receiver oscillator phase stability, and motion of the receiver. Doppler curvature can be partially predicted from assisting almanac or ephemeris data, but its accuracy depends on knowledge of the approximate position of the user. On the other hand, oscillator phase noise is random and unpredictable, and hence resistant to compensation (for this reason, research efforts are currently under way to produce a new generation of low-cost atomic and optical frequency sources). Especially pernicious is motion of a GPS receiver in the user's hand. Because of the short wavelengths of GPS signals, such motion can cause phase variations of more than a full cycle during the time that the receiver is searching for a signal, thus seriously impairing acquisition performance.

6.6.2.3 Signal Processing Losses

There are various forms of processing loss that should be minimized in a high-sensitivity GPS receiver.

Digitization. Losses due to quantization of the ADC digital output must be minimized. The 1-bit quantization often used in low-cost receivers causes almost 2 dB of SNR loss. Hence, it is desirable to use an ADC with at least 2 bits in high-sensitivity applications.

Sampling considerations. The bandwidth of the receiver should be large enough to avoid SNR loss. However, this generally requires higher sampling rates with an attendant increase in power consumption and processing loads, a factor that is detrimental to low-cost, low-power consumer applications.

Correlation losses. Rapid signal acquisition drives the need for coarser quantization of the correlator reference code phase during signal search. However, this causes correlation loss, and an acceptable tradeoff must be made. Correlation loss is further exacerbated if the receiver bandwidth is made small to reduce the required sampling rate.

Doppler compensation losses. One source of these losses is "scalloping loss," caused by the discrete steps of the Doppler frequencies used in searching for the satellites. Scalloping loss can be as large as 2 dB in some receivers. Another source is phase quantization of the Doppler compensation, which can introduce a degradation of as much as 1 dB in the simplest designs.

6.6.2.4 Multipath Fading

It is common in poor signal environments, especially indoors, for the signal to have large and/or numerous multipath components. In addition to causing pseudorange and carrier phase biases, multipath can significantly reduce receiver sensitivity when phase cancellation of the signal occurs.

6.6.2.5 Susceptibility to Interference and Strong Signals

As receiver sensitivity is decreased (i.e. gets more sensitive), so does the susceptibility to various forms of interference. Although this is seldom a problem with receivers of normal sensitivity, in a high-sensitivity receiver, steps must be taken to prevent erroneous acquisition of lower-level PRN code correlation sidelobes from both desired and undesired satellite signals.

6.6.2.6 The Problem of Time Synchronization

In an A-GPS architecture designed for rapid positioning (within a few seconds) using weak signals, the user's receiver does not have time to read the unambiguous GPS time from the received signal itself. The need for the BS to transmit to the user low-latency time accurate enough to estimate the positions of the satellites is a limiting factor.

6.6.2.7 Difficulties in Reliable Sensitivity Assessment

Realistic assessment of receiver sensitivity is a challenging task. At the extremely low signal levels for which a high-sensitivity receiver has been designed, laboratory signal generators often have signal leakage, which causes the signal levels to be higher than indicated by the generator. For this and other reasons, care must be taken in the test and evaluation of high-sensitivity receivers to accurately measure their true sensitivity.

6.7 Software-Defined Radio (SDR) Approach

At the time GPS receivers were first developed in the mid-1970s, most were full 19 in. rack mounted units that contained analog components. It was not until the late 1970s and early 1980s when the first all-digital receiver was manufactured [31]. As technology advanced, the use of application-specific integrated circuit (ASIC) and microprocessors found application in most GNSS receiver designs. Indeed, today, most GNSS receivers and/or chip sets are a combination of ASIC and/or microprocessor-based semiconductor devices. These types of GNSS receiver architectures provide a cost-effective solution, in a small form factor with low power consumption. One disadvantage of these popular configurations is the flexibility in controlling certain aspects of the signal processing of the GNSS signals for advanced users. A more flexible GNSS architecture has emerged that is called a software-defined radio (SDR) approach. While it is true that all-digital GNSS receivers in operation today operate on some form of software, defined by the design programmer, the distinguishing feature in SDRs is that the user can control and program the details of the GNSS receiver's operation. SDRs typically operate on digitized GNSS signal samples. Most often, an RF front end is utilized and digital IF samples are processed by the SDR as illustrated in Figure 6.1. An SDR can operate utilizing a microprocessor, a field-programmable gate array, or a processor commonly found in a personnel computer. Research in user-defined SDR materialized in the mid-1990s [12] and a limited number of SDR products are commercially available today [32]. Most SDRs today are used as research and/or development tools.

6.8 Pseudolite Considerations

Pseudolites (PLs) (i.e. pseudosatellites) have been used in GNSS for a variety of limited applications. For new GNSS development, PLs can provide the range source before SVs can be launched into space. Such a system was used in the initial development of the GPS and was dubbed an "inverted range" [33]. For indoor or underground applications, PLs can provide a ranging source where GNSS SV signals cannot be received. PLs can also have some utility to augment a GNSS when enough SV signals are not available due to signal availability [34].

Some apparent advantages of PLs are their availability since they are typically ground based (vice space based), their increased power available (since they are closer to the user) compared to an SV, and their geometric relationship to the user (where they are typically in a location where an SV is not, and thereby help the dilution of precision in the user solution). However, each of these attributes come with some impact on the GNSS receiver and other non-participating GNSS users that must be dealt with.

Although the signal level received from GNSS SVs is relatively weak, one advantage is that they are all received at approximately the same power level (e.g. -130 dBm $\pm$ 6 dB or so), where signal power level variations largely come from receiving the SV at different aspect angles. Typically, the processing gain of the code-division multiple access (CDMA) spreading code used to encode the GNSS signal can accommodate this power level variation without detrimental effects.

As for the PL, the power received can vary much more dramatically, depending upon application geometry and antenna pattern gains. This is mainly due to the spatial loss factor in propagation that varies as a factor of $(\lambda/4\pi R)^2$, where λ is the carrier wavelength and R is the range between the transmitter and the receiver. Usually, in PL applications, the dynamic range (up to 50 dB or so from 80 m to 20 nmi [35]) needed to support the application exceeds the processing gain of the CDMA spreading code used. When the dynamic range of the application exceeds the processing gain of the CDMA code, interference can occur [36]. The interference that one code (at a strong power level) causes to another code (at a weak power level) in the correlation process is called the *near–far problem*. To mitigate these effects, pulsing the PL signal at a low duty cycle will minimize these effects on nonparticipating GNSS users [37]. Since most CDMA GNSS receivers can tolerate low duty cycle pulsed interference, they can continue to operate with minimal effect. These nonparticipating GNSS receivers will, however, see a slight reduction in their C/N_0, resulting from some GNSS code chips being lost in the correlation process. PLs based on the GPS C/A are limited by the eventual processing gain and cross-correlation performance of the code. Improvements over a C/A-code-based PL can be gained by using a more random code (i.e. wideband [WB] PL [38]).

As the duty cycle of a pulsed PL signal is decreased to minimize its effects on nonparticipating GNSS users, the peak power of the transmitter typically needs to be increased to maintain the same operational range for the PL link (without explicit steps being taken in the GNSS/PL receiver). This, along with the large dynamic range requirement, can cause saturation effects in the GNSS/PL receiver. Saturation in the RF front end (mainly the mixer) and at the digital ADC level can produce measurement biases [39]. Mitigation of these effects can be done by RF AGC (depicted in Figure 6.1) [35].

Since the PL is often placed on the ground, source (i.e. ground) multipath can be a substantial error source for code-based PL applications. This can also be the case for indoor, tunnel, or underground applications of PL systems. To help mitigate these source-induced multipath errors, an advanced multipath limiting antenna (MLA) can be used [35].

While code-based PL systems provide an unambiguous pseudorange, to avoid the code-based multipath and power bias, carrier phase-based PL systems that rely heavily on the carrier phase measurements have been demonstrated. These types of solution techniques either solve for and remove the carrier cycle ambiguity for an absolute position [40, 41] or use a relative, that is, triple difference carrier phase user solution, initialized from an identifiable code-based user solution state (i.e. position) [38].

Problems

6.1 An ultimate limit on the usability of weak GPS signals occurs when the bit error rate (BER) in demodulating the 50 bps navigation message becomes unacceptably large. Find the signal level in dBm at the output of the receiver antenna that will give a BER of 10^{-5}. Assume an effective receiver noise temperature of 513 K and that all signal power has been translated to the baseband I channel with optimal demodulation (integration over the 20 ms bit duration followed by polarity detection).

6.2 Support the claim that a 1-bit ADC provides an essentially linear response to a signal deeply buried in Gaussian noise by solving the following problem. Suppose that the input signal s_{in} to the ADC is a DC voltage embedded in zero-mean additive Gaussian noise $n(t)$ with standard deviation σ_{in} and that the power spectral density of $n(t)$ is flat in the frequency interval $(-W, W)$ and zero outside the interval. Assume that the 1-bit ADC is modeled as a hard limiter that outputs a value $v_{\text{out}} = 1$ if the polarity of the signal plus noise is positive and $v_{\text{out}} = -1$ if the polarity is negative. Define the ADC output signal s_{out} by Eq. (6.20):

$$s_{\text{out}} = E[v_{\text{out}}] \tag{6.20}$$

where E denotes expectation, and let σ_{out} be the standard deviation of the ADC output. The ADC input signal-to-noise ratio SNR_{in} and output signal-to-noise ratio SNR_{out} can then be defined by Eq. (6.21):

$$\text{SNR}_{\text{in}} = \frac{S_{\text{out}}}{\sigma_{\text{in}}}$$

and

$$\text{SNR}_{\text{out}} = \frac{S_{\text{out}}}{\sigma_{\text{out}}} \tag{6.21}$$

where s_{out} and σ_{out}, respectively, are the expected value and the standard deviation of the ADC output. Show that if $s_{\text{in}} \ll \sigma_{\text{in}}$, then $s_{\text{out}} = K s_{\text{in}}$, where K is a constant, and

$$\frac{\text{SNR}_{\text{out}}}{\text{SNR}_{\text{in}}} = \frac{2}{\pi} \tag{6.22}$$

Thus, the signal component of the ADC output is linearly related to the input signal component, and the output SNR is about 2 dB less than that of the input.

6.3 Some GPS receivers directly sample the signal at an IF instead of using mixers for the final frequency shift to baseband. Suppose that you wish to sample a GPS signal with a bandwidth of 1 MHz centered at an IF of 3.5805 MHz. What sampling rates will not result in frequency aliasing? Assuming that a sampling rate of 2.046 MHz was used, show how a digitally sampled baseband signal could be obtained from the samples.

6.4 Instead of forming a baseband signal with I and Q components, a single-component baseband signal can be created simply by multiplying the incoming L1 (or L2) carrier by a sinusoid of the same nominal frequency, followed by low-pass filtering. Discuss the problems inherent in this approach. (*Hint*: Form the product of a sinusoidal carrier with a sinusoidal local oscillator signal; use trigonometric identities to reveal the sum and difference frequency components, and consider what happens to the difference frequency as the phase of the incoming signal assumes various values.)

6.5 Write a computer program using MATLAB®, C or another high-level language that produces the GPS 1023-chip C/A-code used by satellite SV1. The code for this satellite is generated by two 10-stage shift registers called the *GI and G2 registers*, each of which is initialized with all 1 second. The input to the first stage of the G 1 register is the exclusive OR of its 3rd and 10th stages. The input to the first stage of the G2 register is the exclusive OR of its 2nd, 3rd, 6th, 8th, 9th, and 10th stages. The C/A-code is the exclusive OR of stage 10 of Cil, stage 2 of 62, and stage 6 of G2. You may use the GPS IS-200F to help you in this generation.

6.6 For high accuracy of the carrier phase measurements, the most suitable carrier tracking loop will be
(a) PLL with low loop bandwidth
(b) FLL with low loop bandwidth
(c) PLL with high loop bandwidth
(d) FLL with high loop bandwidth

6.7 Which of the following actions does not reduce the receiver noise (code)?
(a) Reducing the loop bandwidth
(b) Decreasing the PDI
(c) Spacing the early–late correlators closer
(d) Increasing the signal strength

6.8 Describe how the time of travel (from satellite to receiver) of the GPS signal is determined and how the pseudorange and carrier phases are computed from it.

References

1 McGraw, G. (2009). Generalized divergence-free carrier smoothing with applications to dual frequency DGPS. *Navigation, Journal of the Institute of Navigation* 56 (2): 115–122.
2 Yang, Y., Sharpe, R.T., and Hatch, R.R. (2002). A fast ambiguity resolution technique for RTK embedded within a GPS receiver. *ION GPS 2002*. Portland, OR (24–27 September 2002), pp. 945–952.
3 Hatch, R.R. (2006). A new three-frequency, geometry-free, technique for ambiguity resolution. In: *Proceedings of the 19th International Technical Meeting of the Satellite Division of The Institute of Navigation (ION GNSS 2006)*, 309–316. Fort Worth, TX: ION (September 2006).
4 Van Nee, R.D.J. and Siereveld, J. (1993). The multipath estimating delay lock loop—approaching theoretical accuracy limits. In: *Proceedings of the 6th International Technical Meeting of the Satellite Division of The Institute of Navigation (ION GPS 1993), Salt Lake City, UT, September 1993*, 921. Salt Lake City, UT: ION.
5 Van Diggelen, F. (2001). Global locate indoor GPS chipset & services. In: *Proceedings of the 14th International Technical Meeting of the Satellite Division of The Institute of Navigation (ION GPS 2001)*, 1515–1521. Salt Lake City, UT: ION (September 2001).
6 Ashjaee, J. and Lorenz, R. (1992). Precision GPS surveying after Y-code, Ashtech, Inc., AN/AFTY/(11/92), Magellan Corporation.
7 RTCA (2011). Radio Technical Commission for Maritime Services Special Committee 104. Recommended Standards for Differential GNSS (Global Navigation Satellite Systems) Service. https://ssl29.pair.com/dmarkle/puborder.php?show=3 (accessed 3 July 2012).
8 3GPP (2012). Requirements for support of Assisted Global Positioning System (A-GPS); Frequency Division Duplex (FDD). http://www.3gpp.org/ftp/Specs/html-info/25171.htm (accessed 4 July 2012).
9 Couch, L.W. III, (2007). *Digital and Analog Communications*, 7e. Upper Saddle River, NJ: Prentice Hall.
10 Pratt, T., Bostian, C., and Allnutt, J. (2003). *Satellite Communications*, 2e. Hoboken, NJ: Wiley.
11 Vaughan, R.G., Scott, N.L., and White, D.R. (1991). The theory of bandpass sampling. *IEEE Transactions on Signal Processing* 39 (9): 1973–1984.
12 Akos, D. (1997). A software radio approach to global navigation satellite system receiver design. PhD. Ohio University. https://etd.ohiolink.edu/!etd.send_file?accession=ohiou1174615606&disposition=inline (accessed 04 September 2019).
13 Gunawardena, S. (2007). Development of a transform-domain instrumentation global positioning system receiver for signal quality and anomalous

event monitoring. Doctoral dissertation. Fritz J. Dolores H. Russ College of Engineering and Technology, Ohio University.

14 Wald, A. (1947). *Sequential Analysis*. New York: Wiley.

15 Uijt de Haag, M. (1999). An investigation into the application of block processing techniques for the global positioning system. PhD dissertation. Ohio University.

16 van Graas, F., Soloviev, A., Uijt de Haag, M. et al. (2005). Comparison of two approaches for GNSS receiver algorithms: batch processing and sequential processing considerations. In: *Proceedings of the 18th International Technical Meeting of the Satellite Division of The Institute of Navigation (ION GNSS 2005)*, 200–211. Long Beach, CA: ION (September 2005).

17 Fenton, P.C., Falkenberg, W.H., Ford, T.J. et al. (1990). NovAtel's GPS receiver—the high performance OEM sensor of the future. In: *Proceedings of the 4th International Technical Meeting of the Satellite Division of The Institute of Navigation (ION GPS 1991)*, 49–58. Albuquerque, NM: ION (September 1990).

18 Van Dierendonck, A.J., Fenton, P., and Ford, T. (1992). Theory and performance of narrow correlator spacing in a GPS receiver. *Navigation, Journal of the Institute of Navigation* 39 (3): 265–284.

19 Costas, J.P. (1956). Synchronous communications. *Proceedings of the IRE* 45: 1713–1718.

20 Ward, P.W. (1998). Performance comparisons between FLL, PLL and a novel FLL-assisted-PLL carrier tracking loop under RF interference conditions. In: *Proceedings of the 11th International Technical Meeting of the Satellite Division of The Institute of Navigation (ION GPS 1998)*, 783–795. Nashville, TN: ION (September 1998).

21 Hatch, R. and Sharpe, T. (2001). A computationally efficient ambiguity resolution technique. In: *Proceedings of the 14th International Technical Meeting of the Satellite Division of The Institute of Navigation (ION GPS 2001), Salt Lake City, UT*, 1558–1564.

22 Hatch, R. and Sharpe, T. (2001). A computationally efficient ambiguity resolution technique. ION GPS.

23 Teunissen, P.J.G., de Jonge, P.J., and Tiberius, C.C.J.M. (1995). The LAMBDA-method for fast GPS surveying. *Presented at the International Symposium*, "GPS Technology Applications," Bucharest, Romania (26–29 September 1995).

24 Hatch, R. (1982). The synergism of GPS code and carrier phase measurements. In: *Proceedings of Third International Geodetic Symposium on Satellite Positioning*, vol. 2, 1213–1231. Las Cruces, NM: New Mexico State University (8–12 February 1982).

25 Hwang, P.Y., McGraw, G.A., and Bader, J.R. (1999). Enhanced differential GPS carrier-smoothed code processing using dual-frequency measurements. *Navigation, Journal of the Institute of Navigation* 46 (2): 127–138.
26 Van Trees, H.L. (1968). *Detection, Estimation, and Modulation Theory,* Part 1. New York: Wiley. ISBN: 471899550.
27 Weill, L. (1994). C/A code pseudoranging accuracy—how good can it get. In: *Proceedings of the 7th International Technical Meeting of the Satellite Division of The Institute of Navigation (ION GPS 1994),* 133–141. Salt Lake City, UT: ION (September 1994).
28 The White House (2000). Office of the Press Secretary, Statement by the President Regarding the United States' Decision to Stop Degrading Global Positioning System Accuracy, for immediate release 1 May 2000.
29 FCC (2005). FCC Amended Report to Congress on the Deployment of E-911 Phase II Services By Tier III Service Providers. Submitted Pursuant to Public Law No. 108-494, Federal Communications Commission, 1 April 2005.
30 u-blox (2012). Technology, Assisted GPS. http://www.u-blox.com/en/assisted-gps.html (accessed 11 July 2012).
31 Institute of Navigation (2012). Virtual navigation museum, systems, SAT-NAV. http://www.ion.org/museum/cat_view.cfm?cid=7&scid=9 (accessed 11 July 2012).
32 IfEN (2012). SX-NSR software receiver. https://www.ion.org/gnss/upload/files/956_SX.NSR.Flyer_Aug2013_Letter.pdf (accessed 04 September 2019).
33 Harrington, R.L. and Dolloff, J.T. (1976). The inverted range: GPS user test facility. *Proceedings of The Institute of Electrical and Electronics Engineers (IEEE), Position, Location, and Navigation, Symposium (PLANS) 1976* (November 1976). New York: IEEE, pp. 204–211.
34 Brown, A.K. (1992). A GPS precision approach and landing system. In: *Proceedings of the 5th International Technical Meeting of the Satellite Division of The Institute of Navigation (ION GPS 1992),* 373–381. Albuquerque, NM: ION (September 1992).
35 Bartone, C. (1998). Ranging airport Pseudolite for local area augmentation using the global positioning system. PhD dissertation. Ohio University. https://etd.ohiolink.edu/!etd.send_file?accession=ohiou1175095346&disposition=inline (accessed 04 September 2019).
36 McGraw, G.A. (1994). Analysis of Pseudolite code interference effects for aircraft precision approaches. In: *Proceedings of the 50th Annual Meeting of The Institute of Navigation,* 433–437. Colorado Springs, CO: ION (June 1994).
37 Winer, B., Mason, W., Manning, P. et al. (1996). GPS receiver laboratory RFI tests. In: *Proceedings of the 1996 National Technical Meeting of The Institute of Navigation, Santa Monica, CA, (January 1996),* 669–676.

38 Kiran, S. (2003). A wideband airport Pseudolite architecture for the local area aug-mentation system. PhD dissertation. Ohio University, November 2003. https://etd.ohiolink.edu > accession=ohiou1081191846 (accessed 26 October 2019).

39 Kiran, S. and Bartone, C. (2004). Verification and mitigation of the power-induced measurement errors for airport Pseudolites in LAAS. *GPS Solutions* 7 (4): 241–252.

40 Cobb, S. (1997). *GPS Pseudolite, Theory, Design, and Applications.* Stanford University http://waas.stanford.edu/~wwu/papers/gps/PDF/Thesis/StewartCobbThesis97.pdf (accessed 04 September 2019).

41 Pervan, B. (1996). *Navigation Integrity for Aircraft Precision Landing Using the Global Positioning System.* Stanford University http://waas.stanford.edu/~wwu/papers/gps/PDF/Thesis/BorisPervanThesis96.pdf (accessed 04 September 2019).

7

GNSS Measurement Errors

7.1 Source of GNSS Measurement Errors

Ranging errors are typically grouped into six classes:

1. *Ionosphere.* Errors in corrections of pseudorange measurements caused by ionospheric effects (free electrons in the ionosphere).
2. *Troposphere.* Errors in corrections of pseudorange measurements caused by tropospheric effects; temperature, pressure, and humidity contribute to variations in the speed of light.
3. *Multipath.* Errors caused by reflected signals entering the receiver antenna.
4. *Ephemeris.* Ephemeris data errors in transmitted parameters in navigation messages for satellites' true positions.
5. *Satellite clock.* Clock errors in the transmitted clock data for global navigation satellite system (GNSS) with respect to the true GNSS time.
6. *Receiver errors.* Errors in the receiver's measurement of range caused by thermal noise, software accuracy, and interchannel biases.

These are described in detail in Sections 7.2–7.8.

7.2 Ionospheric Propagation Errors

The ionosphere, which extends from approximately 50–1000 km above the surface of the Earth, consists of gases that have been ionized by solar radiation. The ionization produces clouds of free electrons that act as a dispersive medium for GNSS signals in which propagation velocity is a function of frequency. A particular location within the ionosphere is alternately illuminated by the Sun and shadowed from the Sun by the Earth in a daily cycle; consequently, the characteristics of the ionosphere exhibit a diurnal variation in which the ionization is usually maximum late in midafternoon and minimum a few hours after midnight. Additional variations result from changes in solar activity.

Global Navigation Satellite Systems, Inertial Navigation, and Integration,
Fourth Edition. Mohinder S. Grewal, Angus P. Andrews, and Chris G. Bartone.

Companion website: www.wiley.com/go/grewal/gnss

The primary effect of the ionosphere on GNSS signals is to change the signal propagation speed as compared with that of free space. A curious fact is that the signal modulation (the code and data stream) is delayed, while the carrier phase is advanced by the same amount. Thus, the measured pseudorange using the code is larger than the correct value, while that using the carrier phase is equally smaller. The magnitude of either error is directly proportional to the total electron content (TEC) in a tube of 1 m^2 cross section along the propagation path. The TEC varies spatially due to spatial nonhomogeneity of the ionosphere. Temporal variations are caused not only by ionospheric dynamics but also by rapid changes in the propagation path due to satellite motion. The path delay for a satellite at zenith typically varies from about 1 m at night to 5–15 m during late afternoon. At low elevation angles, the propagation path through the ionosphere is much longer, so the typical corresponding delays can increase to several meters at night and as much as 50 m during the day.

Since ionospheric error is usually greater at low elevation angles, the impact of these errors could be reduced by not using measurements from satellites below a certain elevation mask angle. However, in difficult signal environments, including blockage of some satellites by obstacles, the user may be forced to use low-elevation satellites. Mask angles of 5–10° offer a good compromise between the loss of measurements and the likelihood of large ionospheric errors.

The L1-only receivers in nondifferential operation can reduce ionospheric pseudorange error by using a model of the ionosphere broadcast by the satellites, which reduces the uncompensated ionospheric delay by about 70% on the average. During the day, errors as large as 10 m at midlatitudes can still exist after compensation with this model and can be much worse with increased solar activity. Other recently developed models offer somewhat better performance. However, they still do not handle adequately the daily variability of the TEC, which can depart from the modeled value by 25% or more.

A dual-frequency receiver in nondifferential operation can take advantage of the dependence of delay on frequency to remove most of the ionospheric error. A relatively simple analysis shows that the group delay varies inversely as the square of the carrier frequency. This can be seen from the following model of the code pseudorange measurements at the L1 and L2 frequencies:

$$\rho_q = \rho_i \pm \frac{k}{f_q^2} \tag{7.1}$$

where ρ is the iono-free pseudorange, ρ_i is the measured pseudorange, and k is a constant that depends on the TEC along the propagation path. The subscript $i = 1, 2$ identifies the measurement at two GNSS frequencies (e.g. L1 and L2 frequencies), respectively, and the plus or minus sign is identified with respective code or carrier phase measurements, respectively. The two equations can be solved for both ρ and k. The solution for ρ for ionosphere free code pseudorange measurements is

$$\rho_q = \frac{f_1^2}{f_1^2 - f_2^2}\rho_1 - \frac{f_2^2}{f_1^2 - f_2^2}\rho_2 \tag{7.2}$$

where, for example, f_1 and f_2 are the L1 and L2 carrier frequencies, respectively, and ρ_1 and ρ_2 are the corresponding pseudorange measurements. Alternatively, the L1 and L5 measurements may be used if the receiver is so equipped.

An equation similar to Eq. (7.2) can be obtained for carrier phase measurements. However, in a nondifferential operation, the residual carrier phase error can be greater than either an L1 or L2 carrier wavelength, making ambiguity resolution difficult.

With a differential operation, ionospheric errors can be nearly eliminated in many applications because ionospheric errors tend to be highly correlated when the base and roving stations are in sufficiently close proximity. With two L1-only receivers separated by 25 km, the unmodeled differential ionospheric error is typically at the 10–20 cm level. At a 100-km separation, this can increase to as much as a meter. Additional error reduction using an ionospheric model can further reduce these errors by 25–50%.

7.2.1 Ionospheric Delay Model

J. A. Klobuchar's model [1, 2] for vertical ionospheric delay in seconds is given by

$$\left. T_g = \mathrm{DC} + A\left[1 - \frac{x^2}{2} + \frac{x^4}{24}\right] \text{for} |x| \le \frac{\pi}{2} \right\} \text{(s)} \tag{7.3}$$

where

$$x = \frac{2\pi(t - T_\rho)}{P} \quad \text{(rad)}$$

$\mathrm{DC} = 5$ ns (constant offset)

T_ρ = phase = 50 400 seconds

A = amplitude

P = period

t = local time of the Earth subpoint of the signal intersection with mean ionospheric height (seconds)

The algorithm assumes this latter height to be 350 km. The DC and phasing T_ρ are held constant at 5 ns and 14 hours (50 400 seconds) local time. Amplitude (A) and period (P) are modeled as third-order polynomials:

$$A = \begin{bmatrix} \sum_{n=0}^{3} \alpha_n \phi_m^n & A \ge 0 \\ \text{if } A < 0, & A = 0 \end{bmatrix} \quad \text{(s)}$$

$$P = \begin{bmatrix} \sum_{n=0}^{3} \beta_n \phi_m^n & P \geq 72\ 000 \\ \text{if } P < 72\ 000, & P = 72\ 000 \end{bmatrix} \quad \text{(s)}$$

where ϕ_m is the geomagnetic latitude of the ionospheric subpoint and α_n, β_n are coefficients selected (from 370 such sets of constants) by the global position system (GPS) master control station and placed in the satellite navigation upload message for downlink to the user. For the legacy GPS navigation (NAV) message, these data are in subframe 4, page 18, so they are seen by the receiver every 12.5 minutes, but get updated fairly infrequently as determined by the master control station.A typical value of coefficients is

$$\alpha_n = [0.8382 \times 10^{-8}, -0.745 \times 10^{-8}, -0.596 \times 10^{-7} - 0.596 \times 10^{-7}]$$
$$\beta_n = [0.8806 \times 10^{+5}, -0.3277 \times 10^{+5}, -0.1966 \times 10^{+6} - 0.1966 \times 10^{+10}]$$

The parameter ϕ_m is calculated as follows with example values:

1. Subtended Earth angle (EA) between user and satellite is given by the approximation

$$\text{EA} \approx \left(\frac{445}{\text{el} + 20}\right)^{-4} (^\circ)$$

   where el is the elevation of the satellite and, with respect to the user, equals 15.5°.
2. Geodetic latitude (lat) and longitude (long) of the ionospheric subpoint are found using the approximations

$$\text{Iono lat } \phi_I = \phi_{\text{user}} + \text{EA} \cos \text{AZ} \ (^\circ)$$
$$\text{Iono long } \lambda_I = \lambda_{\text{user}} + \frac{\text{EA} \sin \text{AZ}}{\cos \phi_I} (^\circ)$$
$$t = 4.32 \times 10^{-4} \lambda_I + \text{GPS time (seconds)}$$

   where ϕ_{user} is geodetic latitude = 41°, λ_{user} is geodetic longitude = −73°, and AZ is azimuth of the satellite with respect to the user = 112.5°.
3. The geodetic latitude is converted to a geomagnetic coordinate system using the approximation

$$\phi_m \approx \phi_I + 11.6^\circ(\lambda_I - 291^\circ) \ (^\circ)$$

4. The final step in the algorithm is to account for elevation angle effect by scaling with an obliquity scale factor (SF):

$$\text{SF} + 1 + 2\left[\frac{96^\circ - \text{el}}{90^\circ}\right]^3 \text{(unitless)}$$

With scaling, time delay due to ionospheric becomes

$$T_g = \begin{cases} \text{SF}\left[(\text{DC}) + A\left(1 - \frac{x^2}{2} + \frac{x^4}{24}\right)\right] & |x| < \frac{\pi}{2} \\ \text{SF(DC)}, & |x| < \frac{\pi}{2} \end{cases}$$

$$T_G = CT_g$$

$$C = \text{speed of light}$$

where T_g is in seconds and T_G is in meters.

The MATLAB® programs Klobuchar `fix.m` and Klobuchar pseudorandom noise (PRN) for computing ionospheric delay (for PRN = satellite number) are described in Appendix A and are programs given at companion website: www.wiley.com/go/grewal/gnss.

7.2.2 GNSS SBAS Ionospheric Algorithms

The ionospheric correction computation (ICC) algorithms enable the computation of the ionospheric delays applicable to a signal on L1 and to the GPS and wide-area reference station (WRS) L1 and L2 interfrequency biases. These algorithms also calculate grid ionospheric vertical errors (GIVEs), empirically derived error bounds for the broadcast ionospheric corrections. The ionospheric delays are employed by the space-based augmentation system (SBAS) user to correct the L1 measurements, as well as internally to correct the WRSs' L1 geostationary Earth orbit (GEO) measurement for orbit determination if dual-frequency corrections are not available from GEOs. The interfrequency biases are needed internally to convert the dual-frequency-derived SBAS corrections to single-frequency corrections for the SBAS users. The vertical ionospheric delay and GIVE information is broadcast to the SBAS user via message types 18 and 26. See the Minimum Operational Performance Standards (MOPs) for details on the content and usage of the SBAS messages [3].

The algorithms used to compute ionospheric delays and interfrequency biases are based on those originated at the Jet Propulsion Laboratory [4]. The ICC models assume that ionospheric electron density is concentrated on a thin shell of height 350 km above the mean Earth surface. The estimates of interfrequency biases and ionospheric delays are derived using a pair of Kalman filters, herein referred to as the L1L2 and ionosphere (IONO) filters. The purpose of the L1L2 filter is to estimate the interfrequency biases, while the purpose of the IONO filter is to estimate the ionosphere delays. The inputs to both filters are leveled WRS receiver slant delay measurements (L2 − L1 differential delay), which are output from the data. Both filters perform their calculations in total electron count units (TECUs) (1 m of L1 ranging delay = 6.16 TECU, and 1 m of L1 − L2 differential delay = 9.52 TECU).

Conceptually, the measurement equation is (neglecting the noise term)

$$\left.\begin{aligned} \tau_{\mathrm{TECU}} &= 9.52 \times \tau_m \\ &= 9.52 \times (t_{\mathrm{L2}_m} - t_{\mathrm{L1}_m}) \\ &= 9.52 \times (b^r_m + b^s_m) + \mathrm{TEC}_{\mathrm{TECU}} \\ &= b^r_{\mathrm{TECH}} + b^s_{\mathrm{TECH}} + \mathrm{TEC}_{\mathrm{TECU}} \end{aligned}\right\} \tag{7.4}$$

where τ is the differential delay, b^r and b^s are the interfrequency biases of the respective receiver and satellite, and TEC is the ionospheric delay. The subscripts m (meters) and TECU denote the corresponding units of each term. The ionospheric delay in meters for a signal on the L1 frequency is

$$\left.\begin{aligned} \tau^{\mathrm{L1}}_m &= 1.5457 \times \frac{1}{9.52}\mathrm{TEC}_{\mathrm{TECU}} \\ &= \frac{1}{6.16}\mathrm{TEC}_{\mathrm{TECU}} \end{aligned}\right\} \tag{7.5}$$

Both Kalman filters contain the vertical delays at the vertices of a triangular spherical grid of height 350 km fixed in the solar-magnetic (SM) coordinate frame as states. The L1L2 filter also contains interfrequency biases as states. In contrast, the IONO filter does not estimate the interfrequency biases, but instead they are periodically forwarded to the IONO filter, along with the variances of the estimates, from the L1L2 filter. Each slant measurement is modeled as a linear combination of the vertical delays at the three vertices surrounding the corresponding measurement pierce point (the intersection of the line of sight and the spherical grid), plus the sum of the receiver and satellite biases, plus noise. The ionospheric delays computed in the IONO filter are eventually transformed to a latitude–longitude grid that is sent to the SBAS users via message type 26. Because SBAS does not have any calibrated ground receivers, the interfrequency bias estimates are all relative to a single receiver designated as a reference, whose L1L2 interfrequency bias filter covariance is initialized to a small value, and to which no process noise is applied.

The major algorithms making up the ICC discussed here are:

Initialization. The L1L2 and IONO filters are initialized using either the Klobuchar model or using previously recorded data.

Estimation. The actual computation of the interfrequency biases and ionospheric delays involves both the L1L2 and IONO filters.

Thread switch. The measurements from a WRS may come from an alternate WRS receiver. In this case, the ICC must compensate for the switch by altering the value of the respective receiver's interfrequency bias state in the L1L2 filter. In the nominal case, an estimate of the L1L2 bias difference is available.

Anomaly processing. The L1L2 filter contains a capability to internally detect when a bias estimate is erroneous. Both thread switch and anomaly processing algorithms may also result in the change of the reference receiver [5].

7.2.2.1 L1L2 Receiver and Satellite Bias and Ionospheric Delay Estimations for GPS

System Model For GPS, the ionospheric delay estimation Kalman filter uses a random walk system model. A state of the Kalman filter at time t_k is modeled to be equal to that state at the previous time t_{k-1}, plus a random process noise representing the uncertainty in the transition from time t_{k-1} to time t_k; that is,

$$\mathbf{x}_k = \mathbf{x}_{k-1} + \mathbf{w}_k$$

where $\mathbf{x}_k$ is the state vector of the Kalman filter at time t_k and $\mathbf{w}_k$ is a white process noise vector with known covariance Q. The state vector $\mathbf{x}_k$ consists of three subgroups of states: the ionospheric vertical delays at triangular tile vertices, the satellite L1L2 biases, and the receiver L1L2 biases; that is,

$$\mathbf{x}_k = \begin{bmatrix} x_{1,k} \\ \vdots \\ x_{\mathrm{NV},k} \\ x_{\mathrm{NV+1},k} \\ \vdots \\ x_{\mathrm{NV+NS},k} \\ x_{\mathrm{NV+NS+1},k} \\ \vdots \\ x_{\mathrm{NV+NS+NR},k} \end{bmatrix}$$

where NV is the number of triangular tile vertices, NS is the number of GPS satellites, and NR is the number of WRSs. The values of NV, NS, and NR must be adjusted to fit the desired configuration. In simulations, one can use 24 GPS satellites in the real orbits generated by GPS Infrared Positioning System (GIPSY) using ephemeris data downloaded from the GPS bulletin board. The number of WRSs is 25 and these WRSs are placed at locations planned for SBAS operations.

Observation Model The observation model or measurement equation establishes the relationship between a measurement and the Kalman filter state vector. For any GPS satellite in view, there is an ionospheric slant delay measurement corresponding to each WRS–satellite pair. Ionospheric slant delay measurement is converted to the vertical delay at its corresponding pierce point through an obliquity factor. At any time t_k, there are approximately 80–200 pierce points, and hence the same number of ionospheric vertical delay measurements that can be used to update the Kalman filter state vector.

Denote the ionospheric vertical delay measurement at t_k for the ith satellite and jth WRS as z_{ijk}. Thus,

$$z_{ijk} = i_{ijk} + \frac{b_{si}}{q_{ijk}} + \frac{b_{sj}}{q_{ijk}} + v_{ijk}$$

where i_{ijk} is the vertical ionospheric delay at the piece point corresponding to satellite i and WRS j; b_{si} and b_{sj} are the L1L2 interfrequency biases for satellite i and WRS j, respectively; q_{ijk} is the obliquity factor; and v_{ijk} is the receiver measurement noise, white with covariance R.To establish an observation model, we need to relate i_{ijk}, b_{si}, and b_{sj} to the state vector of the ionospheric delay estimation Kalman filter. Note that b_{si} and b_{sj} are the elements of the state vector labeled NV + i and NV + NS + j, respectively. The relationship between i_{ijk} and the state vector is established below. The value i_{ijk} is modeled as a linear combination of the vertical delay values at the three vertices of the triangular tile in which the piece point is located, as shown in Figure 7.1. In Figure 7.1, assume a pierce point P is located arbitrarily in the triangular tile ABC. The ionospheric delay at pierce point P is obtained from the vertical delay values at vertices A, B, and C using a bilinear interpolation as follows. Draw a line from point A to point P and find the intersection point D between this line and the line BC. The bilinear interpolation involves two simple linear interpolations – the first yields the vertical delay value at point D from points B and C; the second yields the vertical delay value at point P from points D and A. The result can be summarized as

$$I_P = w_A I_A + w_B I_B + w_C I_C$$

where I_P, I_A, I_B, and I_C are the ionospheric vertical delay values at points P, A, B, and C, respectively, and w_A, w_B, and w_C are the bilinear weighting coefficients from points A, B, and C, respectively, to point P. The values of w_A, w_B, and w_C can be readily calculated from the geometry involved. It is

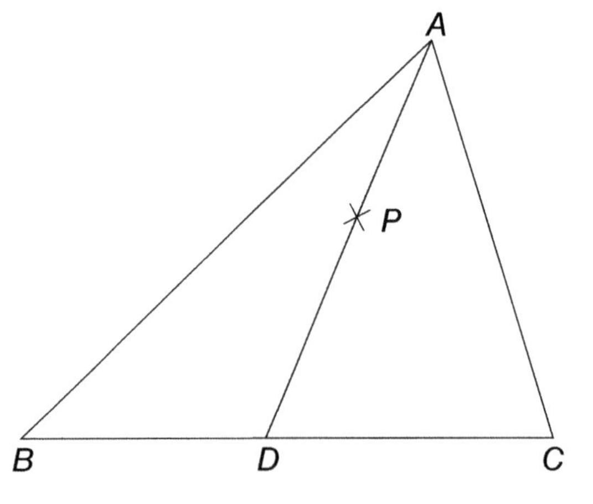

Figure 7.1 Bilinear interpolation.

recognized that I_A, I_B, and I_C are three elements of the Kalman filter state vector. In summary, the measurement equation can be written as

$$z_{ijk} = \mathbf{h}_{ijk}\mathbf{x}_k + v_{ijk}$$

where $\mathbf{h}_{ijk}$ is the measurement matrix and v_{ijk} is the measurement noise, respectively, for the pierce point measurement for the satellite with index i and WRS with index j at time t_k. Here, $\mathbf{h}_{ijk}$ is an (NV + NS + NR) dimension row vector with all elements equal to zeros except five elements. The first three of these five nonzero elements correspond to the vertices of the tile that contains the pierce point under consideration, and the other two correspond to the ith satellite and jth WRS, which yields the ionospheric slant delay measurement z_{ijk}.

UDU^T Kalman filter (See Chapter 10) As noted previously, there are approximately 180–200 pierce points at any time t_k. Each pierce point corresponds to one of the possible combinations of a satellite and a WRS, which further corresponds to an ionospheric vertical delay measurement at that pierce point. The ionospheric estimation Kalman filter is designed so that its state vector is updated upon the reception of each ionospheric vertical delay measurement.

SM-to-Earth-centered, Earth-fixed (ECEF) transformation At the end of each five-minute interval (Kalman filter cycle), the ionospheric vertical delays at the vertices of all tiles are converted from the SM coordinates to the ECEF coordinates. This conversion is completed by first transforming the SBAS ionospheric grid points (IGPs) from the ECEF coordinates to the SM coordinates. For each IGP converted to SM coordinates, the triangular tile that contains this IGP is found. A bilinear interpolation identical to the one described in Figure 5.3 is then used to calculate the ionospheric vertical delay values at this IGP. (Transformations are given in Appendix B.)

New GEOs (3rd, PRN 135 [Galaxy XV], at 133° W longitude; 4th, PRN 138° [Anik F1R], at 107° W longitude) will have L1L5 frequencies (see Chapter 8). Ionospheric delays can be calculated at the WRSs directly instead of using ionospheric delay provided by ionospheric grids from SBAS broadcast messages.

7.2.2.2 Kalman Filter

In estimating the ionospheric vertical delays in the SM coordinate system by the Kalman filter, there are three types of estimation errors:

1. Estimation error due to ionospheric slant delay measurement noise error
2. Estimation error due to the temporal variation of the ionosphere
3. Estimation error due to nonlinear spatial variation of the ionosphere

Each of the three sources of error can be individually minimized by adjusting the values of the covariances **Q** and **R**. However, the requirements to minimize the errors due to noise and temporal variations are often in conflict.

Intuitively, to minimize the measurement noise implies that we want the **Q** and **R** values to result in a Kalman gain that averages out the measurement noise; that is, we want the Kalman gain to take values so that for each new measurement, the value of innovation is small, such that a relatively large noise component of the measurement results in a relatively small estimation error. On the other hand, if we want to minimize the estimation error due to temporal variations, then we want to have a Kalman gain that can produce a large innovation, so that the component in the measurement that represents the actual ionospheric delay variation with time can be quickly reflected in the new state estimate. This suggests that we usually need to compromise in selecting the values of **Q** and **R** when a conventional nonadaptive Kalman filter is used.

Although the Kalman filter estimation error is the dominant source of error, it is not the only source. The nonlinear spatial variation introduces additional error when converting the ionospheric vertical delay estimated by the Kalman filter in the SM coordinate system to the SBAS IGP in the ECEF coordinate system. This is because bilinear interpolation is used and there is an implicit assumption that interpolation is a strictly valid procedure. However, if the actual value of the vertical delay was measured at some location, it would not be equal to the value found by interpolation. Violation of this assumption results in interpolation error during the transformation. It can be shown by simulations that, under certain conditions, this conversion error can be significant and non-negligible.

In order to isolate the sources of errors and to understand how the algorithm responds to various conditions, consider seven scenarios, with each testing one aspect of the possible estimation error, and all their possible combinations.

Scenario 1: Measurement noise. In this scenario, the ionospheric vertical delay is assumed to be a time-invariant constant anywhere over the Earth's surface. The Kalman filter estimation errors due to temporal and spatial variations are zero. For each of the ionospheric slant delay measurements, a zero-mean white Gaussian noise is added. The magnitude of the noise is characterized by its variance. The measurement noise is added to the slant delay rather than the vertical delay because this is where the actual measurement noise is introduced by a GPS receiver.

Scenario 2: Temporal variation. In this scenario, the ionosphere is assumed to be uniformly distributed spatially, but its TEC values change with time; that is, the ionospheric vertical delays vary with time, but these variations are identical everywhere. Various time variation functions, such as a sinusoidal function, a linear ramp, a step function, or an impulse function, can be used to study this scenario. In a simulation using a sinusoidal time variation function, the sinusoidal function is characterized by two parameters – its amplitude and frequency. The values of these two parameters are chosen to produce a time variation that is similar in magnitude to the ionospheric delay variation data published in the literature. The measurement noise is zero. Kalman filter estimation errors due to both the measurement noise and spatial variation are fixed at zero (for this scenario).

Scenario 3: Spatial variation. In this scenario, the ionosphere is assumed to be a constant at any fixed location when observed in the SM coordinate system. The ionospheric delays at different locations in the SM coordinate system, however, are different. Various spatial variation functions can be used to study this scenario. Here, we use a three-dimensional surface constructed from two orthogonal sinusoidal functions of varying amplitudes and frequency to model the values of ionospheric vertical delays over the Earth. The values of the parameters of the two sinusoidal functions are chosen to produce gradients in TEC similar in magnitude to the ionospheric delay variation data published in the literature. The measurement noise is zero. Kalman filter estimation errors due to both the measurement noise and temporal variations are fixed at zero for this scenario.

Scenario 4: Noise + Temporal. Scenarios 1 and 2 are combined, and the Kalman filter estimation error due to spatial variation is zero.

Scenario 5: Noise + Spatial. Scenarios 1 and 3 are combined. In this scenario, the Kalman filter estimation error due to temporal variation is zero.

Scenario 6: Temporal + Spatial. Here, the Kalman filter estimation error due to measurement noise is zero. The combined values of temporal and spatial variations define the "truth ionosphere" in the simulation.

Scenario 7: Noise + Temporal + Spatial. In this scenario, the parameters that define the "true ionosphere" and "measurement noise" can be configured to mimic any ionospheric conditions.

In the simulations, the GPS satellite orbits used are the precise orbits generated by GIPSY using GPS satellite ephemeris data downloaded from the GPS bulletin board. The WRS locations used are those currently recommended by the Federal Aviation Administration (FAA). These locations may be adjusted to evaluate the impact of other WRS locations or additional WRSs.

7.2.2.3 Selection of Q and R

Theoretically, a Kalman filter yields optimal estimation of the states of a system, given a knowledge of the system dynamics and measurement equations, when both the system process noise and measurement noise are zero-mean Gaussian at each epoch and white in time and their variances are known. However, in practice, the system dynamics are often unknown and system modeling errors are introduced when the actual system dynamics differ from the assumptions. In addition, the system process noise and the measurement noise are often non-Gaussian and their variances are not known precisely. To ensure a stable solution, a relatively large value of **Q** is often used, sacrificing estimation accuracy. Careful selection of **Q** and **R** values impacts the performance

Table 7.1 Representative Kalman filter parameter values.

Parameter term	Value	Units
L1L2 filter bias process noise update interval	300	s
L1L2 filter TEC process noise	0.05	TECU/s$^{1/2}$
L1L2 filter TEC process noise update interval	300	s
IONO filter process noise	0.05	TECU/s$^{1/2}$
IONO filter process noise update interval	300	s
IONO measures floor	9	TECU^2
IONO measures scale	0	
L1L2 filter bias process noise	4.25×10^{-4}	TECU/s$^{1/2}$
L1L2 next bias distribution time interval	300	s
L1L2 cold start bias distribution time interval	300	s
L1L2 cold start time interval	86 400	s
Iono a priori covariance matrix	$400 = 20^2$	TECU^2
L1L2 bias a priori covariance matrix	$10\,000 = 100^2$	TECU^2
(ref receiver)	10^{-10}	TECU^2
Maximum initial TEC	1000	TECU
Nominal initial TEC	25	TECU

of the Kalman filter in practical applications, including the SBAS ionospheric estimation filter.

In each phase of the validation, many parameters are tuned. The procedures and rationale involved in selecting the final values of these parameters include an effort to distinguish those parameters for which the performance is particularly sensitive. For many parameters, performance is not particularly sensitive. Table 7.1 shows typical values of the parameters used in two Kalman filters. The L1L2 filter can be eliminated. The IONO filter, including the satellite and receiver biases, may be sufficient to estimate the biases and IONO delays. This reduces the computational load and simplifies the process.

The algorithms must be validated to ensure that the estimation accuracy is good enough to ultimately support downstream precision-approach requirements. Convergence properties of the estimation algorithms must be examined, and the logic associated with restarting the estimation using recorded data must be analyzed. The capabilities to perform thread switches and to detect anomalies must be examined, and the special cases necessitating a change of reference receiver. In each phase of validation, the critical test is whether there is any significant degradation in accuracy as compared with nominal performance, and whether the nominal performance itself is adequate.

7.2.2.4 Calculation of Ionospheric Delay Using Pseudoranges

The calculation of ionospheric propagation delay from P-code and C/A-code can be formulated in terms of the following measurement equalities:

$$\rho_{\mathrm{L1}} = r + l_{\mathrm{1iono}} - c\tau_{\mathrm{RX1}} - c\tau_{\mathrm{GD}} \tag{7.6}$$

$$\rho_{\mathrm{L2}} = r + \frac{l_{\mathrm{1iono}}}{(f_2/f_1)^2} - c\tau_{\mathrm{RX2}} - \frac{c\tau_{\mathrm{GD}}}{(f_2/f_1)^2} \tag{7.7}$$

where

$$\left.\begin{array}{ll}
\rho_{\mathrm{L1}} & = \mathrm{L1\ \ pseudorange} \\
\rho_{\mathrm{L2}} & = \mathrm{L2\ \ pseudorange} \\
\rho & = \text{geometric distance between GPS satellite} \\
 & \quad \text{transmitter and GPS receiver including} \\
 & \quad \text{nondispersive contributions such as} \\
 & \quad \text{tropospheric refraction and user clock error} \\
f_1 & = \mathrm{L1\ \ frequency} \\
 & = 1572.42\ \mathrm{MHz\ \ (GPS)} \\
f_2 & = \mathrm{L2\ \ frequency} \\
 & = 1227.6\ \mathrm{MHz\ \ (GPS)} \\
\tau_{\mathrm{RXL1}} & = \text{receiver noise as manifested in code} \\
 & \quad \text{(receiver and calibration biases) at L1 (ns)} \\
\tau_{\mathrm{RXL2}} & = \text{receiver noise as manifested in code} \\
 & \quad \text{(receiver and calibration biases) at L2 (ns)} \\
T_{\mathrm{GD}} & = \text{satellite group delay (interfrequency bias)} \\
c & = \text{speed of light} \\
 & = 0.2997792458 \dfrac{\mathrm{m}}{\mathrm{ns}} \\
l_{\mathrm{1iono}} & = \text{delay at L1 (m)}
\end{array}\right\} \tag{7.8}$$

Subtracting Eq. (7.6) from Eq. (7.7), we get

$$\mathrm{L1}_{\mathrm{iono}} = \frac{\rho_{\mathrm{RL1}} - \rho_{\mathrm{RL2}}}{1 - (f_{\mathrm{L1}}/f_{\mathrm{L2}})^2} + \frac{c(\tau_{\mathrm{RXL1}} - \tau_{\mathrm{RXL2}})}{1 - (f_{\mathrm{L1}}/f_{\mathrm{L2}})^2} + cT_{\mathrm{GD}} \tag{7.9}$$

What is actually measured in the ionospheric delay is the sum of receiver bias and interfrequency bias. The biases are determined and taken out from the ionospheric delay calculation. These biases may be up to 10 ns (3 m) [6, 7]. However, the presence of ambiguities N_1 and N_2 in carrier phase measurements of L1 and L2 preclude the possibility of calculating the ionosphere delay directly and would involve the solution of the carrier cycle ambiguities iteratively.

The MATLAB® program Iono_delay (PRN#) (described in Appendix A, at www.wiley.com/go/grewal/gnss) uses pseudorange and carrier phase data from L1 and L2 signals.

7.3 Tropospheric Propagation Errors

The lower part of the Earth's atmosphere is composed of dry gases and water vapor, which lengthen the propagation path due to refraction. The magnitude of the resulting signal delay depends on the refractive index of the air along the propagation path and typically varies from about 2.5 m in the zenith direction to 10–15 m at low satellite elevation angles. The troposphere is nondispersive at the GNSS frequencies, so that delay is not frequency dependent. In contrast to the ionosphere, tropospheric path delay is consequently the same for code and carrier signal components. Therefore, this delay cannot be measured by utilizing both L1 and L2 pseudorange measurements, and either models and/or differential techniques must be used to reduce the error.

The refractive index of the troposphere consists of that due to the dry-gas component and the water vapor component, which respectively contributes about 90% and 10% of the total delay. Knowledge of the temperature, pressure, and humidity along the propagation path can determine the refractivity profile, but such measurements are seldom available to the user. However, using standard atmospheric models for the dry delay permits determination of the zenith delay to within about 0.5 m and with an error at other elevation angles that approximately equals the zenith error times the cosecant of the elevation angle. These standard atmospheric models are based on the laws of ideal gases and assume spherical layers of constant refractivity with no temporal variation and an effective atmospheric height of about 40 km. Estimation of dry delay can be improved considerably if surface pressure and temperature measurements are available, bringing the residual error down to within 2–5% of the total.

The component of tropospheric delay due to water vapor (at altitudes up to about 12 km) is much more difficult to model because there is considerable spatial and temporal variation of water vapor in the atmosphere. Fortunately, the wet delay is only about 10% of the total, with values of 5–30 cm in continental midlatitudes. Despite its variability, an exponential vertical profile model can reduce it to within about 2–5 cm.

In practice, a model of the standard atmosphere at the antenna location would be used to estimate the combined zenith delay due to both wet and dry components. Such models use inputs such as the day of the year and the latitude and altitude of the user. The delay is modeled as the zenith delay multiplied by a factor that is a function of the satellite elevation angle. At zenith, this factor is unity, and it increases with decreasing elevation angle as the length of the propagation path through the troposphere increases. Typical values of the multiplication factor are 2 at 30° elevation angle, 4 at 15°, 6 at 10°, and 10 at 5°. The accuracy of the model decreases at low elevation angles, with decimeter level errors at zenith and about 1 m at 10° elevation.

Much research has gone into the development and testing of various tropospheric models. Excellent summaries of these appear in the literature [8–10].

For non-differential users, a simple troposphere model to estimate the delay is represented as:

$$\tau_{\text{tropo}} = \underbrace{2.47e^{-1.33\times10^{-4}h_u}}_{\text{ZTD part}} \underbrace{\left(\frac{1}{\sin(\theta)+0.0121}\right)}_{\text{mapping function part}} \tag{7.10}$$

where,

h_u = the mean sea level (MSL) height of the user (m).

θ = the elevation angle to the satellite.

Other higher fidelity models that have less error for low elevation satellites are the GBAS model [3] or the University of New Brunswick 3 model [ref UNB3 web page, http://gauss.gge.unb.ca/UNB3/frame4.html].

Although a GNSS receiver cannot measure pseudorange error due to the troposphere, differential operation can usually reduce the error to small values by taking advantage of the high spatial correlation of tropospheric errors at two points within 0–100 km on the Earth's surface. However, exceptions often occur when storm fronts pass between the receivers, causing large gradients in temperature, pressure, and humidity.

7.4 The Multipath Problem

Multipath propagation of the GNSS signal is a dominant source of error in positioning, especially in differential GNSS architectures. Objects in the vicinity of a receiver antenna (notably the ground) can easily reflect GNSS signals, resulting in one or more secondary propagation paths. These secondary-path signals, which are superimposed on the desired direct-path signal, always have a longer propagation time and can significantly distort the amplitude and phase of the direct-path signal.

Errors due to multipath cannot be reduced by the use of differential GNSS since they depend on local reflection geometry near each receiver antenna. In a receiver without multipath protection, C/A-code ranging errors of 10 m or more can be experienced. Multipath cannot only cause large code ranging errors but can also severely degrade the ambiguity resolution process required for carrier phase ranging such as that used in precision surveying applications.

Multipath propagation can be divided into two classes: static and dynamic. For a stationary receiver, the propagation geometry changes slowly as the satellites move across the sky, making the multipath parameters essentially constant for perhaps several minutes. However, in mobile applications, there can be rapid fluctuations in fractions of a second. Therefore, different multipath mitigation techniques are generally employed for these two types of multipath environments. For these mobile application smoothing can be used to help mitigate these high frequency multipath errors, which is especially effective with multifrequency GNSS receivers. A significant amount of research has focused on static applications, such as surveying, where the multipath error chances at a low rate, and often there are greater demand for high accuracy (e.g. reference stations, surveying, etc.). For this reason, we will concentrate our attention to the static case.

7.4.1 How Multipath Causes Ranging Errors

To facilitate an understanding of how multipath causes ranging errors, several simplifications can be made that in no way obscure the fundamentals involved. We will assume that the receiver processes only the C/A-code and that the received signal has been converted to complex (i.e. analytic) form at baseband (nominally zero frequency), where all Doppler shift has been removed by a carrier tracking phase-lock loop. It is also assumed that the GNSS navigation data modulation has been removed from the signal, which can be achieved by standard techniques. When no multipath is present, the received waveform is represented by

$$r(t) = ae^{j\phi}c(t-\tau) + n(t) \tag{7.11}$$

where $c(t)$ is the normalized, undelayed C/A-code waveform as transmitted; r is the signal propagation delay; a is the signal amplitude; ϕ is the carrier phase; and $n(t)$ is the Gaussian receiver thermal noise having flat power spectral density. Pseudoranging consists of estimating the delay parameter τ. As we have previously seen, an optimal estimate (i.e. a minimum-variance unbiased estimate) of τ can be obtained by forming the cross-correlation function

$$R(\tau) \int_{T_1}^{T_2} r(t)c_r(t-\tau)dt \tag{7.12}$$

of $r(t)$ with a replica $c_r(t)$ of the transmitted C/A-code and choosing as the delay estimate that value of τ that maximizes this function. Except for an error due to receiver thermal noise, this occurs when the received and replica waveforms are in time alignment. A typical cross-correlation function without multipath for C/A-code receivers having a 2 MHz precorrelation bandwidth is shown by the solid lines Figure 7.2 (these plots ignore the effect of noise, which would add small random variations to the curves).

If multipath is present with a single secondary path, the waveform of Eq. (7.11) changes to

$$r(t) = ae^{j\phi_1}c(t-\tau_1) + be^{j\phi_2}c(t-\tau_2) + n(t) \tag{7.13}$$

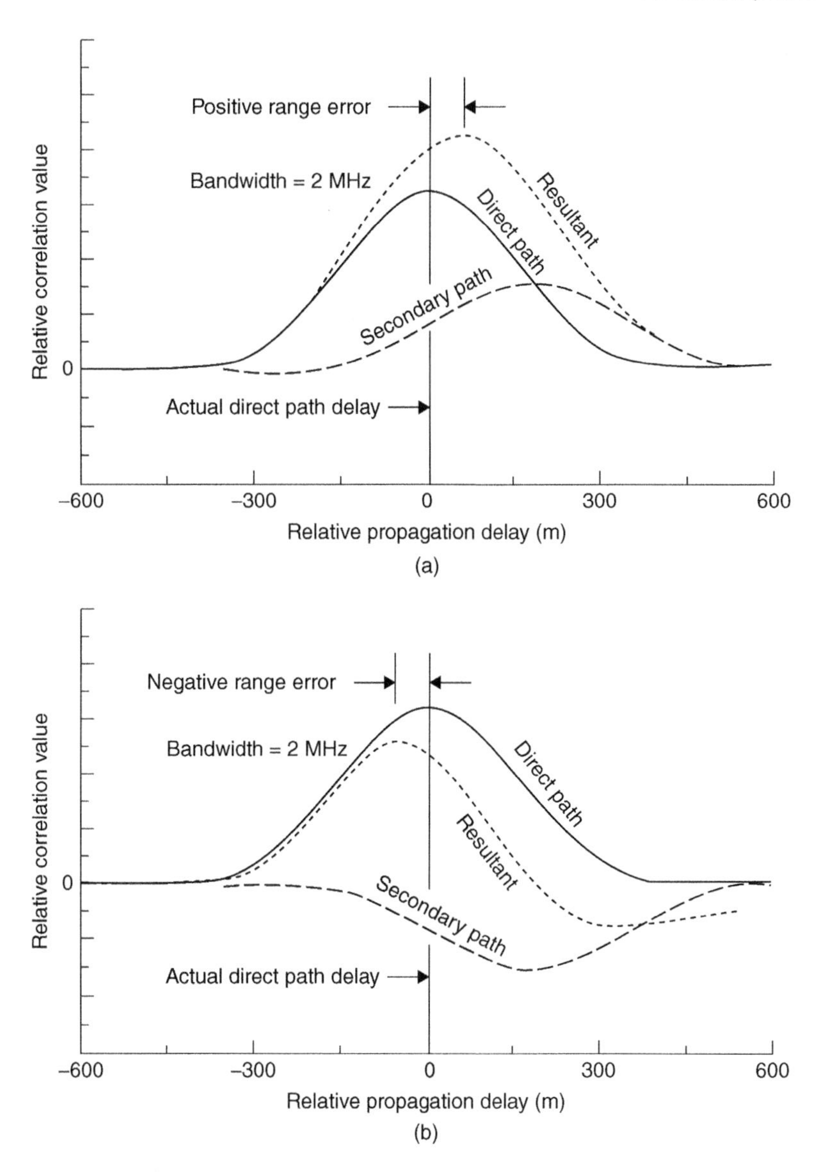

Figure 7.2 Effect of multipath on C/A-code cross-correlation function.

where the direct and secondary paths have respective propagation delays τ_1 and τ_2, amplitudes a and b, and carrier phases ϕ_1 and ϕ_2. In a receiver not designed expressly to handle multipath, the resulting cross-correlation function will now have two superimposed components, one from the direct path and one from the secondary path. The result is a function with a distortion depending on the relative amplitude, delay, and phase of the secondary-path signal, as illustrated at the top of Figure 7.2 for an in-phase secondary path and at the bottom of Figure 7.2 for an out-of-phase secondary path. Most importantly, the location of the peak of the function has been displaced from its correct position, resulting in a pseudorange error.

In vintage receivers employing standard code tracking techniques (early and late codes separated by one C/A-code chip), the magnitude of pseudorange error caused by multipath can be quite large, reaching 70–80 m for a secondary-path signal one-half as large as the direct-path signal and having a relative delay of approximately 250 m. Further details can be found in Ref. [11].

7.5 Methods of Multipath Mitigation

Processing against slowly changing multipath can be broadly separated into two classes: spatial processing and time-domain processing. Spatial processing uses antenna design in combination with known or partially known characteristics of signal propagation geometry to isolate the direct-path received signal. In contrast, time domain processing operates on the multipath-corrupted signal within the receiver to mitigate the multipath on the measurements before they are applied to the position, velocity and time (PVT) solution method.

7.5.1 Spatial Processing Techniques

7.5.1.1 Antenna Location Strategy

Two of the main factors that affect the multipath error are the magnitude and delay of the multipath signal, relative to the direct signal received at the user antenna from the GNSS satellite. A simple, but often unpractical technique that minimizes ever-present ground signal reflections is to place the receiver antenna directly at ground level. This causes the point of ground reflection to be essentially coincident with the antenna location so that the secondary path has very nearly the same delay as the direct path. Clearly, such antenna location strategies may not always be possible but can be very effective when feasible. In generally, the antenna should be placed at a high enough location so that multipath reflections do not come into the antenna at positive elevation angles. It should be kept in mind that the higher the antenna is mounted from reflecting surfaces, the greater the multipath delay, which will intern, increase the multipath error, but at the same time, the multipath reflection will come into the antenna at a more negative elevation angle, which will decrease the multipath error.

7.5.1.2 Ground Plane Antennas

The most common form of spatial processing is an antenna designed to attenuate signals reflected from the ground. A simple design uses a metallic ground plane disk centered at the base of the antenna to shield the antenna from below. A deficiency of this design is that when the signal wave fronts arrive at the disk edge from below, they induce surface waves on the top of the disk that then travel to the antenna. The surface waves can be minimized by replacing the ground plane with a choke ring, which is essentially a ground plane containing a series of concentric circular troughs one-quarter wavelength deep. These troughs act as transmission lines shorted at the bottom ends so that their top ends exhibit very high impedance at the GNSS carrier frequency. Therefore, induced surface waves cannot form, and signals that arrive from below the horizontal plane are significantly attenuated. However, the size, weight, and cost of a choke ring antenna are significantly greater than those of simpler designs. Most importantly, the choke ring cannot effectively attenuate secondary-path signals arriving from above the horizontal, such as those reflecting from buildings or other structures. Nevertheless, such antennas have proven to mitigate multipath when signal ground bounce is the dominant source of multipath, particularly in GNSS surveying applications.

7.5.1.3 Directive Antenna Arrays

A more advanced form of spatial processing uses antenna arrays to form a highly directive spatial response pattern with high gain in the direction of the direct-path signal and attenuation in directions from which secondary-path signals arrive. However, inasmuch as signals from different satellites have different directions of arrival and different multipath geometries, many directivity patterns must be simultaneously operative, and each must be capable of adapting to changing geometry as the satellites move across the sky. For these reasons, highly directive arrays seldom are practical or affordable for most applications, which would require a large number of antenna elements within the array or implement a digital beam forming array and digital receiver, whereby each satellite signal is sampled, and individually processed [12].

7.5.1.4 Long-Term Signal Observation

If a GNSS signal is observed for sizable fractions of an hour to several hours, one can take advantage of changes in multipath geometry caused by satellite motion. This motion causes the relative delays between the direct and secondary paths to change, resulting in measurable variations in the received signal. For example, a periodic change in signal level caused by alternate phase reinforcement and cancellation by the reflected signals is often observable. Although a variety of algorithms have been proposed for extracting the direct-path signal component from measurements of the received signal, the need for long observation times rules out this technique for most applications. However, it can be an effective method of multipath mitigation at a fixed site, such as at a differential GNSS base station. In this case, it is even possible to observe the same satellites from one day to the next, looking for patterns of pseudorange or phase measurements that repeat daily.

Multipath calculation from long-term observations. Delays can be computed as follows by using pseudoranges and carrier phases over long signal observations (one day to next). This technique may be ruled out for most applications. Ambiguities and cycle slips have been eliminated or mitigated.

Let

$$\left.\begin{array}{ll} \lambda_1 & = 19.03\ \text{cm, wavelength of L1 (GPS)} \\ \lambda_1 & = 24.42\ \text{cm, wavelength of L2 (GPS)} \\ \phi_{\text{L1}} & = \text{carrier phase for L1 (cycles)} \\ \phi_{\text{L2}} & = \text{carrier phase for L2 (cycles)} \\ f_1 & = \text{L1 frequency} = 1575.42\ \text{MHz (GPS)} \\ f_2 & = \text{L2 frequency} = 1227.6\ \text{MHz (GPS)} \\ \rho & = \text{error free pseudorange} \\ \rho_{\text{L1}} & = \text{pseudorange L1 (m)} \\ \rho_{\text{L2}} & = \text{pseudorange L2 (m)} \\ I & = \text{ionospheric delay} \\ I_{\text{L1}} & = \text{ionospheric delay at L1 (m)} \\ \text{MP}_{\text{L1}} & = \text{multipath in L1} \\ \text{MP}_{\text{L2}} & = \text{multipath in L2} \end{array}\right\} \tag{7.14}$$

For dual-frequency GNSS receivers, one obtains

$$\lambda_1\phi_{\text{L1}} = r - \frac{I}{(f_1)^2} \tag{7.15}$$

$$\lambda_2\phi_{\text{L2}} = r - \frac{I}{(f_2)^2} \tag{7.16}$$

Subtracting Eq. (7.16) from Eq. (7.15), one can obtain

$$\left.\begin{array}{r} \lambda_1\phi_{\text{L1}} - \lambda_2\phi_{\text{L2}} = \dfrac{I(f_1)^2 - I(f_2)^2}{(f_1)^2(f_2)^2} \\ I_{\text{L1}} = \dfrac{(\lambda_1\phi_{\text{L1}} - \lambda_2\phi_{\text{L2}})(f_2)^2}{(f_1)^2 - (f_2)^2} \end{array}\right\} \tag{7.17}$$

$$\left.\begin{array}{r} K = \dfrac{(f_2)^2}{(f_1)^2 - (f_2)^2} \\ I_{\text{L1}} = K(\lambda_1\phi_{\text{L1}} - \lambda_2\phi_{\text{L2}}) \\ \rho_{\text{L1}} = r + \dfrac{1}{(f_1)^2} \end{array}\right\} \tag{7.18}$$

Subtracting Eq. (7.15) from Eq. (7.18), one obtains the multipath as

$$\text{MP}_{\text{L1}} = \rho_{\text{L1}} - \lambda_1\phi_{\text{L1}} - 2I_{\text{L1}} \tag{7.19}$$

where

$$I_{L1} = \frac{I}{(f_1)^2}$$

Substitute Eq. (7.17) into Eq. (7.19) to obtain

$$\left.\begin{aligned} \mathrm{MP}_{L1} &= \rho_{L1} - \lambda_1\phi_{L1} - 2K(\lambda_1\phi_{L1} - \lambda_2\phi_{L2}) \\ &= \rho_{L1} - [(1+2K)\lambda_1\phi_{L1} - 2K\lambda_2\phi_{L2}] \end{aligned}\right\} \tag{7.20}$$

7.5.2 Time-Domain Processing

Although time-domain processing against GNSS multipath errors has been the subject of active research for at least two decades, there is still much to be learned, both at theoretical and practical levels. Most of the practical approaches have been developed by receiver manufacturers, who are often reluctant to explicitly reveal their methods. Nevertheless, enough information about multipath processing exists to gain insight into its recent evolution.

7.5.2.1 Narrow-Correlator Technology (1990–1993)

The first significant means to reduce GPS multipath effects by receiver processing made its debut in the early 1990s. Until that time, most receivers had been designed with a 2 MHz precorrelation bandwidth that encompassed most, but not all, of the GPS C/A spread-spectrum signal power. These receivers also used one-chip spacing between the early and late reference C/A-codes in the code tracking loops. However, the 1992 paper [13] makes it clear that using a significantly larger bandwidth combined with much closer spacing of the early and late reference codes would dramatically improve the ranging accuracy both with and without multipath. It is somewhat surprising that these facts were not recognized earlier by the GNSS community, given that they had been well known in radar circles for many decades.

A 2 MHz precorrelation bandwidth causes the peak of the direct-path cross-correlation function to be severely rounded, as illustrated in Figure 7.2. Consequently, the sloping sides of a secondary-path component of the correlation function can significantly shift the location of the peak, as indicated in the figure. The result of using an 8 MHz bandwidth is shown in Figure 7.3, where it can be noted that the sharper peak of the direct-path cross-correlation function is less easily shifted by the secondary-path component. It can also be shown that at larger bandwidths, the sharper peak is more resistant to disturbance by receiver thermal noise, even though the precorrelation signal-to-noise ratio is increased.

Another advantage of a larger precorrelation bandwidth is that the spacing between the early and late reference codes in a code tracking loop can be made smaller without significantly reducing the gain of the loop, hence the term narrow correlator. It can be shown that this causes the noises on the early and late correlator outputs to become more highly correlated, resulting in less noise on the loop error signal. An additional benefit is that the code tracking loop will be affected only by the multipath-induced distortions near the peak of the correlation function.

7.5.2.2 Leading-Edge Techniques

Because the direct-path signal always precedes secondary-path signals, the leading (left-hand) portion of the correlation function is uncontaminated by multipath, as is illustrated in Figure 7.3. Therefore, if one could measure the location of just the leading part, it appears that the direct-path delay could be determined with no error due to multipath. Unfortunately, this seemingly happy state of affairs is illusory. With a small direct-/secondary-path separation, the uncontaminated portion of the correlation function is a minuscule piece at the extreme left, where the curve just begins to rise. In this region, not only is the signal-to-noise ratio relatively poor, but also the slope of the curve is relatively small, which severely degrades the accuracy of delay estimation.

For these reasons, the leading-edge approach best suits situations with a moderate to large direct-/secondary-path separation. However, even in these cases, there is the problem of making the delay measurement insensitive to the slope of the correlation function leading edge, which can vary with signal strength. Such a problem does not occur when measuring the location of the correlation function peak.

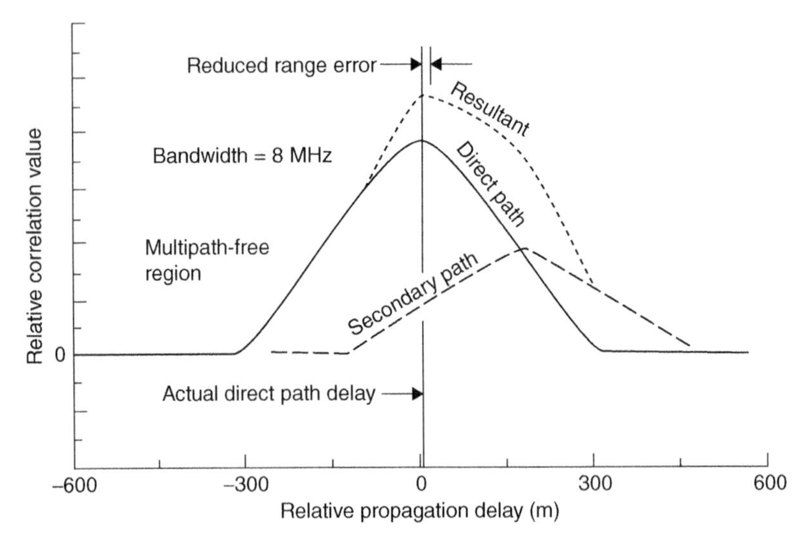

Figure 7.3 Reduced multipath error with larger precorrelation bandwidth.

7.5.2.3 Correlation Function Shape-Based Methods

Some GNSS receiver designers have attempted to determine the parameters of the multipath model from the shape of the correlation function. The idea has merit, but for best results, many correlations with different values of reference code delay are required to obtain a sampled version of the function shape. Another practical difficulty arises in attempting to map each measured shape into a corresponding direct-path delay estimate. Even in the simple two-path model (Eq. (7.12)), there are six signal parameters, so that a very large number of correlation function shapes must be handled. An example of a heuristically developed shape-based approach called the early–late slope (ELS) method can be found in Ref. [14], while a method based on maximum-likelihood estimation (MLE) called the multipath-estimating delay-lock loop (MEDLL) is described in Ref. [15].

7.5.2.4 Modified Correlator Reference Waveforms

Another new approach to multipath mitigation alters the waveform of the correlator reference PRN code to provide a cross-correlation function with inherent resistance to errors caused by multipath. Examples include the strobe correlator [16], the use of special code reference waveforms to narrow the correlation function developed in Refs. [17, 18], and the gated correlator developed in Ref. [19]. These techniques take advantage of the fact that the range information in the received signal resides primarily in the chip transitions of the C/A-code. By using a correlator reference waveform that is not responsive to the flat portions of the C/A-code, the resulting correlation function can be narrowed down to the width of a chip transition, thereby being almost immune to multipath having a primary/secondary-path separation greater than 30–40 m. An example of such a reference waveform and the corresponding correlation function are shown in Figure 7.4.

7.5.3 Multipath Mitigation Technology (MMT)

Yet another approach to time-domain multipath mitigation is called MMT and incorporated a number of GNSS receivers manufactured by NovAtel Corporation of Canada. The MMT technique not only reaches theoretical performance limits described in Section 7.6 for both code and carrier phase ranging but also, compared with existing approaches, has the advantage that its performance improves as the signal observation time is lengthened. A description of MMT follows and also appears in a patent [20].

7.5.3.1 Description

MMT is based on MLE. Although the theory of MLE is well-developed, its application to GNSS multipath mitigation has not been feasible until now due to the large amount of computation required. However, recent mathematical

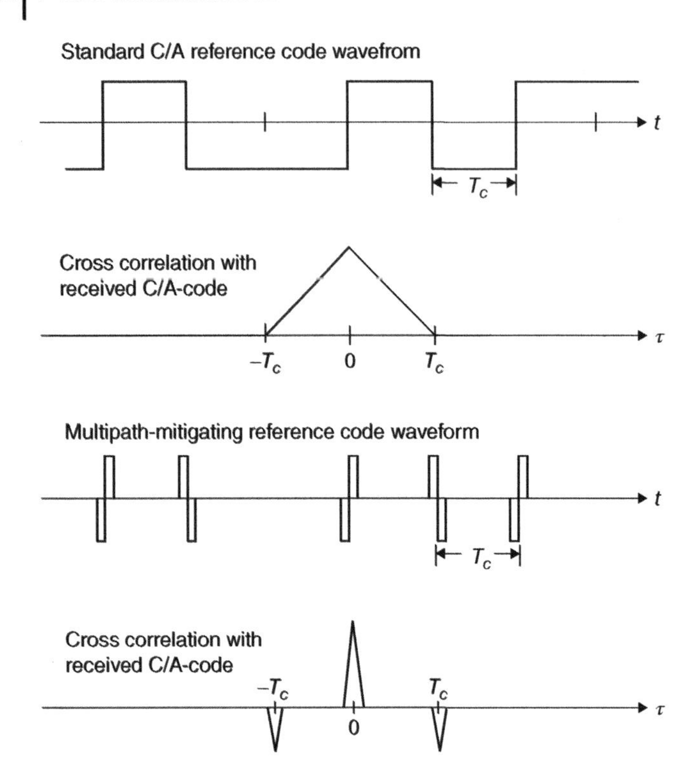

Figure 7.4 Multipath-mitigating reference code waveform.

breakthroughs have solved this problem. Before introducing the MMT algorithm, we first briefly describe the process of MLE in the context of the multipath problem.

7.5.3.2 Maximum-Likelihood (ML) Multipath Estimation

MLE is described in detail in Chapter 10. Its application to multipath mitigation is described in the following text.

7.5.3.3 The Two-Path ML Estimator (MLE)

The simplest ML estimator designed for multipath is based on a two-path model (one direct path and one secondary delayed path). For simplicity in describing MMT, we consider only this model, although generalization to additional paths is straightforward, and the MMT algorithm can be implemented for such cases. It is assumed that the received signal has been frequency-shifted to baseband, and the navigation data have been stripped off. The two-path signal model is

$$r(t) = A_1 e^{j\phi_1} m(t - \tau_1) + A_2 e^{j\phi_2} m(t - \tau_2) + n(t) \tag{7.21}$$

In this model, the parameters A_1, ϕ_1, τ_1, respectively, are the direct-path signal amplitude, phase, and delay, and the parameters A_2, ϕ_2, and τ_2 are the corresponding parameters for the secondary path. The code modulation is denoted by $m(t)$, and the noise function $n(t)$ is an additive zero-mean complex Gaussian noise process with a flat power spectral density. It will be convenient to group the multipath parameters into the vector

$$\overline{\theta} = [A_1, \phi_1, \tau_1, A_2, \phi_2, \tau_1] \tag{7.22}$$

Observation of the received signal $r(t)$ is accomplished by sampling it on the time interval $[0, T]$ to produce a complex observed vector $\overline{r}$.

The ML estimate of the multipath parameters is the vector $\overline{\theta}$ of parameter values that maximizes the likelihood function $p(\overline{r} \mid \overline{\theta})$, which is the probability density of the received signal vector conditioned on the values of the multipath parameters. In this maximization, the vector $\overline{r}$ is held fixed at its observed value. Within the vector $\hat{\overline{\theta}}$, the estimates $\hat{\tau}_1$ and $\hat{\phi}_1$ of direct-path delay and carrier phase are normally the only ones of interest. However, the ML estimate of these parameters requires that the likelihood function $p(\overline{r} \mid \overline{\theta})$ be maximized over the six-dimensional (6D) space of *all* multipath parameters (components of 0). For this reason, the unwanted parameters are called *nuisance parameters.*

Since the natural logarithm is a strictly increasing function, the maximization of $p(\overline{r} \mid \overline{\theta})$ is equivalent to maximization of $L(\overline{r}; \overline{\theta}) = \ln p(\overline{r} \mid \overline{\theta})$, which is called the log-likelihood function. The log-likelihood function is often simpler than the likelihood function itself, especially when the noise in the observations is additive and Gaussian. In our application this is the case.

Maximization of $L(\overline{r}; \overline{\theta})$ by standard techniques is a daunting task. A brute-force approach is to find the maximum by a search over the 6D multipath parameter space, but it takes too long to be of practical value. Reliable gradient-based or hill climbing methods are too slow to be useful. Finding the maximum using differential calculus is difficult because of the nonlinearity of the resulting equations and the possibility of local maxima that are not global maxima. Iterative solution techniques are often difficult to analyze and may not converge to the correct solution in a timely manner, if they converge at all. As we shall see, the MMT algorithm solves these problems by reducing the dimensionality of the search space.

7.5.3.4 Asymptotic Properties of ML Estimators

MLE is used by MMT not only because it can be made computationally simple enough to be practical but also because ML estimators have desirable asymptotic properties (Asymptotic refers to the behavior of an estimator when the error becomes small. In GNSS, this occurs when E/N_0 is sufficiently large.):

1. The ML estimate of a parameter asymptotically converges in probability to the true parameter value.

2. The ML estimate is asymptotically efficient; that is, the ratio of the variance of the estimation error to the Cramer–Rao bound approaches unity.
3. The ML estimate is asymptotically Gaussian.

7.5.3.5 The MMT Multipath Mitigation Algorithm

The MMT algorithm uses several mathematical techniques to solve what would otherwise be intractable computational problems. The first of these is a nonlinear transformation on the multipath parameter space to permit rapid computation of a log-likelihood function that has been partially maximized with respect to all of the multipath parameters except for the path delays. Thus, final maximization requires a search in only two dimensions for the two-path case, aided by acceleration techniques.

A new method of signal compression, described in Section 7.5.3.10, is used to transform the received signal into a very small vector on which MMT can operate very rapidly.

A major advantage of the MMT algorithm is that its performance improves with increasing E/N_0, the ratio of signal energy E to noise power spectral density. This is not true for most GNSS multipath mitigation methods because their estimation error is in the form of an irreducible bias. Additionally, the MMT algorithm provides ML estimates of all parameters in the multipath model and can utilize known bounds on the magnitudes of the secondary paths, if available, to improve performance.

7.5.3.6 The MMT Baseband Signal Model

In the complex baseband signal $r(t)$ given by Eq. (7.21), it is assumed that the signal has been Doppler-compensated and stripped of the 50 bps navigation data modulation. In developing the MMT algorithm, it is useful to separate $r(t)$ into its real component, $x(t)$, and imaginary component, $y(t)$:

$$\left.\begin{aligned} x(t) &= A_1 \cos\phi_1 m(t-\tau_1) + A_2 \cos\phi_2 m(t-\tau_2) + n_x(t) \\ y(t) &= A_1 \sin\phi_1 m(t-\tau_1) + A_2 \sin\phi_2 m(t-\tau_2) + n_y(t) \end{aligned}\right] \tag{7.23}$$

where $n_x(t)$ and $n_y(t)$ are independent, real-valued, zero-mean Gaussian noise processes with flat power spectral density.

7.5.3.7 Baseband Signal Vectors

The real and imaginary signal components are synchronously sampled on $[0, T]$ at the Nyquist rate $2\,W$, corresponding to the low-pass baseband bandwidth W, to produce the vectors

$$\left.\begin{aligned} \bar{x} &= (x_1,\ x_2, \dots, x_M) \\ \bar{y} &= (y_1,\ y_2, \dots, x_M) \end{aligned}\right] \tag{7.24}$$

in which the noise components of distinct samples are essentially uncorrelated (hence independent, since the noise is Gaussian).

7.5.3.8 The Log-Likelihood Function

The ML estimates of the six parameters in the vector $\overline{\theta}$ given by Eq. (7.22) are obtained by maximizing the log-likelihood function with respect to these parameters. For MMT, the log-likelihood function is

$$\left.\begin{aligned} L(\overline{x},\overline{y}\mid\overline{\theta}) &= \ln[p(\overline{x},\overline{y}\mid\overline{\theta})] \\ &= \ln C_1 \\ -C_2\sum_{k=1}^{M} &\begin{bmatrix} x_k - A_1\cos\theta_1 m_k(\tau_1) \\ -A_2\cos\theta_2 m_k(\tau_2) \end{bmatrix}^2 \\ -C_2\sum_{k=1}^{M} &\begin{bmatrix} y_k - A_1\cos\theta_1 m_k(\tau_1) \\ -A_2\cos\theta_2 m_k(\tau_2) \end{bmatrix}^2 \end{aligned}\right\} \tag{7.25}$$

where

$$\begin{aligned} C_1 &= \left(\frac{1}{\sqrt{2\pi\sigma}}\right)^M \\ C_2 &= \frac{1}{2\pi\sigma^2} \end{aligned} \tag{7.26}$$

σ^2 = noise variance of $x(t)$ and $y(t)$

$m_k(\tau_1)$ = kth sample of $m(t-\tau_1)$

$m_k(\tau_2)$ = kth sample of $m(t-\tau_2)$

Replacing the summations in Eq. (7.25) by integrals and utilizing the fact that C_1 and $-C_2$ are negative constants that do not depend on the multipath parameters, maximization of Eq. (7.25) is equivalent to *minimization* of

$$\left.\begin{aligned} \Gamma = \int_0^T &\begin{bmatrix} x(t) - A_1\cos\phi_1 m(t-\tau_1) \\ -A_2\cos\phi_2 m_k(t-\tau_2) \end{bmatrix}^2 dt \\ +\int_0^T &\begin{bmatrix} y(t) - A_1\sin\phi_1 m(t-\tau_1) \\ -A_2\sin\phi_2 m_k(t-\tau_2) \end{bmatrix}^2 dt \end{aligned}\right\} \tag{7.27}$$

with respect to the six multipath parameters. This is a highly coupled, nonlinear minimization problem on the 6D space spanned by the parameters A_1, ϕ_1, τ_1, A_2, ϕ_2, and τ_2. Standard minimization techniques such as a gradient search on this space or ad hoc iterative approaches are either unreliable or too slow to be useful.However, a major breakthrough results by using the invertible transformation

$$\left.\begin{aligned} a &= A_1\cos\phi_1 \quad & c &= A_1\sin\phi_1 \\ b &= A_2\cos\phi_2 \quad & c &= A_2\sin\phi_2 \end{aligned}\right\} \tag{7.28}$$

When this transformation is applied and the integrands in Eq. (7.27) are expanded, the problem becomes one of minimizing

$$\left.\begin{aligned}\Gamma = & \int_0^T [x^2(t) + y^2(t)]\ dt \\ & +(a^2 + b^2 + c^2 + d^2)R_{mm}(0) \\ & -2aR_{xm}(\tau_1) - 2bR_{xm}(\tau_2) + 2abR_{mm}(\tau_1 - \tau_2) \\ & -2cR_{ym}(\tau_1) - 2dR_{ym}(\tau_2) + 2cdR_{mm}(\tau_1 - \tau_2)\end{aligned}\right\} \tag{7.29}$$

Note that Γ in Eq. (7.29) is quadratic in a, b, c, and d and uses the correlation functions

$$\left.\begin{aligned}R_{xm}(\tau) &= \int_0^T x(t)m(t-\tau)\ dt \\ R_{ym}(\tau) &= \int_0^T y(t)m(t-\tau)\ dt \\ R_{mm}(\tau) &= \int_0^T m(t)m(t-\tau)\ dt\end{aligned}\right\} \tag{7.30}$$

Thus, minimization of Eq. (7.29) with respect to a, b, c, and d can be accomplished by taking partial derivatives with respect to these parameters, resulting in the linear system

$$\left.\begin{aligned}0 &= \frac{\partial\Gamma}{\partial a} = 2aR_{mm}(0) - 2R_{xm}(\tau_1) + 2bR_{mm}(\tau_1 - \tau_2) \\ 0 &= \frac{\partial\Gamma}{\partial b} = 2bR_{mm}(0) - 2R_{xm}(\tau_2) + 2aR_{mm}(\tau_1 - \tau_2) \\ 0 &= \frac{\partial\Gamma}{\partial c} = 2cR_{mm}(0) - 2R_{ym}(\tau_1) + 2dR_{mm}(\tau_1 - \tau_2) \\ 0 &= \frac{\partial\Gamma}{\partial d} = 2dR_{mm}(0) - 2R_{ym}(\tau_2) + 2cR_{mm}(\tau_1 - \tau_2)\end{aligned}\right] \tag{7.31}$$

For each pair of values of τ_1 and τ_2, this linear system can be explicitly solved for the minimizing values of a, b, c, and d. Thus, the space to be searched for a minimum of (7.29) (i.e. Eq. (7.29)) is now 2D instead of 6D. The minimization procedure is as follows. Search the (τ_1, τ_2) domain. At each point (τ_1, τ_2), compute the values of the correlation functions in the system (Eq. (7.31)) and then solve the system to find the values of a, b, c, and d that minimize Γ at that point. Identify the point $(\hat{\tau}_1, \hat{\tau}_2,)_{\text{ML}}$ where the smallest of all such minima is obtained, as well as the associated minimizing values of a, b, c, and d. Transform these values of a, b, c, and d back to the estimates $\widehat{A}_{1\text{ML}}, \widehat{A}_{2\text{ML}}, \ \widehat{\phi}_{1\text{ML}}, \ \widehat{\phi}_{2\text{ML}}$ by using the inverse of transformation (7.26), which is

$$\left.\begin{aligned}A_1 &= \sqrt{a^2 + c^2} & A_2 &= \sqrt{b^2 + d^2} \\ \phi_1 &= \arctan\ 2(a, c) & \phi_2 &= \arctan\ 2(b, d)\end{aligned}\right\} \tag{7.32}$$

7.5.3.9 Secondary-Path Amplitude Constraint

In the majority of multipath scenarios, the amplitudes of secondary-path signals are smaller than that of the direct path. The multipath mitigation performance of MMT can be significantly improved by minimizing Γ in Eq. (7.29) subject to the constraint

$$\frac{A_2}{A_1} \leq \alpha \tag{7.33}$$

where α is a positive constant (a typical value is 0.7). The constraint in terms of the transformed parameters a, b, c, and d is

$$b^2 + d^2 \leq \alpha^2(a^2 + c^2) \tag{7.34}$$

The constrained minimization of Eq. (7.29) uses the method of Lagrange multipliers.

7.5.3.10 Signal Compression

In the MMT algorithm, the correlation functions $R_{xm}(\tau)$, $R_{ym}(\tau)$, and $R_{mm}(\tau)$ defined by Eq. (7.30) and appearing in Eq. (7.31) are computed very rapidly by first using a process called signal compression, in which the large number of signal samples (on the order of 10^8–10^9) that would normally be involved is reduced to only a few tens of samples (the exact number depends on which type of GNSS signal is being processed). This processing is easily done in real time.

The correlation functions appearing in Eq. (7.30) have the form

$$R(\tau) = \int_0^T r(t)m(t-\tau)\,dt \tag{7.35}$$

where $r(t)$ is a given function and $m(t)$ is a replica of the code modulation, which includes the effects of filtering in the satellite and receiver. The calculation of $R(\tau)$ in a conventional receiver is ordinarily not computationally difficult because in such receivers, $m(t)$ can be an ideal chipping sequence with only the values ± 1, and the multiplications of samples of the integrand of Eq. (7.33) then become trivial. Furthermore, conventional receivers track only the peak of the correlation function so that $R(\tau)$ needs to be computed for only a few values of τ (usually for early, punctual, and late correlations). However, the MMT algorithm cannot employ these simplifications. The function $m(t)$ used by MMT must include the aforementioned effects of filtering, thus requiring multibit multiplications (typically numbering in the millions) in the calculation of $R(\tau)$. Furthermore, $R(\tau)$ must be calculated for many values of τ to obtain high resolution for accurate estimation of direct-path delay in the presence of multipath.

These difficulties are circumvented by using signal compression. To simplify its description, we assume that the correlation function $R(\tau)$ in Eq. (7.33) is a cyclic correlation over one period T of the replica code $m(t)$, in which $m(t-\tau)$ is a rotation by τ (right for positive τ and left for negative τ). However, compression can be accomplished over an arbitrary interval of observation of the function $r(t)$, in which many periods of a received position navigation (PN) code occur, and furthermore, the correlation function need not be cyclic.

A single period of replica code can be written as

$$m(t) = \sum_{k=0}^{N-1} \varepsilon_k c(t - kT_c) \tag{7.36}$$

where T_c is the duration of each chip, ε_k is the chip polarity (either +1 or −1), and N is the number of chips in one period of the code. The function $c(t)$ is the response of the combined satellite and receiver filtering to a single ideal chip of the code. This ideal chip has a constant value of 1 on the interval $0 \le t \le T_c$. Because the filtering is linear and time invariant, it follows that $m(t)$ is the filter response to the entire code sequence. The index k identifies the individual chips of the code, where $k = 0$ identifies the epoch chip, defined as the first chip of the chipping sequence.

The compressed signal is defined by

$$\tilde{r}(t) = \sum_{k=0}^{N-1} \varepsilon_k r(t + kT_c) \tag{7.37}$$

In this expression $\varepsilon_k r(t - kT_c)$ is $r(t)$ weighted by ε_k and left-rotated by kT_c. In GNSS applications, the compressed signal has the very nice property that essentially all of its energy (excluding noise) is concentrated into a pulse of one filtered chip in duration. This is made evident by noting that the received signal $r(t)$ without multipath can be expressed as

$$r(t) = am(t - \tau_0) + n(t) = a\left[\sum_{j=0}^{N-1} \varepsilon_j c(t - \tau_0 - jT_c)\right] + n(t) \tag{7.38}$$

where a is the signal amplitude, τ_0 is the signal delay, $n(t)$ is noise, and all time shifts are rotations (i.e. cyclic over one code period). Substitution of this expression into Eq. (7.37) gives

$$\left.\begin{aligned}
\tilde{r}(t) &= \sum_{k=0}^{N-1} \varepsilon_k r(t + kT_c) \\
&= \sum_{k=0}^{N-1} \varepsilon_k \left\{\left[\sum_{j=0}^{N-1} \varepsilon_j c(t - \tau_0 + kT_c - jT_c)\right] + n(t + kT_c)\right\} \\
&= \sum_{k=0}^{N-1}\sum_{j=0}^{N-1} \varepsilon_k \varepsilon_j c[t - \tau_0 + (k - j)T_c] + \sum_{k=0}^{N-1} \varepsilon_k n(t + kT_c) \\
&= \sum_{k=0}^{N-1}\sum_{j=0}^{N-1} \varepsilon_k \varepsilon_j c[t - \tau_0 + (k - j)T_c] + \tilde{n}(t)
\end{aligned}\right\} \tag{7.39}$$

where the double summation is the compressed signal component and the single summation is the compressed noise function $\widetilde{n}(t)$. The terms in the double summation can be grouped into N groups such that each group contains N terms having the same value of $k - j$ modulo N. Thus, $\widetilde{r}(t)$ will be the summation of N group sums plus $\widetilde{n}(t)$. The group sum corresponding to particular value p of $k - j$ modulo N is $c[t - \tau_0 + pT_c]$ weighted by the sum of terms $\varepsilon_j \varepsilon_k$, which satisfy $k - j = p$ modulo N. Since T_c is the duration of $c(t)$ before filtering, it can be seen that $\widetilde{r}(t)$ consists of a concatenation of N weighted and translated copies of $c(t)$, which do not overlap, except for a trailing transient from each copy due to filtering.

7.5.3.11 Properties of the Compressed Signal

If the number of chips N is sufficiently large (on the order of 10_3 or more), the autocorrelation function of the GNSS chipping sequence has the property that the group sums in which $k - j \neq 0$ modulo N are negligible compared to the group sum in which $k - j = 0$ modulo N. Furthermore, the sum of all of these small group sums is also negligible because the translations of the weighted copies of $c(t)$ prevent the small group sums from accumulating to large values. Thus, to a very good approximation, the double summation in Eq. (7.39) is just the sum of the terms where $k - j = 0$ modulo N:

$$\widetilde{r}(t) \cong \left[\sum_{k=0}^{N-1} \varepsilon_k^2 c(t - \tau_0) \right] + \widetilde{n}(t) = Nc(t - \tau_0) + \widetilde{n}(t) \tag{7.40}$$

This is a very significant result because it tells us that the compressed received signal is essentially just the single weighted filtered chip $Nc(t - \tau_0) + \widetilde{n}$ plus noise, with small "sidelobe" chips to either side. Furthermore, the compression process provides a processing gain of $10 \log N$ dB. Since a receiver can measure the delay τ_0, a window can be constructed that need be long enough only to contain $Nc(t - \tau_0)$, and the sidelobe chips as well as all noise outside this window can be rejected. The required length of the window is $T_c + \delta$, where δ is large enough to accommodate the measurement uncertainty of τ_0, the trailing transient due to filtering, and any multipath components with delays larger than τ_0 (almost certainly the only multipath components having significant amplitude are found within one chip of the direct-path delay). Thus, the window length is somewhat larger than the one-chip duration of the code, a quantity much smaller than the length T of the observed signal $r(t)$, which must include all N chips of the code. It is because of this result that $\widetilde{r}(t)$ can justifiably be called a compressed signal. An illustration of the compressed signal is shown in Figure 7.5.

If N is sufficiently large, the processing gain is great enough to make the compressed signal within the window visible with very little noise, so that small

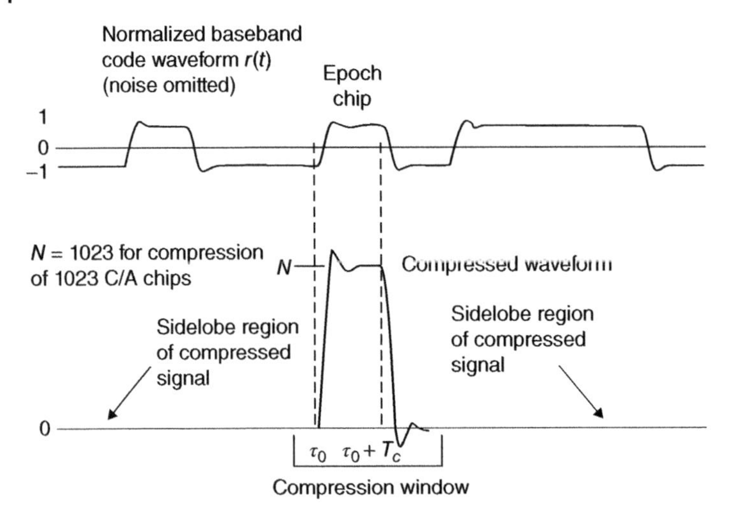

Figure 7.5 Compression of the received signal.

subtleties in the chip waveshape due to multipath or other causes can easily be seen. This property is very beneficial for signal integrity monitoring. It has been put to practical use in GNSS receivers sold by the NovAtel Corporation, which calls its implementation the *vision correlator*.

The compressed signal also enjoys a *linearity property*: If $r(t) = a_1 r_1(t) + a_2 r_2(t)$, then $\widetilde{r}(t) = a_1 \widetilde{r}_1(t) + a_2 \widetilde{r}_2(t)$. The linearity property is essential for the MMT to properly process a multipath-corrupted signal.

7.5.3.12 The Compression Theorem

Most importantly, the compressed signal can be used to drastically reduce the amount of computation of the correlation function $R(\tau)$ in Eq. (7.35). The basis for this assertion is the following theorem:

The correlation function

$$R(\tau) = \int_0^T r(t)m(t-\tau)\,du \tag{7.41}$$

can be computed by the alternate method

$$R(\tau) = \int_0^T \widetilde{r}(t)c(t-\tau)\,du \tag{7.42}$$

Proof.

$$\left.\begin{aligned}
R(\tau) &= \int_0^T r(t)m(t-\tau)\,dt \\
&= \int_0^T r(t)\left[\sum_{k-1}^{N-1} \varepsilon_k c(t-kT_c-\tau)\right] dt \\
&= \sum_{k=0}^{N-1} \int_0^T \varepsilon_k r(t)\, c(t-kT_c-\tau)\,dt \\
&= \sum_{k=0}^{N-1} \int_0^T \varepsilon_k r(u+kT_c)c(u-\tau)\,du\,(\text{using } u=t-kT_c) \\
&= \int_0^T \left[\sum_{k=0}^{N-1} \varepsilon_k r(u+kT_c)\right] c(u-\tau)\,du \\
&= \int_0^T \tilde{r}(u)\, c(u-\tau)\,du
\end{aligned}\right\} \tag{7.43}$$

This theorem shows that $R(\tau)$ can be computed by cross correlating the compressed signal $\tilde{r}(t)$ with the very short function $c(t)$. Furthermore, since we have already noted that the significant portion of $\tilde{r}(t)$ also spans a short time interval, the region surrounding the correlation peak of $R(\tau)$ can be obtained with far less computation than the original correlation (Eq. (7.44)). The bottom line is that the cross correlations in Eq. (7.30) used by MMT can be calculated very efficiently by using the compressed versions of the signals $x(t)$, $y(t)$, and $m(t)$.

7.5.4 Performance of Time-Domain Methods

7.5.4.1 Ranging with the C/A-Code

Typical C/A-code ranging performance curves for several multipath mitigation approaches are shown in Figure 7.6 for the case of an in-phase secondary path with amplitude one-half that of the direct path. Even with the best available methods (other than MMT), peak range errors of 3–6 m are not uncommon. It can be observed that the error tends to be largest for "close-in" multipath, where the separation of the two paths is on the order of 10 m. Indeed, this region poses the greatest challenge in multipath mitigation research because the extraction of direct-path delay from a signal with small direct/secondary-path separation is an ill-conditioned parameter estimation problem.

A serious limitation of most existing multipath mitigation algorithms is that the residual error is mostly in the form of a bias that cannot be removed by further filtering or averaging. On the other hand, the aforementioned MMT

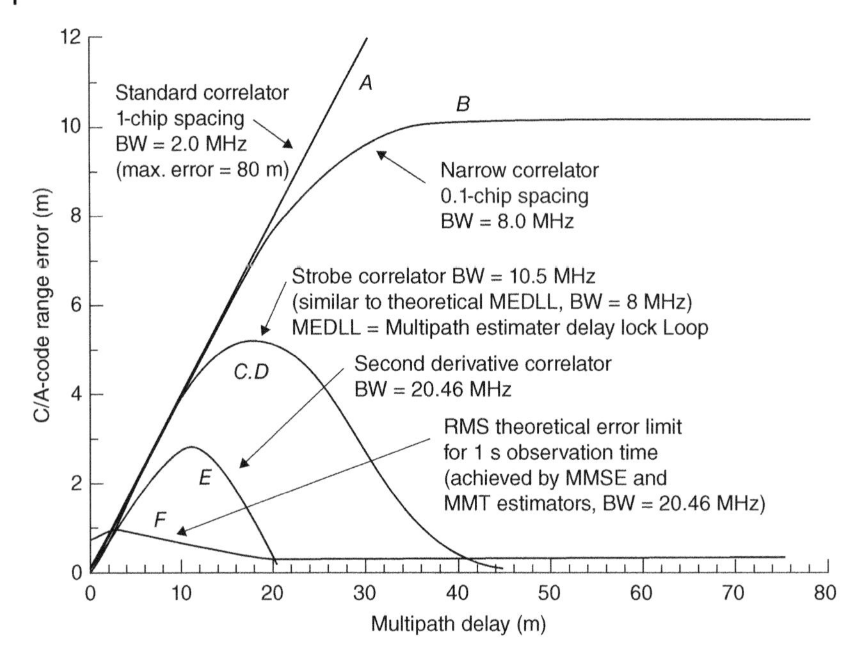

Figure 7.6 Performance of various multipath mitigation approaches.

algorithm overcomes this limitation and also appears to have significantly better performance than other published algorithms, as is indicated by curve *F* of Figure 7.6.

7.5.4.2 Carrier Phase Ranging

The presence of multipath also causes errors in estimating carrier phase, which limits the performance in surveying and other precision applications, particularly with regard to carrier phase ambiguity resolution. Not all current multipath mitigation algorithms are capable of reducing multipath-induced phase error. The most difficult situation occurs at small separations between the direct and secondary paths (less than a few meters). It can be shown that, under such conditions, essentially no mitigation is theoretically possible. Typical phase error curves for the MMT algorithm, which appears to have the best performance of published methods, is shown in Figure 7.7 [17].

7.5.4.3 Testing Receiver Multipath Performance

Conducting tests of receiver multipath mitigation performance on either an absolute or a comparative basis is often done in two ways. For very controlled and repeatable multipath error testing, an advanced GNSS simulator is used to represent an operational scenario whereby multiple signals (direct) and multipath are simulated and fed into the GNSS receiver. The results can then be compared with the known simulated measurements and position of the user GNSS receiver.

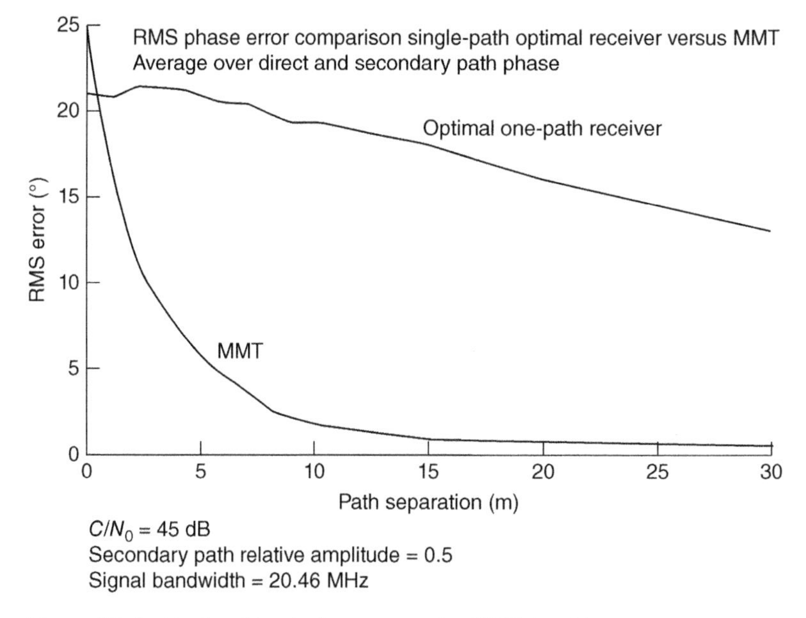

Figure 7.7 Residual multipath phase error using MMT algorithm.

Another way to perform multipath performance analysis is in the real operational environment of the receiver. This type of testing will evaluate the entire system and processing performed, i.e. antenna, receiver correlation, postcorrelation processing. An analysis technique commonly referred to as a code-minus-carrier (CMC) technique is used. This technique is most often applied to analyze the code multipath in a postprocessing fashion whereby the code measurement is detrended by the carrier phase measurement and all other error sources are removed, except for the multipath to be analyzed. The predicted multipath error can then be used to analyze the performance of the GNSS receiver systems or compared with other prediction methods for validation.

7.6 Theoretical Limits for Multipath Mitigation

7.6.1 Estimation-Theoretic Methods

Relatively little has been published on multipath mitigation from the fundamental viewpoint of statistical estimation theory despite the power of its methods and its ability to reach theoretical performance limits in many cases. Knowledge of such limits provides a valuable benchmark in receiver design by permitting an accurate assessment of the potential payoff in developing techniques that are better than those in current use. Of equal importance is the revelation of the signal processing operations that can reach performance bounds. Although it may not be feasible to implement the processing directly, its revelation often leads to a practical method that achieves nearly the same performance.

7.6.1.1 Optimality Criteria

In discussing theoretical performance limits, it is important to define the criterion of optimality. In GPS, the optimal range estimator is traditionally considered to be the minimum-variance unbiased estimator (MVUE), which can be realized by properly designed receivers. However, in Ref. [21], it is shown that the standard deviation of a MVUE designed for multipath becomes infinite as the primary-to-secondary-path separation approaches zero. For this reason, it seems that a better criterion of optimality would be the minimum root mean square (RMS) error, which can include both random and bias components. Unfortunately, it can be shown that no estimator exists having minimum RMS error for every combination of true multipath parameters.

7.6.2 Minimum Mean-Squared Error (MMSE) Estimator

There is an estimator that can be claimed optimal in a weaker sense. The MMSE estimator has the property that no other estimator has a uniformly smaller RMS error. In other words, if some other estimator has smaller RMS error than the MMSE estimator for some set of true multipath parameter values, then that estimator must have a larger RMS error than the MMSE estimator for some other set of values.

The MMSE estimator also has an important advantage not possessed by most current multipath mitigation methods in that the RMS error decreases as the length of the signal observation interval is increased.

7.6.3 Multipath Modeling Errors

Although a properly designed estimation-theoretic approach such as the MMSE estimator will generally outperform other methods, the design of such estimators requires a mathematical model of the multipath-contaminated signal containing parameters to be estimated. If the actual signal departs from the assumed model, performance degradation can occur. For example, if the model contains only two signal propagation paths but in reality the signal is arriving via three or more paths, large bias errors in range estimation can result. On the other hand, poorer performance (usually in the form of random error cause by noise) can also occur if the model has too many degrees of freedom. Striking the right balance in the number of parameters in the model can be difficult if little information exists about the multipath reflection geometry.

7.7 Ephemeris Data Errors

Small errors in the ephemeris data transmitted by each satellite cause corresponding errors in the computed position of the satellite. Satellite ephemerides are determined by the master control station of the GNSS ground segment based on monitoring of individual signals by four monitoring stations. Because the locations of these stations are known precisely (e.g. at the meter level in real-time), an "inverted" positioning process can calculate the orbital parameters of the satellites as if they were users. This process is aided by precision clocks at the monitoring stations and by tracking over long periods of time with optimal filter processing. Based on the orbital parameter estimates thus obtained, the master control station uploads the ephemeris data to each satellite, which then transmits the data to users via the navigation data message. Errors in satellite position when calculated from the ephemeris data typically result in range errors on the order of 1–2 m. Improvements in satellite tracking by the ground segment will undoubtedly reduce this error further.

7.8 Onboard Clock Errors

Timing of the signal transmission from each satellite is directly controlled by its own atomic clock without any corrections applied. This time frame is called *space vehicle* (SV) *time*. A schematic of a rubidium atomic clock is shown in Figure 7.8. Although the atomic clocks in the satellites are highly accurate, errors are large enough to require correction. Correction is needed partly because it would be difficult to directly synchronize the clocks closely in all the satellites in real time. Instead, the clocks are allowed some degree of relative drift that is estimated by ground station observations and is used to generate clock correction data in the GNSS navigation message. When SV time is corrected using this data, the result is called GNSS *time*. The time of transmission used in calculating pseudoranges must be GNSS time, which is common to all satellites.

The onboard clock error is typically less than 1 ms and varies slowly. This permits the correction to be specified by a quadratic polynomial in time whose coefficients are transmitted in the navigation message. The correction has the form

$$\Delta t^i_{\text{sv}} = a_{f0} + a_{f1}(t^i_{\text{sv}} - t^i_{\text{oc}}) + a_{f2}(t^i_{\text{sv}} - t^i_{\text{oc}})^2 + \Delta t^i_r, \ [\text{S}] \tag{7.44}$$

with

$$t_{\text{GNSS}} = t_{\text{sv}} - \Delta t_{\text{sv}} \tag{7.45}$$

where a_{f0}, a_{f1}, a_{f2} are the correction coefficients, t_{sv} is SV time, and Δt_r is a small relativistic clock correction caused by the orbital eccentricity. The clock

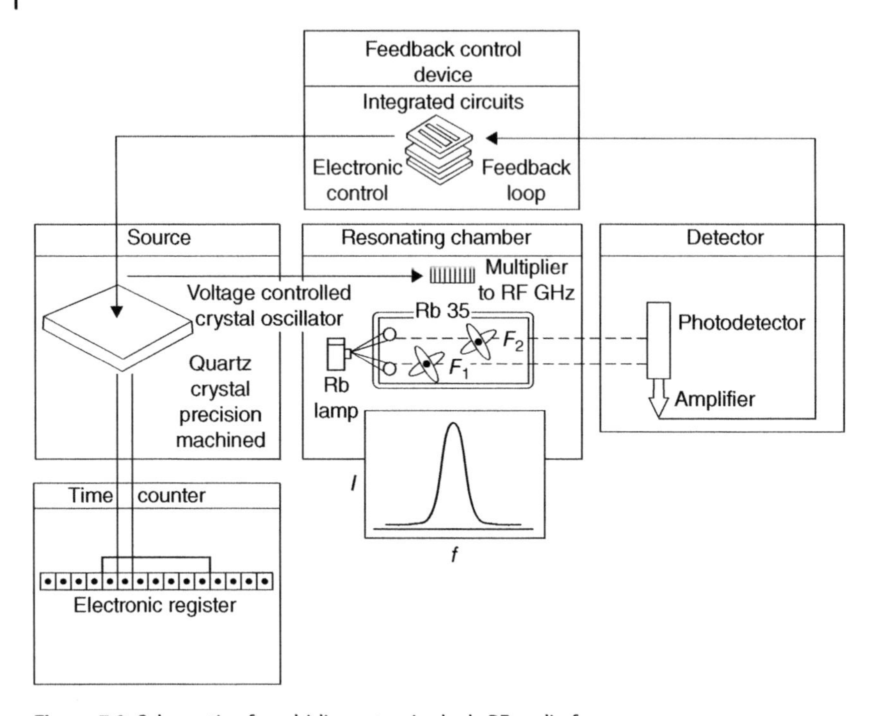

Figure 7.8 Schematic of a rubidium atomic clock. RF: radio frequency.

data reference time t_{oc}, in seconds is broadcast in the navigation data message. The stability of the atomic clocks permits the polynomial correction given by Eq. (7.44) to be valid over a time interval of four to six hours. After the correction has been applied, the residual error in GNSS time is typically less than a few nanoseconds, or about 1 m in range. Complete calculations of GNSS time are given as exercises in Section 4.1.3.4.

7.9 Receiver Clock Errors

Because the navigation solution includes a solution for receiver clock error, the requirements for accuracy of receiver clocks is far less stringent than for GNSS satellite clocks. In fact, for receiver clocks, short-term stability over the pseudorange measurement period is usually more important than absolute frequency accuracy. In almost all cases, such clocks are quartz crystal oscillators with absolute accuracies in the 1–10 ppm range over typical operating temperature ranges. When properly designed, such oscillators typically have stabilities of 0.01–0.05 ppm over a period of a few seconds.

Receivers that incorporate receiver clock error in the Kalman filter state vector need a suitable mathematical model of the crystal clock error. A typical

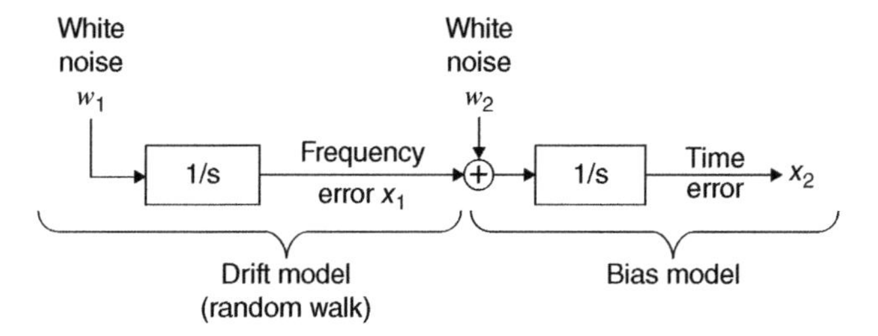

Figure 7.9 Crystal clock error model.

model in the continuous-time domain is shown in Figure 7.9, which is easily changed to a discrete version for the Kalman filter. In this model, the clock error consists of a bias (frequency) component and a drift (time) component. The frequency error component is modeled as a random walk produced by integrated white noise. The time error component is modeled as the integral of the frequency error after additional white noise (statistically independent from that causing the frequency error) has been added to the latter. In the model, the key parameters that need to be specified are the power spectral densities of the two noise sources, which depend on characteristics of the specific crystal oscillator used.

The continuous-time model has the form

$$\dot{x}_1 = w_1 \tag{7.46}$$

$$\dot{x}_2 = x_1 + w_2 \tag{7.47}$$

where $w_1(t)$ and $w_2(t)$ are independent zero-mean white-noise processes with known variances.

The equivalent discrete-time model has the state vector

$$\mathbf{x} = \begin{bmatrix} x_1 \\ x_2 \end{bmatrix} \tag{7.48}$$

and the stochastic sequence model

$$\mathbf{x}_k = \begin{bmatrix} 1 & 0 \\ \Delta t & 1 \end{bmatrix} \mathbf{x}_{k-1} + \begin{bmatrix} w_{1,k-1} \\ w_{2,k-1} \end{bmatrix} \tag{7.49}$$

where Δt is the discrete-time step and $\{w_{1,k-1}\}$, $\{w_{2,k-1}\}$ are independent zero-mean white-noise sequences with known variances.

7.10 Error Budgets

For purposes of analyzing the effects of the errors discussed earlier, it is convenient to convert each error into an equivalent range error experienced by a user,

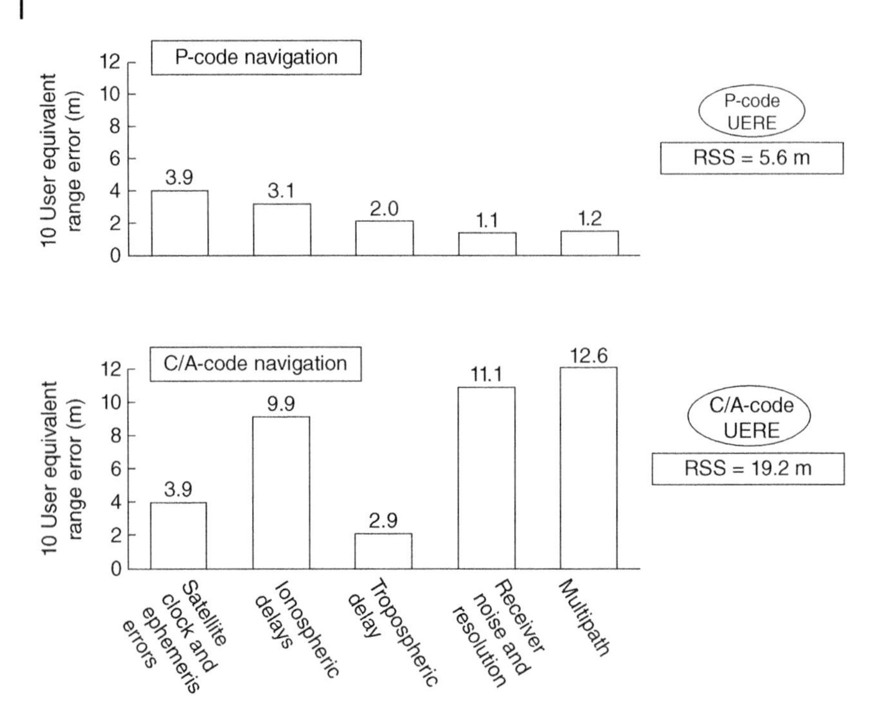

Figure 7.10 GPS UERE budget.

which is called the *user-equivalent range error* (UERE). In general, the errors from different sources will have different statistical properties. For example, satellite clock and ephemeris errors tend to vary slowly with time and appear as biases over moderately long time intervals, perhaps hours. On the other hand, errors due to receiver noise and quantization effects may vary much more rapidly, perhaps within seconds. Nonetheless, if sufficiently long time durations over many navigation scenarios are considered, all errors can be considered as zero-mean random processes that can be combined to form a single UERE. This is accomplished by forming the root sum square (RSS) of the UERE errors from all sources:

$$\text{UERE} = \sqrt{\sum_{i=1}^{n} (\text{UERE})_i^2} \tag{7.50}$$

Figure 7.10 depicts the various GPS UERE errors and their combined effect for both C/A-code and P(Y)-code navigation at the $1 - \sigma$ level.

The UERE for the C/A-code user is typically about 19 m. It can be seen that, for such a user, the dominant error sources in non-differential operations are multipath, receiver noise/resolution, and ionospheric and tropospheric delay (however, recent advances in receiver technology have in some cases significantly reduced receiver noise/resolution errors). On the other hand, the P(Y)-code user has a significantly smaller UERE of about 6 m for the following reasons:

1. The full use of the L1 and L2 signals permits significant reduction of ionospheric error.
2. The wider bandwidth of the P(Y)-codes greatly reduces errors due to multipath and receiver noise.

Problems

7.1 Using the values provided for Klobuchar's model in Section 7.2.1, calculate the ionospheric delay and plot the results.

7.2 Assume that a direct-path GNSS L1 C/A-code signal arrives with a phase such that all of the signal power lies in the baseband I channel, so that the baseband signal is purely real. Further assume an infinite signal bandwidth so that the cross correlation of the baseband signal with an ideal C/A reference code waveform will be an isosceles triangle 600 m wide at the base.

(a) Suppose that in addition to the direct-path signal there is a secondary-path signal arriving with a relative time delay of precisely 250 L1 carrier cycles (so that it is in phase with the direct-path signal) and with an amplitude one-half that of the direct path. Calculate the pseudorange error that would result, including its sign, under noiseless conditions. Assume that pseudorange is measured with a delay-lock loop using 0.1-chip spacing between the early and late reference codes.
(*Hint*: The resulting cross-correlation function is the superposition of the cross-correlation functions of the direct- and secondary-path signals.)

(b) Repeat the calculations of part (a) but with a secondary-path relative time delay of precisely 250.5 carrier cycles. Note that in this case, the secondary-path phase is 180° out of phase with the direct-path signal but still lies entirely in the baseband I channel.

7.3 (a) Using the discrete matrix version of the receiver clock model given by Eq. (7.49), find the standard deviation σ_{w1} of the white-noise sequence $w_{1,k}$ needed in the model to produce a frequency standard deviation σ_{x1} of 1 Hz after 10 minutes of continuous oscillator operation. Assume that the initial frequency error at $t = 0$ is zero and that the discrete-time step Δt is one second.

(b) Using the assumptions and the value of σ_{w1} found in part (a), find the standard deviation σ_{x2} of the bias error after 10 minutes. Assume that $\sigma_{w2} = 0$

(c) Show that σ_{x1} and σ_{x2} approach infinity as the time t approaches infinity. Will this cause any problems in the development of a Kalman filter that includes estimates of the clock frequency and bias error?

7.4 The peak electron density in the ionosphere occurs in a height range of
(a) 50–100 km
(b) 250–400 km
(c) 500–700 km
(d) 800–1000 km

7.5 The refractive index of the gaseous mass in the troposphere is
(a) Slightly higher than unity
(b) Slightly lower than unity
(c) Unity
(d) Zero

7.6 If the range measurements for two simultaneously tracking satellites in a receiver are differenced, then the differenced measurement will be free of
(a) Receiver clock error only
(b) Satellite clock error and orbital error only
(c) Ionospheric delay error and tropospheric delay error only
(d) Ionospheric delay error, tropospheric delay error, satellite clock error, and orbital error only

7.7 Zero baseline test (code) can be performed to estimate
(a) Receiver noise and multipath
(b) Receiver noise
(c) Receiver noise, multipath, and atmospheric delay errors
(d) None of the above

7.8 What is the purpose of antispoofing (AS)?

7.9 Derive the multipath formula equivalent to Eq. (7.20) for L2 using the same notation as in Eq. (7.14).

7.10 Calculate the ionospheric delay using dual-frequency carrier phases.

References

1 Feess, W.A. and Stephens, S.G. (1986). Evaluation of GPS ionospheric time delay algorithm for single frequency users. *Proceedings of the IEEE Position, Location, and Navigation Symposium (PLANS '86)*, Las Vegas, NV (4–7 November 1986). New York, pp. 206–213.
2 Klobuchar, J.A. (1976). Ionospheric time delay corrections for advanced satellite ranging systems. *Propagation Limitations of Navigation and Positioning Systems*, NATO AGARD Conference Proceedings 209. Paris: NATO AGARD.
3 RTCA (1999). Minimum operational performance standards for global positioning system, wide area augmentation system. *Document RTCA/D0–229.* Washington, DC: Radio Technical Commission for Aeronautics.
4 Mannucci, A., Wilson, B., and Edwards, C.D. (1993). A new method for monitoring the Earth's total electron content using the GPS Global Network. *Proceedings of ION GPS-93*, Salt Lake City, UT (September 1993), pp. 22–24.
5 Grewal, M.S. (2012). Space-based augmentation for global navigation satellite systems. *IEEE Transactions on Ultrasonics, Ferroelectrics, and Frequency Control* 59 (3): 497–504.
6 El-Arini, M.B., Conker, R.S., Albertson, T.W. et al. (1994/1995). Comparison of real-time ionospheric algorithms for a GPS wide-area augmentation system. *NAVIGATION, Journal of the Institute of Navigation* 41 (4): 393–413.
7 Moreno, R. and Suard, N. (1999). Ionospheric delay using only LI: validation and application to GPS receiver calibration and to inter-frequency bias estimation. In: *Proceedings of The Institute of Navigation (ION)* (25–27 January 1999), 119–129. Alexandria, VA: ION.
8 Hofmann-Wellenhof, B., Lichtenegger, H., and Collins, J. (1997). *GPS: Theory and Practice.* Vienna: Springer-Verlag.
9 Janes, H.W., Langley, R.B., and Newby, S. (1991). Analysis of tropospheric delay prediction models: comparisons with ray-tracing and implications for GPS relative positioning. *Bulletin Geodisique* 65 (3): 151–161.
10 Seeber, G. (1993). *Satellite Geodesy: Foundation, Methods, and Applications.* Berlin: Walter de Gruyter.
11 Hagerman, L. (1973). Effects of multipath on coherent and noncoherent PRN ranging receiver. *Aerospace Report TOR-0073(3020-03)-3.* El Segundo, CA: Aerospace Corporation, Development Planning Division.
12 Dickman, J., Bartone, C., Zhang, Y., and Thornburg, B. (2003). Characterization and performance of a prototype wideband airport pseudolite multipath limiting antenna for the local area augmentation system. In: *Proceedings*

of the 2003 National Technical Meeting of The Institute of Navigation, 783–793. Anaheim, CA: ION (January 2003).

13 Van Dierendonck, A.J., Fenton, P., and Ford, T. (1992). Theory and performance of narrow correlator spacing in a GPS receiver. In: *Proceedings of the National Technical Meeting,* 115–124. San Diego, CA: Institute of Navigation.

14 Townsend, B. and Fenton, P. (1994). A practical approach to the reduction of pseudorange multipath errors in a L1 GPS receiver. In: *Proceedings of ION GPS-94, 7th International Technical Meeting of the Satellite Division of the Institute of Navigation,* 143–148. Salt Lake City, UT, Alexandria, VA: ION.

15 Townsend, B., Van Nee, D.J.R., Fenton, P., and Van Dierendonck, K. (1995). Performance evaluation of the multipath estimating delay lock loop. In: *Proceedings of the National Technical Meeting,* 277–283. Anaheim, CA: Institute of Navigation.

16 Garin, L., van Diggelen, F., and Rousseau, J. (1996). Strobe and edge correlator multipath mitigation for code. In: *Proceedings of ION GPS-96, 9th International Technical Meeting of the Satellite Division of the Institute of Navigation, Kansas City, MO,* 657–664. Alexandria, VA: ION.

17 Weill, L. (1997). GPS multipath mitigation by means of correlator reference waveform design. In: *Proceedings of the National Technical Meeting, Institute of Navigation, Santa Monica, CA,* 197–206. Alexandria, VA, ION.

18 Weill, L. (1998). Application of superresolution concepts to the GPS multipath mitigation problem. In: *Proceedings of the National Technical Meeting, Institute of Navigation, Long Beach, CA,* 673–682. Alexandria, VA: ION.

19 McGraw, G. and Braasch, M. (1999). GNSS multipath mitigation using gated and high resolution correlator concepts. In: *Proceedings of the 1999 National Technical Meeting and 19th Biennial Guidance Test Symposium,* 333–342. San Diego, CA: Institute of Navigation.

20 Weill, L. and Fisher, B. (2000). Method for mitigating multipath effects in radio ranging systems. US Patent 6, 031, 881, 29 February 2000.

21 Weill, L. (1995). Achieving theoretical accuracy limits for pseudoranging in the presence of multipath. In: *Proceedings of ION GPS-95, 8th International Technical Meeting of the Satellite Division of the Institute of Navigation, Palm Springs, CA,* 1521–1530. Alexandria, VA: ION.

8

Differential GNSS

8.1 Introduction

Differential global navigation satellite system (differential GNSS [DGNSS]) is a technique for reducing the errors in GNSS-derived positions by using additional data from one or more reference GNSS receivers at known locations. One of the most common forms of DGNSS involves determining the combined effects of navigation message ephemeris and satellite clock errors (including the effects of propagation) at a reference station and transmitting corrections, in real time, to a user's receiver. The receiver applies the corrections in the process of determining its user solution (i.e. position, velocity, and time [PVT]) [1]. These include corrections for satellite ephemeris orbit and clock errors, and atmospheric delay errors. Common systematic errors will cancel, to the extent they do not decorrelate with respect to the distance (and time) between the user and where the location(s) where the reference measurements were formed. Still other error sources, such as multipath and receiver noise are non-common and cannot be corrected with DGNSS; these errors should be mitigated at each location.

While there are various ways to implement DGNSS, most can be categorized as correction-based DGNSS and measurement/relative-based DGNSS. Correction-based DGNSS typically involves a reference/monitor station that is most often fixed and surveyed, whereby pseudorange corrections are generated with respect to the "true range" from the reference station to the space vehicle (SV). When all of the error terms are put into a single correction, per SV, and matched with the ephemeris set, it is often referred to as a lumped pseudorange correction. Some correction-based DGNSS architectures decompose the error sources and provide corrections for specific error terms (e.g. orbit, ionosphere). In measurement/relative-based DGNSS architectures, the emphasis is to send monitor/reference station measurements to the rover so that the user can perform difference processing and solve for the relative range vector (i.e. baseline) between the reference station and the mobile user.

Global Navigation Satellite Systems, Inertial Navigation, and Integration,
Fourth Edition. Mohinder S. Grewal, Angus P. Andrews, and Chris G. Bartone.

Companion website: www.wiley.com/go/grewal/gnss

8.2 Descriptions of Local-Area Differential GNSS (LADGNSS), Wide-Area Differential GNSS (WADGNSS), and Space-Based Augmentation System (SBAS)

8.2.1 LADGNSS

Local-area differential GNSS (LADGNSS) is a form of DGNSS in which the user's GNSS receiver receives real-time measurement corrections from a reference receiver generally located within the line of sight (LOS). The corrections account for the combined effects of navigation message ephemeris and satellite clock errors and, usually, atmospheric propagation delay errors between the reference station and the mobile user. With the assumption that these errors are also common between the measurements made by the reference receiver and the user's receiver, the application of the DGNSS corrections will result in more accurate PVT user solutions [2].

8.2.2 WADGNSS

Wide-area differential GNSS (WADGNSS) is a form of DGNSS in which the user's GNSS receiver receives corrections determined from a network of reference stations distributed over a wide geographic area. Separate corrections are usually determined for specific error sources, such as satellite ephemeris (orbit), and ionospheric propagation delay. The corrections are applied in the user's receiver or attached computer in computing the receiver's PVT solution. The corrections are typically supplied in real time by way of a geostationary communications satellite or through a network of ground-based transmitters. Corrections may also be provided at a later date for postprocessing collected data [2].

8.2.3 SBAS

8.2.3.1 Wide-Area Augmentation System (WAAS)

Wide-area augmentation system (WAAS) enhances the global positioning system (GPS) standard positioning service (SPS) and is available over a wide geographic area. The WAAS, developed by the Federal Aviation Administration (FAA) together with other agencies, provides WADGPS corrections, additional ranging signals from geostationary Earth orbit (GEO) satellites, and integrity data on the GPS and GEO satellites [2]. Improvements to WAAS are still being made under a contract from the FAA. This section describes the current state of WAAS with three GEO satellites [3, 4].

The FAA is currently implementing WAAS Phase IV-A to support operational system maintenance and future incorporation of dual frequency capability. Significant infrastructure modifications and technical refresh include replacement of all operational processors, inclusion of L2C and L5 GPS data in reference station data structures, and incorporation of three additional GEO satellites and associate ground facilities to replace aging satellites. WAAS is also preparing for the start of Phase IV-B, when dual frequency capability will be implemented. The FAA intends to replace the use of the GPS L2 P(Y) semicodeless signal with the use of the GPS L5 civil signal since Department of Defense (DOD) support of GPS L2 codeless/semicodeless capability is planned to be discontinued in approximately 2026, two years after 24 operational GPS satellites are broadcasting L5. In addition to the WAAS ground system transition from L2 to the L5 civil signal in WAAS Phase IV-B, the FAA also plans to introduce a new dual frequency SBAS navigation service, while retaining legacy single-frequency user services.

Each GEO Uplink Subsystem (GUS) includes a closed-loop control algorithm and special signal generator (SigGen) hardware. These ensure that the downlink signal to the users is controlled adequately to be used as a ranging source to supplement the GPS satellites in view.

The primary mission of WAAS is to provide a means for air navigation for all phases of flight in the National Airspace System (NAS) from departure, en route, and through approach and landing. GPS augmented by WAAS offers the capability for both nonprecision approach (NPA) and precision approach (PA) procedures within a specific service volume. A secondary mission of the WAAS is to provide a WAAS network time (WNT) offset between the WNT and Coordinated Universal Time (UTC) for non-navigation users.

WAAS provides improved en route navigation and PA capability to WAAS-certified avionics. The safety critical WAAS system consists of the equipment and software necessary to augment the DOD-provided GPS SPS. WAAS provides a signal in space (SIS) to WAAS-certified aircraft avionics using the WAAS for any FAA-approved phase of flight. The SIS provides two services: (i) data on GPS and GEO satellites and (ii) a ranging capability.

The GPS satellite data are received and processed at widely dispersed Wide-Area Reference Stations (WRSs), which are strategically located to provide coverage over the required WAAS service volume. Data are forwarded to Wide-Area Master Stations (WMSs), which process the data from multiple WRSs to determine the integrity, differential corrections, and residual errors for each monitored satellite and for each predetermined ionospheric grid point (IGP). Multiple WMSs are provided to eliminate single-point failures within the WAAS network. Information from all WMSs is sent to each GUS and uplinked along with the GEO navigation message to GEO satellites. The GEO satellites downlink these data to the users via the GPS SPS L-band ranging signal (L1) frequency with GPS-type modulation. Each ground-based station/subsystem communicates via a terrestrial communications subsystem (TCS) (see Figure 8.1).

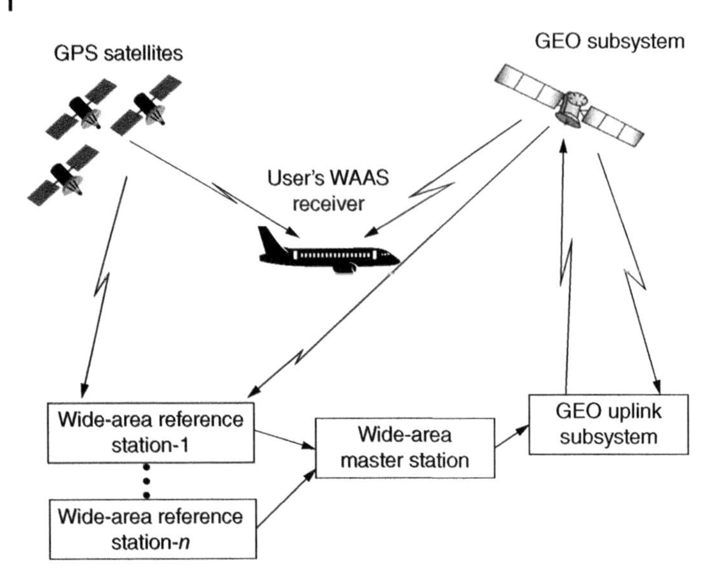

Figure 8.1 WAAS top-level view.

In addition to providing augmented GPS data to the users, WAAS verifies its own integrity and takes any necessary action to ensure that the system meets the WAAS performance requirements. WAAS also has a system operation and maintenance function that provides status and related maintenance information to FAA airway facilities (AFs) NAS personnel.

Correction and verification (C&V) processes data from all WRSs to determine integrity, differential corrections, satellite orbits, and residual error bounds for each monitored satellite. It also determines ionospheric vertical delays and their residual error bounds at each of the IGPs. C&V schedules and formats WAAS messages and forwards them to the GUSs for broadcast to the GEO satellites.

C&V's capabilities are as follows:

1. Control C&V operations and maintenance (COM) supports the transfer of files, performs remotely initiated software configuration checks, and accepts requests to start and stop execution of the C&V application software.
2. Control C&V modes manage mode transitions in the C&V subsystem while the application software is running.
3. Monitor C&V (MCV) reports line replaceable unit (LRU) faults and configuration status. In addition, it monitors software processes and provides performance data for the local C&V subsystems.

4. Process input data (PID) selects and monitors data from the wide-area reference equipments (WREs). Data that pass PID screening are repackaged for other C&V capabilities. PID performs clock and L1 GPS Precise Positioning Service L-band ranging signal (L2) receiver bias calculations, cycle slip detection, outlier detection, data smoothing, and data monitoring. In addition, PID calculates and applies the windup correction to the carrier phase, accumulates data to estimate the pseudorange to carrier phase bias, and computes the ionosphere corrected carrier phase and measured slant delay.
5. Satellite orbit determination (SOD) determines the GPS and GEO satellite orbits and clock offsets, WRE receiver clock offsets, and troposphere delay.
6. Ionosphere correction computation (ICC) determines the L1 IGP vertical delays, grid ionosphere vertical error (GIVE) for all defined IGPs, and L1–L2 interfrequency bias for each satellite transmitter and each WRS receiver.
7. Satellite correction processing (SCP) determines the fast and long-term satellite corrections, including the user differential range error (UDRE). It determines the WNT and the GEO and WNT clock steering commands [3].
8. Independent data verification (IDV) compares satellite corrections, GEO navigation data, and ionospheric corrections from two independent computational sources, and if the comparisons are within limits, one source is selected from which to build the WAAS messages. If the comparisons are not within limits, various responses may occur, depending on the data being compared, all the way from alarms being generated to the C&V being faulted.
9. Message output processing (MOP) transmits messages containing independently verified results of C&V calculations to the GUS processing (GP) for broadcast.
10. C&V playback (PLB) processes the playback data that have been recorded by the other C&V capabilities.
11. Integrity data monitoring (IDM) checks both the broadcast and the to-be-broadcast UDREs and GIVEs to ensure that they are properly bounding their errors. In addition, it monitors and validates that the broadcast messages are sent correctly. It also performs the WAAS time-to-alarm validation [5, 6].

WRS algorithms. Each WRS collects raw pseudorange (PR) and accumulated Doppler range (ADR) measurements from GPS and GEO satellites selected for tracking. Each WRS performs smoothing on the measurements and corrects for atmospheric effects, that is, ionospheric and tropospheric delays. These smoothed and atmospherically corrected measurements are provided to the WMS.

WMS foreground (fast) algorithms. The WMS foreground algorithms are applicable to real-time processing functions, specifically the computation of fast correction, determination of satellite integrity status, and WAAS message formatting. This processing is done at a 1 Hz rate.

WMS background (slow) algorithms. The WMS background processing consists of algorithms that estimate slowly varying parameters. These algorithms consist of WRS clock error estimation, grid ionospheric delay computation, broadcast ephemeris computation, SOD, satellite ephemeris error computation, and satellite visibility computation.

IDV and validation algorithms. This includes a set of WRS and at least one WMS, which enable monitoring the integrity status of GPS and the determination of wide-area DGPS correction data. Each WRS has three dual-frequency GPS receivers to provide parallel sets of measurement data. The presence of parallel data streams enables independent data verification and validation (IDV&V) to be employed to ensure the integrity of GPS data and their corrections in the WAAS messages broadcast via one or more GEOs. With IDV&V active, the WMS applies the corrections computed from one stream to the data from the other stream to provide verification of the corrections prior to transmission. The primary data stream is also used for the validation phase to check the active (already broadcast) correction and to monitor their SIS performance. These algorithms are continually being improved [5, 7–11].

8.2.3.2 European Global Navigation Overlay System (EGNOS)

European Global Navigation Overlay System (EGNOS) is a joint project of the European Space Agency, the European Commission, and the European Organization for the Safety of Air Navigation (Eurocontrol). Its primary service area is the European Civil Aviation Conference (ECAC) region. However, several extensions of its service area to adjacent and more remote areas are under study. An overview of the EGNOS system architecture is presented in Figure 8.2, where

RIMS are the Ranging and Integrity Monitoring Stations
MCC is the Mission and Control Center
NLES are the Navigation Land Earth Stations
PACF is the Performance Assessment and Checkout Facility
DVP is the Development and Verification Platform
ASQF is the Application-Specific Qualification Facility
EWAN is the EGNOS Wide Area (communication) Network

8.2.3.3 Other SBAS

Service areas of current and future SBAS systems are mapped in Figure 8.3, and the acronyms are listed in Table 8.1.

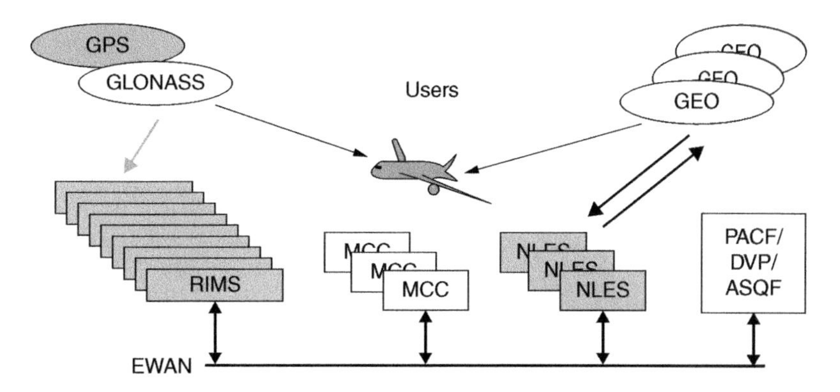

Figure 8.2 European Global Navigation Overlay System architecture.

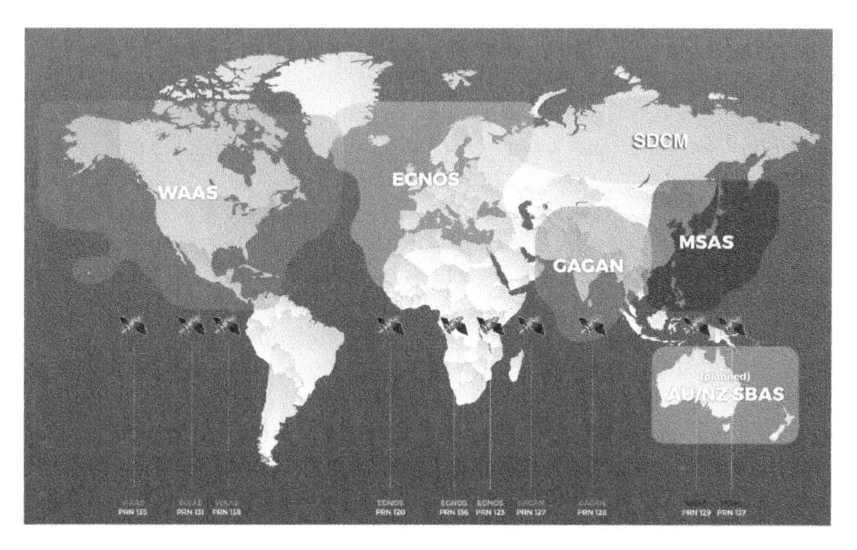

Figure 8.3 Current and planned SBAS service areas.

8.3 GEO with L1L5 Signals

The Space-Based Augmentation System (SBAS) uses GEO satellites to relay correction and integrity information to users. A secondary use of the GEO signal is to provide users with a GPS-like ranging source. Two independent ranging signals are generated on the ground and provided via C-band uplink to the GEO, where the navigation payload translates the uplinked signals to L1 downlink frequencies. A GEO satellite and its corresponding pair of GUS sites comprise a GEO Communication and Control Segment (GCCS).

Table 8.1 Worldwide SBAS system coverage.

Country	Acronym	Title
United States	WAAS	Wide-Area Augmentation System
Europe	EGNOS	European Geostationary Navigation Overlay System
Japan	MSAS	MTSAT Satellite-Based Augmentation System
Japan	QZSS	Quasi-Zenith Satellite System (Sub-meter Level Augmentation Service [SLAS])
China	SNAS	Satellite Navigation Augmentation System
India	GAGAN	GPS and GEO Augmented Navigation (per ISRO site)
Russia	SDCM	System for Differential Corrections and Monitoring
Korea	KAS	Korean Augmentation System

Closed-loop control of the GEO's L1 and L5 broadcast signals in space is necessary to ensure that the algorithms compensate for various sources of uplink divergence between the code and carrier, including uplink ionospheric delay, uplink Doppler, and divergence due to carrier frequency translation errors induced by the GEO's transponder.

The GUS receives SBAS messages from the WMS via terrestrial communication circuits, adds forward error correction (FEC) encoding, and transmits the messages via C-band uplink to the GEO satellite for broadcast to SBAS users. The GUS uplink signal uses the GPS standard positioning service waveform (C/A code, binary phase-shift keying [BPSK] modulation); however, the data rate is higher (250 bps). The 250 bits of data are encoded with a one-half rate convolutional code, resulting in a 500 sps transmission rate.

A key feature of WAAS GCCS is that satellite broadcasts are available at both the GPS L1 and L5 frequencies. Figure 8.4 provides a top-level view of the GCCS architecture.

For the L1 loop, each symbol is modulated by the C/A code, a 1.023 × 106 cps pseudorandom sequence to provide a spread-spectrum signal. This signal is then BPSK modulated by the GUS onto an intermediate frequency (IF) carrier, up-converted to a C-band frequency, and uplinked to the GEO. The satellite's navigation transponder translates the signal in frequency to an L-band (GPS L1) downlink frequency. The GUS monitors the L1 downlink signal from the GEO to provide closed-loop control of the code and L1 carrier. When properly controlled, the SBAS GEO provides ranging signals, as well as GPS corrections and integrity data, to end users.

The L5 spread-spectrum signal is generated by modulating each message symbol with a 10.23 × 106 cps pseudorandom code, which is an order of magnitude longer than that of the L1 C/A-code. As with L1, the L5 signal is then BPSK modulated onto an IF carrier, up-converted to a C-band frequency,

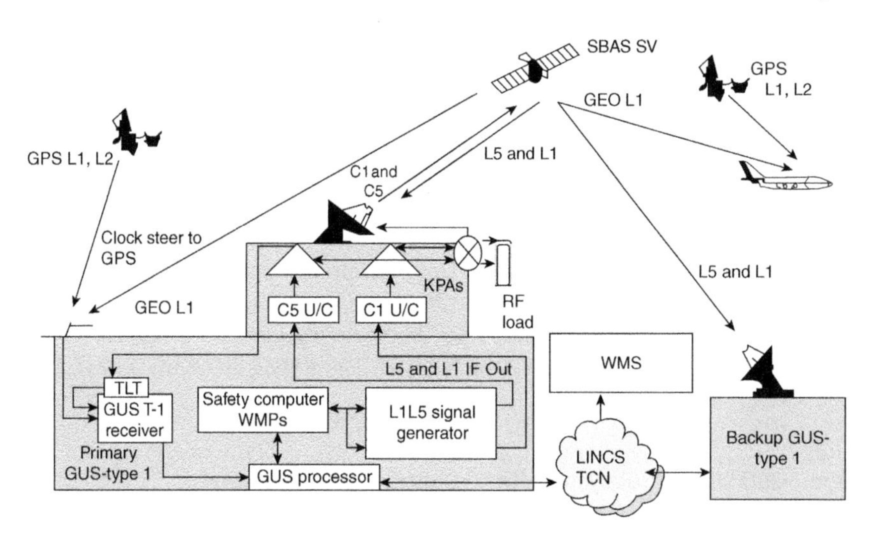

Figure 8.4 GCCS top-level view. TLT: test loop translator; KPA: Klystron power amplifier; LINCS: local information network communication system; TCN: terrestrial communication network; RF: radio frequency.

and uplinked to the GEO. The GEO transponder independently translates the second uplink signal to the L-band for broadcast to SBAS end users. Use of two independent broadcast signals creates unique challenges in estimating biases and maintaining coherency between the two signals.

An important aspect of the downlink signals is coherence between the code and carrier frequency. To ensure code-carrier coherency, closed-loop control algorithms, implemented in the safety computer's WAAS Message Processors (WMPs), are used to maintain the code chipping rate and carrier frequency of the received L1 signal at a constant ratio of 1 : 1540. The second L-band (L5) downlink is used by the control algorithms to estimate and correct for ionospheric delay on the uplink signal. Control algorithms also correct for other uplink effects such as Doppler, equipment delays, and transponder offsets in order to maintain the correct Doppler and ionospheric divergence as observed by the user.

Closed-loop control of each signal is required to maintain coherence between its code and carrier frequency, as described earlier. With two independent signal paths, it is also required that coherence between the two carriers be maintained for correct ionospheric delay estimation. The control-loop algorithms "precorrect" the code phase, carrier phase, and carrier frequency of the L1 and L5 signals to remove uplink effects such as ionospheric delays, uplink Doppler, equipment delays, and frequency offsets. In addition, differential biases between the L1 and L5 signals must be estimated and corrected.

Each control algorithm contains two Kalman filters and two control loops. One Kalman filter estimates the ionospheric delay and its rate of change from L1 and L5 pseudorange measurements. The second Kalman filter estimates range, range rate, range acceleration, and acceleration rate from raw pseudorange measurements. Range estimates are adjusted for ionospheric delay, as estimated by the first Kalman filter. Each code control loop generates a code chip rate command and chip acceleration command to compensate for uplink ionospheric delay and for the uplink Doppler effect. Each frequency control loop generates a carrier frequency command and a frequency rate command. A final estimator is used to calculate bias between the L1 and L5 signals.

Results of laboratory tests utilizing live L1L5 hardware elements and simulated satellite effects follow.

8.3.1 GEO Uplink Subsystem (GUS) Control Loop Overview

The primary GUS control loop functional block diagram is shown in Figure 8.5. The backup GUS control loop is similar to the primary GUS control loop except that the uplink signal is radiated into a dummy load. The operation of the backup GUS control loop is different from the primary GUS because of the latter.

Each of the L1 and L5 control loops in the primary GUS consists of an iono Kalman filter, a range Kalman filter, a code control function, and a frequency control function. In addition, there is an L1L5 bias estimation function. These control loop functions reside inside the safety computer. The external inputs to the control loop algorithm are the pseudorange, carrier phase, Doppler, and carrier-to-noise ratio from the receiver.

8.3.1.1 Ionospheric Kalman Filters

The L1 and L5 ionospheric (iono) Kalman filters are two-state filters:

$$\mathbf{x} = \begin{bmatrix} \text{iono delay} \\ \text{iono delay rate} \end{bmatrix}$$

During every one-second timeframe in the safety computer, the ionospheric Kalman filter states and the covariance are propagated. The equations for Kalman filter propagation are given in Table 10.1.

The L1 filter measurement is formulated as follows:

$$z = \frac{(\rho_{L1} - d_{L1}) - (\rho_{L5} - d_{L5})}{(1 - f_{L1})^2/(f_{L5})^2}$$

where ρ_{L1} is the L1 pseudorange, ρ_{L5} is the L5 pseudorange, d_{L1} and d_{L5} are the predetermined L1 and L5 downlink path hardware delays, f_{L1} is the L1 nominal frequency of 1575.42 MHz, and f_{L5} is the L5 nominal frequency of 1176.45 MHz.

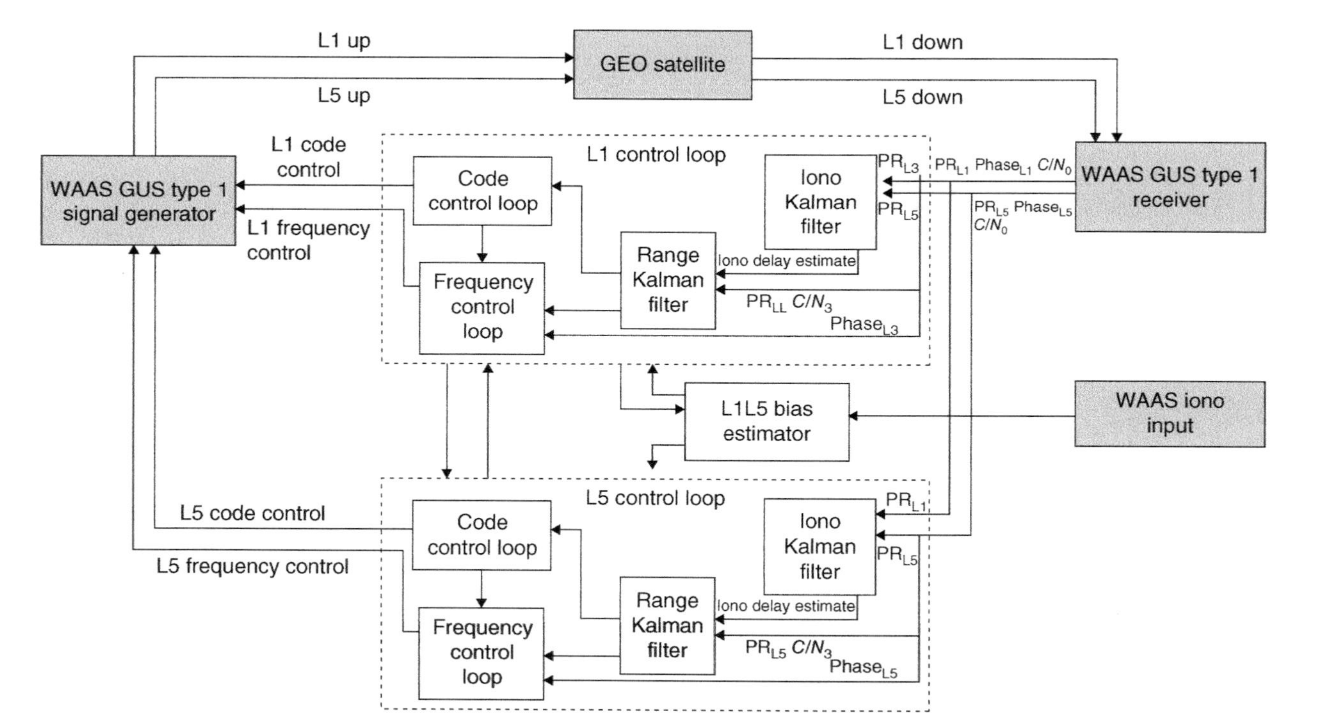

Figure 8.5 Primary GEO uplink subsystem type (GUST) control loop functional block diagram.

The L5 ionospheric Kalman filter design is similar to that for L1, with the filter measurement as follows:

$$z = \frac{(\rho_{L1} - d_{L1}) - (\rho_{L5} - d_{L5})}{(f_{L5})^2/(f_{L1})^2 - 1} \tag{8.1}$$

8.3.1.2 Range Kalman Filter

The L1 and L5 range Kalman filters use four state variables:

$$\mathbf{x} \stackrel{\text{def}}{=} \begin{bmatrix} \text{range} \\ \text{range rate} \\ \text{acceleration} \\ \text{acceleration rate} \end{bmatrix} \tag{8.2}$$

During every one-second timeframe in the safety computer, the range Kalman filter states and their covariance of uncertainty are propagated (predicted) in the Kalman filter.

After the filter propagation, if L1 pseudorange is valid, the L1 range estimate and covariance are updated in the Kalman filter using the L1 pseudorange measurement correction. Likewise, if L5 pseudorange is valid, the L5 range estimate and covariance are updated in the Kalman filter using the L5 pseudorange measurement correction.

The L1 range Kalman filter measurement is

$$z = \rho_{L1} - d_{L1} - I_{L1}$$

where ρ_{L1} is the L1 pseudorange, d_{L1} is the predetermined L1 downlink path hardware delays, and I_{L1} is the L1 ionospheric delay estimate.

Likewise, the L5 range Kalman filter measurement is

$$z = \rho_{L5} - d_{L5} - I_{L5}$$

where ρ_{L5} is the L5 pseudorange, d_{L5} is the predetermined L5 downlink path hardware delays, and I_{L5} is the L5 ionospheric delay estimate. The required Kalman filter equations are given in Table 10.1.

8.3.1.3 Code Control Function

The L1 and L5 code control functions compute the corresponding code chip rate commands and the chip acceleration commands to be sent to the signal generator. The signal generator adjusts its L1 and L5 chip rates according to these commands. The purpose of code control is to compensate for any initial GEO range estimation error, the iono delay on the uplink C-band signal, and the Doppler effects due to the GEO movement on the uplink signal code chip rate. This compensation will ensure that the GEO signal code phase deviation is within the required limit.

The receiver and signal generator timing 1 pps (pulse per second) errors also affect the GEO signal code phase deviation. These errors are compensated separately by the clock steering algorithm [12].

Measurement errors in the predetermined hardware delays of the two signal paths (both uplink and downlink) will result in additional code phase deviation for the GEO signal due to the closed-loop control. This additional code phase deviation will be interpreted as GEO satellite clock error by the master station's GEO orbit determination (OD). Since the clock steering algorithm will use the SBAS broadcast type 9 message GEO clock offset as part of the input to the clock steering controller [12], the additional code phase deviation due to common measurement errors will be compensated for by the clock steering function.

There are several inputs to the code control function: the uplink range, the projected range of the GEO for the next one-second timeframe, the estimated iono delay, and so on. The uplink range is the integration of the commanded chip rate, and this integration is performed in the safety computer. The commanded chip acceleration is computed on the basis of the estimated acceleration from the Kalman filter (see Table 10.1).

8.3.1.4 Frequency Control Function

The L1 and L5 frequency control functions compute the corresponding carrier frequency commands and the frequency change rate (acceleration) commands to be sent to the signal generator. The signal generator adjusts the L1 and L5 IF outputs according to these commands. The purpose of frequency control is to compensate for the Doppler effects due to the GEO movement on the carrier of the uplink signal, the effect of iono rate on the uplink carrier, and the frequency offset of the GEO transponders. This function also continuously estimates the GEO transponder offset, which could drift during the lifetime of the GEO satellite.

8.3.1.5 L1L5 Bias Estimation Function

This function estimates the bias between the L1 and L5 that is due to differential measurement errors in the predetermined hardware delays of the two signal paths. If not estimated and compensated, the bias between L1 and L5 will be indistinguishable from iono delay, as shown in the equations (8.3) and (8.4). L1 and L5 pseudorange can be expressed as

$$\rho_{L1} = r + I_{L1} + \text{true } d_{L1} + \text{clock error} + \text{tropo delay} \tag{8.3}$$

$$\rho_{L5} = r + I_{L5} + \text{true } d_{L5} + \text{clock error} + \text{tropo delay} \tag{8.4}$$

where r is the true range, I_{L1} is the true L1 iono delay, I_{L5} is the true L5 iono delay, true d_{L1} is the true L1 downlink path hardware delay, and true d_{L5} is the true L5 downlink path hardware delay.

This becomes

$$\rho_{L1} - d_{L1} = r + I_{L1} + \text{true } d_{L1} + \text{clock error} + \text{tropo delay} - d_{L1} \tag{8.5}$$

$$\rho_{L5} - d_{L5} = r + I_{L5} + \text{true } d_{L5} + \text{clock error} + \text{tropo delay} - d_{L5} \tag{8.6}$$

where d_{L1} is the predetermined (measured) L1 downlink path hardware delay and d_{L5} is the predetermined (measured) L5 downlink path hardware delay.

Let Δd_{L1} = true $d_{L1} - d_{L1}$ and Δd_{L5} = true $d_{L5} - d_{L5}$. The measurement for the L1 iono Kalman filter becomes

$$\begin{aligned} z &= \frac{(\rho_{L1} - d_{L1}) - (\rho_{L5} - d_{L5})}{(1 - f_{L1})^2/(f_{L5})^2} \\ &= I_{L1} + \frac{(\Delta d_{L1} - \Delta d_{L5})}{(1 - f_{L1})^2/(f_{L5})^2} \end{aligned} \tag{8.7}$$

The term $(\Delta d_{L1} - \Delta d_{L5})/(1 - f_{L1})^2/(f_{L5})^2$ is the differential L1L5 bias term, and it becomes an error in the L1 iono delay estimation. The L5 iono Kalman filter is similarly affected by the L1L5 bias term.

8.3.1.6 Code-Carrier Coherence

The GEO's broadcast code-carrier coherence (CCC) requirement is specified in the WAAS System Specification and in appendix A of Ref. [13]. The WAAS System Specification states the requirement as:

> The lack of coherence between the broadcast SIS L1 carrier phase and its respective code phase shall be limited in accordance with the following equation such that the standard deviation over T seconds of the error due to 100-second carrier-smoothing of the code-based pseudorange is less than one carrier wavelength. This equation does not include code-carrier divergence due to ionospheric refraction in the downlink propagation path.
>
> $\sigma_{(T\,\mathrm{sec})}[\rho_{L1}(t) - \overline{\rho_{L1}(t)}] > 0.19\ \mathrm{m}$

The term ρ_{L1} is the L1 pseudorange that would be measured by a noiseless receiver, and $\overline{\rho_{RL1}(t)}$ is the carrier smoothed L1 pseudorange. WAAS CCC is defined over an interval of T – 86 400 seconds (1 day). Carrier smoothing is performed over a 100-second interval. Note that a noiseless receiver, as used here, means the values that would be measured by a hypothetical noiseless receiver in parallel with the ground station receiver. Alternately, because code-carrier divergence due to ionospheric refraction in the downlink signal path is excluded, the noiseless receiver may be considered to be at the focus of the GEO transmit antenna.

The WAAS CCC equation earlier is interpreted from the Radio Technical Commission for Aeronautics (RTCA) Minimum Operational Performance Standards (MOPSs) equation [13] and is based on the need to limit errors associated with carrier smoothing of the code-based pseudorange. CCC results for two WAAS GEOs are provided in Table 8.2. The results indicate that the control loop algorithm performance meets WAAS requirements [3].

Table 8.2 Code-carrier coherence results.

GEO satellite	Date	Code-carrier coherence requirement <1 cycle
CRW	25 August 2009	0.65
CRE	25 August 2009	0.99

8.3.1.7 Carrier Frequency Stability

Carrier frequency stability is a function of the uplink frequency standard, GUS signal generator, and GEO satellite transponder. The GEO's short-term carrier frequency stability requirement is specified in the WAAS System Specification and appendix A of Ref. [13]. It states: "The short term stability of the carrier frequency (square root of the Allan variance) at the input of the user's receiver antenna shall be better than 5×10^{-11} over 1 to 10 s, excluding the effects of the ionosphere and Doppler."

The Allan variance [14] is calculated on the double difference of L1 phase data divided by the center frequency over 1–10 seconds. Effects of smoothed ionosphere and Doppler are compensated for in the data prior to this calculation.

8.4 GUS Clock Steering Algorithm

Presently, the SBAS WMS calculates SBAS network time (WNT) and estimates clock parameters (offset and drift) for each satellite. The GEO Uplink System (GUS) clock is an independent free running clock. However, the GUS clock must track WNT (GPS time) to enable accurate ranging from the GEO SIS. Therefore, a clock steering algorithm is necessary. The GUS clock steering algorithms reside in the SBAS message processor (WMP). The SBAS type 9 message (GEO navigation message) is used as input to the GUS WMP, provided by the WMS.

The GUS clock is steered to the GPS time epoch (see also Figure 8.6). The GUS receiver clock error is the deviation of its one-second pulse from the GPS epoch. The clock error is computed in the GUS processor by calculating the user position error by combining (in the least-squares sense, weighted with expected error statistics) multiple satellite data (pseudorange residuals called MOPS *residuals*) [13] into a position error estimate with respect to surveyed GUS position. The clock steering algorithm is initialized with the SBAS type 9 message (GEO navigation message). This design keeps the GUS receiver clock 1 pps synchronized with the GPS time epoch. Since the 10 MHz frequency standard is the frequency reference for the receiver, its frequency output needs to be

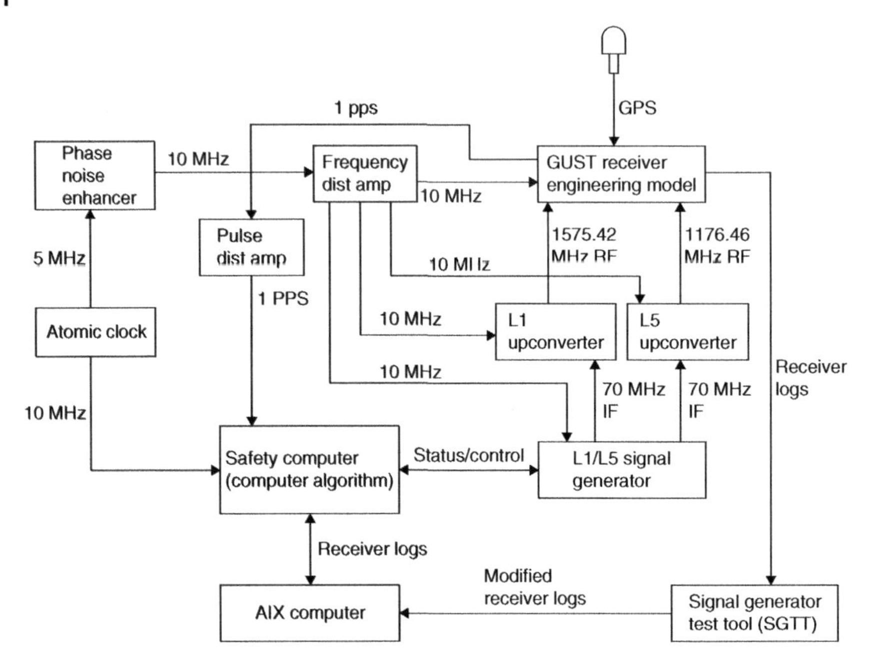

Figure 8.6 Control loop test setup. AIX = IBM Advanced Interactive eXecutive Processor.

controlled so that the 1 pps is adjusted. A proportional, integral, and differential controller has been designed to synchronize to the GPS time at GUS locations.

This algorithm also decouples the GUS clock from orbit errors and increases the observability of orbit errors in the OD filter in the correction processor of the WMS. It also synchronizes GUS clocks at all GUS locations to GPS time.

In the initial 24 hours after a GUS becomes primary, the clock steering algorithm uses SBAS type 9 messages from the WMS to align the GEO's epoch with the GPS epoch. The SBAS type 9 message contains a term referred to as a_{Gf0} or clock offset. This offset represents a correction, or time difference, between the GEO's epoch and SBAS network time (WNT). WNT is the internal time reference scale of SBAS and is required to track the GPS timescale, while at the same time providing users with the translation to UTC. Since GPS master time is not directly obtainable, the SBAS architecture requires that WNT be computed at multiple WMSs using potentially differing sets of measurements from potentially differing sets of receivers and clocks (SBAS reference stations). WNT is required to agree with GPS to within 50 ns. At the same time, the WNT to UTC offset must be provided to the user, with the offset being accurate to 20 ns. The GUS calculates local clock adjustments. On the basis of these clock adjustments, the frequency standard can be made to speed up or slow the GUS clock. This will keep the total GEO clock offset within the range allowed by the

SBAS type 9 message so that users can make the proper clock corrections in their algorithms [15, 16].

After the initial 24 hours, once the GUS clock is synchronized with WNT, a second steering method of clock steering is used. The algorithm now uses the composite of the MOPS [13] solution for the receiver clock error, and the average of the a_{Gf0}, and the average of the MOPS solution as the input to the clock steering controller.

8.4.1 Receiver Clock Error Determination

Determination of receiver clock error is based on the user position solution algorithm described in the SBAS MOPS. The clock bias (C_b) is a resultant of the MOPS weighted least-squares solution.

Components of the weighted least-squares solution are the observation matrix (**H**), the measurement weighting matrix (**W**), and the MOPS residual column vector ($\boldsymbol{\Delta\rho}$). The weighted gain matrix (**K**) is calculated using **H** and **W** (see Eq. (2.30)):

$$\mathbf{K} = (\mathbf{H}^T\mathbf{W}\mathbf{H})^{-1}\mathbf{H}^T\mathbf{W} \tag{8.8}$$

From this, the column vector for the user position error and the clock bias solution is

$$\boldsymbol{\Delta}\mathbf{x} = \mathbf{K}\boldsymbol{\Delta\rho} \tag{8.9}$$

$$\boldsymbol{\Delta}\mathbf{x} = (\mathbf{H}^T\mathbf{W}\mathbf{H})^{-1}\mathbf{H}^T\mathbf{W}\boldsymbol{\Delta\rho} \tag{8.10}$$

where

$$\boldsymbol{\Delta}\mathbf{x} = \begin{pmatrix} \Delta X(U) \\ \Delta X(E) \\ \Delta X(N) \\ C_b \end{pmatrix} \tag{8.11}$$

where $\Delta X(U)$ is the up error, $\Delta X(E)$ is the east error, $\Delta X(N)$ is the north error, and C_b is the clock bias or receiver clock error.

The $n \times 4$ observation matrix (**H**) is computed in up–east–north (UEN) reference frame using the LOS azimuth (Az_i) and LOS elevation (El_i) from the GUS omni antenna to the SV. The value n is the number of satellites in view. The formula for calculating the observation matrix is

$$\mathbf{H} = \begin{bmatrix} \cos(\mathrm{E1}_1)\cos(\mathrm{Az}_1) & \cos(\mathrm{E1}_1)\sin(\mathrm{Az}_1) & \sin(\mathrm{E1}_1) & 1 \\ \cos(\mathrm{E1}_2)\cos(\mathrm{Az}_2) & \cos(\mathrm{E1}_2)\sin(\mathrm{Az}_1) & \sin(\mathrm{E1}_2) & 1 \\ \vdots & \vdots & \vdots & \vdots \\ \cos(\mathrm{E1}_n)\cos(\mathrm{Az}_n) & \cos(\mathrm{E1}_n)\sin(\mathrm{Az}_n) & \sin(\mathrm{E1}_n) & 1 \end{bmatrix} \tag{8.12}$$

The $n \times n$ weighting matrix (**W**) is a function of the total variance (σ_i^2) of the individual satellites in view. The inverse of the weighting matrix is

$$\mathbf{W}^{-1} = \begin{bmatrix} \sigma_1^2 & 0 & 0 & \vdots & 0 \\ 0 & \sigma_2^2 & 0 & \vdots & 0 \\ 0 & 0 & \sigma_3^2 & \vdots & 0 \\ \vdots & \vdots & \vdots & \ddots & \vdots \\ 0 & 0 & 0 & 0 & \sigma_n^2 \end{bmatrix}. \tag{8.13}$$

The equation to calculate the total variance (σ_i^2) is

$$\sigma_i^2 = \left(\frac{\text{UDRE}_i}{3.29}\right)^2 + \left(\frac{F_{\text{pp}i} \times \text{GIVE}_i}{3.29}\right)^2 + \sigma_{\text{L1nmp},i}^2 + \frac{\sigma_{\text{tropo},i}^2}{\sin^2 \text{El}_i} \tag{8.14}$$

The algorithms for calculating user differential range error (UDRE_i), user grid ionospheric vertical error (GIVE_i), LOS obliquity factor ($F_{\text{pp}i}$), standard deviation of uncertainty for the vertical troposphere delay model (a tropo_i), and the standard deviation of noise and multipath on the L1 omni pseudorange $\sigma_{\text{L1nmp},i}$ are found in the SBAS MOPS [13].

The MOPS residuals ($\Delta\rho$) are the difference between the smoothed MOPS measured pseudorange ($\text{PR}_{M,i}$) and the expected pseudorange ($\text{PR}_{\text{corr},i}$)

$$\boldsymbol{\Delta\rho} = \begin{bmatrix} \text{PR}_{M,1} - \text{PR}_{\text{corr},1} \\ \text{PR}_{M,2} - \text{PR}_{\text{corr},2} \\ \text{PR}_{M,3} - \text{PR}_{\text{corr},3} \\ \vdots \end{bmatrix} \tag{8.15}$$

The MOPS measured pseudorange ($\text{PR}_{M,i}$) in Earth-centered, Earth-fixed (ECEF) reference is corrected for Earth rotation, for SBAS clock corrections, for ionospheric effects, and for tropospheric effects. The equation to calculate ($\text{PR}_{M,i}$) is

$$\text{PR}_{M,i} = \text{PR}_{L,i} + \Delta\text{PR}_{\text{CC},i} + \Delta\text{PR}_{\text{FC},i} + \Delta\text{PR}_{\text{ER},i} - \Delta\text{PR}_{T,i} - \Delta\text{PR}_{I,i} \tag{8.16}$$

The algorithms used to calculate smoothed L1 omni pseudorange ($\text{PR}_{L,i}$), pseudorange clock correction ($\Delta\text{PR}_{\text{CC},i}$), pseudorange fast correction ($\Delta\text{PR}_{\text{FC},i}$), pseudorange Earth rotation correction ($\Delta\text{PR}_{\text{ER},i}$), pseudorange troposphere correction ($\Delta\text{PR}_{T,i}$), and pseudorange ionosphere correction ($\Delta\text{PR}_{I,i}$) are found in the SBAS MOPS [13].

Expected pseudorange ($PR_{corr,i}$) ECEF, at the time of GPS transmission is computed from broadcast ephemeris corrected for fast and long-term corrections. The calculation is

$$PR_{corr,i} = \sqrt{(X_{corr,i} - X_{GUS})^2 + (Y_{corr,i} - Y_{GUS})^2 + (Z_{corr,i} - Z_{GUS})^2} \quad (8.17)$$

The fixed-position parameters of the WRE (X_{GUS}, Y_{GUS}, Z_{GUS}) are site specific.

8.4.2 Clock Steering Control Law

In the primary GUS, the clock steering algorithm is initialized with SBAS type 9 message (GEO navigation message). After the initialization, composite of MOPS solution and type 9 message for the receiver clock error is used as the input to the control law (see Figure 8.7). For the backup GUS, the MOPS solution for the receiver clock error is used as the input to the control law (see Figure 8.8).

For both the primary and backup clock steering algorithm, the control law is a PID controller. The output of the control law will be the frequency adjustment command. This command is sent to the frequency standard to adjust the

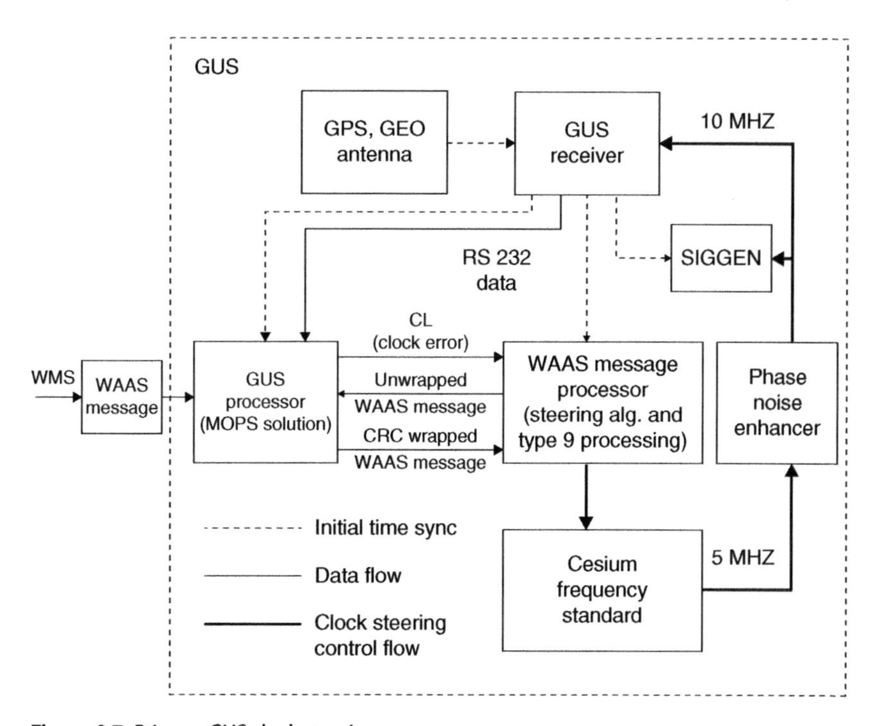

Figure 8.7 Primary GUS clock steering.

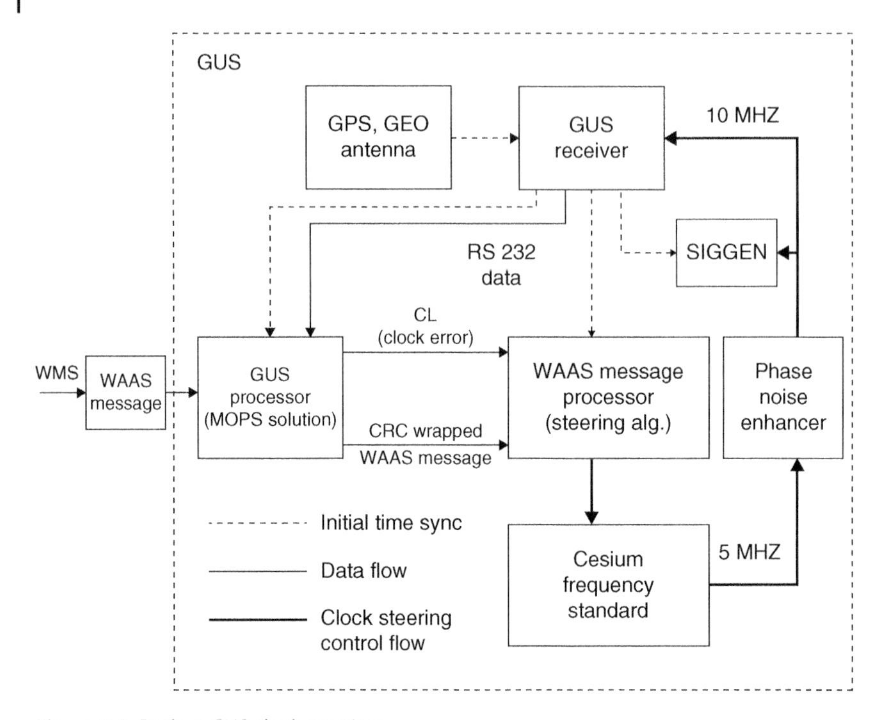

Figure 8.8 Backup GUS clock steering.

atomic clock frequency. The output frequency to the receiver causes the 1 pps to approach the GPS epoch. Thus, a closed-loop control of the frequency standard is established.

8.5 GEO Orbit Determination (OD)

The purpose of WAAS is to provide pseudorange and ionospheric corrections for GPS satellites to improve the accuracy for the GPS navigation user and to protect the user with "integrity." Integrity is the ability to provide timely warnings to the user whenever any navigation parameters estimated using the system are outside tolerance limits. WAAS may also augment the GPS constellation by providing additional ranging sources using GEO satellites that are being used to broadcast the WAAS signal.

The two parameters having the most influence on the integrity bounds for the broadcast data are UDRE for the pseudorange corrections and GIVE for the ionospheric corrections. With these, the onboard navigation system estimates the horizontal protection limit (HPL) and the vertical protection limit (VPL), which are then compared to the horizontal alert limit (HAL) and the vertical alert limit (VAL) requirements for the particular phase of flight involved, that is, oceanic/remote, en route, terminal, NPA, and PA. If the estimated protection limits are greater than the alert limits, the navigation system is declared unavailable. Therefore, the UDRE and GIVE values obtained by the WAAS (in concert with the GPS and GEO constellation geometry and reliability) essentially determine the degree of availability of the WADGPS navigation service to the user.

The WAAS algorithms calculate the broadcast corrections and the corresponding UDREs and GIVEs by processing the satellite signals received by the network of ground stations. Therefore, the expected values for UDREs and GIVEs are dependent on satellite and station geometries, satellite signal and clock performance, receiver performance, environmental conditions (such as multipath and ionospheric storms), and algorithm design [17, 18].

8.5.1 OD Covariance Analysis

A full WAAS algorithm contains three Kalman filters – an OD filter, an ionospheric corrections filter, and a fast corrections filter. The fast corrections filter is a Kalman filter that estimates the GEO, GPS, and ground station clock states every second. In this section, we derive an estimated lower bound of the GEO UDRE for a WAAS algorithm that contains only the OD Kalman filter, called the UDRE (OD), where OD refers to orbit determination.

A method is proposed to approximate the UDRE obtained for a WAAS including both the OD filter and the fast corrections filter from UDRE (OD). From case studies of the geometries studied in Table 8.3, we obtain the essential dependence of UDRE on ground station geometry.

A covariance analysis on the OD is performed using a simplified version of the OD algorithms. The performance of the ionospheric corrections filter is treated as perfect, and therefore, the ionospheric filter model is ignored. The station clocks are treated as if perfectly synchronized using the GPS satellite measurements. Therefore, the station clock states are ignored. This allows the decoupling of the ODs for all the satellites from each other, simplifying the OD problem to that for one satellite with its corresponding ground station geometry and synchronized station clocks. Both of these assumptions are liberal; therefore, the UDRE(OD) obtained here is a lower bound for the actual UDRE(OD). Finally, we consider only users within the service volume covered by the stations and, therefore, ignore any degradation factors depending on user location.

Table 8.3 Cases used in geometry-per-station analysis.

Case	UDRE	GDOP	Satellite	Geometry
1	17.9	905	AOR-W	WAAS stations (25), 21 in view
2	45.8	2516	AOR-W	4 WAAS stations (CONUS)
3	135.0	56 536	AOR-W	4 WAAS stations (NE)
4	4.5	254	AOR-W	WAAS stations + Santiago
5	5.8	212	AOR-W	WAAS stations + London
6	4.0	154	AOR-W	WAAS stations + Santiago + London
7	7.5	439	AOR-W	4 WAAS stations (CONUS) + Santiago
8	8.6	337	AOR-W	4 WAAS stations (CONUS) + London
9	6.6	271	AOR-W	4 WAAS stations (CONUS) + Santiago + London
10	47.7	2799	AOR-W	4 WAAS stations (NE) + Santiago
11	21.5	1405	AOR-W	4 WAAS stations (NE) + London
12	16.4	1334	AOR-W	4 WAAS stations (NE) + Santiago + London
13	28.5	1686	POR	WAAS stations (25), 8 in view
14	45.4	3196	POR	WAAS stations, Hawaii
15	31.1	1898	POR	WAAS stations, Cold Bay
16	55.0	4204	POR	WAAS stations, Hawaii, Cold Bay
17	6.7	257	POR	WAAS stations + Sydney
18	8.3	338	POR	WAAS stations + Tokyo
19	6.7	257	POR	WAAS stations + Sydney + Tokyo
20	21.0	1124	MTSAT	MSAS stations, 8 in view
21	22.0	1191	MTSAT	MSAS stations, Hawaii
22	24.9	1407	MTSAT	MSAS stations, Australia
23	54.6	4149	MTSAT	MSAS stations, Hawaii, Australia
24	22.0	1198	MTSAT	MSAS stations, Ibaraki
25	29.0	1731	MTSAT	MSAS stations—Ibaraki, Australia
26	54.8	4164	MTSAT	MSAS stations, Ibaraki, Australia, Hawaii
27	13.2	609	MTSAT	MSAS stations + Cold Bay
A		139	TEST	0 = 75′
B		422	TEST	0 = 30′
C		3343	TEST	0 = 10′
D		13 211	TEST	0 = 5′
E		67	TEST	41 stations
F		64	TEST	41 + 4 stations

The four WAAS stations (CONUS) are Boston, Miami, Seattle, and Los Angeles.
The four WAAS stations (NE) are Boston, New York, Washington, DC, and Cleveland.

To simulate the Kalman filter for the covariance matrix P, the following four matrices are necessary (Table 10.1):

$$\begin{aligned}\mathbf{\Phi} &= \text{state transition matrix}\\ \mathbf{H} &= \text{measurement sensitivity matrix}\\ \mathbf{Q} &= \text{process noise covariance matrix}\\ \mathbf{R} &= \text{measurement noise covariance matrix}\end{aligned}$$

The methods used to determine these matrices are described in the following text.

The state vector for the satellite is

$$\mathbf{x} = \begin{bmatrix} \mathbf{r} \\ \dot{\mathbf{r}} \\ C_b \end{bmatrix}$$

where

$$\mathbf{r} \equiv [x, y, z]^T$$

is the satellite position in the Earth-centered inertial (ECI) frame;

$$\dot{\mathbf{r}} \equiv [\dot{x}, \dot{y}, \dot{z}]^T$$

is the satellite velocity in the ECI frame; and C_b is the satellite clock offset relative to the synchronized station clocks. Newton's second and third (gravitational) laws provide the equations of motion for the satellite:

$$\dot{\mathbf{r}} \equiv \frac{d^2 r}{dt^2} = -\frac{\mu_E \mathbf{r}}{|\mathbf{r}|^3} + M$$

where $\dot{\mathbf{r}}$ is the acceleration in the EC1 frame, μ_{E} is the gravitational constant for the Earth, and M is the total perturbation vector in the ECI frame containing all the perturbing accelerations. For this analysis, only the perturbation due to the oblateness of the Earth is included. The effect of this perturbation on the behavior of the covariance is negligible, and therefore higher-order perturbations are ignored. (Note that although the theoretical model is simplified, the process noise covariance matrix Q is chosen to be consistent with a far more sophisticated orbital model.)

Therefore,

$$M = -\frac{3}{2} J_2 \frac{\mu_E}{|\mathbf{r}|^3} \frac{a_E^2}{|\mathbf{r}|^2} [\mathbf{I}_{3\times3} \widehat{\mathbf{z}}\widehat{\mathbf{z}}^T] \mathbf{r}$$

where a_E is the semimajor axis of the Earth-shape model, J_2 is the second zonal harmonic coefficient of the Earth-shape model, and $\widehat{\mathbf{z}} \equiv [0, 0, 1]^T$ [19].

The second-order differential equation of motion can be rewritten as a pair of first-order differential equations:

$$\dot{\mathbf{r}} = \mathbf{r}_2, \dot{\mathbf{r}}_2 = \frac{\mu_E r_1}{|\mathbf{r}|^3} + M \tag{8.18}$$

where $\mathbf{r}_1$ and $\mathbf{r}_2$ are vectors, which therefore gives a system of six first-order equations.

The variational equations are differential equations describing the rates of change of the satellite position and velocity vectors as functions of variations in the components of the estimation state vector. These lead to the state transition matrix Φ used in the Kalman filter. The variational equations are

$$\ddot{Y}(t) = A(t)Y(t) + B(t)\dot{Y}(t) \tag{8.19}$$

where

$$Y(t_k)_{3\times6} \equiv \left[\left(\frac{\partial \mathbf{r}(t_k)}{\partial \mathbf{r}(t_{k-1})} \right)_{3\times3} \left(\frac{\partial \mathbf{r}(t_k)}{\partial \dot{\mathbf{r}}(t_{k-1})} \right)_{3\times3} \right] \tag{8.20}$$

$$\dot{Y}(t_k)_{3\times6} \equiv \left[\left(\frac{\partial \dot{\mathbf{r}}(t_k)}{\partial \mathbf{r}(t_{k-1})} \right)_{3\times3} \left(\frac{\partial \mathbf{r}(t_k)}{\partial \dot{\mathbf{r}}(t_{k-1})} \right)_{3\times3} \right] \tag{8.21}$$

$$\begin{aligned} A(t)_{3\times3} &\equiv \frac{\partial \ddot{\mathbf{r}}}{\partial \mathbf{r}} \\ &= \frac{-\mu_E}{|\mathbf{r}|^3}[\mathbf{I}_{3\times3} - 3\hat{\mathbf{r}}\hat{\mathbf{r}}^T] - \frac{3}{2}J_2 \frac{\mu_E}{|\mathbf{r}|^3}\frac{a_E^2}{|\mathbf{r}|^2} \\ &\quad \times [\mathbf{I}_{3\times3} + 2\hat{\mathbf{z}}\hat{\mathbf{z}}^T - 10(\hat{\mathbf{r}}^T\hat{\mathbf{z}}^T)(\hat{\mathbf{z}}\hat{\mathbf{r}}^T + \hat{\mathbf{r}}\hat{\mathbf{z}}^T) + (10(\hat{\mathbf{r}}^T\hat{\mathbf{z}})^2 - 5)(\hat{\mathbf{r}}\hat{\mathbf{r}}^T)] \end{aligned} \tag{8.22}$$

$$B(t)_{3\times3} \equiv \frac{\partial \ddot{\mathbf{r}}}{\partial \dot{\mathbf{r}}} = \mathbf{0}_{3\times3} \tag{8.23}$$

where $\hat{\mathbf{r}} = \mathbf{r}/|\mathbf{r}|$.

Equations (8.20)–(8.23) are substituted into Eqs. (8.19) and (8.18), and the differential equations are solved using the fourth-order Runge–Kutta method. The time step used is a 5five-minute interval. The initial conditions for the GEO are specified for the particular case given and propagated forward for each time step, whereas the initial conditions for the Y terms are

$$Y(t_{k-1})_{3\times6} = [\mathbf{I}_{3\times3} \ \mathbf{0}_{3\times3}], \ \dot{Y}(t_k)_{3\times6} = [\mathbf{0}_{3\times3} \ \mathbf{I}_{3\times3}]$$

and are reset for each time step. This is due to the divergence of the solution of the differential equation used in this method to calculate the state transition matrix for the Kepler problem.This gives the state $\mathbf{x}_1^T = [\mathbf{r}_1^T \ \mathbf{r}_2^T]$ and the state transition matrix

$$\Phi_{k,k-1_{7\times7}} = \begin{bmatrix} Y(t_k)_{3\times6} & \mathbf{0}_{3\times1} \\ \dot{Y}(t_k)_{3\times6} & \mathbf{0}_{3\times1} \\ \mathbf{0}_{1\times6} & \mathbf{I}_{1\times1} \end{bmatrix} \tag{8.24}$$

for the Kalman filter.

The measurement sensitivity matrix is given by

$$H_{N\times 7} \equiv \left[\left(\frac{\partial \rho}{\partial \mathbf{r}}\right)_{N\times 3} \left(\frac{\partial \rho}{\partial \dot{\mathbf{r}}}\right)_{N\times 3} \left(\frac{\partial \rho}{\partial (ct)}\right)_{N\times 1} \right]$$

where ρ is the pseudorange for a station and N is the number of stations in view of the satellite. Note that this is essentially the same H as in Section 8.4.1. Ignoring relativistic corrections and denoting the station position by the vector $r_s \equiv [x_s\ y_s\ z_s]^T$, the matrices earlier are given by

$$\frac{\partial \rho}{\partial \mathbf{r}} = \frac{[\mathbf{r} - r_s]^T}{\mathbf{r} - r_s} \frac{\partial \mathbf{r}(t_k)}{\partial \mathbf{r}(t_{k-1})}$$

$$\frac{\partial \rho}{\partial \dot{\mathbf{r}}} = \frac{[\mathbf{r} - r_s]^T}{\mathbf{r} - r_s} \frac{\partial \mathbf{r}(t_k)}{\partial \dot{\mathbf{r}}(t_{k-1})}$$

and

$$\frac{\partial \rho}{\partial (ct)} = 1 \tag{8.25}$$

The station position is calculated with the WGS84 model for the Earth and converted to the ECI frame using the J2000 epoch (see Appendix B).

These are then combined with the measurement noise covariance matrix R and the process noise covariance matrix Q to obtain the Kalman filter equations for the covariance matrix P, as shown in Table 10.1.

The initial condition, $P_0(+)$, and Q are chosen to be consistent with the WAAS algorithms. The value of R is chosen by matching the output of the GEO covariance for AOR-W with $R = \sigma^2 \mathbf{I}$ and is used as the input R for all other satellites and station geometries (note that this therefore gives approximate results). This corresponds to carrier phase ranging for the stations. The results corresponding to the value of R for code ranging are also presented.

From this covariance, the lower bound on the UDRE is obtained by

$$\text{UDRE} \geq \text{EMRBE} + K_{\text{ss}} \sqrt{\text{tr}(p)}$$

where EMRBE is the estimated maximum range and bias error. EMRBE $= 0$, $K_{\text{ss}} = 3.29$ will bring the 0.999 level of bounding for the UDRE. Finally, since the message is broadcast every second, $\Delta t = 1$, so the trace can be used for the velocity components as well.

Figure 8.9 shows the relationship between UDRE and geometric dilution of precision (GDOP) for various GEO satellites and WRS locations. Table 8.3 describes the various cases considered in this analysis.

The numerical values used for the filter are as follows (all units are Système International [SI]):

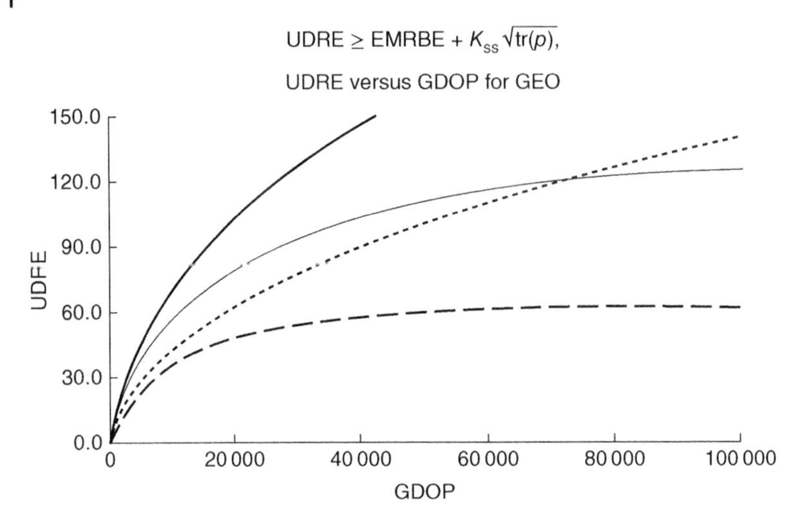

Figure 8.9 Relationship between UDRE and GDOP.

Earth parameters:

$$\mu_E = 3.986\,004\,41 \times 10^{14}$$
$$a_E = 6\,378\,137.0$$
$$J_2 = 1082.63 \times 10^{-6}$$
$$b_E = 6\,356\,752.3142$$

8.6 Ground-Based Augmentation System (GBAS)

8.6.1 Local-Area Augmentation System (LAAS)

The local-area augmentation system (LAAS) (near airports) is being designed to provide DGPS corrections in support of navigation and landing systems. The system provides monitoring functions via LAAS ground facility (LGF) and includes individual measurements, ranging sources, reference receivers, navigation data, data broadcast, environment sensors, and equipment failures. Each identified monitor has a corresponding system response including alarms, alerts, and service alerts (see Figure 8.10.)

8.6.2 Joint Precision Approach and Landing System (ALS)

The joint precision approach landing system (ALS) is an all-weather military landing system based on differential GPS for land-based and sea-based aircraft.

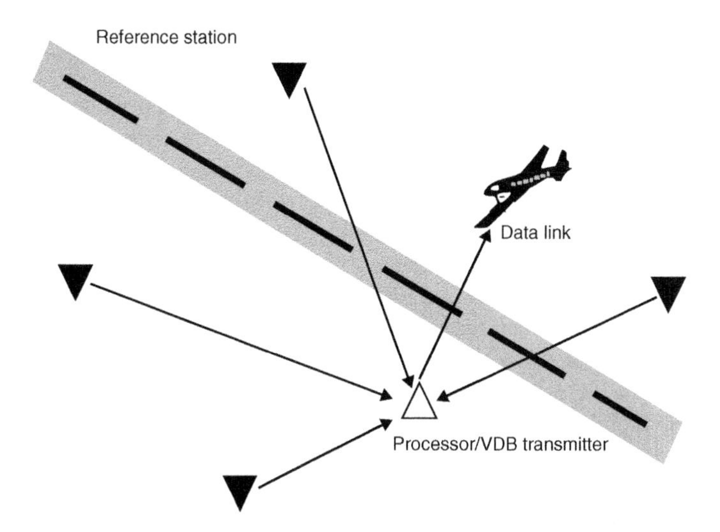

Figure 8.10 Local-area augmentation system (LAAS). VDB: very high frequency data broadcast.

One of the two main types of ALS is Shipboard Relative GPS (SRGPS), which is used for approaches and types of ALS is SRGPS, which is used for approaches and landings on ships. SRGPS uses a variant of carrier phase differential GPS to obtain accuracies within 10–20 cm of a moving point on a ship. The location is part of the information broadcast to users. The other type of ALS is known as Local Differential GPS (LDGPS). LDGPS is land-based and is used for military airfields. A tactical variant of LDGPS is portable and can be set up for relatively short-term airfield operations [20].

ALS is completing the System Development and Demonstration phase and preparing for a production manufacturing, with initial operating capability projected by the end of 2020. Full operational capability is expected to take another decade or more.

8.6.3 Enhanced Long-Range Navigation (eLORAN)

While enhanced long-range navigation (eLORAN) is not a GNSS and is not currently operational in the United States, other parts of the world have been using eLORAN as a backup to GNSS. eLORAN was developed to enhance the capabilities provided by the legacy long range navigation-version C (LORAN-C) for data channel messaging, all-in-view, and traceable time reference to UTC. Data messaging formats have been established within eLORAN to provide differential eLORAN capabilities [21, 22].

8.7 Measurement/Relative-Based DGNSS

DGNSS is a technique for improving the performance of GNSS positioning. The basic idea of DGNSS is to compute the spatial displacement vector of the user's receiver (sometimes called the roving or remote receiver) relative to another receiver (usually called the reference receiver or base station). In most DGNSS applications, the coordinates of the reference receiver are precisely known from highly accurate survey information; thus, the accurate location of the roving receiver can be determined by vector addition of the reference receiver coordinates and the reference-to-rover displacement vector. For relative DGNSS applications, the reference receiver may be mobile as well, but the paramount importance is the relative position and/or velocity between the mobile user and the mobile user reference receiver.

The positioning accuracy of DGNSS depends on the error in estimating the reference-to-rover displacement vector. This error can be made considerably smaller than the positioning error of a stand-alone receiver because major components of pseudorange measurement errors are common to the roving and reference receivers and can be canceled out by using the difference between the reference and rover measurements to compute the displacement vector.

There are basically two ways that errors common to the roving and reference receiver can be canceled. The first method is called the *measurement* or *solution-domain* technique, in which both receivers individually compute their positions and the reference-to-rover displacement vector is simply the difference of these positions. However, the two receivers must use exactly the same set of satellites for this method to be effective. Since this requirement is often impossible to fulfill (e.g. due to blockage of signals at the roving receiver), this method is seldom used. A far better method, which offers more flexibility, is to use only the difference of the measurements from the set of satellites that are viewed in common by both receivers. Therefore, only this method will be described.

The two primary types of differential measurements are code measurements and carrier phase measurements.

8.7.1 Code Differential Measurements

To obtain code differential measurements, the roving and reference receivers each make a pseudorange measurement of the following form for each satellite:

$$\begin{aligned} \rho^j_{q,A}(t) &= r^j_A(t) + ct^j(t) + ct^j_A(t) + \varepsilon^j_{q,A}(t) \\ \rho^j_{q,B}(t) &= r^j_B(t) + ct^j(t) + ct^j_B(t) + \varepsilon^j_{q,B}(t) \end{aligned} \tag{8.26}$$

where

$\rho^j_{q,A}$ = code phase measurements for frequency q (m)

u = user receiver A or B

j = SV number

where $\rho^j_{q,u}(t)$ is the measured pseudorange in meters, $r^j_A(t)$ is the true receiver-to-satellite j geometric range in meters, c is the speed of light in meters per second, t^j is the clock error for each satellite in seconds, t_u(for user receiver A or B) is the receiver clock error in seconds, and $\varepsilon_{q,u}(t)$ is the combined error due to orbit ephemeris, atmospheric, multipath, receiver, antenna group delay, and code bias (SV and user) errors. The pseudorange measurements made by both receivers must occur at a common GNSS time, or if not, corrections must be applied to extrapolate the measurements to a common time.

8.7.1.1 Single-Difference Observations

A code single-difference (SD) observation is determined by subtracting an equation of the form (Eq. (8.26)) for the reference receiver from a similar equation for the roving receiver, where both equations relate to the same satellite. The result is

$$\rho^j_{q,AB}(t) = r^j_{AB}(t) + t^j_{AB}(t) + \varepsilon^j_{q,AB} \tag{8.27}$$

where the symbols $\rho^j_{q,AB}(t), r^j_{AB}(t), t^j_{AB}(t), \varepsilon^j_{q,AB}$ denote the difference between the corresponding terms in the two equations of the form (Eq. (8.26)). The term t^j representing the satellite clock error has disappeared since the satellite clock error is the same for the pseudorange measurements made by each receiver.

8.7.1.2 Double-Difference Observations

A code double-difference (DD) measurement is formed by subtraction of the single-difference observation of the form (Eq. (8.27)) for one satellite (j) from a similar single-difference observation for another satellite (k). Thus, if there are M single-difference observations corresponding to M satellites, there will be $(M-1)$ independent double-difference observations that can be formed. The double-difference observations have the form

$$\rho^{jk}_{q,AB}(t) = r^{jk}_{AB}(t) + \varepsilon^{jk}_{q,AB} \tag{8.28}$$

where the symbol $\rho^{jk}_{AB}(t)$ denotes difference between the corresponding difference terms in the two equations of the form (Eq. (8.27)). Note that the double-difference error term $t^{jk}_{AB}(t)$ involving receiver clock error has been largely cancelled out since receiver clock error is constant across all satellite measurements made in the reference and roving receivers. The residual code phase error $\varepsilon^{jk}_{q,AB}$ term remains for any error term that is not common between the reference receiver and rover receiver that has not been mitigated. For LADGNSS type architectures, this residual error term is mostly multipath and receiver measurement error. For WADGNSS this may also include residual orbit ephemeris and residual atmospheric errors.

Although DGNSS is effective in removing satellite and receiver clock errors, ionospheric and tropospheric errors, and ephemeris orbit errors, it cannot remove errors due to multipath, receiver interchannel biases, and thermal noise since these errors are not common to the roving and reference receivers. These types of non-common errors should be mitigated at each receiver location to minimize user receiver PVT solution error.

8.7.2 Carrier Phase Differential Measurements

Because carrier phase measurements have significantly less noise than the pseudorange code phase measurement, positioning accuracy is potentially much more accurate. However, since only the fractional and not the integer part of a carrier cycle can be observed, some method of finding the integer part must be employed. This is the classic *ambiguity resolution* problem.

Single- and double-difference observations can be obtained from carrier phase measurements having the form

$$\begin{aligned}\Phi^j_{q,A}(t) &= \lambda_q\varphi^j_A = r^j_A(t) + ct^j(t) + ct^j_A(t) + \lambda_q N^j_{q,A} + v^j_A(t)\\ \Phi^j_{q,B}(t) &= \lambda_q\varphi^j_B = r^j_B(t) + ct^j(t) + ct^j_B(t) + \lambda_q N^j_{q,B} + v^j_B(t)\end{aligned} \tag{8.29}$$

where

$\Phi^j_{q,A}$ = carrier phase measurements for frequency q (m)

λ_q = GNSS wavelength (m)

$\varphi^j_{q,A}$ = carrier phase measurement for frequency q (cycles)

u = user receiver A or B

j = SV number

where the variable λ_q is the carrier wavelength at the respective GNSS frequency (e.g. 0.1903 m for L1 and 0.2442 m for L2), the measured carrier phase φ_u in cycles, the carrier phase ambiguity $N^j_{q,u}$ in cycles, and for station A or B. The other carrier phase residual error term $v^j_A(t)$ is the combined error due to orbit ephemeris, atmospheric, multipath, receiver, antenna phase delay (SV and user) errors.

8.7.2.1 Single-Difference Observations

Each carrier phase single-difference observation is determined in the same manner as for the code by subtracting two equations of the form (Eq. (8.30)) for the reference receiver from a similar equation for the roving receiver, where both equations relate to the same satellite. The result is

$$\Phi^{j}_{q,AB}(t) = \lambda_q \varphi^{j}_{q,AB} = r^{j}_{AB}(t) + ct^{j}_{AB}(t) + \lambda_q N^{j}_{q,AB} + v^{j}_{AB}(t) \tag{8.30}$$

where, as before, the satellite clock error term has cancelled because it is common to both reference and user receiver measurements.

8.7.2.2 Double-Difference Observations

A double-difference carrier phase measurement is formed by subtraction of the single-difference observation of the form (Eq. (8.31)) for one satellite from a similar single-difference observation for another satellite. The double-difference observations have the form

$$\Phi^{jk}_{q,AB}(t) = \lambda_q \varphi^{jk}_{q,AB} = r^{jk}_{AB}(t) + \lambda_q N^{jk}_{q,AB} + v^{jk}_{AB}(t) \tag{8.31}$$

Again, note that the double-difference error term $t^{jk}_{AB}(t)$ involving receiver clock error has been largely cancelled out since receiver clock error is constant across all satellite measurements made in the reference and roving receivers. The residual carrier phase error term v^{jk}_{AB} remains for any error term that is not common between the reference receiver and rover receiver that has not been mitigated. For LADGNSS type architectures, this residual error term is mostly multipath and receiver measurement error. For WADGNSS this may also include residual orbit ephemeris and residual atmospheric errors. The $N^{jk}_{q,AB}$ term is referred to the DD carrier cycle ambiguity and is the term that needs to be accurately estimated to enable high-accuracy PVT solutions using carrier phase measurements. When this term is estimated as a real quality, the term "float" solution is used for PVT estimation. If this term can be estimated as an integer, with high levels of confidence, then the ambiguity may be declared as "fixed" and high accuracy PVT estimation of the user PVT solution is possible.

8.7.2.3 Triple-Difference Observations

Triple-difference carrier observations are sometimes used in DGNSS. The triple difference typically involves taking DD carrier phase measurements, and differencing them in time. These observations then have the form:

$$\Phi^{jk}_{q,AB}(t_1 - t_2) = \lambda_q \varphi_{q,AB}(t_1 - t_2) + v^{jk}_{AB}(t_1 - t_2) \tag{8.32}$$

Note that the DD carrier cycle ambiguity term, $N^{jk}_{q,AB}$ is subtracted out (assuming no cycle slips from $t_2 - t_1$, where $t_2 > t_1$). Also, because the residual error terms v^{jk}_{AB}, change very little in a short period of time, from one measurement epoch to another, the errors often very small. The triple difference carrier phase of Eq. (8.32) can be useful in several applications. This can be useful for high precision PVT projection form one PVT solution to the next, but this process requires initialization. This is extremely useful for high accuracy velocity estimation with the carrier phase. Additionally, the triple difference carrier phase formulation of Eq. (8.32) can be used for cycle slips detection by observing the deviation of successive triple difference observations from their predicted values as the carrier is tracked [23–25].

8.7.2.4 Combinations of Code and Carrier Phase Observations

Various code and carrier combinations can be formed for to enable various performance features. For example, taking the difference between two GNSS measurements at two different frequencies (e.g. L1 and L2) can produce a measurement at a different "effective wavelength"; this formulation is often called a "widelane" measurement. When GNSS carrier phase measurements are used, this may allow for faster and more reliable carrier phase ambiguity resolution, but less accuracy than other combinations. Summing two different GNSS measurements at two different frequencies (e.g. L1 and L2), called a "narrowlane" measurement, can result in increased accuracy; however, because the effective wavelength is much smaller than either of the single frequency, or the widelane wavelength, carrier cycle ambiguity resolution is much more challenging.

8.7.3 Positioning Using Double-Difference Measurements

8.7.3.1 Code-Based Positioning

The linearized matrix equation for positioning using code double-difference measurements from four satellites has the form

$$\overbrace{\delta Z^{\rho}_{q,AB}}^{3\times1} = \overbrace{H^{[1]}_{q,AB}}^{3\times3} \overbrace{\delta \mathbf{x}}^{3\times1} + \overbrace{\mathbf{v}_{\rho}}^{3\times1} \tag{8.33}$$

which is the same form as shown in Section 2.3.3 (Eq. 2.27). However, because the double-difference measurements have eliminated receiver clock error as an unknown, the unknowns are simply the X, Y, and Z coordinates of the roving receiver, constituting the components of the 3×1 vector $\delta\mathbf{x}$. Thus, the measurement matrix $H^{[1]}$ is 3×3 and the partial derivatives in it are partial derivatives of the double differences $\rho^{jk}_{q,AB}$ with respect to user position coordinates X, Y, Z instead of partial derivatives of the pseudorange measurements. Accordingly, the measurement vector δZ^{ρ} and the measurement noise vector $\mathbf{v}_{\rho}$ are 3×1. As indicated in Section 2.2.3, a solution for position can be found by computing the measurement vector associated with an assumed initial position $\mathbf{x}$, finding the difference δZ^{ρ} between the computed and actual measurement vectors, solving (Eq. (8.34)) (omitting the measurement noise vector) for the position correction δx, and obtaining the new value $x + \delta\mathbf{x}$ for X. Iteration of this process is used to produce a sequence of positions that converges to the position solution.

8.7.3.2 Carrier Phase-Based Positioning

For positioning using carrier phase double-difference measurements, the linearized matrix equation from four satellites used for iterative position solution has the form

$$\overbrace{\delta Z^{\Phi}_{q,AB}}^{3\times1} = \overbrace{H^{[1]}_{q,AB}}^{3\times3} \overbrace{\delta \mathbf{x}}^{3\times1} + \overbrace{\mathbf{v}_{\rho}}^{3\times1} \tag{8.34}$$

where the measurement matrix $H^{[1]}$ contains the partial derivatives of the double differences $\Phi^{jk}_{q,AB}$ with respect to user position coordinates X, Y, Z. As compared with code-based positioning, the measurement noise term $\mathbf{v}_{\rho}$ is much smaller, often in the centimeter range. However, the major difference is that the ambiguity in the phase measurements can cause convergence to any one of many possible positions in a spatial grid of points. Only one of these points is the correct position. Various techniques for resolving the ambiguity have been developed. A simple method is to use the position solution from the code double-difference measurements as the initial position X in the carrier phase iterative position solution. If this initial position is sufficiently accurate, convergence to the correct solution will be obtained.

8.7.3.3 Real-Time Processing Versus Postprocessing

Since double differencing combines measurements made in the roving and reference receivers, these measurements must be brought together for processing. Often the processing site is at the roving receiver, although in other applications it can be at the reference station or at another off-site location. In real-time processing, measurements are transmitted to the processing site using wireless communication system, cellular link, or the Internet. In postprocessing, the data can be physically carried to the processing site in a storage medium.

8.8 GNSS Precise Point Positioning Services and Products

The cost and inconvenience of setting up one's own DGNSS system can be eliminated because there are numerous services and software packages available to the user, some of which are free. There are too many to describe completely, so only a few of them are described in this section.

8.8.1 The International GNSS Service (IGS)

Many of the DGNSS services are subsumed under the International GNSS Service (IGS), which is a voluntary federation of more than 200 worldwide agencies that pool resources and permanent GPS and Global Orbiting

Navigation Satellite System (GLONASS) station data to generate precise DGNSS positioning services. The IGS is committed to providing the highest-quality data and products as the standard for global navigation satellite systems (GNSSs) in support of Earth science research, multidisciplinary applications, and education. The IGS network of stations tracks GPS, GLONASS, Galileo, and BeiDou constellations, as well as SBAS signals [26].

8.8.2 Continuously Operating Reference Stations (CORSs)

The National Geodetic Survey (NGS), an office of National Oceanic and Atmospheric Administration NOAA's, National Ocean Service, manages two networks of Continuously Operating Reference Station (CORS): the National CORS network and the Cooperative CORS network. These networks consist of numerous base stations containing DGPS reference receivers that operate continuously to generate pseudorange and other DGPS data for postprocessing. The data are disseminated to a wide variety of users. Surveyors, geographic information system (GIS)/land information system (LIS) professionals, engineers, scientists, and others can apply CORS data to their own GPS measurements to obtain positioning accuracies approaching a few centimeters relative to the National Spatial Reference System (NSRS), both horizontally and vertically. The CORS program is a multipurpose cooperative endeavor involving more than 130 governments, academic, and private organizations, each of which operates at least one CORS site. In particular, it includes all existing National Differential GPS (NDGPS) sites and all existing FAA WAAS sites. New sites are continually being evaluated according to established criteria.

Typical uses of CORS include land management, coastal monitoring, civil engineering, boundary determination, mapping and GISs, geophysical and infrastructure modeling, as well as future improvements to weather prediction and climate modeling.

CORS data from various sites are available from NGS at various locations throughout the Globe for varying installation configurations and sampling rates. A large concentration of CORS sites is within the Continental US.

8.8.3 GPS Inferred Positioning System (GIPSY) and Orbit Analysis Simulation Software (OASIS)

The GPS Inferred Positioning System and Orbit Analysis Simulation Software (GIPSY-OASIS II [GOA II]) package consists of extremely versatile software that can be used for GPS positioning, and satellite orbit prediction and analysis. Developed by the Caltech Jet Propulsion Laboratory (L), it can provide centimeter-level DGPS positioning accuracy over short to intercontinental baselines. It is capable of unattended, automated, low-cost operation in near real time for precise positioning and time transfer in ground, sea, air, and space applications.

GOA II also includes many force models useful for OD, such as the Earth/Sun/moon/planet (and tidal) gravity perturbations, solar pressure, thermal radiation, and drag, which make it useful in non-GPS satellite positioning applications. To augment its potential accuracy, models are included for Earth characteristics, such as tides, ocean/atmospheric loading, and crustal plate motion.

Parameter estimation for positioning and time transfer is state of the art. A general estimator can be used for GPS and non-GPS data. Matrix factorization is used to maintain robustness of solutions, and the estimator can intelligently identify, correct, or exclude questionable data. A general and flexible noise model is included.

8.8.4 Scripps Coordinate Update Tool (SCOUT)

Scripps Coordinate Update Tool (SCOUT), managed by the Scripps Institute of Oceanography, is also a system that provides precise positioning for users who submit GPS RINEX data from their receiver via the Internet. The reference stations are by default the three nearest sites for which data have been collected and are available for the specific day the user's data are taken. However, the user can specify the reference stations if desired. Station maps are provided to assist the user in specifying nearby reference sites. When SCOUT has finished determining a DGPS position solution, it sends a report of the results to the user via the Internet. The report contains both Cartesian and geodetic coordinates, standard deviations, and the locations of the reference sites that were used. The reported Cartesian coordinates are referenced to the International Terrestrial Reference Frame 2000 (ITRF2000), and the geodetic coordinates are referenced to both ITRF2000 and the World Geodetic System 1984 (WGS84) ellipsoid.

8.8.5 The Online Positioning User Service (OPUS)

The NGS operates Online Positioning User Service (OPUS) as a means to provide GPS users easier access to the NSRS. OPUS users submit their GPS data files to the NGS Internet site. The NGS computers and software determine a position by using reference receivers from three CORS sites. The position is reported back to the user by e-mail with reference to various datums: ITRF; North American Datum 1983 (NAD83); Universal Transverse Mercator (UTM); and State Plain Coordinate (SPC) northing and easting. OPUS is intended for use in the coterminous United States and in most US territories. It is NGS policy not to publish geodetic coordinates outside the United States without the agreement of the affected countries. In other parts of the Globe, other countries offer their own static or mobile positioning services that may take advantage of NGS or other countries' specific data, such as a country's own datum.

8.8.6 Australia's Online GPS Processing System (AUPOS)

The Australia's Online GPS Processing System (AUPOS) provides users with the facility to submit via the Internet dual-frequency geodetic quality GPS RINEX data observed in a "static" mode and receive rapid-turnaround precise position coordinates. The service is free and provides both International Terrestrial Reference Frame (ITRF) and Geocentric Datum of Australia (GDA94) coordinates. This Internet service takes advantage of both IGS products and the IGS GPS network and can handle GPS data collected anywhere on Earth.

8.8.7 National Resources Canada (NRCan)

The NRCan provides a free on-line PPP tool using single or dual-frequency GNSS measurements for static and kinematic data. The tool can accept RINEX and text file formats. Positioning results are sent back to the user with reference to the Canadian Spatial Reference System and ITRF [27].

Problems

8.1 Determine the CCC at the GUS location using L1 code and carrier.

8.2 Determine the frequency stability of the GEO transponder using Allan variance for the L1 using 1–10 second intervals.

8.3 What are GNSS?
(a) Single difference
(b) Double difference
(c) Triple difference
(d) Wide lane
(e) Narrow lane

8.4 What is a CORS site?

References

1 Institute of Navigation, Global Positioning System (1999). Selected papers on satellite based augmentation systems (SBASs) ("Redbook"), Vol. VI. Alexandria, VA: ION.
2 Langley, R.B. (1995). A GPS glossary. *GPS World* (October 1995), pp. 61–63.

3 Cheung, L.A. and Hsu, P.H. (2011). WAAS single frequency GEO operation field test and L1 signal code-carrier coherence. *ION GNSS 2011 Conference Proceedings* (20–23 September 2011).

4 Grewal, M.S. (2012). Space-based augmentation for global navigation satellite systems. *IEEE Transactions on Ultrasonics, Ferroelectrics, and Frequency Control* 59 (3): 497–504.

5 Peck, S., Griffith, C., Reinhardt, V. et al. (1998). WAAS network time performance and validation results. In: *Proceedings of the Institute of Navigation, Santa Monica, CA* (January 1998). Alexandria, VA: ION.

6 Ahmadi, R., Becker, G.S., Peck, S.R. et al. (1998). Validation analysis of the WAAS GIVE and UIVE algorithms. In: *Proceedings of the Institute of Navigation, ION '98, Santa Monica, CA* (January 1998). Alexandria, VA: ION.

7 Grewal, M.S., Brown, W., and Lucy, R. (1999). *Test Results of Geostationary Satellite (GEO) Uplink Sub-System (GUS) Using GEO Navigation Payloads,* Monographs of the Global Positioning System: Papers Published in Navigation ("Redbook"), vol. VI, 339–348. Alexandria, VA: Institute of Navigation, ION.

8 Parkinson, B.W. and Spilker, J.J. Jr., (eds.) (1996). *Global Positioning System: Theory and Applications,* Progress in Astronautics and Aeronautics, vol. 1. Washington, DC: American Institute of Aeronautics and Astronautics.

9 Parkinson, B.W. and Spilker, J.J. Jr., (eds.) (1996). *Global Positioning System: Theory and Applications,* Progress in Astronautics and Aeronautics, vol. 2. Washington, DC: American Institute of Aeronautics and Astronautics.

10 Parkinson, B.W., O'Connor, M.L., and Fitzgibbon, K.T. (1995). Aircraft automatic approach and landing using GPS. In: *Global Positioning System: Theory & Applications,* Vols. II and 164, Chapter 14 (eds. B.W. Parkinson and J.J. Spilker) Jr.) , 397–425. Washington, DC: American Institute of Aeronautics and Astronautics.Progress in Astronautics and Aeronautics (P. Zarchan editor-in-chief).

11 Yunck, T. W. I. Bertiger, S. M. Lichten, A. J. Mannucci, R. J. Muellerschoen, S. C. Wu (1995). A robust and efficient new approach to real time wide area differential GPS navigation for civil aviation. NASA/L Internal Report L D-12584. Pasadena, CA.

12 Grewal, M.S., Hsu, P., and Plummer, T.W. (2003). A new algorithm for SBAS GEO uplink subsystem (GUS) clock steering. *ION GPS/GNSS Proceedings* (September 2003), pp. 2712–2719.

13 RTCA (1996). Minimum operational performance standards (MOPS) for global positioning system/wide area augmentation system airborne equipment. RTCA/DO-229, January 16, 1996, and subsequent changes, Appendix A, "WAAS System Signal Specification," RTCA, Washington, DC.

14 Allan, D.W. (1975). The measurement of frequency and frequency stability of precision oscillators. *NBS Technical Note 669*. pp. 1–27.

15 Carolipio, E.N., Pandya, N., and Grewal, M.S. (2002). GEO orbit determination via covariance analysis with a known clock error. *Navigation, Journal of the Institute of Navigation* 48 (4): 255–260.
16 Tran, P. and DiLellio, J. (2000). Impacts of GEOs as ranging sources on precision approach category I availability. In: *Proceedings of ION Annual Meeting* (June 2000). Alexandria, VA: Institute of Navigation.
17 Grewal, M.S., Pandya, N., Wu, J., and Carolipio, E. (2000). Dependence of user differential ranging error (UDRE) on augmentation systems—ground station geometries. In: *Proceedings of the Institute of Navigation's (ION) 2000 National Technical Meeting*, Anaheim, CA (26–28 January 2000), 80–91. Alexandria, VA: ION.
18 Pandya, N. (1999). Dependence of GEO UDRE on ground station geometries. In: *WAAS Engineering Notebook*. Fullerton, CA: California State University, Fullerton (1 December 1999).
19 Bate, R.R., Mueller, D.D., and White, J.E. (1971). *Fundamentals of Astrodynamics*. New York: Dover.
20 Naval Air Systems Command. Joint precision approach & landing system. https://www.flightglobal.com/news/articles/analysis-us-navy-precision-landing-system-to-enter-457458/ (accessed 01 September 2019).
21 General lighthouse authority, research & radionavigation. https://www.gla-rad.org (visited November 21 November , 2012).
22 International Loran Associations (2007). Enhanced Loran *(eLoran)* definition document, Report Version 1.0, Report Version Date: 16 October 2007. https://www.loran.org/otherarchives/2007%20eLoran%20Definition%20Document-1.0.pdf (accessed 01 September 2019).
23 van Graas, F. and Lee, S.-W. (1995). High-accuracy differential positioning for satellite-based systems without using code-phase measurements. *Proceedings of the 1995 National Technical Meeting of The Institute of Navigation*, Anaheim, CA (January 1995), pp. 231–239.
24 Kiran, S. and Bartone, C. (2003). A viable airport pseudolite architecture for the local area augmentation system. *Proceedings of the 16th International Technical Meeting of the Satellite Division of The Institute of Navigation (ION GPS/GNSS 2003)*, Portland, OR (September 2003), pp. 2326–2336.
25 van Graas, F. and Soloviev, A. Precise velocity estimation using a stand-alone GPS receiver. *Navigation, Journal of the Institute of Navigation* 51 (4): 283–292.
26 International GNSS Service. http://www.igs.org/ (accessed 3 April 2019).
27 National Resources Canada Precise point positioning. https://webapp.geod.nrcan.gc.ca/geod/tools-outils/ppp.php?locale=en (accessed 3 April 2019).

9

GNSS and GEO Signal Integrity

9.1 Introduction

Navigation system integrity refers to the ability of the system to provide timely warnings to users when the system should not be used for navigation. Global Navigation Satellite Systems (GNSSs) have both internal and independent methods to maintain integrity. Satellites monitor for some of the anomalies, but not all. Clock failures, data errors, selective availability (SA) (currently discontinued), and antispoof (AS) are checked internally. The master control station monitors the constellations. In the case of Global Positioning System (GPS), data are collected from five monitoring stations distributed around the Earth. GPS performance is checked every 15 minutes by conducting tolerance and validation checks of the measured pseudoranges, using a Kalman filter, error management process [1].

The basic GNSS (as described in Chapter 4) provides integrity information to the user via the navigation message, but this may not be timely enough for some applications, such as civil aviation. Therefore, additional methods of providing integrity are necessary.

Two different methods will be discussed – GNSS-only receiver (TSO-C129-compliant) autonomous integrity monitoring (RAIM) and use of ground monitoring stations to monitor the health of the satellites, as is done via space-based augmentation system (SBAS) and ground-based augmentation system (GBAS) (TSO-C145-compliant receivers).

The analytic structure of RAIM is stochastic detection theory. Two hypothesis testing questions are raised. First, has the failure occurred? Second, if so, which satellite failed? In the case of no backup navigation system, both questions must be answered. The bad satellite must be identified and eliminated from the navigation solution, so that the vehicle can proceed safely without the bad GNSS solution. However, if there is a backup navigation system available, then it can be used when a failure occurs.

Global Navigation Satellite Systems, Inertial Navigation, and Integration,
Fourth Edition. Mohinder S. Grewal, Angus P. Andrews, and Chris G. Bartone.

Companion website: www.wiley.com/go/grewal/gnss

Determining which satellite has failed is more difficult than failure detection, and it requires more measurement redundancy [2–5].

Three RAIM methods have been proposed in recent papers on GPS integrity:

1. Range comparison method
2. Least-squares residual method
3. Parity method

The three methods are called "snapshot methods" because the detection algorithms assume that noisy redundant range-type measurements are available at a given point in time. The basic measurement equation is derived in Eq. (2.27). The following measurement equation is used in all three methods:

$$\overbrace{\delta Z_\rho}^{n\times 1} = \overbrace{\mathbf{H}}^{n\times 4}\ \overbrace{\delta \mathbf{x}}^{4\times 1} + \overbrace{\nu_\rho}^{n\times 1}$$

where n is the number of satellites and $\mathbf{H}$ is the linearized sensitivity matrix about nominal user position and clock bias.

9.1.1 Range Comparison Method

For the GNSS navigation problem described in Section 2.3.3, there are four unknowns (three position coordinates $[X, Y, Z]$ and clock bias C_b) and more than four satellites in view (e.g. six satellites). One can solve the position and time equations for the first four satellites, ignoring noise, and find the user position. This solution can then be used to predict the remaining two pseudorange measurements, and the predicted values could be compared with actual measured values. If the two differences (residuals) are small, we have near consistency in the measurements and the detection algorithm can declare "no failure." It only remains to quantify what we mean by "small" or "large" and then assess the decision rule performance on actual data.

There are six satellites in view. With the range comparison method, two range residuals $(\widetilde{\delta Z}_\rho^1, \widetilde{\delta Z}_\rho^2)$ represent a point in a statistical plane as shown in Figure 9.1.

If the statistics of the noise (ν_ρ) are Gaussian (normal), the contour will be elliptical as shown in Figure 9.1. The particular contour chosen is the one that sets the alarm rate at the desired value. The alarm rate could be set at 1/15 000 as specified in the RTCA MOPS [6, 7].

9.1.2 Least-Squares Method

The basic measurement equation with noise (Eq. (2.27)) is

$$\delta Z_\rho = \mathbf{H}\delta\mathbf{x} + \nu_\rho \tag{9.1}$$

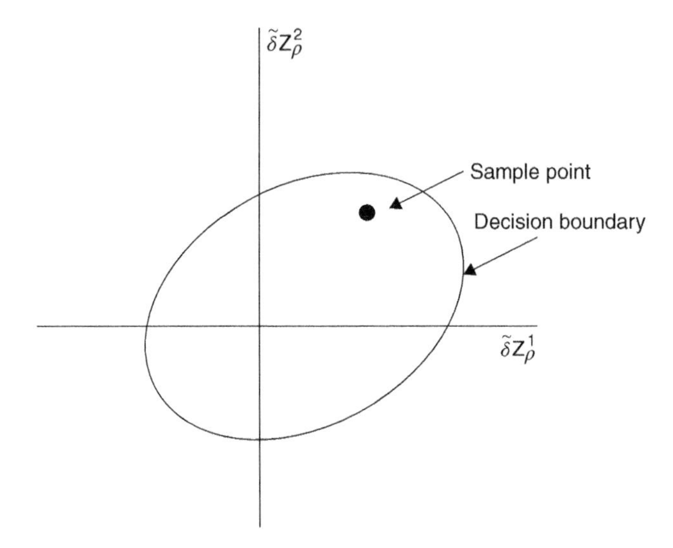

Figure 9.1 Test statistic plane for the six satellites in view.

where the additive white noise $v_\rho \in N(0, \sigma^2)$.

Let us suppose six satellites are in view and four unknowns, as in Section 9.1.1, and solve for the four unknowns by the least-squares method.

The least-squares solution is given by Eq. (2.30):

$$\widehat{\delta \mathbf{x}} = (\mathbf{H}^T\mathbf{H})^{-1}\mathbf{H}^T\delta Z_\rho \tag{9.2}$$

The least-squares solution can be used to predict the six measurements, in accordance with

$$\widehat{\delta Z_\rho}(\text{predicted}) = \mathbf{H}\widehat{\delta \mathbf{x}} \tag{9.3}$$

We can get a formula for the sum-squared residual error S by substituting $\widehat{\delta \mathbf{x}}$ from Eq. (9.2) into Eq. (9.3):

$$\begin{aligned}\Delta Z_\rho &= \delta Z_\rho - \widehat{\delta Z_\rho}(\text{residual error})\\ &= [\mathbf{I} - \mathbf{H}(\mathbf{H}^T\mathbf{H})^{-1}\mathbf{H}^T]\delta Z_\rho\end{aligned} \tag{9.4}$$

$$S = \Delta Z_\rho^T \Delta Z_\rho, \text{the sum-squared error} \tag{9.5}$$

This sum of squared error has three properties that are important in the decision rule:

1. S is a nonnegative scalar quantity. Choose a threshold value τ of S such that $S < \tau$ will be considered safe and that $S \geq \tau$ will be declared a failure.
2. If the v_ρ have the same independent zero-mean Gaussian distribution, then the statistical distribution of S is completely independent of the satellite

geometry for any number of satellites (n). Thresholds are precalculated, which results in the desired alarm rate for the various anticipated values of n. Then the real-time algorithm sets the threshold appropriately for the number of satellites in view at the moment.

3. With the v_ρ, from above, S has an unnormalized chi-square (χ^2) distribution with ($n-4$) degrees of freedom (see Section 10.9.4.1). Parkinson and Axelrad [2] use $\sqrt{S/n-4}$ as the test statistic. Calculating the test statistic involves the same matrix manipulation, but these are no worse than calculating the dilution of precision (DOP) [3].

9.1.3 Parity Method

The parity RAIM method is somewhat similar to the range comparison method except that the way in which the test statistic is formed is different. In the parity method, perform a linear transformation on the measurement vector as follows:

$$\begin{bmatrix} \delta \mathbf{x} \\ p \end{bmatrix} = \begin{bmatrix} (\mathbf{H}^T\mathbf{H})^{-1}\mathbf{H}^T \\ \mathbf{P} \end{bmatrix} \delta Z_\rho \tag{9.6}$$

The lower portion of Eq. (9.6), which yields p, is the result of operating on δZ_ρ with the special $(n-4)\times n$ matrix $\mathbf{P}$, whose rows are mutually orthogonal, unity magnitude and orthogonal to the columns of $\mathbf{H}$.

Under the same assumptions about the noise v_ρ as the previous, the following statements can be made:

$$\left.\begin{aligned} &E\langle p\rangle = 0 \\ &E\langle pp^T\rangle = \sigma^2\mathbf{I}\,(\text{covariance of}\,p) \end{aligned}\right\} \tag{9.7}$$

where σ^2 is the variance associated with v_ρ. Use p as the test statistic in this method. For detection, obtain all the information needed about p from its magnitude or magnitude squared. Thus, in the parity method, the test statistic for detection reduces to a scalar, as in the least-squares method [8, 9].

These RAIM protection levels have assumed that there is at most one ranging source with bias. Extending RAIM protection to multiple bias is given in Refs. [10–12].

9.2 SBAS and GBAS Integrity Design

The objectives of SBAS and the GBAS are to provide integrity, accuracy, availability, and continuity for GPS, Global Orbiting Navigation Satellite System (GLONASS), and Galileo Standard Positioning Service (SPS). Integrity is defined as the ability of the system to provide timely warnings to the user when individual corrections or certain satellites should not be used for navigation, specifically, the prevention of hazardously misleading information (HMI) data transmission to the user. The system should not be used for navigation when hardware, software, or environmental errors directly pose a threat to the user or indirectly pose a threat by obscuring HMI from the integrity monitors. SBAS integrity is based on the premise that errors not detected or corrected in the operational environment can become threats to integrity and, if not mitigated, can become hazards to the user.

An SBAS design should mitigate the majority of these data errors with corrections that are proved to bound the integrity hazard to an acceptable level. The leftover data errors (referred to as *residual errors*) are mitigated by the transmission of residual error bounding information. The threat of potential underbounding of integrity information is mitigated by integrity monitors. This section examines both the faulted and unfaulted cases and mitigation strategies for these cases. These SBAS corrections improve the accuracy of satellite signals. The integrity data ensure that the residual errors are bounded. The SBAS integrity monitors help ensure that the integrity data have not been corrupted by SBAS failures.

The section addresses the data errors, error detection and correction pitfalls, and how such threats can become HMI to the user, as well as fault conditions, failure conditions, threats, and mitigation, and how safety integrity requirements are satisfied. Safety integrity assurance rules will be evaluated. Results from real signal in space (SIS) data, a high-level overview of the required SBAS safety architecture, and a data processing path protection approach are included.

This section provides information that defines how a safety-of-life-critical SBAS system should be designed and implemented in order to ensure mitigation of the entire International Civil Aviation Organization (ICAO) threat space to the required level less than 10^{-7}. It provides as an example, the rationale, background, and references to show that the SBAS can be used as a trusted navigational aid to augment the GPS for lateral positioning with vertical guidance (LPV).

Rail integrity is one of the most stringent operational requirements, as evidenced by the European Rail Traffic Management System (ERTMS) required integrity levels, which are in the order of 10^{-11}. Train detection will require an equally high level of positive integrity.

The section addresses the hazardous/severe–major integrity failure condition using LPV as an illustrative example. The ICAO integrity requirement is based on the premise that errors not detected or corrected in the operational environment system can become threats to the integrity and, if not mitigated, can become hazards to the user. These errors in the operational environment (referred to as data errors) can affect both the user and the SBAS system. Integrity in this context is defined as the ability of the system to provide timely warning to users when individual corrections or satellites should not be used for navigation, that is, the prevention of HMI data transmission to the user. The system should not be used for navigation when data errors in the environment, such as the ionosphere, and data processing, such as multipath, render the integrity data erroneous. The user must be protected from residual errors that can become threats to the integrity data that could result in HMI being transmitted to the user [13–15].

An SBAS design mitigates the majority of these data errors with "corrections." The leftover data errors (referred to as residual errors) are mitigated by the transmission of residual error bounding information. The threat of potential underbounding of the integrity information is mitigated by integrity monitors and point design features that protect the integrity of the information within the SBAS system. Additionally, analytic safety analyses are required to provide evidence and proof that the residual errors are acceptable (i.e. that the probability of HMI transmission to the user is sufficiently low).

Table 9.1 lists the SBAS error sources. Mitigation of these errors when they become integrity threats are presented in Section 9.2.8. Section 9.3 gives an application of these techniques to SBAS for threat mitigation. GNSS data integrity channel (GIC) is discussed in Section 9.5.

9.2.1 SBAS Error Sources and Integrity Threats

The SBAS operational environment contains data errors. The SBAS ensures that these data errors do not become threats to the integrity data, so that HMI is not broadcasted to the user with a P_{HMI} greater than 10^{-7}.

The data used by an SBAS to calculate the correction and/or integrity data are assumed to contain errors, such as GNSS satellite clock offset, which must be sufficiently mitigated. The errors discussed are inherent in any SBAS design that utilizes GPS, Galileo, GLONASS, or geostationary Earth orbit (GEO) satellites; reference receivers; corrections; and integrity bounds. Depending on the system architecture, other error sources may also exist. Table 9.1 summarizes the error sources that every SBAS system must address [16].

The integrity threats associated with each of these error sources generally have two cases, shown in Figure 9.2. The *fault-free case* addresses the nominal errors associated with each error source, and the *faulted case* represents the

Table 9.1 List of SBAS error sources.

GNSS satellite	Integrity bound associated
GEO satellites	Message uplink
Reference receiver	Environment (ionosphere and troposphere)
Estimation	

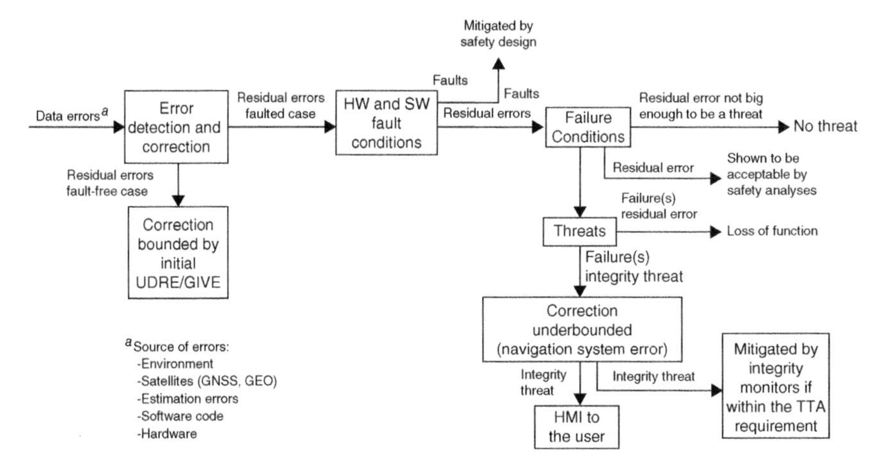

Figure 9.2 Integrity mitigation within an SBAS.

errors when one or more of the system's components cause errors. The defining quality of an SBAS system that meets the ICAO standards is the mitigation of the faulted case and the fault-free case.

9.2.2 GNSS-Associated Errors

GNSS error sources are mitigated in an SBAS system by using corrections and integrity bounds. Generally, the SBAS system corrects the errors as well as possible and then bounds the residual errors with integrity bounds that are broadcast to the user. The nominal GNSS satellite errors are well understood. The literature includes many techniques for mitigating these errors. GNSS failure modes are not as well understood and often require careful study to define the threat, which must be accounted for in threat models.

9.2.2.1 GNSS Clock Error

Each GNSS satellite broadcasts a navigation data message containing an estimate of its clock offset (relative to GNSS time) and drift rate. The GNSS satellite clock value is utilized to correct the satellite's pseudorange, the measurement used to calculate the distance (range) from the satellite to the receiver (either the user's receiver or the reference receiver).

Under fault-free conditions, the SBAS can accurately compute these corrections and mitigate this error source. Simple statistical techniques can be used to characterize these errors. The SBAS must also address satellite failures that cause the clock to rapidly accelerate, rendering the corrections suddenly invalid. As a result, the error bounds may not be bounding the residual error in the corrections. These types of failures have been observed many times in the history of GNSS.

9.2.2.2 GNSS Ephemeris Error

Each GNSS also broadcasts a navigation data message containing a prediction of its orbital parameters – Keplerian orbital parameters. The satellite's ephemeris data enable determination of the satellite position and velocity. Any difference between the satellite's calculated position and velocity and the true position is a potential source of error.

Under fault-free conditions, the SBAS provides corrections relative to the GNSS broadcast ephemeris data. The SBAS can accurately compute these corrections and mitigate this error source using standard statistical techniques. The satellite may experience an unexpected maneuver, rendering the corrections suddenly invalid. This threat includes geometric constraints that may be insufficient for the SBAS to adequately detect the orbit error.

Under fault-free conditions, the SBAS provides corrections relative to the GNSS broadcast ephemeris data. The SBAS can accurately compute these corrections and mitigate this error source using standard statistical techniques. The satellite may experience an unexpected maneuver, rendering the corrections suddenly invalid. This threat includes geometric constraints that may be insufficient for the SBAS to adequately detect the orbit error.

9.2.2.3 GNSS Code and Carrier Incoherence

The GNSS signal consists of a radiofrequency carrier encoded with a pseudorandom spread-spectrum code. The user's receiver performs smoothing of its pseudorange measurements using the carrier phase measurements. If the code and carrier are not coherent, there will be an error in this pseudorange smoothing process. This error is caused by a satellite failure. Incoherence between the code and carrier phase can increase the range error, ultimately resulting in the user incorrectly determining the code/carrier ambiguity.

9.2.2.4 GNSS Signal Distortion

A satellite may fail in a manner that distorts the pseudorange portion (pseudorandom noise [PRN] encoding) of the GPS transmission. This causes an error in the user's pseudorange measurements that may be different from the error that the SBAS receiver experiences. In 1993, the SV-19 GPS satellite experienced a failure that fits into this category. This error is caused by a satellite failure. If a satellite experiences this type of failure, the SBAS may not be able to estimate the satellite clock corrections that are aligned with the user's measurements, which could result in HMI.

9.2.2.5 GNSS L1L2 Bias

The GNSS L1 and L2 signals are utilized together to compute the ionospheric delay so the delay can be removed from the range calculations. The satellite has separate signal paths for these two frequencies; therefore, the signals can have different delays. The difference in the delays must be modeled accurately to be able to properly calibrate and use L1 and L2 signals together.

Under nominal conditions, the SBAS estimation process is very accurate, and this error is easily modeled with standard statistical techniques. If a satellite experiences a fault, the L1L2 bias can suddenly change, resulting in a large estimation error. A large estimation error can lead to excessive errors in correction processing.

9.2.2.6 Environment Errors: Ionosphere

As the GNSS L1 and L2 signals propagate through the ionosphere, the signals are delayed by charged particles. The density of the charged particles, and therefore the delay, varies with location, time of day, angle of transmission through the ionosphere, and solar activity. This delay will cause an error in range measurements and must be corrected and properly accounted for in the SBAS measurement error models. As discussed earlier in Chapter 7, during calm ionospheric conditions, modeling errors are well understood and can be handled using standard statistical techniques. Ionospheric storms pose a multitude of threats for SBAS users. The model used in the error estimation may become invalid. The user may experience errors that are not observable to the SBAS due to the geometry of the reference station pierce points. The error in the corrections may increase over time due to rapid fluctuations in the ionosphere.

9.2.2.7 Environment Errors: Troposphere

As the GNSS L1 and L2 signals propagate through the troposphere, the signals are delayed. This delay is dependent on temperature, humidity, angle of transmission through the atmosphere, and atmospheric pressure. This delay will cause an error in range measurements and must be corrected and properly accounted for in the measurement error models. Tropospheric modeling errors manifest themselves in the algorithms that generate the corrections. The user utilizes a separate tropospheric model that may have errors due to tropospheric modeling.

9.2.3 GEO-Associated Errors

9.2.3.1 GEO Code and Carrier Incoherence

The GEO signal consists of a radiofrequency carrier encoded with a pseudorandom spread-spectrum code. The user's receiver performs smoothing of its GEO pseudorange measurements using the carrier phase measurements. If the code and carrier are not coherent, there will be an error in this pseudorange smoothing process. Under fault-free conditions, some incoherence is possible (due to environmental effects). This will be a very small error that is easily modeled by the ground system. Under faulted conditions, severe divergence and potentially large errors are theoretically possible if the GEO uplink subsystem fails.

9.2.3.2 GEO-Associated Environment Errors: Ionosphere

As the GEO L1 signal propagates through the ionosphere, the signal is delayed by charged particles. The density of the charged particles, and therefore the delay, varies with location, time of day, angle of transmission through the ionosphere, and solar activity. GEO satellites that are available today broadcast single-frequency (L1) signals that do not allow a precise determination of the ionospheric delay at a reference station. Without dual-frequency measurements, uncertainty in the calculated ionospheric delay estimates bleed into the corrections. New GEO satellites [PRN 135,138] have two frequencies L1 and L5. Ionospheric delay is calculated using those frequencies (see Chapter 7).

9.2.3.3 GEO-Associated Environment Errors: Troposphere

Like GNSS satellite signals, the GEO L1 signal is delayed as it propagates through the troposphere. This delay will cause an error in range measurements and must be corrected and properly accounted for in the measurement error models.

9.2.4 Receiver and Measurement Processing Errors

Measurement errors affect an SBAS system in two ways. They can corrupt or degrade the accuracy of the corrections. They can also mask other system errors and result in HMI slipping through to the user. The errors given below must all be mitigated and residual errors bounded.

9.2.4.1 Receiver Measurement Error

The receiver outputs pseudorange and carrier phase measurements for all satellites that are in view. The receiver and antenna characteristics limit the measurement accuracy. Under fault-free conditions, these errors can be addressed using well-documented processes. A receiver could fault and output measurement data that are in error for any or all of the satellites in view. A latent common-mode failure in the receiver firmware could cause all measurements in the system to simultaneously fail. Erroneous measurements pose two threats. They cannot only result in correction errors; they can also fool the integrity monitors and let HMI slip through to the user.

9.2.4.2 Intercard Bias

For receiver designs that include multiple correlators, the internal delays in the subreceivers are different. This creates a different apparent clock for each subreceiver, called an *intercard bias*. Under nominal conditions, the intercard bias estimate is extremely accurate and the intercard bias error is easily accounted for. Any failure condition in the receiver or the algorithm computing the bias will result in an increase in the measurement data error.

9.2.4.3 Multipath

Under nominal conditions, the dominant source of noise is multipath. Multipath is caused by reflected signals arriving at the receiver delayed relative to the direct signal. The amount of error is dependent on the delay time and the receiver correlator type. (See discussion of *multipath mitigation methods* in Chapter 7.)

9.2.4.4 L1L2 Bias

The GNSS L1 and L2 signals are utilized together to compute the ionospheric delay so that the delay can be removed from the range calculations. The receivers and antenna will experience different delays in the electronics when monitoring these two frequencies. The difference in the delays must be accurately modeled to be able to remove the bias and to use the L1 and L2 signals together. If a receiver fails, the L1L2 bias can suddenly change, resulting in large estimation errors. Under nominal conditions, the estimation process is very accurate and the error is not significant.

9.2.4.5 Receiver Clock Error

A high-quality receiver generally utilizes a (cesium) frequency standard that provides a long-term stable time reference (clock). This clock does drift. If a receiver fails, the clock bias can suddenly change, resulting in a large estimation error. Under nominal conditions, the SBAS is able to accurately account for receiver clock bias and drift. If a receiver fails, the clock may accelerate, introducing errors into the corrections and the integrity monitoring algorithms.

9.2.4.6 Measurement Processing Unpack/Pack Corruption

The measurement processing software that interfaces with the receiver needs to unpack and repack the GNSS ephemeris. A software failure or network transmission failure could corrupt the GNSS ephemeris data and result in the SBAS using an incorrect ephemeris.

9.2.5 Estimation Errors

The SBAS system provides corrections to improve the accuracy of the GNSS measurements and to mitigate the GPS/GEO error sources. Estimations of parameters and corrections described in Sections 9.2.5.1–9.2.5.4 cause these errors, which must be accounted for.

9.2.5.1 Reference Time Offset Estimation Error

The difference between the SBAS and GNSS reference time must be less than 50 ns. If the user is en route and mixing SBAS-corrected satellite data with non-SBAS-corrected satellite data, then the offset (error) between the SBAS reference time and the GNSS reference time could affect the user's receiver position solution. Under fault-free conditions, this error varies slowly. If one or more GNSS satellites fail, the offset between GNSS time and the SBAS reference time could vary rapidly.

9.2.5.2 Clock Estimation Error

The SBAS system must compute estimates of the reference receiver clocks and GNSS/GEO satellite clock errors. An error in this estimation results in errors in the user's position solution. The error in the estimation process must be accounted for in the integrity bounds.

9.2.5.3 Ephemeris Correction Error

The SBAS computes estimates of each satellite's orbit (ephemeris) and then uses these estimates to compute corrections. Error in the orbit (ephemeris) estimation process will result in erroneous corrections. Sources of error include measurement noise, troposphere modeling error, and orbital parameter modeling error. The error in the estimation process must be accounted for in the integrity bounds.

9.2.5.4 L1L2 Wide-Area Reference Equipment (WRE) and GPS Satellite Bias Estimation Error

The L1L2 bias of the satellites and the receivers is used to generate the SBAS corrections. SBAS users utilize single-frequency corrections while corrections are generated using dual-frequency measurements that are unaffected by ionospheric delay errors. An error in the estimation process will result in erroneous corrections. Sources of estimation error include measurement error, time in view, ionospheric storms, and receiver/satellite malfunctions. The error in the estimation process must be accounted for in the integrity bounds.

9.2.6 Integrity-Bound Associated Errors

The integrity monitoring functionality in an SBAS system ensures that the system meets the allocated integrity requirement. This processing includes functionality that must be performed on a "trusted" platform with software developed to the proper RTCA/D0178-B safety level.

The ICAO HMI hazard has been evaluated to be a "hazardous/severe–major" failure condition. This requires all software responsible for preventing HMI to be developed using a process that meets all the RTCA/DO-178B Level B objectives.

A critical aspect of mitigating an integrity threat is the determination of the threat model. Threats originating in the RTCA/DO-178B Level B software can be characterized using observed performance, provided all the inputs originate from Level B software and the algorithms have been designed in an analytic methodology.

9.2.6.1 Ionospheric Modeling Errors

The SBAS system uses an underlying characterization to transmit ionospheric corrections to the user. During periods of high solar activity, the ionospheric decorrelation can be quite rapid and large, and the true delay variation around the grid point may not match the underlying characterization. In this case, the SBAS-estimated delay measurement and the associated error bound may not be accurate or the SBAS may not sample a particular ionospheric event that is affecting a user.

9.2.6.2 Fringe Area Ephemeris Error

Errors may be present in the SBAS GNSS position estimates that are not observable from the reference receivers. These errors could cause position errors in a user's position solution that are not observable to the reference receivers.

9.2.6.3 Small-Sigma Errors

It is possible that any quantity of satellites could contain small- or medium-sized errors that combine in such a manner that creates an overall position error that is unbounded to a user.

9.2.6.4 Missed Message: Old but Active Data (OBAD)

The user could have missed one or more messages and is allowed to use old corrections and integrity data. The use of these old data could result in an increased error compared to users that have not missed messages.

9.2.6.5 Time to Alarm (TTA) Exceeded

If there is an underbound condition, the SBAS is required to correct that condition within a specified period of time. This is called the time to alarm (TTA). This alarm is a series of messages that contain the new information, such as an increased error bound or new corrections that are needed to correct the situation and prevent HMI. Different types of failure, such as hardware, software, or network transmission delay, could occur and cause the alarm messages to be delayed in excess of the required time.

9.2.7 GEO Uplink Errors

Errors caused by the uplink system can also be a source of HMI to the user.

9.2.7.1 GEO Uplink System Fails to Receive SBAS Message

Any hardware or software along the path to the satellite could fault, causing the message to be delayed or not broadcast at all.

9.2.8 Mitigation of Integrity Threats

This section describes some approaches that may be used to eliminate and minimize data errors, mitigate integrity threats, and satisfy the safety integrity requirements.

Safety design and safety analyses are utilized to protect the data transmission path into the integrity monitors and out to the user through the geostationary satellite.

Such integrity monitors, written to DO-178B Level B standards to provide adjustments to the integrity bounds, must test the associated integrity data, user differential range error (UDRE), or grid ionosphere vertical error (GIVE) in an analytically tractable manner. The test prevents HMI by passing the integrity data with no changes, increasing the integrity data to bound the residual error in the corrections, or setting the integrity data to "not monitored" or "don't use." Each integrity monitor must carefully account for the uncertainty in each component of a calculation. Noisy measurements or poor quality corrections will result in large integrity bounds.

The examples given are for a system that utilizes either a "calculate then monitor" or "monitor then calculate" design. Both techniques are used in the examples to fully illustrate the types of mitigation needed to meet the general SBAS integrity requirements. Under the "calculate then monitor" design, corrections and error bounds are computed assuming that the inputs to the system follow some observed or otherwise predetermined model. A monitoring system then verifies the validity of these corrections and error bounds against the integrity threats. With the "monitor then calculate design," the measurements inputs to the monitor are carefully screened and forced to meet strict integrity requirements. The corrections and the error bounds are then computed in an analytically tractable manner and no further testing is required. Both designs must address all of the errors associated with an SBAS system in an analytically tractable manner.

9.2.8.1 Mitigation of GNSS Associated Errors

GNSS Clock Error

Fault-free case. The clock corrections are computed in a Kalman filter. The broadcast UDRE should be constructed using standard statistical techniques to ensure that the nominal errors in the fast corrections and long-term clock corrections are bounded.

Faulted case. A monitor is designed to ensure that the probability of a large fast correction error and/or long-term clock correction error is less than the allocation on the fault tree. The monitor must use measurements that are independent of the measurements used to compute the corrections. Error models for each input into the monitor must be determined and validated. The monitor either passes the UDRE or increases the UDRE or sets it to "not monitored" or "don't use" depending on the size of the GNSS clock error.

GPS Ephemeris Error

Fault-free case. The orbit corrections are computed in a Kalman filter. The broadcast UDRE would be constructed using standard statistical techniques to ensure that the nominal errors in the long-term position corrections are bounded.

Faulted case. Clock errors are easily observed by a differential GNSS system. The ability of an SBAS to observe orbit errors is dependent on the location of the system's reference stations. The SBAS can generate a covariance matrix and package it in SBAS message type 28. This message provides a location-specific multiplier for the broadcast UDRE. The covariance matrix must take into account the quality of the measurements from the reference stations and the quality of the ephemeris corrections broadcast from the SBAS. When the GNSS ephemeris is grossly in error, the SBAS must either detect and correct the problem or increase the uncertainty in the UDRE. Under faulted conditions, the SBAS must account for the situation where clock error cancels with the ephemeris error at one or more of the reference stations.

GNSS Code and Carrier

Fault-free case. GNSS code-carrier divergence results from a failure on the GNSS satellite and errors do not need to be mitigated in the fault-free case.

Faulted case. A monitor must be developed to detect and alarm if the GNSS code and carrier phase become incoherent. The monitor must account for differences in the SBAS measurement smoothing algorithm and the user's measurement smoothing algorithm. The most difficult threat to detect and mitigate is one where the code-carrier divergence occurs shortly (within seconds) after the user acquires the satellite. In this case, the error has an immediate effect on the user and a gradual effect on the SBAS.

GNSS Signal Distortion

Fault-free case. GNSS signal distortion results from a failure on the GNSS satellite and errors do not need to be mitigated in the fault-free case.

Faulted case. A monitor can be developed to mitigate the errors from GNSS signal distortion. The measurement error incurred from signal distortion is receiver dependent. The monitor must mitigate the errors regardless of the type of equipment the user is employing.

GNSS L1L2 Bias

Fault-free case. L1L2 bias errors can be computed with a Kalman filter. These corrections are not sent to the user but are used in the other monitors. Nominal error bounds are computed with standard statistical techniques.

Faulted case. If the SBAS design utilizes the L1L2 bias corrections in the integrity monitors, then they must account for the faulted case. The L1L2 bias can suddenly change due to an equipment failure on board the GNSS satellite. The SBAS must be designed so that this type of failure does not "blind" the monitors. One approach to this design is to form a single-frequency integrity monitor that tests the corrections without using the L1L2 bias corrections.

Environment (Ionosphere) Errors

Fault-free case. Under calm ionospheric conditions, the GIVE is computed in a fashion that accounts for measurement uncertainty, L1L2 bias errors, and nominal fluctuations in the ionosphere.

Faulted case. The integrity monitors must ensure that an ionospheric storm cannot cause HMI. One approach to this problem is to create an ionospheric storm detector that is sensitive to spatial and/or temporal changes in the ionospheric delay. Proving such a detector mitigates HMI is a difficult endeavor since the ionosphere is unpredictable during ionospheric storms. It is possible for ionospheric storms to exist in regions where the SBAS does not sample the event. An additional factor can be added to the GIVE to account for unobservable ionospheric storms. In some cases (when a reference receiver is out or the grid point is on the edge of the service volume), this term can be quite large. The GIVE must also account for rapid fluctuations in the ionosphere between ionospheric correction updates. One way to mitigate such errors is to run the monitor frequently and to send alarm messages if such an event occurs.

Environment (Troposphere) Errors

Both cases. Tropospheric delay errors are built into many of the SBAS corrections. The SBAS must determine error bounds on the tropospheric delay error and build them into the UDRE.

9.2.8.2 Mitigation of GEO-Associated Errors

GEO Code and Carrier and Environment Errors For GEO code-associated errors, fault-free and faulted, see section "GNSS Code and Carrier."

Fault-free case. Since GEO measurements are single frequency, the dual-frequency techniques utilized for GNSS integrity monitoring have to be modified. One approach to working with single-frequency measurements is to compensate for the iono delay using the broadcast ionospheric grid delays. The uncertainty of the iono corrections (GIVE) needs to be accounted for in the integrity monitors.

Faulted case. During ionospheric storms, the GIVE is likely to be substantially inflated. The inflated values will "blind" the other integrity monitors from detecting small GEO clock and ephemeris errors, resulting in a large GEO UDRE.

For both faulted and fault-free cases, of environment (troposphere) errors, see section "Environment (Troposphere) Errors" – Both cases.

9.2.8.3 Mitigation of Receiver and Measurement Processing Errors

Receiver Measurement Error

Fault-free case. The integrity monitors must account for the noise in the reference station measurements. A bound on the noise can be computed and utilized in the integrity monitors. In the "calculate then monitor" approach, integrity monitors must use measurements that are uncorrelated with the measurements used to compute the corrections. Otherwise, error cancellation may occur.

Faulted case. In the faulted case, one or more receivers may be sending out erroneous measurements. An integrity monitor must be built to detect such events and to ensure that erroneous measurements do not blind the integrity monitors.

Intercard bias both cases. Intercard bias errors appear to be measurement errors and are mitigated by the methods discussed in Section 9.2.4.1.

Code Noise and Multipath (CNMP)

Fault-free case. Small multipath errors are accounted for in the receiver measurement error discussed in Section 9.2.8.3.

Faulted case. Large multipath errors must be detected and screened from the integrity monitors or accounted for in the measurement noise error bounds.

WRE L1L2 Bias

Fault-free case. The WRE L1L2 bias can be computed in a manner similar to that for the GNSS L1L2 bias. The nominal errors in this computation must be bounded and accounted for in the integrity monitors.

Faulted case. A receiver can malfunction, causing the L1L2 bias to suddenly change. The L1L2 bias is used in the correction and integrity monitoring functions and such a change must be detected and corrected to prevent HMI. A single-frequency monitor can be created that tests the corrections without using L1L2 bias as an input.

WRE Clock Error

Fault-free case. The receiver clock error can be computed using a Kalman filter. Standard statistical techniques can be used to determine the error in the wide-area reference equipment (WRE) clock estimates. This error bound can be utilized by the integrity monitors.

Faulted case. If bad data are received in the Kalman filter, erroneous WRE clock corrections could result. An integrity monitor can be built that does not utilize the WRE clock estimates from the Kalman filter to test the corrections when the WRE clock estimates are bad.

9.2.8.4 Mitigation of Estimation Errors

Reference Time Offset Estimation Error

Fault-free case. In the fault-free case, the difference between the GPS reference time and the SBAS reference time is accounted for by the user, provided the difference is less than 50 ns.

Faulted case. In the faulted case, due to some system fault or GPS anomaly, the difference in the SBAS reference time and the GPS reference time exceeds 50 ns. A simple monitor can be constructed to measure the difference between the two references. The monitor would respond to a large offset by setting all satellites not monitored, stopping the user from mixing corrected and uncorrected satellites.

Clock Estimation Error, Ephemeris Correction Error, L1L2 WRE, and GNSS Satellite Bias Estimation Error See sections "GNSS Clock Error," "GNSS Ephemeris Error," "GNSS L1L2 Bias," and "WRE L1L2 Bias."

9.2.8.5 Mitigation of Integrity-Bound-Associated Errors

Ionospheric Modeling Error

Fault-free case. Extensive testing of the models used in the SBAS will provide assurance that the iono model error is properly bounding under quiet ionospheric conditions.

Faulted case. During an ionospheric storm, the validity of the model is in question. A monitor can be constructed to test the validity of the model and to increase the GIVE when the model is in question.

Fringe Area Ephemeris Error

Fault-free case. This error is mitigated by message type 28 as discussed in section "GNSS Ephemeris Error."

Faulted case. Special considerations must be taken to ensure that the integrity monitors are sensitive to satellite ephemeris errors on the fringe of coverage. Errors in the satellite ephemeris are not well viewed by the SBAS on the edge of the service region. A specific proof of the monitors' sensitivity to errors of this nature is required. Additional inflation factors may be needed to adjust the UDRE for this error.

Small-Sigma Errors

Fault-free case. Tests can easily be performed on individual corrections; the user, however, must be protected from the combination of all error sources. An analysis can be performed to demonstrate that any combination of errors observed in the fault-free case is bounded by the broadcast integrity bounds. An example of this analysis is discussed in Ref. [15].

Faulted case. Under faulted conditions, small biases may occur, which can "add" in the user position solution to cause HMI. This threat can be mitigated by monitoring the accuracy of the user position solution at the reference stations.

Missed Message: OBAD

Fault-free case. The old but active data (OBAD) deprivation factors broadcast by the SBAS account for aging data.

Faulted case. The integrity monitors must ensure that every combination of active SBAS messages meets the integrity requirements. Two methods are suggested for this threat. First, the integrity monitors can run on every active set of broadcast messages to check their validity after broadcast. If a large error is detected, an alarm will be sent. A second, preferable, approach is to test the messages against every active data set before broadcast and to adjust the corrections/integrity bounds accordingly.

TTA Exceeded

Fault-free case. The system is designed to meet the TTA requirement by continually monitoring the satellite signals and responding to integrity faults with alarms.

Faulted case. A monitor can be designed to test the "loop back" time in the system and continually ensure that the TTA requirement is met. The monitor sends a test message every minute and measures the time it takes for the message to loop back through the system.

9.3 SBAS Example

The process for identifying, characterizing, and mitigating a failure condition is illustrated by the following SBAS example. SBAS broadcasts corrections to compensate for range errors incurred as the signal passes through the ionosphere. The uncertainty in these corrections is computed and sent to the user along with the corrections. HMI would result if the SBAS broadcasts erroneous integrity data (error bounds) and does not alert the user to the erroneous integrity data within a specified time limit. This time limit is referred to as the TTA.

1. *Identify error conditions that can cause HMI*. Error conditions can be caused by internal or external hardware or software failures or fluctuations in environmental conditions. The onset of an ionospheric storm represents a failure condition that could result in large errors in the ionospheric corrections, ultimately resulting in an increased probability of HMI.
2. *Precisely characterize the threat*. On days with nominal ionospheric behavior, the ionospheric threats are well understood and are reasonably easy to quantify. Scientists are not yet able to characterize the ionosphere during storm conditions. For these reasons, SBAS has generated specific threat models for the ionosphere based on real data collected during the worst ionospheric activity from the solar maximum period (an 11-year solar cycle). An important aspect of this model is the ionospheric irregularity detector, which assures the validity of the model and inflates the error bounds if the validity of the model is in question.
3. *Identify error detection mechanisms*. In the SBAS, errors in ionospheric corrections are mitigated by a monitor located in a "safety processor" and a special detector called the "ionospheric irregularity detector."
4. *Analytically determine that the threat is mitigated*. It is tempting to take an reliability, maintainability, availability (RMA) approach to dealing with ionospheric storms:

   Ionospheric storms are "infrequent events."
   "We haven't seen them cause HMI yet ...."
   "They don't last very long."
   "The system has other margins ...."

The a priori probability of a storm is not the mitigation of the threat. SBAS must meet its 10^{-7} integrity allocation during ionospheric storms. The analysis must account for worst-case events, like storms that are not well sampled by the ground system. Furthermore, it is not necessarily the storms with the highest magnitude that are the hardest to detect or are most likely to cause HMI. Extensive analysis is needed to characterize the threat.

In general, every requirement in a system's specification is tested by some type of formal demonstration. Most of the SBAS system-level requirements fall into this category; however, the SBAS integrity requirement does not. Testing fault-tree allocations of 10^{-7} and smaller requires on the order of 100 000 000 independent points (I sample every 5 minutes for 950 years). Integrity can only be demonstrated where reference stations exist.

Integrity must be proved for every satellite/user geometry. Every user at every point in space must be protected at all times. Demonstrations cannot be conducted where data are not available. In addition, every satellite geometry (subset) must be tested. Since GPS orbits repeat, then, if at a specific airport a satellite/user geometry exists with an increased probability of HMI, the situation will repeat every day at the same time until the constellation changes. It is because of these considerations that analytic proofs are required to satisfy integrity requirements.

The identification, characterization, and mitigation of a threat to the SBAS user should be carefully scrutinized by a panel of experts in the SBAS field. The analysis supporting claims is formally documented, scrutinized, and approved by this panel. This four-step process should be completed for every error identified in the system.

9.4 Summary

The data used by an SBAS to calculate the corrections and integrity data are assumed to contain errors that have been sufficiently mitigated. The errors discussed are inherent in any SBAS design that utilizes GPS satellites. An SBAS design mitigates the majority of these errors with "corrections," thereby making it a trusted navigation aid. The leftover errors, referred to as residual errors, are mitigated by the transmission of residual error bounding information. The threat of potential underbounding of integrity information is mitigated by integrity monitors. Both faulted and unfaulted cases are examined and mitigation strategies have been discussed. These SBAS corrections improve the accuracy of satellite signals. The integrity data ensure that the residual errors are bounded. The SBAS integrity monitors ensure that the integrity data have not been corrupted by SBAS failures. Following the integrity design guidelines given in this chapter is an important factor in obtaining certification and approval for use of the SBAS system.

SBAS integrity concepts may be applied to GBAS. In GBAS, the integrity will be broadcast from the ground.

9.5 Future: GIC

A GIC will be provided in the next generation of GPS satellites such as GPS IIF and GPS III. In addition, the next generation will include airborne monitoring by using redundant measurements (RAIM). The GIC consists of a network of ground-based GNSS signal monitoring stations, located at known reference stations that cover a wide geographical area over which signal integrity is guaranteed by navigation providers, such as the Federal Aviation Agency (FAA).

These monitors will be connected to a central control station where the integrity decision will be made. The integrity message will be broadcast through GEO stationary satellites [14].

Problem

9.1 Use the data from GPS_Position(PRN#) MATLAB program from Appendix A at www.wiley.com/go/grewal/gnss for six satellites (position), pseudoranges, and user position to find the integrity using three RAIM methods.

References

1 Langley, R.B. (1999). The integrity of GPS. *GPS World*.

2 Parkinson, B.W. and Axelrad, P. (1988). Autonomous GPS integrity monitoring using the pseudorange residual. *NAVIGATION, Journal of the Institute of Navigation* 35 (2): 255–274.

3 Parkinson, B.W. and Spilker, J.J. Jr., (eds.) (1996). *Global Positioning System: Theory and Applications*, Progress in Astronautics and Aeronautics, vol. 1. Washington, DC: American Institute of Aeronautics and Astronautics.

4 Parkinson, B.W. and Spilker, J.J. Jr., (eds.) (1996). *Global Positioning System: Theory and Applications*, Progress in Astronautics and Aeronautics, vol. 2. Washington, DC: American Institute of Aeronautics and Astronautics.

5 Parkinson, B.W., O'Connor, M.L., and Fitzgibbon, K.T. (1995). Aircraft automatic approach and landing using GPS. In: *Global Positioning System: Theory & Applications*, Progress in Astronautics and Aeronautics, Chapter 14, vols. II and 164 (ed. B.W. Parkinson, J.J. Spilker, and editor-in-chief P. Zarchan), 397–425. Washington, DC: American Institute of Aeronautics and Astronautics.

6 Lee, Y.C. (1986). Analysis of range and position comparison methods as a means to provide GPS integrity in the user receiver. *Proceedings of the Annual Meeting of the Institute of Navigation*, Seattle, WA (24–26 June 1986), pp. 1–4.

7 RTCA (1996). Minimum Operational Performance Standards (MOPS) for Global Positioning System/Wide Area Augmentation System Airborne Equipment. RTCA/DO-229, January 16, 1996, and subsequent changes, Appendix A, "WAAS System Signal Specification". Washington, DC: RTCA.

8 Sturza, M.A. (1988/89). Navigation system integrity monitoring using redundant measurements. *NAVIGATION, , Journal of the Institute of Navigation* 35 (4): 483–501.

9 Sturza, M.A. and Brown, A.K. (1990). Comparison of fixed and variable threshold RAIM algorithms. *Proceedings of the 3rd International Technical Meeting of the Institute of Navigation*, Satellite Division, ION GPS-90, Colorado Springs, CO (19–21 September 1990), pp. 437–443.

10 Angus, J.E. (2006). RAIM with multiple faults. *NAVIGATION, Journal of the Institute of Navigation* 53 (4): 249–257.

11 Shively, C.A. and O'Laughlin, D.G. (2009). Detailed analysis of RAIM performance for ADP-B separation error. *NAVIGATION, Journal of the Institute of Navigation* 56 (4): 261–274.

12 Larson, C., Raquet, J.F., and Veth, M.J. (2010). The impact of altitude on image based integrity. *NAVIGATION, Journal of the Institute of Navigation* 57 (4): 249–262.

13 Schempp, T. Stephen R. Peck and Robert M. Fries (2001). WAAS algorithm contribution to hazardously misleading information (HMI). *14th International Technical Meeting of the Satellite Division of the Institute of Navigation*, Salt Lake City, Utah (11–14 September 2001).

14 Watt, G. R. M. Fries, H. L Habereder, D.R. Heine, T. L. McKendree (2003). Lessons learned in the certification of integrity for a satellite-based navigation system. *Proceedings of the ION NTM 2003*, Anaheim, CA (January 2003).

15 Grewal, M.S. and Andrews, A.P. (2013). *Application of Kalman Filtering to GPS, INS, & Navigation, Short Course Notes*. Anaheim, CA: Kalman Filtering Consulting Associates.

16 Schempp, T. Arthur L. Rubin (2002). An application of Gaussian overbounding for the WAAS fault-free error analysis. *Proceedings of ION GPS*, Portland, OR (September 2002), pp. 766–772.

10

Kalman Filtering

Once you get the physics right, the rest is mathematics.

Rudolf E. Kalman (1930–2016) Kailath Lecture,
Stanford University, 11 May 2008

10.1 Chapter Focus

The primary purpose of this chapter is to provide a working familiarity with Kalman filtering – both the theoretical and practical aspects of it, and especially those features essential for global navigation satellite system (GNSS) navigation, inertial navigation performance analysis (in Chapter 11), and performance analysis of integrated GNSS/INS navigation (in Chapter 12).

The next section (10.2) is a heuristic introduction to Kalman filtering, primarily intended to give the uninitiated an appreciation of its place in the history of technology. This is followed by a section defining some mathematical notation needed in the rest of the chapter, beginning with a section on the mathematical foundations of the Kalman filter. Although people who like Kalman filters and sausages need not watch them being made, they should benefit from understanding some of their ingredients. In the case of the Kalman filter, a key ingredient is the linear least mean squares estimator (LLMSE) first developed by Carl Friedrich Gauss (1777–1855) in the 1820s [1] and extended to include real-time estimation (i.e. filtering) by the late Rudolf Emil Kalman in the 1950s [2]. Understanding the LLMSE is key to understanding many essential attributes of Kalman filtering, including the fact that it does not depend on any statistics of the underlying error probability density functions (PDFs) beyond their means and covariances. We start with the LLMSE to establish these attributes, then present a brief overview of how Kalman (and others) extended this to develop his famous filter.

We also present alternative but equivalent mathematical representations of the Kalman filter that have been found to be more robust against computer

Global Navigation Satellite Systems, Inertial Navigation, and Integration,
Fourth Edition. Mohinder S. Grewal, Angus P. Andrews, and Chris G. Bartone.

Companion website: www.wiley.com/go/grewal/gnss

roundoff errors, especially for applications with large numbers of unknowns to be estimated. This can be critical for maintaining accuracy in large-scale applications such as GNSS system state estimation at the constellation level.

Some of the more successful approaches used for approximating nonlinear transformations of means and covariances in Kalman filtering are also presented. Appendix C (on www.wiley.com/go/grewal/gnss) examines the "PDF ambiguity errors" this introduces with only means and covariances to start with.

For GNSS receiver applications, we show how to represent the random dynamical characteristics of various host vehicles in state-space form for using Kalman filters in vehicle tracking, and assess the relative statistical significance of linearization errors in those applications.

We conclude with some methods used for monitoring Kalman filter operation to detect and diagnose anomalous behaviors.

10.2 Frequently Asked Questions

1. **What is a Kalman filter?**
   In this context, a *filter* is a real-time estimator. The Kalman filter uses noisy measurements to track the status of dynamic systems with random disturbances.
   Its popularity is due, in large part, to the fact that so many practical estimation problems can be modeled and solved in this way, and the fact that the Kalman filter has performed so well in so many of these applications.
   It has been justifiably labeled "navigation's integration workhorse" [3] because it has become an essential part of modern navigation systems – especially for integrating navigation systems as disparate as GNSS and INS (inertial navigation system). Kalman filtering is the connective tissue binding satellite navigation systems together.
2. **What is it used for?**
   Its principal uses in navigation are:
   (a) *In performance-predictive design of sensor systems* used for estimating the current state of a particular dynamic system in a proposed application. A subset of the Kalman filter implementation, called the *matrix riccati equations*, determines how well the state of the dynamic system can be estimated, given certain attributes of the sensors to be used. For example, it has been used for comparing the relative performance of various configurations of GNSS satellite constellations, locations and operations of supportive ground stations, and receiver hardware and software configurations. It is also used to determine *if* the unknown state variables can be uniquely determined – and how well – from a given set of measurements. This is the issue of *observability*.

(b) *As an embedded real-time estimator* for a particular application. Kalman filters in one form or another are in current use throughout GNSS systems to maintain their accuracy, and in receivers with Kalman model parameters representing the host vehicle dynamics, receiver signal characteristics, and any auxiliary sensors used.

3. **How does it work?**

The Kalman filter has been called "ideally suited to digital computer implementation" [4], in part because it represents the estimation problem using algebraic expressions with a finite number of variables and parameters. It does, however, assume that these variables are *real numbers* – with *infinite* precision. Some of the problems encountered in its use arise from this distinction. These are all issues on the practical side of Kalman filtering that must be considered along with the theory:

(a) *Representing uncertainty.*

Generally, a *random number* or *variate* is a variable whose value may be random but whose probability distribution is known to some degree. The truly remarkable thing about Gauss's LLMSE is that the solution depends only on two parameters of the probability distributions of the random variates involved. It is otherwise independent of other attributes of the underlying error distributions.

i. The first of these parameters is the *mean value* of a probability distribution. If the variate is a vector, the mean will also be a vector.
ii. The second parameter is the *mean-squared deviation about the mean*, which will be a matrix if the variate is a vector. It is also called the *covariance* about the mean of the distribution or the *covariance of uncertainty* in the estimate, because it characterizes all the cross correlations among the components of vector variates.

These two parameters are used for characterizing each of three probability distributions in Kalman filtering:

i. The distributions of *measurement errors*, in which case the means are usually assumed to be zero, but can otherwise be treated as known measurement biases and subtracted from the sensor outputs.
ii. The distributions of *random dynamic disturbances*, the means of which are also assumed to be zero. Otherwise, any nonzero mean can be treated as a known dynamic disturbance.
iii. The distribution of *estimation errors*, in which case equations for maintaining means and covariances can be further subdivided into two classes:
   A. Matrix equations for maintaining the covariance of estimation errors are collectively called *the riccati equations*, a reference to a nonlinear differential equation studied by the Italian mathematician Jacopo Riccati (1676–1754) and shown to be equivalent to a pair of linear differential equations – an attribute shared

with its matrix counterpart. It was Kalman's colleague Richard S. Bucy who recognized that the nonlinear differential equations for covariance propagation in the Kalman filter had the form of Riccati's equation. In Kalman filtering, the solution to the riccati equations does not depend on the estimate, which is why the riccati equations can be used independently for assessing the expected performance of a proposed application defined by its stochastic dynamic model and the performance specifications and measurement schedules for the proposed sensor suite.

B. *The Kalman filter is essentially an estimator for the mean of the distribution of the unknown variate,* in which case the estimation errors will have zero mean. The filter uses the covariance of estimation errors from the solution to the riccati equations described earlier to maintain the linear least mean squares estimate of the unknown variate. This part of the Kalman filter implementation depends on the solution of the riccati equations, but *the riccati equations do not depend on the estimate.*

(b) *Dependence on linearity.*
Transformations of the means and covariances in its implementation depend on linearity of the dynamics and measurements. Although several approximations for not-quite-linear applications have been derived, applied, and evaluated, the resulting means and covariances from nonlinear transformations of the variates generally depend on other attributes of the starting probability distributions beyond just the mean and covariance. Even assuming that the starting distributions are Gaussian is inadequate, because nonlinear transformations of Gaussian distributions do not preserve gaussianity.

(c) *Linear algebra.*
The Kalman filter is based on linear algebraic procedures for the addition and multiplication of vectors and matrices, and inversion of matrices. This makes it particularly easy to implement in MATLAB®.

4. **How is it implemented?**
It is called a "*predictor–corrector*" algorithm because it propagates the linear least mean squares estimate $\hat{\mathbf{x}}$ and its covariance of estimation errors $\mathbf{P}$ (also called the covariance of estimation *uncertainty*) forward in the time between measurements, predicting the estimate and its covariance of estimation uncertainty before the next measurement is used. Then the results of the measurement(s) are used to correct the predicted values to reflect the influence of the information gained from the new measurements.
The top-level data flow structure of the implementation is shown in block form in Figure 10.1, where the box titled "Measurement update (corrector)" uses as its inputs the estimated state variable $\hat{\mathbf{x}}$ prior to using the measurements and the associated covariance matrix $\mathbf{P}$ of prior state

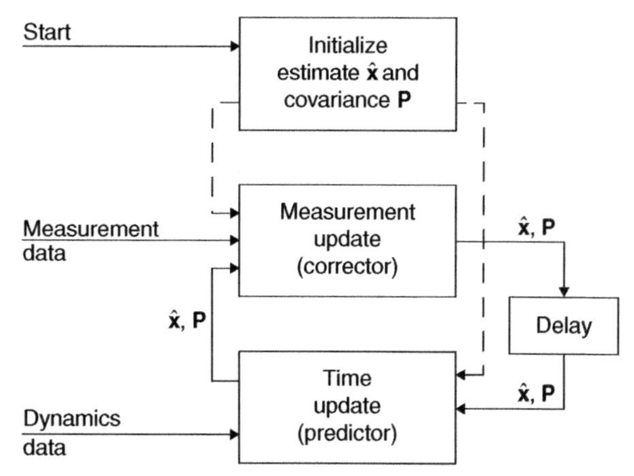

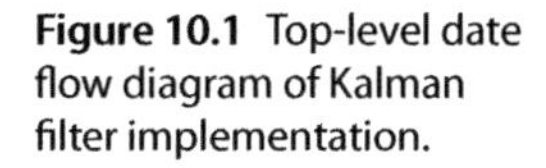
Figure 10.1 Top-level date flow diagram of Kalman filter implementation.

estimation uncertainty. Inputs also include measurement data in the form of the vector of measurement values $\mathbf{z}$, its associated matrix $\mathbf{H}$ of linear sensitivities to the unknown dynamic state variable $\mathbf{x}$, and the associated covariance matrix $\mathbf{R}$ of measurement noise. Outputs include the estimated state variable $\hat{\mathbf{x}}$ after using the measurements and the covariance matrix $\mathbf{P}$ of state estimation uncertainty corrected to reflect the improvement in estimation uncertainty resulting from the used of measurements. If no usable measurements are available, no correction is made, which allows for testing and rejecting anomalous measurements.

The box titled "Delay" represents the time delay between one set of measurements and the next, which is not necessarily constant and can be zero.

The box titled "Time update[1] (predictor)" uses as its inputs the estimated state variable $\hat{\mathbf{x}}$ after using the last set of measurements, and the associated covariance matrix $\mathbf{P}$ of estimation uncertainty. Inputs also include parameters defining the changes in the estimate and its associated uncertainty due to the passage of time. These input parameters include the linear transformation matrix $\boldsymbol{\Phi}$ modeling the deterministic dynamics of the state vector $\mathbf{x}$ over the time period since the last measurement update and the matrix $\mathbf{Q}$ representing the covariance matrix of additional dynamic uncertainty resulting from random disturbances of the state vector during that period. Outputs of the predictor include the time-updated estimate $\hat{\mathbf{x}}$ of the dynamic state vector $\mathbf{x}$ and its associated covariance matrix $\mathbf{P}$ of estimation uncertainty.

Activities required for starting the Kalman filter are represented by the box title "Initialize estimate $\hat{\mathbf{x}}$ and covariance $\mathbf{P}$," which gives initial values of the estimate $\hat{\mathbf{x}}$ and its associated covariance matrix of estimation uncertainty $\mathbf{P}$ to either the corrector or predictor – but not both. After initialization, program control loops between the predictor and corrector, the derivations of which are outlined in the following sections.

1 Also called the "temporal update."

10.3 Notation

The Kalman filter makes extensive use of linear algebra, so it will be necessary to define some related mathematical conventions before we can start to use them.

10.3.1 Real Vectors and Matrices

10.3.1.1 Notation

We will use font, case, and face to distinguish between printed symbols used for different types of mathematical objects:

1. The symbol $\mathbb{R}$ represents the field of *real numbers*, also called *scalars*. Real numbers or scalars $r \in \mathbb{R}$ will be denoted by lowercase unbolded letters.
2. The symbol $\mathbb{R}^n$ represents n-dimensional Euclidean space, the set of n-dimensional real *vectors*, also called n-vectors. Unless specified otherwise, they are assumed to be *column vectors*. Vectors $\mathbf{v} \in \mathbb{R}^n$ will be denoted by lowercase **bold** letters.
   (a) The scalar element in the ith row of a column vector $\mathbf{v}$ is represented as v_i, unbolded and post-subscripted. If $\mathbf{v}$ is a row vector, v_i represents the element in the ith column of $\mathbf{v}$.
   (b) The post-superscript ^{T} denotes the *transpose* of a vector (represented by the post-apostrophe' in MATLAB). The transpose of a column vector is a row vector, and *vice versa*.
   (c) The letter $\mathbf{x}$ will be used to denote the state vector of a dynamic system being tracked by a Kalman filter.
   (d) The overbar $\overline{\mathbf{x}}$ will be used to denote the mean of the probability distribution of $\mathbf{x}$.
   (e) The "hatted" letter $\hat{\mathbf{x}}$ will be used to denote the *estimate* (or estimated mean) of the probability distribution of the state vector.
3. The symbol $\mathbb{R}^{\ell \times n}$ represents the set of $\ell \times n$ real matrices, where ℓ and n are positive integers. ℓ representing the number of rows and n representing the number of columns. Matrices $\mathbf{M} \in \mathbb{R}^{\ell \times n}$ will be denoted by uppercase **bold** letters.
   (a) The letter $\mathbf{P}$ will be used to denote the covariance of uncertainty (defined hereafter) of the estimate $\hat{\mathbf{x}}$ (defined earlier).
   (b) The post-superscript ^{T} denotes the *transpose* (apostrophe' in MATLAB) of a matrix:

$$\{\mathbf{M}^T\}_{ij} = \{\mathbf{M}\}_{ji} = m_{ji}$$

   (c) The post-superscript $^{-1}$ denotes the *inverse* of a matrix – if it exists.
   (d) The post-superscript$^{1/2}$ in $\mathbf{P}^{1/2}$ denotes a *square root* of a symmetric positive-definite matrix $\mathbf{P}$, defined as a solution to the matrix equation

$$\mathbf{P}^{1/2}\mathbf{P}^{T/2} = \mathbf{P} \quad \text{for} \quad \mathbf{P}^{T/2} \stackrel{\text{def}}{=} [\mathbf{P}^{1/2}]^T \tag{10.1}$$

(e) The post-superscript$^{-1/2}$ in $\mathbf{P}^{-1/2}$ denotes a *square root* of $\mathbf{P}^{-1}$.
(f) The scalar element in the ith row and jth column of a matrix **M** is represented as m_{ij}, where the m is lowercase and unbolded.
(g) The "dot" ($\cdot$) will be used to indicate the full range of a subscript, so $\mathbf{m}_{\cdot j}$ will denote the jth column vector of the matrix **M** and $\mathbf{m}_{i\cdot}$ will denote the ith row vector of **M**. (Its equivalent in MATLAB is the colon, ":")

4. Vectors and matrices in mathematical expressions are assumed to be *conformably dimensioned* for the operations shown. A matrix product **AB**, for example, assumes the number of columns in **A** equals the number of rows in **B**.

10.3.1.2 Vector and Matrix Properties

Properties useful in Kalman filtering include the following:

Unit vectors **u** have unit length

$$\sqrt{\sum_i u_i^2} = 1$$

Zero matrices have zero (0) in all elements and are denoted by the zero symbol (0).

Square matrices have the same number of rows and columns. Non-square matrices are called *rectangular*.

Symmetric matrices are square matrices **M** such that $\mathbf{M}^T = \mathbf{M}$.

Positive definite matrices are symmetric $n \times n$ matrices **M** such that, for any nonzero column n-vector **v**,

$$\mathbf{v}^T\mathbf{M}\mathbf{v} > 0$$

For **M** *positive semi-definite*, $\mathbf{v}^T\mathbf{M}\mathbf{v} \geq 0$.

The **main diagonal** of a matrix **M** is that subset of elements m_{ii} with the same row and column index i. (Note that this applies to non-square matrices.)

The **trace** of a square matrix **M** is defined as the sum of its diagonal elements and denoted as Tr(**M**). The trace operator is linear, in that

$$\mathrm{Tr}(a\mathbf{A} + b\mathbf{B}) = a\mathrm{Tr}(\mathbf{A}) + b\mathrm{Tr}(\mathbf{B})$$

The trace of a matrix product is also invariant under commutation:

$$\mathrm{Tr}(\mathbf{AB}) = \mathrm{Tr}(\mathbf{BA}) \tag{10.2}$$

The trace of a matrix **M** is computed in MATLAB by the function `trace(M)`.

Diagonal matrices are square matrices **D** in which only the diagonal elements d_{ii} are nonzero. The notation diag(**v**) will denote the square diagonal matrix with diagonal elements specified by the vector **v**.

Identity matrices are diagonal matrices in which all diagonal elements equal 1. The symbol **I** will be used to denote an identity matrix.

Triangular matrices have zeros on one side or the other of the main diagonal.
Upper triangular matrices have zeros below the main diagonal.
Lower triangular matrices have zeros above the main diagonal.
Strictly triangular matrices have zeros on the main diagonal, as well.
Unit triangular matrices have ones on their main diagonal.

Determinants of $n \times n$ matrices **M** are real numbers equal to their *volumetric scaling*, which is the volume of the unit n-cube after transformation by **M**. The determinant of **M** will be denoted by det **M** and is computed in MATLAB by the function `det(M)`.

Eigenvalues or **characteristic values** of a matrix **M** are the the solutions λ of its *characteristic polynomial*

$$\det(\mathbf{M} - \lambda \mathbf{I}) = 0$$

Eigenvectors of a matrix **M** are the unit vector solutions of $\mathbf{Mu} = \lambda_i \mathbf{u}_i$ for the eigenvalues λ_i of **M**.

Decomposition – in mathematics – refers to representing something as a sum or product. A square matrix **M**, for example, can be represented as a sum of its symmetric and antisymmetric parts as

$$\mathbf{A} = \underbrace{(\mathbf{M} + \mathbf{M}^T)/2}_{\text{symmetric}} + \underbrace{(\mathbf{M} - \mathbf{M}^T)/2}_{\text{antisymmetric}}$$

Factorization refers to matrix decomposition as a product, perhaps done to distinguish it from biological terminology.

The **eigenvalue–eigenvector decomposition** of a symmetric matrix **M** is the representation

$$\mathbf{M} = \sum_i \lambda_i \mathbf{u}_i \mathbf{u}_i^T$$

where the λ_i are the eigenvalues of **M** and the $\mathbf{u}_i$ are the corresponding unit eigenvectors.

Eigendecomposition – or spectral decomposition – of a square matrix **M** refers to its factorization in the form

$$\mathbf{M} = \mathbf{E}\mathbf{D}_\lambda \mathbf{E}^{-1},$$

where the columns of **E** are the eigenvectors of **M** and the diagonal entries of $\mathbf{D}_\lambda$ are the corresponding eigenvalues.

Orthogonal matrices are nonsingular square matrices Ω such that

$$\boldsymbol{\Omega}^T = \boldsymbol{\Omega}^{-1} \quad \text{and} \quad \boldsymbol{\Omega}^T\boldsymbol{\Omega} = \boldsymbol{\Omega}\boldsymbol{\Omega}^T = \mathbf{I}$$

These have properties which are very useful in matrix computations:

- Products of $n \times n$ orthogonal matrices are also $n \times n$ orthogonal matrices and

$$[\boldsymbol{\Omega}_1\boldsymbol{\Omega}_2\boldsymbol{\Omega}_3 \cdots \boldsymbol{\Omega}_m]^T = \boldsymbol{\Omega}_m^{-1}\boldsymbol{\Omega}_{m-1}^{-1}\boldsymbol{\Omega}_{m-2}^{-1} \cdots \boldsymbol{\Omega}_1^{-1}$$

- If $\mathbf{P}^{1/2}$ is a square root of $\mathbf{P}$ (defined in Eq. (10.1)) and $\boldsymbol{\Omega}$ is a conformable orthogonal matrix, then $\mathbf{P}^{1/2}\boldsymbol{\Omega}$ is also a square root of $\mathbf{P}$, because

$$\mathbf{P}^{1/2}\boldsymbol{\Omega}[\mathbf{P}^{1/2}\boldsymbol{\Omega}]^T = \mathbf{P}^{1/2}\boldsymbol{\Omega}\boldsymbol{\Omega}^T\mathbf{P}^{T/2} = \mathbf{P}^{1/2}\mathbf{I}\mathbf{P}^{T/2} = \mathbf{P} \tag{10.3}$$

  This implies that matrix square roots are not unique and are related by orthogonal matrices, which will have important consequences in square root implementations of the Kalman filter.
- Determinants of orthogonal matrices are ± 1 and the absolute values of all eigenvalues are $+1$.

Singular values of a matrix $\mathbf{M}$ are the square roots of the eigenvalues of $\mathbf{M}\mathbf{M}^T$ (left singular values) and $\mathbf{M}^T\mathbf{M}$ (right singular values).

Singular value decomposition of an $n \times n$ symmetric positive-definite matrix $\mathbf{M}$ is a representation in the form

$$\mathbf{M} = \boldsymbol{\Omega}\mathbf{D}\boldsymbol{\Omega}^T$$

where the unit eigenvectors of $\mathbf{M}$ are the columns of the $n \times n$ orthogonal matrix $\boldsymbol{\Omega}$ and $\mathbf{D}$ is a diagonal matrix with the corresponding eigenvalues on its diagonal, ordered such that

$$d_{11} \geq d_{22} \geq d_{33} \geq \cdots \geq d_{nn} > 0$$

For any symmetric positive-definite matrix $\mathbf{M}$ the MATLAB function `[OmegaL,D,OmegaR] = svd(M)` returns $\boldsymbol{\Omega}_L$, $\mathbf{D}$, $\boldsymbol{\Omega}_R$ such that $\boldsymbol{\Omega}_L = \boldsymbol{\Omega}_R$.

Condition numbers for computational procedures generally refer to sensitivities of outputs to inputs. The condition number for inverting a square matrix $\mathbf{M}$ is the ratio of the largest singular value of $\mathbf{M}$ to the smallest singular value of $\mathbf{M}$. Larger condition numbers generally indicate poorer conditioning for inversion. The condition number for an orthogonal matrix is 1 (one), which is as good as it gets.

10.3.2 Probability Essentials

Kalman filtering is not about probabilities. It is about expected values of estimation errors and their squared values, but we need to use probabilities to explain what it is about.

10.3.2.1 Basic Concepts

Derivation of the Kalman filter equations requires only a few concepts from probability theory. To wit:

1. All probability distributions are assumed to be defined on n-dimensional Euclidean spaces $\mathbb{R}^n$ by integrable PDFs $p(\cdot)$ such that
   (a) $p(\mathbf{x}) \geq 0$ for all $\mathbf{x} \in \mathbb{R}^n$
   (b) $\int_{\mathbb{R}^n} p(\mathbf{x})dx = 1$

2. The *expected value* $\mathrm{E}_p\langle F\rangle$ of any real-, vector-, or matrix-valued function F of $\mathbf{x} \in \mathbb{R}^n$ is defined as

$$\mathrm{E}_p\langle F\rangle \stackrel{\text{def}}{=} \int_{\mathbb{R}^n} F(\mathbf{x})\, p(\mathbf{x})\, d\mathbf{x}$$

   where $d\mathbf{x} = dx_1\, dx_2\, dx_3\, \cdots\, dx_n$ and the expected value of a vector or matrix is the vector or matrix of the expected values of its elements. The subscript p on E represents the probability density function defining the operative expectancy operator, although this notation can be dropped after the quasi-independence of the LLMSE from full specification of PDFs has been established.
3. *Raw moments* of probability distributions on $\mathbb{R}^n$ are the expected values of outer-product powers of the variate. The Nth such raw moment is an N-dimensional data structures $\mathcal{D}$ with elements

$$\{\mathcal{D}\}_{j_1,\, j_2,\, j_3, \ldots,\, j_N} \stackrel{\text{def}}{=} \mathrm{E}\langle x_{j_1}\, x_{j_2}\, x_{j_3}\, \cdots\, x_{j_N}\,\rangle, \quad 1 \le j_i \le n, \quad 1 \le i \le N$$

4. The first-order raw moment is called the *mean* of the probability distribution defined by a PDF $p(\cdot)$ as the vector

$$\overline{\mathbf{x}} = \mathrm{E}_p\langle \mathbf{x}\rangle$$

5. *Central moments* are defined with respect to the mean as the raw moments of $\mathbf{x} - \overline{\mathbf{x}}$.
6. The second central moment of the probability distribution defined by $p(\cdot)$ would then be the $n \times n$ matrix
   (a) $\mathbf{P} = \mathrm{E}_p\langle[\mathbf{x} - \overline{\mathbf{x}}]\, [\mathbf{x} - \overline{\mathbf{x}}]^T\rangle$, and its trace (from Eq. (10.2))
   (b) $\mathrm{Tr}(\mathbf{P}) = \mathrm{E}_p\langle|\mathbf{x} - \overline{\mathbf{x}}|^2\rangle$, the mean squared value.

It should be noted that there are so-called "pathological" probability distributions (e.g. the Cauchy distribution) for which the mean and covariance are not defined. Otherwise, in the LLMSE, the underlying probability distribution p does not need to be specified beyond its mean and covariance about the mean. This is because the theory depends only on the mean and covariance about the mean and is otherwise independent of any other particulars of the probability distribution.

10.3.2.2 Linearity of the Expectancy Operator E⟨·⟩

An important property of the expectancy operator $\mathrm{E}\langle\cdot\rangle$ is that it is *linear*, in the sense that for any real numbers a and b, scalar functions f and g, vector-valued functions $\mathbf{f}$ and $\mathbf{g}$, real matrices $\mathbf{A}$ and $\mathbf{B}$, and matrix-valued functions $\boldsymbol{F}$ and $\boldsymbol{G}$,

$$\mathrm{E}\langle af + bg\rangle = a\mathrm{E}\langle f\rangle + b\mathrm{E}\langle g\rangle \tag{10.4}$$

$$\mathrm{E}\langle \mathbf{A}\mathbf{f} + \mathbf{B}\mathbf{g}\rangle = \mathbf{A}\mathrm{E}\langle \mathbf{f}\rangle + \mathbf{B}\mathrm{E}\langle \mathbf{g}\rangle \tag{10.5}$$

$$\mathrm{E}\langle \mathbf{A}\boldsymbol{F} + \mathbf{B}\boldsymbol{G}\rangle = \mathbf{A}\mathrm{E}\langle \boldsymbol{F}\rangle + \mathbf{B}\mathrm{E}\langle \boldsymbol{G}\rangle \tag{10.6}$$

10.3.2.3 Means and Covariances of Linearly Transformed Variates

If **A** is any matrix, $\overline{\mathbf{x}}$ is the mean of a probability distribution on $\mathbf{x} \in \mathbb{R}^n$ and **P** is its covariance, then the mean and covariance of **Ax** will be

$$\begin{aligned} \text{E}\langle \mathbf{Ax} \rangle &= \mathbf{A}\text{E}\langle \mathbf{x} \rangle \\ &= \mathbf{A}\overline{\mathbf{x}} \quad \text{(mean)} \end{aligned} \tag{10.7}$$

$$\begin{aligned} \text{E}\langle [\mathbf{Ax} - \mathbf{A}\overline{\mathbf{x}}][\mathbf{Ax} - \mathbf{A}\overline{\mathbf{x}}]^T \rangle &= \mathbf{A}\text{E}\langle [\mathbf{x} - \overline{\mathbf{x}}][\mathbf{x} - \overline{\mathbf{x}}]^T \rangle \mathbf{A}^T \\ &= \mathbf{APA}^T \quad \text{(covariance)} \end{aligned} \tag{10.8}$$

where **P** is the covariance of **x**. *This will be very important in the derivation of the Kalman filter.*

10.3.3 Discrete Time Notation

10.3.3.1 Subscripting

The Kalman filter is defined in terms of a discrete time sequence of events,

$$t_0,\ t_1,\ t_2,\ t_3, \dots,$$

where the real time sequence is nondecreasing,

$$t_0 \le t_1 \le t_2 \le t_3 \le \cdots$$

and the time interval between successive events is not necessarily constant. These discrete time values t_k generally represent the times at which measurements can be made but may also represent times when there is no measurement used but estimates are needed. These conditions may occur when a measurement is deemed to be in error and not usable, for example.

As a measure for mitigating subscript confusion, sub-arrays or elements of time-subscripted arrays can be represented by braces, so that $\{\mathbf{M}_k\}_{ij}$ denotes the element in the *i*th row and *j*th column of the time-subscripted matrix $\mathbf{M}_k$.

10.3.3.2 A Priori and A Posteriori Values

In the discrete time instance t_k of each Kalman filter measurement update there can be two values of the estimate $\hat{\mathbf{x}}$ and its associated covariance matrix of estimation uncertainty **P**:

1. The *a priori* values of $\hat{\mathbf{x}}$ and **P** before the information in the measurement(s) is used, which will be denoted as $\hat{\mathbf{x}}_{k(-)}$ and $\mathbf{P}_{k(-)}$.
2. The *a posteriori* values of $\hat{\mathbf{x}}$ and **P** after the information in the measurement(s) has been used, which will be denoted as $\hat{\mathbf{x}}_{k(+)}$ and $\mathbf{P}_{k(+)}$.

10.3.3.3 Allowing for Testing and Rejecting Measurements

There is the possibility that a measurement vector $\mathbf{z}_k$ or some elements thereof at time t_k can be tested and rejected as being too anomalous to be acceptable without significant risk.

In the case that the entire measurement vector $\mathbf{z}_k$ is unacceptable, the a posteriori values of the estimate $\hat{\mathbf{x}}_{k(+)}$ and its covariance of uncertainty $\mathbf{P}_{k(+)}$ will be equal to their respective a priori values, $\hat{\mathbf{x}}_{k(-)}$ and $\mathbf{P}_{k(-)}$.

In the case that only some components of a measurement vector are deemed unusable, the suspect components of the measurement vector $\mathbf{z}_k$, the corresponding rows of $\mathbf{H}_k$, and the corresponding rows and columns of the associated covariance matrix $\mathbf{R}_k$ of measurement noise can be eliminated and their dimension ℓ_k reduced accordingly.

Methods for monitoring and testing are discussed in Section 10.7.

10.4 Kalman Filter Genesis

Kalman played a major role in a mid-twentieth century paradigm shift in engineering mathematics for estimation and control problems – from probability theory and spectral characteristics in the frequency domain to linear first-order differential equations in "state space" (Euclidean space) and matrix theory. In 1955, when the Glenn L. Martin Company founded its Research Institute for Advanced Studies (RIAS) just outside Baltimore, one of its major research areas would be in the mathematical theory for control of dynamic systems. Kalman joined RIAS in 1956, and in November of 1958 he posed for himself the problem of transforming Wiener–Kolmogorov "filtering" (real-time estimation) in the frequency domain to state-space form in the time domain. The effort would come to rely on mathematics beyond historical engineering curricula, but its result – about a year later – would be the Kalman filter [2], followed by major discoveries about the relationships between linear estimation and control problems in what has come to be called *mathematical systems theory* [5].

A brief historical overview of the mathematical origins of that transformation is presented in the following subsections 10.4.1–10.4.3, the main purpose of which is to demonstrate the special properties of Gauss's LLMSE that have been inherited by Kalman filtering. This also honors some of the other major contributors to the mathematical development of Kalman filtering.

10.4.1 Measurement Update (Corrector)

The measurement update (also called the "observational update") is where measurements related to the state vector are used to update the estimated state vector.

To really understand why the Kalman filter depends only on the first two moments of probability distributions it is necessary to start with one of its progenitors, the *LLMSE* of Gauss [1], and follow through some of its transformations to become the so-called "corrector" implementation of the Kalman filter.

10.4.1.1 Linear Least Mean Squares Estimation: Gauss to Kalman

Least Squares Gauss discovered his *method of least squares* in 1795, at the age of 18 [6], but he did not apply it to an estimation problem until 1801 and did not publish his method until 1809 [7]. In the meantime, his method of least squares had been discovered independently and published by Andrien-Marie Legendre (1752–1833) [8] in France and Robert Adrian (1775–1843) in the United States [9].

Gauss's method of least squares finds a solution to the overdetermined linear equation $\mathbf{z} = \mathbf{H}\mathbf{x}$ for $\mathbf{x}$, given $\mathbf{z}$ and $\mathbf{H}$. The solution for $\mathbf{x}$ that minimizes the squared error $|\mathbf{z} - \mathbf{H}\mathbf{x}|^2$ can be found by setting its derivative to 0:

$$\frac{d}{d\mathbf{x}}|\mathbf{z} - \mathbf{H}\mathbf{x}|^2 = 0$$

and solving the resulting *normal equation*

$$\underbrace{\mathbf{H}^T\mathbf{H}}_{\text{Gramian}}\ \mathbf{x} = \mathbf{H}^T\mathbf{z} \tag{10.9}$$

for the least-squares estimate

$$\hat{\mathbf{x}} = [\mathbf{H}^T\mathbf{H}]^{-1}\mathbf{H}^{\mathrm{T}}\mathbf{z} \tag{10.10}$$

if the *Gramian*[2] *matrix* $\mathbf{H}^T\mathbf{H}$ is nonsingular.

Gauss–Markov Theorem and Homoscedasticity In [1], Gauss showed that the least squares solution is also the linear least mean squares solution if the *measurement error*

$$\mathbf{v} \stackrel{\text{def}}{=} \mathbf{z} - \mathbf{H}\mathbf{x} \tag{10.11}$$

is zero-mean and *homoskedastic*. That is, it has covariance

$$\begin{aligned} \mathrm{E}\langle \mathbf{v}\mathbf{v}^T \rangle &= \begin{bmatrix} \sigma^2 & 0 & 0 & \cdots & 0 \\ 0 & \sigma^2 & 0 & \cdots & 0 \\ 0 & 0 & \sigma^2 & \cdots & 0 \\ \vdots & \vdots & \vdots & \ddots & \vdots \\ 0 & 0 & 0 & \cdots & \sigma^2 \end{bmatrix} \\ &= \sigma^2\mathbf{I} \end{aligned} \tag{10.12}$$

where $\mathbf{I}$ is an identity matrix. This implies that the individual components of the measurement error vector $\mathbf{v}$ are statistically independent and all have the same variance σ^2. Note that *homoskedasticity* is a weaker constraint than that of *independent identically distributed*, a not-uncommon assumption in statistics.

2 Named for Jørgen Pedersen Gram (1850–1916), also codiscoverer (with Erhard Schmidt (1876–1959)) of Gram–Schmidt orthogonalization, a modified version of which is used in the unit triangular diagonal (UD) square root implementation of the Kalman filter.

Gauss showed that under the constraints on the unknown measurement error vector **v** to be zero-mean and satisfy Eq. (10.12), the least squares solution also minimizes the mean squared error

$$\mathrm{E}\langle |\mathbf{z} - \mathbf{H}\mathbf{x}|^2 \rangle$$

Gauss published his results in New Latin,[3] which eventually fell out of favor as a language for mathematical discourse. As a consequence, his work was largely forgotten until essentially the same result was obtained by Andrey Andreyevich Markov (1856–1922) nearly a century later [11]. It was called the *Markov theorem* until the earlier publication by Gauss was rediscovered. It is now called the *Gauss–Markov theorem*.

Generalization by Aitken Distributions that are not homoskedastic are called *heteroskedastic*. The Gauss–Markov theorem was generalized to the heteroskedastic (i.e. general) case by Alexander Craig Aitken (1895–1967) in a 1935 publication [12]. In this case, the covariance of the zero-mean measurement errors

$$\mathrm{E}\langle \mathbf{v}\mathbf{v}^T \rangle = \mathbf{R} \tag{10.13}$$

which is not necessarily in the form of a scalar matrix, as in Eq. (10.12).

A weighted least squares solution. The trick is to convert the least squares problem of $\mathbf{z} = \mathbf{H}\mathbf{x} + \mathbf{v}$ to a *weighted least squares* problem,

$$\underbrace{\mathbf{W}\mathbf{z}}_{\mathbf{z}^\star} = \underbrace{\mathbf{W}\mathbf{H}\mathbf{x}}_{\mathbf{H}^\star} + \underbrace{\mathbf{W}\mathbf{v}}_{\mathbf{v}^\star} \tag{10.14}$$

such that the weighted noise vector $\mathbf{v}^\star$ is homoskedastic. That is, such that a scalar matrix

$$\begin{aligned} \sigma^2 \mathbf{I} &= \mathrm{E}\langle \mathbf{v}^\star \mathbf{v}^{\star T} \rangle \\ &= \mathrm{E}\langle (\mathbf{W}\mathbf{v})(\mathbf{W}\mathbf{v})^T \rangle \\ &= \mathbf{W}\mathrm{E}\langle \mathbf{v}\mathbf{v}^T \rangle \mathbf{W}^T \\ &= \mathbf{W}\mathbf{R}\mathbf{W}^T \end{aligned} \tag{10.15}$$

or, for $\sigma = 1$,

$$\begin{aligned} \mathbf{R} &= \mathbf{W}^{-1}\mathbf{I}\mathbf{W}^{-T} \\ &= \mathbf{W}^{-1}\mathbf{W}^{-T} \end{aligned}$$

3 Also called Modern Latin or Neoclassical Latin, it was a revival of Latin as a language for scientific publications between the late fourteenth and late nineteenth centuries. Another of Gauss's discoveries, the fast Fourier transform, was also published in New Latin, in 1805, later to be discovered independently by James W. Cooley (1926–2016) and John W. Tukey (1915–2000) and published in 1965 [10].

$$= [\mathbf{W}^T\mathbf{W}]^{-1} \tag{10.16}$$

$$\mathbf{R}^{-1} = \mathbf{W}^T\mathbf{W} \tag{10.17}$$

which constrains the values of the weighting matrix $\mathbf{W}$ such that the weighted least squares problem of Eq. (10.14) is homoskedastic. In that case, the *Gauss–Markov theorem* guarantees that the linear least mean squares solution for the unknown $\mathbf{x}$ will be the least squares solution of Eq. (10.14), given by Eq. (10.10) with $\mathbf{H}$ replaced by $\mathbf{H}^\star$ and $\mathbf{z}$ replaced by $\mathbf{z}^\star$:

$$\begin{aligned}\hat{\mathbf{x}} &= [\mathbf{H}^{\star T}\mathbf{H}^\star]^{-1}\mathbf{H}^{\star T}\mathbf{z}^\star \\ &= [(\mathbf{W}\mathbf{H})^T\mathbf{W}\mathbf{H}]^{-1}(\mathbf{W}\mathbf{H})^T\mathbf{W}\mathbf{z} \\ &= [\mathbf{H}^T\mathbf{W}^T\mathbf{W}\mathbf{H}]^{-1}\mathbf{H}^T\mathbf{W}^T\mathbf{W}\mathbf{z} \\ &= [\mathbf{H}^T\mathbf{R}^{-1}\mathbf{H}]^{-1}\mathbf{H}^T\mathbf{R}^{-1}\mathbf{z}\end{aligned} \tag{10.18}$$

– the general linear least mean squares solution. Note that it does not depend on choosing a particular solution $\mathbf{W}$ to Eq. (10.17), but only on the covariance matrix $\mathbf{R}$ of measurement errors.

Means and Mean Squared Estimation Error If $\hat{\mathbf{x}}$ is an estimated value of a variate with mean $\overline{\mathbf{x}}$ and covariance $\mathbf{P}$, then the squared estimation error for $\hat{\mathbf{x}}$ as a function of $\mathbf{x} \in \mathbb{R}^n$ is $|\hat{\mathbf{x}} - \mathbf{x}|^2$ and the mean squared estimation error over the probability distribution of $\mathbf{x}$ is its expected value,

$$\begin{aligned}\mathrm{E}\langle|\hat{\mathbf{x}} - \mathbf{x}|^2\rangle &= \mathrm{E}\langle[\hat{\mathbf{x}} - \mathbf{x}]^T[\hat{\mathbf{x}} - \mathbf{x}]\rangle \\ &= |\hat{\mathbf{x}}|^2 - 2\hat{\mathbf{x}}^T\mathrm{E}\langle\mathbf{x}\rangle + \mathrm{E}\langle|\mathbf{x}|^2\rangle \\ &= |\hat{\mathbf{x}}|^2 - 2\hat{\mathbf{x}}^T\mathrm{E}\langle\mathbf{x}\rangle + \mathrm{E}\langle|\overline{\mathbf{x}} + (\mathbf{x} - \overline{\mathbf{x}})|^2\rangle \\ &= |\hat{\mathbf{x}}|^2 - 2\hat{\mathbf{x}}^T\overline{\mathbf{x}} + |\overline{\mathbf{x}}|^2 + \mathrm{E}\langle|\mathbf{x} - \overline{\mathbf{x}}|^2\rangle \\ &= |\hat{\mathbf{x}}|^2 - 2\hat{\mathbf{x}}^T\overline{\mathbf{x}} + |\overline{\mathbf{x}}|^2 + \sum_i \mathrm{E}\langle(x_i - \overline{x}_i)^2\rangle\end{aligned} \tag{10.19}$$

$$\mathrm{E}\langle|\hat{\mathbf{x}} - \mathbf{x}|^2\rangle = |\hat{\mathbf{x}} - \overline{\mathbf{x}}|^2 + \mathrm{Tr}(\mathbf{P}) \tag{10.20}$$

Equation (10.20) is a quadratic equation in the estimated value $\hat{\mathbf{x}}$ that reaches its minimum value at $\hat{\mathbf{x}} = \overline{\mathbf{x}}$, its mean, and that minimum value is the trace of $\mathbf{P}$, which the sum of the diagonal elements of $\mathbf{P}$. That is, in the LLMSE, *the least mean squares estimate is always the mean of the probability distribution* – independent of any other attributes of the probability distribution in question. That is why – in the LLMSE – only the means and covariances of the distributions matter. The linear least mean squares estimate is otherwise independent of other attributes of the probability distributions. This is a very important attribute of Kalman filtering, as well.

See Appendix C on www.wiley.com/go/grewal/gnss of a demonstration showing that Eq. (10.20) is independent of the PDF of the variates involved.

Covariance of Estimation Error *Estimation error* is defined as the difference between the estimate $\hat{\mathbf{x}}$ and $\mathbf{x}$:

$$\begin{aligned}
\delta &\stackrel{\text{def}}{=} \hat{\mathbf{x}} - \mathbf{x} \\
&= [\mathbf{H}^T\mathbf{R}_\mathbf{v}^{-1}\mathbf{H}]^{-1}\mathbf{H}^T\mathbf{R}_\mathbf{v}^{-1}\mathbf{z} - \mathbf{x} \\
&= [\mathbf{H}^T\mathbf{R}_\mathbf{v}^{-1}\mathbf{H}]^{-1}\mathbf{H}^T\mathbf{R}_\mathbf{v}^{-1}(\mathbf{H}\mathbf{x} + \mathbf{v}) - \mathbf{x} \\
&= \underbrace{\left[\underbrace{[\mathbf{H}^T\mathbf{R}_\mathbf{v}^{-1}\mathbf{H}]^{-1}\mathbf{H}^T\mathbf{R}_\mathbf{v}^{-1}\mathbf{H}}_{\mathbf{I}} - \mathbf{I}\right]}_{0}\mathbf{x} + [\mathbf{H}^T\mathbf{R}_\mathbf{v}^{-1}\mathbf{H}]^{-1}\mathbf{H}^T\mathbf{R}_\mathbf{v}^{-1}\mathbf{v} \\
&= [\mathbf{H}^T\mathbf{R}_\mathbf{v}^{-1}\mathbf{H}]^{-1}\mathbf{H}^T\mathbf{R}_\mathbf{v}^{-1}\mathbf{v}
\end{aligned} \tag{10.21}$$

Hence its mean,

$$\begin{aligned}
\underset{\mathbf{v}}{\mathrm{E}}\langle\delta\rangle &= \mathrm{E}_\mathbf{v}\langle[\mathbf{H}^T\mathbf{R}_\mathbf{v}^{-1}\mathbf{H}]^{-1}\mathbf{H}^T\mathbf{R}_\mathbf{v}^{-1}\mathbf{v}\rangle \\
&= [\mathbf{H}^T\mathbf{R}_\mathbf{v}^{-1}\mathbf{H}]^{-1}\mathbf{H}^T\mathbf{R}_\mathbf{v}^{-1}\mathrm{E}_\mathbf{v}\langle\mathbf{v}\rangle \\
&= 0
\end{aligned} \tag{10.22}$$

if $\mathbf{v}$ is zero-mean, and its covariance about the mean

$$\begin{aligned}
\mathbf{P} &\stackrel{\text{def}}{=} \mathrm{E}\langle\delta\delta^T\rangle \\
&= \mathrm{E}_\mathbf{v}\langle[\mathbf{H}^T\mathbf{R}_\mathbf{v}^{-1}\mathbf{H}]^{-1}\mathbf{H}^T\mathbf{R}_\mathbf{v}^{-1}\mathbf{v}\,\mathbf{v}^T\mathbf{R}_\mathbf{v}^{-1}\mathbf{H}\,[\mathbf{H}^T\mathbf{R}_\mathbf{v}^{-1}\mathbf{H}]^{-1}\rangle \\
&= [\mathbf{H}^T\mathbf{R}_\mathbf{v}^{-1}\mathbf{H}]^{-1}\mathbf{H}^T\mathbf{R}_\mathbf{v}^{-1}\mathrm{E}_\mathbf{v}\langle\mathbf{v}\,\mathbf{v}^T\rangle\mathbf{R}_\mathbf{v}^{-1}\mathbf{H}\,[\mathbf{H}^T\mathbf{R}_\mathbf{v}^{-1}\mathbf{H}]^{-1} \\
&= [\mathbf{H}^T\mathbf{R}_\mathbf{v}^{-1}\mathbf{H}]^{-1}\mathbf{H}^T\,\underbrace{\mathbf{R}_\mathbf{v}^{-1}\mathbf{R}_\mathbf{v}}_{\mathbf{I}}\,\mathbf{R}_\mathbf{v}^{-1}\mathbf{H}\,[\mathbf{H}^T\mathbf{R}_\mathbf{v}^{-1}\mathbf{H}]^{-1} \\
&= \underbrace{[\mathbf{H}^T\mathbf{R}_\mathbf{v}^{-1}\mathbf{H}]^{-1}\mathbf{H}^T\mathbf{R}_\mathbf{v}^{-1}\mathbf{H}}_{\mathbf{I}}[\mathbf{H}^T\mathbf{R}_\mathbf{v}^{-1}\mathbf{H}]^{-1} \\
&= [\mathbf{H}^T\mathbf{R}_\mathbf{v}^{-1}\mathbf{H}]^{-1}
\end{aligned} \tag{10.23}$$

the covariance of estimation errors from the linear least-mean-squared estimator.

Recursive Linear Least Mean Squares Estimation If $\mathbf{z}_{k-1}$ and $\mathbf{z}_k$ are two independent measurements of the same variate $\mathbf{x}$ with statistically independent zero-mean measurement errors, then their respective measurement errors $\mathbf{v}_{k-1}$ and $\mathbf{v}_k$ are zero-mean with covariances

$$\mathrm{E}\langle\mathbf{v}_{k-1}\mathbf{v}_{k-1}^T\rangle = \mathbf{R}_{k-1} \tag{10.24}$$

$$\mathrm{E}\langle\mathbf{v}_k\mathbf{v}_k^T\rangle = \mathbf{R}_k \tag{10.25}$$

$$\mathrm{E}\langle\mathbf{v}_{k-1}\mathbf{v}_k^T\rangle = 0, \quad \text{the zero matrix} \tag{10.26}$$

and respective measurement sensitivity matrices $\mathbf{H}_{k-1}$ and $\mathbf{H}_k$.

Consequently the first measurement $\mathbf{z}_{k-1}$ will yield the initial estimate

$$\hat{\mathbf{x}}_{k-1(+)} = (\mathbf{H}_{k-1}^T \mathbf{R}_{k-1}^{-1} \mathbf{H}_{k-1})^{-1} \mathbf{H}_{k-1}^T \mathbf{R}_{k-1}^{-1} \mathbf{z}_{k-1} \tag{10.27}$$

with covariance of estimation error

$$\mathbf{P}_{k-1(+)} = (\mathbf{H}_{k-1}^T \mathbf{R}_{k-1}^{-1} \mathbf{H}_{k-1})^{-1} \tag{10.28}$$

from which we can extract two useful formulas:

$$\mathbf{H}_{k-1}^T \mathbf{R}_{k-1}^{-1} \mathbf{z}_{k-1} = \mathbf{P}_{k-1(+)}^{-1} \hat{\mathbf{x}}_{k-1(+)} \tag{10.29}$$

$$\mathbf{H}_{k-1}^T \mathbf{R}_{k-1}^{-1} \mathbf{H}_{k-1} = \mathbf{P}_{k-1(+)}^{-1} \tag{10.30}$$

The combined first and second measurements should then yield the linear least mean squares solution to

$$\begin{bmatrix} \mathbf{z}_{k-1} \\ \mathbf{z}_k \end{bmatrix} = \begin{bmatrix} \mathbf{H}_{k-1} \\ \mathbf{H}_k \end{bmatrix} \mathbf{x}_{k(+)} + \begin{bmatrix} \mathbf{v}_{k-1} \\ \mathbf{v}_k \end{bmatrix} \tag{10.31}$$

with measurement noise covariance

$$\mathrm{E}\left\langle \begin{bmatrix} \mathbf{v}_{k-1} \\ \mathbf{v}_k \end{bmatrix} \begin{bmatrix} \mathbf{v}_{k-1} \\ \mathbf{v}_k \end{bmatrix}^T \right\rangle = \begin{bmatrix} \mathbf{R}_{k-1} & 0 \\ 0 & \mathbf{R}_k \end{bmatrix} \tag{10.32}$$

and solution

$$\begin{aligned} \hat{\mathbf{x}}_{k(+)} &= \left\{ \begin{bmatrix} \mathbf{H}_{k-1} \\ \mathbf{H}_k \end{bmatrix}^T \begin{bmatrix} \mathbf{R}_{k-1} & 0 \\ 0 & \mathbf{R}_k \end{bmatrix}^{-1} \begin{bmatrix} \mathbf{H}_{k-1} \\ \mathbf{H}_k \end{bmatrix} \right\}^{-1} \\ &\quad \times \begin{bmatrix} \mathbf{H}_{k-1} \\ \mathbf{H}_k \end{bmatrix}^T \begin{bmatrix} \mathbf{R}_{k-1} & 0 \\ 0 & \mathbf{R}_k \end{bmatrix}^{-1} \begin{bmatrix} \mathbf{z}_{k-1} \\ \mathbf{z}_k \end{bmatrix} \\ &= \{ \mathbf{H}_{k-1}^T \mathbf{R}_{k-1}^{-1} \mathbf{H}_{k-1} + \mathbf{H}_k^T \mathbf{R}_k^{-1} \mathbf{H}_k \}^{-1} \\ &\quad \times \{ \mathbf{H}_{k-1}^T \mathbf{R}_{k-1}^{-1} \mathbf{z}_{k-1} + \mathbf{H}_k^T \mathbf{R}_k^{-1} \mathbf{z}_k \} \end{aligned} \tag{10.33}$$

Using Eqs. (10.29) and (10.30), this last result can be transformed into

$$\begin{aligned} \hat{\mathbf{x}}_{k(+)} &= \{ \mathbf{P}_{k-1(+)}^{-1} + \mathbf{H}_k^T \mathbf{R}_k^{-1} \mathbf{H}_k \}^{-1} \\ &\quad \times \{ \mathbf{P}_{k-1(+)}^{-1} \hat{\mathbf{x}}_{k-1} + \mathbf{H}_k^T \mathbf{R}_k^{-1} \mathbf{z}_k \} \end{aligned} \tag{10.34}$$

Equation (10.34) expresses the kth linear least mean squares estimate $\hat{\mathbf{x}}_{k(+)}$ in terms of the $(k-1)$ th estimate $\hat{\mathbf{x}}_{k-1(+)}$ and the kth measurement $\mathbf{z}_k$. This can be put into the form used in Kalman filtering by using a matrix inversion formula from the mid-1940s.

Duncan–Guttman Formula This is a matrix inversion formula published first by William Jolly Duncan (1984–1960) in 1945 [13], followed by Louis Eliyahu Guttman (1916–1987) in 1946 [14]:

$$(\mathbf{A} - \mathbf{U}\mathbf{D}^{-1}\mathbf{V})^{-1} = \mathbf{A}^{-1} + \mathbf{A}^{-1}\mathbf{U}\,(\mathbf{D} - \mathbf{V}\mathbf{A}^{-1}\mathbf{U})^{-1}\mathbf{V}\mathbf{A}^{-1} \tag{10.35}$$

which, with the substitutions

$$\mathbf{A} = \mathbf{P}_{k-1(+)}^{-1} \tag{10.36}$$

$$\mathbf{U} = \mathbf{H}_k^T \tag{10.37}$$

$$\mathbf{D} = -\mathbf{R}_k \tag{10.38}$$

$$\mathbf{V} = \mathbf{H}_k \tag{10.39}$$

becomes

$$(\mathbf{P}_{k-1(+)}^{-1} + \mathbf{H}_k^T\mathbf{R}_k^{-1}\mathbf{H}_k)^{-1} = \mathbf{P}_{k-1(+)} - \underbrace{\mathbf{P}_{k-1(+)}\mathbf{H}_k^T(\mathbf{R}_k + \mathbf{H}_k\mathbf{P}_{k-1(+)}\mathbf{H}_k^T)^{-1}}_{\mathbf{K}_k} \mathbf{H}_k\mathbf{P}_{k-1(+)} \tag{10.40}$$

changing Eq. (10.34) to

$$\begin{aligned}
\hat{\mathbf{x}}_{k(+)} &= \{\mathbf{P}_{k-1(+)} - \mathbf{P}_{k-1(+)}\mathbf{H}_k^T(\mathbf{R}_k + \mathbf{H}_{k(+)}\mathbf{P}_{k-1(+)}\mathbf{H}_k^T)^{-1}\mathbf{H}_k\mathbf{P}_{k-1(+)}\} \\
&\quad \times\{\mathbf{P}_{k-1(+)}^{-1}\hat{\mathbf{x}}_{k-1(+)} + \mathbf{H}_k^T\mathbf{R}_k^{-1}\mathbf{z}_k\}. \\
&= \{\mathbf{I} - \mathbf{P}_{k-1(+)}\mathbf{H}_k^T(\mathbf{R}_k + \mathbf{H}_k\mathbf{P}_{k-1(+)}\mathbf{H}_k^T)^{-1}\mathbf{H}_k\} \\
&\quad \times\{\hat{\mathbf{x}}_{k-1(+)} + \mathbf{P}_{k-1(+)}\mathbf{H}_k^T\mathbf{R}_k^{-1}\mathbf{z}_k\} \\
&= \hat{\mathbf{x}}_{k-1(+)} + \mathbf{P}_{k-1(+)}\mathbf{H}_k^T(\mathbf{R}_k + \mathbf{H}_k\mathbf{P}_{k-1(+)}\mathbf{H}_k^T)^{-1} \\
&\quad \times\{[(\mathbf{R}_k + \mathbf{H}_k\mathbf{P}_{k-1(+)}\mathbf{H}_k^T) - \mathbf{H}_k\mathbf{P}_{k-1(+)}\mathbf{H}_k^T]\mathbf{R}_k^{-1}\mathbf{z}_k - \mathbf{H}_k\hat{\mathbf{x}}_{k-1(+)}\} \\
&= \hat{\mathbf{x}}_{k-1(+)} + \underbrace{\mathbf{P}_{k-1(+)}\mathbf{H}_k^T(\mathbf{R}_k + \mathbf{H}_k\mathbf{P}_{k-1(+)}\mathbf{H}_k^T)^{-1}}_{\mathbf{K}_k} \\
&\quad \times\{[\mathbf{R}_k]\mathbf{R}_k^{-1}\mathbf{z}_k - \mathbf{H}_k\hat{\mathbf{x}}_{k-1(+)}\} \\
&= \hat{\mathbf{x}}_{k-1(+)} + \mathbf{K}_k\{\mathbf{z}_k - \mathbf{H}_k\hat{\mathbf{x}}_{k-1(+)}\}
\end{aligned} \tag{10.41}$$

Kalman Gain Matrix Equation (10.41) is almost the form used in the Kalman measurement update, except for subscription substitutions to correct the assumptions that

1. The discrete times $t_k = t_{k-1}$.
2. The state vector $\mathbf{x}$ remained unmolested by random dynamics between t_{k-1} and t_k, so that

$$\hat{\mathbf{x}}_{k(-)} = \hat{\mathbf{x}}_{k-1(+)}$$

$$\mathbf{P}_{k(-)} = \mathbf{P}_{k-1(+)}$$

With these substitutions, the measurement update has the form

$$\mathbf{K}_k = \mathbf{P}_{k(-)}\mathbf{H}_k^T(\mathbf{R}_k + \mathbf{H}_k\mathbf{P}_{k(-)}\mathbf{H}_k^T)^{-1} \tag{10.42}$$

$$\hat{\mathbf{x}}_{k(+)} = \hat{\mathbf{x}}_{k(-)} + \mathbf{K}_k\{\mathbf{z}_k - \mathbf{H}_k\hat{\mathbf{x}}_{k(-)}\} \tag{10.43}$$

$\mathbf{K}_k$ is called the *Kalman gain matrix*. It is the weighting matrix applied to the difference between the kth measurement $\mathbf{z}_k$ and the *expected measurement* $\mathbf{H}_k\hat{\mathbf{x}}_{k(-)}$ based on the a priori estimate $\hat{\mathbf{x}}_{k(-)}$. It is considered to be the *crown jewel* of the Kalman filter derivation.

The fact that the Kalman gain matrix does not depend on the measurement can be exploited by computing it in the time between measurements.

10.4.1.2 Kalman Measurement Update Equations

The Kalman measurement correction equations account for the changes in the estimate $\hat{\mathbf{x}}_k$ and its associated covariance matrix of estimation uncertainty to reflect the information gained from the most recent measurement $\mathbf{z}_k$. It can be computed more efficiently by reusing the Kalman gain in the covariance update:

$$\mathbf{K}_k \stackrel{\text{def}}{=} \mathbf{P}_{k(-)}\mathbf{H}_k^T(\mathbf{R}_k + \mathbf{H}_k\mathbf{P}_{k(-)}\mathbf{H}_k^T)^{-1} \tag{10.44}$$

$$\hat{\mathbf{x}}_{k(+)} = \hat{\mathbf{x}}_{k(-)} + \mathbf{K}_k\{\mathbf{z}_k - \mathbf{H}_k\hat{\mathbf{x}}_{k(-)}\} \tag{10.45}$$

$$\mathbf{P}_{k(+)} = \mathbf{P}_{k(-)} - \mathbf{K}_k\mathbf{H}_k\mathbf{P}_{k(-)} \tag{10.46}$$

where the subscripting indicates the sequencing of the measurements $\mathbf{z}_1, \mathbf{z}_2, \mathbf{z}_3, \dots$.

This completes the derivation of the Kalman corrector equations.

10.4.2 Time Update (Predictor)

The time update is also called the "temporal update."

The Kalman predictor uses the same unknown vector $\mathbf{x}$ from the corrector, but now addresses the question of how the estimate $\hat{\mathbf{x}}$ and its associated covariance of estimation uncertainty $\mathbf{P}$ are affected by known and unknown (i.e. random) dynamics. Like the measurement update, it also uses linearity to characterize the estimation problem using only means and covariances.

10.4.2.1 Continuous-Time Dynamics

State space is a representation of a dynamic system of n variables x_i as components of a vector $\mathbf{x}$ in n-dimensional Euclidean space $\mathbb{R}^n$. The dimension of this state space is determined by the number of degrees of freedom of the dynamic system. In the Kalman *predictor*, the same vector $\mathbf{x} \in \mathbb{R}^n$ used in the Kalman corrector becomes the vector of variables subject to changes over time – as governed by linear differential equations of the sort

$$\frac{d}{dt}\mathbf{x}(t) = \mathbf{F}(t)\,\mathbf{x}(t) + \mathbf{w}(t) \tag{10.47}$$

where

$\mathbf{x}(t)$ is a real n-vector-valued function of time (t), where n is the dimension of $\mathbf{x}$.

$\mathbf{F}(t)$ is an $n \times n$ real matrix-valued function of time.
$\mathbf{w}(t)$ is a real vector-valued function of time (t).

Homogeneous and Nonhomogeneus Differential Equations If the function $\mathbf{w}(t) = 0$, the resulting differential equation (10.47) assumes the form

$$\frac{d}{dt}\mathbf{x}(t) = \mathbf{F}(t)\,\mathbf{x}(t) \tag{10.48}$$

which is called *homogeneous*, meaning that the dependent variable $\mathbf{x}$ appears in every term of the equation. Otherwise, the differential equation (10.47) is considered to be *nonhomogeneous*.

Example 10.1 ***(mth Order Nonhomogeneous Ordinary Linear Differential Equation)***
This is a differential equation of the sort

$$\sum_{i=0}^{m} a_i(t)\left(\frac{d}{dt}\right)^i y(t) = b(t)$$

in the scalar function $y(t)$. For pedagogical purposes it can be rewritten in the form

$$\left(\frac{d}{dt}\right)^m y(t) = \frac{b(t)}{a_m(t)} - \sum_{i=0}^{m-1} \frac{a_i(t)}{a_m(t)}\left(\frac{d}{dt}\right)^i y(t) \tag{10.49}$$

so that the first-order linear differential equation for the vectorized dependent variable

$$\mathbf{x}(t) \stackrel{\text{def}}{=} \begin{bmatrix} x_1(t) \\ x_2(t) \\ x_3(t) \\ \vdots \\ x_m(t) \end{bmatrix} \stackrel{\text{def}}{=} \begin{bmatrix} y(t) \\ \frac{d}{dt}y(t) \\ \left(\frac{d}{dt}\right)^2 y(t) \\ \vdots \\ \left(\frac{d}{dt}\right)^{m-1} y(t) \end{bmatrix}$$

$$\frac{d}{dt}\mathbf{x}(t) = \begin{bmatrix} \frac{d}{dt}y(t) \\ \left(\frac{d}{dt}\right)^2 y(t) \\ \left(\frac{d}{dt}\right)^3 y(t) \\ \vdots \\ \left(\frac{d}{dt}\right)^m y(t) \end{bmatrix} = \begin{bmatrix} x_2(t) \\ x_3(t) \\ x_4(t) \\ \vdots \\ \frac{b(t)}{a_m(t)} - \sum_{i=0}^{m-1} \frac{a_i(t)}{a_m(t)}\left(\frac{d}{dt}\right)^i y(t) \end{bmatrix},$$

where we have used Eq. (10.49) so that the result can be expressed in state-space form as

$$\frac{d}{dt}\mathbf{x}(t) = \mathbf{F}(t)\mathbf{x}(t) + \mathbf{w}(t)$$

$$\mathbf{F} \stackrel{\text{def}}{=} \begin{bmatrix} 0 & 1 & 0 & \cdots & 0 \\ 0 & 0 & 1 & \cdots & 0 \\ 0 & 0 & 0 & \cdots & 0 \\ \vdots & \vdots & \vdots & \ddots & \vdots \\ -\frac{a_0(t)}{a_m(t)} & -\frac{a_1(t)}{a_m(t)} & -\frac{a_2(t)}{a_m(t)} & \cdots & -\frac{a_{m-1}(t)}{a_m(t)} \end{bmatrix}$$

$$\mathbf{w}(t) \stackrel{\text{def}}{=} \begin{bmatrix} 0 \\ 0 \\ 0 \\ \vdots \\ \frac{b(t)}{a_m(t)} \end{bmatrix}$$

Homogeneous Equation Solutions The key to solving the nonhomogeneous differential equation (10.47) depends on the solution to the corresponding homogeneous equation (10.48). In either case, the solution on a time interval $t_{k-1} \leq t \leq t_k$ can be expressed in terms of its initial value $\mathbf{x}(t_{k-1})$ on that interval, and the solution will involve the matrix exponential function.

Matrix Exponentials The exponential function exp $(\mathbf{M})$ of an $n \times n$ square matrix $\mathbf{M}$ is defined by the exponential power series

$$\exp(\mathbf{M}) \stackrel{\text{def}}{=} \sum_{m=0}^{\infty} \frac{1}{m!} \mathbf{M}^m \tag{10.50}$$

which always converges – but not fast enough to compute it that way. It is implemented more efficiently in MATLAB as the function `expm`.

If the square matrix $\mathbf{M}(t)$ is a differentiable function of t, the derivative of its exponential will be

$$\begin{aligned} \frac{d}{dt} \exp(\mathbf{M}(t)) &= \frac{d\,\mathbf{M}(t)}{dt} \exp(\mathbf{M}(t)) \\ &= \exp(\mathbf{M}(t)) \frac{d\,\mathbf{M}(t)}{dt} \end{aligned} \tag{10.51}$$

Consequently, if we let the matrix $\mathbf{M}$ be defined as an integral of $\mathbf{F}(t)$ from the homogeneous equation (10.48), then

$$\mathbf{M}(s,t) \stackrel{\text{def}}{=} \int_s^t \mathbf{F}(\tau) d\tau \tag{10.52}$$

$$\mathbf{\Phi}(s,t) \stackrel{\text{def}}{=} \exp(\mathbf{M}(s,t)) \tag{10.53}$$

$$\mathbf{\Phi}(s,s) = \mathbf{\Phi}(t,t) = \mathbf{I} \tag{10.54}$$

$$\mathbf{\Phi}(t,\tau)\mathbf{\Phi}(s,t) = \mathbf{\Phi}(s,\tau) \tag{10.55}$$

$$\mathbf{\Phi}^{-1}(s,t) = \mathbf{\Phi}(t,s) \tag{10.56}$$

$$\frac{d}{dt}\mathbf{\Phi}(s,t) = \mathbf{F}(t)\ \mathbf{\Phi}(s,t) \tag{10.57}$$

$$\frac{d}{ds}\mathbf{\Phi}(s,t) = -\mathbf{F}(s)\ \mathbf{\Phi}(s,t) \tag{10.58}$$

The matrix $\mathbf{\Phi}(s,\ t)$ is what is called the *fundamental solution matrix* for Eq. (10.48), in that it solves the associated boundary value problem as

$$\mathbf{x}(t) = \mathbf{\Phi}(s,\ t)\ \mathbf{x}(s) \tag{10.59}$$

State Transition Matrices Kalman filtering uses the discrete-time notation

$$\mathbf{\Phi}_k \stackrel{\text{def}}{=} \mathbf{\Phi}(t_{k-1},\ t_k) \tag{10.60}$$

for the *state transition matrix* $\mathbf{\Phi}_k$ between the $(k-1)$ th and kth discrete time epochs.

Nonhomogeneous Solutions Equation (10.47) with a known integrable dynamic disturbance function $\mathbf{w}(t)$ has a closed-form solution for the initial value problem in terms of the state transition matrix function as

$$\mathbf{x}(t) = \mathbf{\Phi}(s,\ t)\ \mathbf{x}(s) + \int_s^t \mathbf{\Phi}(\tau,\ t)\ \mathbf{w}(\tau)d\tau \tag{10.61}$$

where $\mathbf{x}(s)$ is the initial value on the interval $[s,\ t]$ and the integral is the ordinary (Riemann) integral.

Linear Stochastic Differential Equations One might like to generalize Eq. (10.61) for a *white noise process* $\mathbf{w}(t)$ such that

$$\mathrm{E}\langle \mathbf{w}(t)\rangle = 0 \quad \text{for all} \quad t \quad \text{(zero mean)}$$

$$\mathrm{E}\langle \mathbf{w}(t)\mathbf{w}^T(s)\rangle = \begin{cases} 0, & s \neq t \quad \text{(uncorrelated in time)} \\ \mathbf{Q}(t), & s = t \end{cases}$$

where the covariance matrix $\mathbf{Q}(t)$ of random noise uncertainty is an $n \times n$ symmetric positive definite matrix and n is the dimension of $\mathbf{w}(t)$. If that could be done, then a solution might be obtained as

$$\mathbf{x}_k = \mathbf{\Phi}_k\mathbf{x}_{k-1} + \mathbf{w}_k \tag{10.62}$$

where the discrete-time additive random disturbance vector

$$\mathbf{w}_k = \int_{t_{k-1}}^{t_k} \mathbf{\Phi}(\tau,\ t_k)\ \mathbf{w}(\tau)d\tau \tag{10.63}$$

has zero mean

$$\mathrm{E}\langle \mathbf{w}_k\rangle = \int_{t_{k-1}}^{t_k} \mathbf{\Phi}(\tau,\ t_k)\ \mathrm{E}\langle \mathbf{w}(\tau)\rangle d\tau = 0 \tag{10.64}$$

and covariance

$$\begin{aligned}
\mathbf{Q}_k &= \mathrm{E}\langle \mathbf{w}_k \mathbf{w}_k^T \rangle \\
&= \mathrm{E}\left\langle \left[\int_{t_{k-1}}^{t_k} \boldsymbol{\Phi}(\tau,\ t_k)\ \mathbf{w}(\tau) d\tau\right] \left[\int_{t_{k-1}}^{t_k} \boldsymbol{\Phi}(\sigma,\ t_k)\ \mathbf{w}(\sigma) d\sigma\right]^T \right\rangle \\
&= \int_{t_{k-1}}^{t_k} \int_{t_{k-1}}^{t_k} \boldsymbol{\Phi}(\tau,\ t_k)\ \mathrm{E}\langle \mathbf{w}(\tau)\ \mathbf{w}^T(\sigma)\rangle \boldsymbol{\Phi}^T(\sigma,\ t_k) d\tau\ d\sigma \\
&= \int_{t_{k-1}}^{t_k} \boldsymbol{\Phi}(\tau,\ t_k)\ \mathbf{Q}(\tau) \boldsymbol{\Phi}^T(\tau,\ t_k) d\tau
\end{aligned} \tag{10.65}$$

and the corresponding discrete time update for the covariance of state estimation uncertainty could then be expressed as

$$\mathbf{P}_{k(-)} = \boldsymbol{\Phi}_k \mathbf{P}_{k-1(+)} \boldsymbol{\Phi}_k^T + \mathbf{Q}_k \tag{10.66}$$

The problem with this approach is that Eq. (10.47) then becomes what is called a *stochastic*[4] *differential equation* because it includes a white noise process, which is not an integrable function in the Riemann calculus. A new calculus would be required for modeling random disturbance noise $\mathbf{w}(t)$ in Eq. (10.47).

Stochastic Integrals and Markov Processes The problem of integrating white noise processes had been studied for more than half a century [15] until the first stochastic integral was defined by Kyosi Itô (1915–2008) in 1944 [16]. The resulting stochastic integral equation is not in the same form, but it does result in what is called a **Markov process** in discrete time. This is a sequence of random vectors

$$\dots,\ \mathbf{w}_{k-1},\ \mathbf{w}_k,\ \mathbf{w}_{k+1},\ \dots$$

such that

$$\mathrm{E}\langle \mathbf{w}_k \rangle = 0 \tag{10.67}$$

$$\mathrm{E}\langle \mathbf{w}_j \mathbf{w}_k^T \rangle = \begin{cases} 0, & j \neq k \\ \mathbf{Q}_k, & j = k \end{cases} \tag{10.68}$$

where the $\mathbf{Q}_k$ are $n \times n$ symmetric positive definite covariance matrices and n is the dimension of $\mathbf{x}_k$ and $\mathbf{w}_k$. This is a somewhat generalized Markov process, in that we only care about the sequence of means and covariances of the $\mathbf{w}_k$, and their covariances can vary with k. For a derivation including the stochastic calculus step, see [17]. That is all that is needed for Kalman filtering.

10.4.2.2 Discrete-Time Dynamics

From Eqs. (10.62) and (10.68),

$$\hat{\mathbf{x}}_k = \boldsymbol{\Phi}_k\ \hat{\mathbf{x}}_{k-1(+)} \tag{10.69}$$

---

4 From the Greek word for *guess* or *aim at*.

$$\mathbf{P}_{k(-)} = \mathbf{\Phi}_k \mathbf{P}_{k-1(+)} \mathbf{\Phi}_k^T + \mathbf{Q}_k \tag{10.70}$$

where the parameters $\mathbf{\Phi}_k$ and $\mathbf{Q}_k$, and the inputs $\hat{\mathbf{x}}_{k-1(+)}$ and $\mathbf{P}_{k-1(+)}$ are given.

10.4.3 Basic Kalman Filter Equations

The top-level processing flow of the Kalman filter is shown in Figure 10.1.

The essential implementation equations for the Kalman filter are summarized in Table 10.1. In most cases, these follow the same symbolic notation used by Kalman in his original paper [2]. The main exception is that Kalman used the symbol Δ for what is now called the Kalman gain matrix and – in his honor – represented by the letter **K**.

10.4.4 The Time-Invariant Case

If the set of Kalman filter model parameters $\{\mathbf{\Phi}_k, \mathbf{Q}_k, \mathbf{H}_k, \mathbf{R}_k\}$ have the same values for every time t_k, the system is called *time-invariant* or *linear time-invariant* (LTI).

Perhaps the most common use of time-invariant Kalman filtering is for solving the steady-state riccati equations to calculate the steady-state Kalman gain $\mathbf{K}_\infty$, which can be used as the measurement feedback gain without having to solve the riccati equation. This approach can be much more efficient than full-blown Kalman filtering. Time-invariance is otherwise not that common in practice.

10.4.5 Observability and Stability Issues

Observability is the issue of whether an unknown variable can be determined from the relationships between the unknowns and knowns (i.e. the data) of an estimation problem. For linear problems, this does not generally depend on the data, but only on its relationship to the unknowns.

Least squares problems are deemed observable if and only if the associated *Gramian matrix* of Eq. (10.9) is nonsingular. This is a *qualitative* characterization of observability that does not depend on the data **z**, but may require infinite precision to be calculated precisely.

Kalman filtering observability is characterized by the resulting covariance matrix **P** of estimation uncertainty, which is a *quantitative* measure of observability depending on the problem parameters $\mathbf{H}_k$, $\mathbf{R}_k$, $\mathbf{\Phi}_k$, and $\mathbf{Q}_k$. The final value of $\mathbf{P}_k$ characterizes *how well* the state vector **x** can be determined, and it does not depend on the actual measurements $\mathbf{z}_k$. This is very important in Kalman filtering problems, because it allows the degree of observability of the estimation problem to be determined just from the problem parameters, without requiring the actual measured values $\mathbf{z}_k$.

Stability of a linear stochastic system is the issue of whether its solutions are bounded over time. For time-invariant systems, this can be characterized by the eigenvalues of the dynamic coefficient matrix **F** being in the left half of the complex plane (i.e. having negative real parts) or those of the associated state transition matrix $\mathbf{\Phi}$ being inside the unit circle in the complex plane. However, the accuracy of numerically calculated eigenvalues near the boundaries of stability may leave room for ambiguity.

Table 10.1 Basic Kalman filter equations.

Implementation equations		
Inputs	Processing	Outputs
Time update (predictor)		
$\hat{\mathbf{x}}_{k-1(+)}$, $\mathbf{P}_{k-1(+)}$	$\hat{\mathbf{x}}_{k(-)} = \mathbf{\Phi}_k \hat{\mathbf{x}}_{k-1(+)}$	$\hat{\mathbf{x}}_{k(-)}$, $\mathbf{P}_{k(-)}$
$\mathbf{\Phi}_k$, $\mathbf{Q}_k$	$\mathbf{P}_{k(-)} = \mathbf{\Phi}_k \mathbf{P}_{k-1(+)} \mathbf{\Phi}_k^T + \mathbf{Q}_k$	
Measurement update (corrector)		
$\hat{\mathbf{x}}_{k(-)}$, $\mathbf{P}_{k(-)}$	$\mathbf{K}_k = \mathbf{P}_{k(-)} \mathbf{H}_k^T (\mathbf{R}_k + \mathbf{H}_k \mathbf{P}_{k(-)} \mathbf{H}_k^T)^{-1}$	$\hat{\mathbf{x}}_{k(+)}$, $\mathbf{P}_{k(+)}$
$\mathbf{z}_k$, $\mathbf{H}_k$, $\mathbf{R}_k$	$\hat{\mathbf{x}}_{k(+)} = \hat{\mathbf{x}}_{k(-)} + \mathbf{K}_k \{\mathbf{z}_k - \mathbf{H}_k \hat{\mathbf{x}}_{k(-)}\}$	
	$\mathbf{P}_{k(+)} = \mathbf{P}_{k(-)} - \mathbf{K}_k \mathbf{H}_k \mathbf{P}_{k(-)}$	
Parameters and variables		
Symbol	Description	Dimension
n	Dimension of state vector (integer)	1
ℓ_k	Dimension of kth measurement vector (integer)	1
k	Discrete measurement time index (integer)	1
t_k	Discrete time of kth measurement (scalar)	1
$\hat{\mathbf{x}}_{k-1(+)}$	A posteriori estimate at time t_{k-1}	$n \times 1$
$\mathbf{P}_{k-1(+)}$	A posteriori estimation covariance at timet_{k-1}	$n \times n$
$\mathbf{\Phi}_k$	State transition matrix from t_{k-1} to t_k	$n \times n$
$\mathbf{Q}_k$	Uncertainty accumulated between t_{k-1} and t_k	$n \times n$
$\hat{\mathbf{x}}_{k(-)}$	A priori estimate at time t_k	$n \times 1$
$\mathbf{P}_{k(-)}$	A priori estimation covariance at time t_k	$n \times n$
$\mathbf{z}_k$	Measurement vector at time t_k	$\ell_k \times 1$
$\mathbf{H}_k$	Measurement sensitivity matrix at time t_k	$\ell_k \times n$
$\mathbf{R}_k$	Measurement error covariance at timet_k	$\ell_k \times \ell_k$
$\mathbf{K}_k$	Kalman gain matrix at time t_k	$n \times \ell_k$
$\hat{\mathbf{x}}_{k(+)}$	A posteriori estimate at time t_k	$n \times 1$
$\mathbf{P}_{k(+)}$	A posteriori estimation covariance at time t_k	$n \times n$

One of the triumphs of Kalman filtering was showing that *dynamically unstable* estimation problems can still be observable. That is, the riccati equations (characterized by $\mathbf{H}_k$, $\mathbf{R}_k$, $\mathbf{\Phi}_k$, and $\mathbf{Q}_k$) can be stable even if the dynamic equations (characterized by $\mathbf{\Phi}_k$ and $\mathbf{Q}_k$ only) are unstable.

10.5 Alternative Implementations

The first working implementation of a Kalman filter was directed by Stanley F. Schmidt (1926–2015) when he was chief of the Dynamics Analysis Branch at the NASA Ames Research Center (ARC) in Mountain View, California in the late 1950s and early 1960s [18]. Schmidt was searching for navigation methods suitable for what would become the Apollo project to send American astronauts to the moon and back. Schmidt's approach included linearization of the navigation problem about nominal trajectories, and he consulted with Kalman about Kalman's "new approach to linear filtering and prediction problems" [2]. Schmidt's efforts resolved many of the essential implementation issues, but they also exposed some remaining vulnerabilities related to accumulated computer roundoff errors, especially in solving the riccati equations for updating the covariance matrix **P** of estimation uncertainty. Something better would be needed for shoehorning the implementation into a space-qualified flight computer for the Apollo missions.

10.5.1 Implementation Issues

Performance-related issues for the implementation of Kalman filters on digital computers include:

1. *Computational complexity*, which determines the numbers of arithmetic operations required to execute one cycle of the predictor and corrector equations. This generally scales as a polynomial in the dimensions of the state vector and measurement vector. Computational complexity formulas for most of the algorithms mentioned in this chapter can be found in the book [19] by Gilbert Stewart. Complexity is often quoted as an "order," $O(n^p \ell_k^q)$, in terms of the highest powers (p and q) of the dimensions (n and ℓ_k) of the matrices involved. Inversion of an $n \times n$ square matrix, for example, is of $O(n^3)$ complexity, as is multiplication of two $n \times n$ matrices.
2. *Wordlength requirements*, which determine the number of bits of precision needed to control to acceptable levels the roundoff errors in the digital implementation of a Kalman filters. This may depend on such model attributes as the condition numbers of matrices used in the Kalman filter model, or other algebraic properties of the matrices used. This issue has been exploited greatly by what are called "square root" implementations

of the Kalman filter. Better algorithms for matrix methods robust against computer roundoff may be found in the book by Golub and Van Loan [20].

Although digital processing technologies have improved by orders of magnitude since the introduction of the Kalman filter in 1960, these issues are still important for very large-scale implementations such as GNSS system state tracking at the constellation level.

10.5.2 Conventional Implementation Improvements

These are minor modifications of the implementation shown in Table 10.1 that have been shown to reduce computational complexity.

10.5.2.1 Measurement Decorrelation by Diagonalization

Kaminski [21] discovered that the computational complexity of the conventional Kalman measurement update implementation can be reduced significantly by processing the components of vector-valued observations sequentially after factoring the measurement covariance matrix $\mathbf{R}_k$ into a form such as

$$\mathbf{R}_k = \mathbf{U}_k \mathbf{D}_k \mathbf{U}_k^T \quad \text{(factorization)} \tag{10.71}$$

$$\mathbf{D}_k = \mathbf{U}_k^{-1} \mathbf{R}_k \mathbf{U}_k^{-T} \quad \text{(diagonalization)} \tag{10.72}$$

where $\mathbf{D}_k$ is a diagonal matrix with positive elements and $\mathbf{U}_k$ is easily inverted.

The modified Cholesky decomposition algorithm implemented by the MATLAB m-file `modchol.m`, for example, results in an $\ell_k \times \ell_k$ unit upper triangular $\mathbf{U}_k$. Inversion of $\ell_k \times \ell_k$ triangular matrices is of $O(\ell_k^2)$ computational complexity, whereas general $\ell_k \times \ell_k$ matrix inversion is of $O(\ell_k^3)$ complexity. Then the alternative measurement vector and model parameters

$$\mathbf{z}_k^\star \stackrel{\text{def}}{=} \mathbf{U}_k^{-1} \mathbf{z}_k \tag{10.73}$$

$$\mathbf{H}_k^\star \stackrel{\text{def}}{=} \mathbf{U}_k^{-1} \mathbf{H}_k \tag{10.74}$$

$$\mathbf{R}_k^\star \stackrel{\text{def}}{=} \mathbf{D}_k \tag{10.75}$$

represent a measurement vector $\mathbf{z}_k^\star$ with uncorrelated errors, with corresponding measurement sensitivity matrices equal to the rows of $\mathbf{H}_k^\star$ and corresponding variances equal to the diagonal elements of $\mathbf{D}_k$. Individual elements of the vector $\mathbf{z}_k^\star$ can then be processes serially as uncorrelated scalar measurements.

The computational advantage of Kaminski's decorrelation approach [21] is

$$\frac{1}{3}\ell_k^3 - \frac{1}{2}\ell_k^2 + \frac{7}{6}\ell_k - \ell_k n + 2\ell_k^2 n + \ell_k n^2 \text{ floating-point operations (flops)}$$

where ℓ_k is the dimension of $\mathbf{z}_k$ and n is the dimension of the state vector. That is, it requires that many fewer flops to decorrelate vector-valued measurements and process the components serially.

10.5.2.2 Exploiting Symmetry

Computation requirements for computing symmetric matrices can be reduced by computing only the diagonal elements and the off-diagonal elements on one side of the main diagonal. For example, this would apply to the following terms from the conventional Kalman filter implementation in Table 10.1:

$$\mathbf{P}_{k(-)} = \mathbf{\Phi}_k \mathbf{P}_{k-1(+)} \mathbf{\Phi}_k^T + \mathbf{Q}_k, \ (\mathbf{R}_k + \mathbf{H}_k \mathbf{P}_{k(-)} \mathbf{H}_k^T)^{-1}, \ \mathbf{P}_{k(+)} = \mathbf{P}_{k(-)} - \mathbf{K}_k \mathbf{H}_k \mathbf{P}_{k(-)}$$

Furthermore, Verhaegen and Van Dooren [22] have shown that this "forced symmetry" also mitigates some cumulative effects of computer roundoff errors.

10.5.2.3 Information Filter

The inverse of the covariance matrix of estimation uncertainty is called an *information matrix*:

$$\mathbf{Y} \stackrel{\text{def}}{=} \mathbf{P}^{-1} \tag{10.76}$$

a concept first defined in a broader context by Ronald Aylmer Fisher (1890–1962) in 1925 [23]. This alternative representation can be useful in some situations – such as those in which there is essentially *no* information in some subspace of state space. This can occur during start-up with incomplete state estimates, for example. In that case the linear least mean squares estimate $\hat{\mathbf{x}}$ satisfying

$$\mathbf{z} = \mathbf{H}\mathbf{x} + \mathbf{v}$$
$$\mathrm{E}\langle \mathbf{v}\mathbf{v}^T \rangle = \mathbf{R}$$

is *underdetermined* because its Gramian $\mathbf{H}^T\mathbf{H}$ is singular. However, the associated information matrix

$$\mathbf{Y} = \mathbf{H}\mathbf{R}^{-1}\mathbf{H}^T$$

is computable, even though some of its eigenvalues are 0. In this case, the corresponding eigenvalues of $\mathbf{P}$ would be infinite.

The Kalman filter implementation in terms of $\mathbf{Y}$ greatly simplifies the measurement update but complicates the time update. It is also less transparent because the information matrix needs to be inverted to understand and evaluate estimation uncertainties quantitatively, and because the state estimate for the information filter is redefined as

$$\hat{\mathbf{y}} \stackrel{\text{def}}{=} \mathbf{Y}\hat{\mathbf{x}} \tag{10.77}$$

The information filter implementation in terms of $\hat{\mathbf{y}}$ and $\mathbf{Y}$ is listed in Table 10.2. Note that this implementation assumes the state transition matrices $\mathbf{\Phi}_k$ and dynamic disturbance covariance matrices $\mathbf{Q}_k$ are nonsingular.

Table 10.2 Information filter equations.

Information filter variables

$$\mathbf{Y} \stackrel{\text{def}}{=} \mathbf{P}^{-1} \qquad \mathbf{y} \stackrel{\text{def}}{=} \mathbf{Y}\,\mathbf{x}$$

Implementation equations

Predictor	Corrector
$\mathbf{C}_k = \mathbf{\Phi}_k^{-1}\mathbf{Y}_{k-1(+)}\mathbf{\Phi}_k^{-T}$	$\mathbf{A}_k = \mathbf{H}_k^T\mathbf{R}_k^{-1}$
$\mathbf{D}_k = \mathbf{C}_k[\mathbf{C}_k + \mathbf{Q}_k^{-1}]^{-1}$	$\mathbf{B}_k = \mathbf{A}_k^T\mathbf{H}_k^T$
$\mathbf{E}_k = \mathbf{I} - \mathbf{D}_k$	$\hat{\mathbf{y}}_{k(+)} = \hat{\mathbf{y}}_{k(-)} + \mathbf{A}_k\mathbf{z}_k$
$\hat{\mathbf{y}}_{k(-)} = \mathbf{E}_k\mathbf{\Phi}_k^{-T}\hat{\mathbf{y}}_{k-1(+)}$	$\mathbf{Y}_{k(+)} = \mathbf{Y}_{k(-)} + \mathbf{B}_k$
$\mathbf{Y}_{k(-)} = \mathbf{E}_k\mathbf{C}_k\mathbf{E}_k^T + \mathbf{D}_k\mathbf{Q}_k^{-1}\mathbf{D}_k^T$	

10.5.2.4 Sigma Rho Filtering

This implementation by Grewal and Kain [24] uses the alternative covariance factorization

$$\mathbf{C} = \mathbf{D}_\sigma \boldsymbol{P} \mathbf{D}_\sigma \tag{10.78}$$

$$\mathbf{D}_\sigma = \text{diag}(\sigma_1, \sigma_2, \sigma_3, \ldots, \sigma_n) \tag{10.79}$$

$$\sigma_i > 0 \quad \text{for all} \quad 1 \le i \le n \tag{10.80}$$

$$\boldsymbol{P} = \begin{bmatrix} 1 & \rho_{12} & \rho_{13} & \cdots & \rho_{1n} \\ \rho_{21} & 1 & \rho_{23} & \cdots & \rho_{2n} \\ \rho_{31} & \rho_{32} & 1 & \cdots & \rho_{3n} \\ \vdots & \vdots & \vdots & \ddots & \vdots \\ \rho_{n1} & \rho_{n2} & \rho_{n3} & \cdots & 1 \end{bmatrix} \tag{10.81}$$

where $\mathbf{C} = \mathbf{R}$ or $\mathbf{P}$, not to be confused with the capital of ρ ("rho"), which is $\boldsymbol{P}$ ("Rho").

The σ_i are the RMS uncertainties of the measurement or state vector components and the ρ_{ij} are the correlation coefficients between the errors of the ith and jth components. For more detail see [25].

10.5.3 James E. Potter (1937–2005) and Square Root Filtering

A major breakthrough came in 1962, when Potter was a staff member at the MIT Instrumentation Laboratory (later reorganized and renamed *The Charles Stark Draper Laboratory*), working on the MIT-designed navigation system for the Apollo missions. Studies on mainframe computers with 36-bit floating point arithmetic had shown that a Kalman filter implemention was adequate for the task, but it had to be implemented on the Apollo Computer, a beyond-state-of-the-art special-purpose flight computer with 15-bit fixed-point arithmetic. Potter took the problem home with him on a Friday and returned with a solution the following Monday.

Potter's idea was to redefine the Kalman filter implementation in terms of a square root $\mathbf{P}^{1/2}$ of the covariance matrix $\mathbf{P}$ of estimation uncertainty,

$$\mathbf{P} = \mathbf{P}^{1/2}\mathbf{P}^{T/2} \tag{10.82}$$

Potter's definition of a matrix square root is at odds with standard mathematical nomenclature, which defines the square of a matrix $\mathbf{S}$ as $\mathbf{S}^2 = \mathbf{SS}$ – without the second $\mathbf{S}$ being transposed. After the success of Potter's approach, however, the term "square root" for a solution $\mathbf{P}^{1/2}$ of Eq. (10.82) has become common usage in Kalman filtering.

Potter's implementation of the Kalman filter in terms of $\mathbf{P}^{1/2}$ has come to be called *the Potter square root filter*. It has been characterized as "achieving the same accuracy with half as many bits" [26], and it has spawned a host of alternative Kalman filter implementations[5] with improved robustness against computer roundoff errors. These have come to utilize a variety of methods for creating and manipulating matrix square roots.

10.5.4 Square Root Matrix Manipulation Methods

These are digital computation methods for linear algebra problems in square root filtering that have been found to be particularly robust against computer roundoff error.

10.5.4.1 Cholesky Decomposition

Factorization or **decomposition** refers to methods for factoring matrices as products of two or more matrices with useful properties. "Decomposition" is the older established term in mathematics, but some objected to its meaning outside mathematics and substituted "factorization" instead.

André-Louis Cholesky (1875–1918) was a French mathematician, geodesist, and artillery officer killed in WWI and credited [29] with the discovery of an improved solution to the linear least-squares problem in geodesy (determining the shape of geopotential surfaces). Cholesky's algorithms factored the $n \times n$ Gramian matrix

$$\mathbf{G} = \mathbf{H}^T\mathbf{H} \tag{10.83}$$

5 See, e.g. [27] and [28] for a sampling of alternative implementations, including some with cross correlation between $\mathbf{w}_k$ and $\mathbf{v}_k$.

of Eq. (10.9) as a symmetric product of a triangular[6] matrix $\mathbf{\Delta}$ *with reversal of the order of the transposed factor*[7]:

$$\mathbf{G} = \mathbf{\Delta}\mathbf{\Delta}^T \tag{10.84}$$

$$\mathbf{\Delta} = \begin{bmatrix} \delta_{11} & \delta_{12} & \delta_{13} & \cdots & \delta_{1n} \\ 0 & \delta_{22} & \delta_{23} & \cdots & \delta_{2n} \\ 0 & 0 & \delta_{33} & \cdots & \delta_{3n} \\ \vdots & \vdots & \vdots & \ddots & \vdots \\ 0 & 0 & 0 & \cdots & \delta_{nn} \end{bmatrix} \quad \text{(upper triangular)} \tag{10.85}$$

or

$$= \begin{bmatrix} \delta_{11} & 0 & 0 & \cdots & 0 \\ \delta_{21} & \delta_{22} & 0 & \cdots & 0 \\ \delta_{31} & \delta_{32} & \delta_{33} & \cdots & 0 \\ \vdots & \vdots & \vdots & \ddots & \vdots \\ \delta_{n1} & \delta_{n2} & \delta_{n3} & \cdots & \delta_{nn} \end{bmatrix} \quad \text{(lower triangular)} \tag{10.86}$$

Cholesky's algorithm for computing the triangular factors in Eq. (10.84) came to be called *Cholesky decomposition* or *Cholesky factorization*, and the triangular factors $\mathbf{\Delta}$ are called *Cholesky factors*.

The built-in MATLAB function `chol` returns the upper triangular Cholesky factor $\mathbf{\Delta}^\star$ for the alternative factorization

$$\mathbf{G} = \mathbf{\Delta}^{\star T}\mathbf{\Delta}^\star$$

which is the lower triangular solution to Eq. (10.84). Therefore, `Delta = chol(G)';` should return the **lower triangular** solution of Eq. (10.84).

The **upper triangular** solution of Eq. (10.84) is returned by the MATLAB function `utchol` in the m-file `utchol.m` on www.wiley.com/go/grewal/gnss.

10.5.4.2 Modified Cholesky Decomposition

This is a factoring of the form

$$\mathbf{P} = \mathbf{U}\mathbf{D}\mathbf{U}^T \tag{10.87}$$

where $\mathbf{D}$ is a diagonal matrix with positive diagonal elements and $\mathbf{U}$ is a *unit triangular* matrix.

---

6 Linear solutions of triangular systems require $O(n^2)$ arithmetic operations, compared with $O(n^3)$ for non-triangular systems, which tends to favor using triangular square roots of matrices wherever it is possible.

7 Reversal of the order of the transposed factor is required here for cleaving to Potter's definition of his matrix square root. Cholesky had things the other way around.

10.5.4.3 Nonuniqueness of Matrix Square Roots

Matrix square roots (defined in Eq. (10.1)) are not unique and can be transformed into one another by orthogonal transformations (Eq. (10.3)). This relationship with orthogonal transformations is exploited in square root filter implementations.

10.5.4.4 Triangularization by QR Decomposition

This is a factorization of a matrix $\mathbf{M}$ into a matrix product $\mathbf{M} = \mathbf{QR}$, where $\mathbf{Q}$ is orthogonal and $\mathbf{R}$[8] is triangular. It is usually traced to the Gram–Schmidt orthogonalization methods of Jørgen Pedersen Gram [30] and Erhard Schmidt [31], although it may have had its roots in works by Cauchy and Laplace, and its numerical stability has been improved by methods due to Wallace Givens [32] and Alston Householder [33]. Also, a modified form of Gram–Schmidt orthogonalization was used by Catherine L. Thornton [34] for implementing the time update of the Bierman–Thornton UD filter (described in the following text).

QR decomposition was originally derived with the orthogonal matrix factor coming first but has been generalized to allow it to come last – the form needed in square root filtering. The MATLAB m-file `housetri.m` on www.wiley.com/go/grewal/gnss performs the triangularization of a matrix $\mathbf{M}$ through a series of Householder transformations $\mathbf{\Omega}_j$ such that the result

$$\begin{bmatrix} \mathbf{\Delta} & 0 \end{bmatrix} = \mathbf{M} \times \mathbf{\Omega}_1 \times \mathbf{\Omega}_2 \times \cdots \times \mathbf{\Omega}_N \tag{10.88}$$

where $\mathbf{\Delta}$ is a triangular matrix. Applied to the matrix $\mathbf{P}^{1/2}_{k(-)}$ from Eq. (10.89), it returns the upper triangular square root of $\mathbf{P}_{k(-)}$ for the Potter square root filter with dynamic disturbance noise.

10.5.4.5 Householder Triangularization

Alston S. Householder (1904–1993) devised a robust method for triangularizing rectangular matrices using symmetric orthogonal matrices of the sort

$$\mathbf{\Omega}_H(\mathbf{u}) = \mathbf{I} - \frac{2\mathbf{u}\mathbf{u}^T}{|\mathbf{u}|^2}$$

with carefully selected vectors $\mathbf{u}$ [33]. His algorithm is implemented in the m-file `housetri.m`, on www.wiley.com/go/grewal/gnss.

10.5.5 Alternative Square Root Filter Implementations

10.5.5.1 Potter Implementation

Potter's derivation of the Kalman filter in terms of a square root of the covariance matrix assumed scalar measurements and did not include dynamic disturbance noise (not an issue for space travel without Klingons). Measurement vector decorrelation (described in Section 10.5.2.1) takes care of

---

8 We will use different notation to avoid confusion with Kalman's use of $\mathbf{Q}$ for the covariance of dynamic disturbance noise and $\mathbf{R}$ for the covariance of measurement noise.

the scalar measurement issue and triangularization methods (Section 10.5.4.4) take care of the dynamic disturbance noise issue.

Measurement Update Potter's algorithm is implemented in the m-file `potter.m` on www.wiley.com/go/grewal/gnss. It does not preserve triangularity of the matrix square root, but that is corrected by the time update.

Time Update This starts with the block matrix[9]

$$\mathbf{B}_{k(-)} = \left[\mathbf{\Phi}_k \mathbf{P}^{1/2}_{k-1(+)} \ \ \mathbf{Q}^{1/2}_k \right] \tag{10.89}$$

the "Potter square" of which

$$\begin{aligned}\mathbf{B}_{k(-)}\mathbf{B}^T_{k(-)} &= \mathbf{\Phi}_k \mathbf{P}^{1/2}_{k-1(+)} \mathbf{P}^{T/2}_{k-1(+)} \mathbf{\Phi}^T_k + \mathbf{Q}^{1/1}_k \mathbf{C}^{T/2}_k \\ &= \mathbf{\Phi}_k \mathbf{P}_{k-1(+)} \mathbf{\Phi}^T_k + \mathbf{Q}_k \\ &= \mathbf{P}_{k(-)}\end{aligned} \tag{10.90}$$

Therefore, triangularization of $\mathbf{B}_{k(-)}$ by multiplying it on the right by orthogonal transformations will produce

$$\mathbf{B}_{k(-)}\mathbf{\Omega} = \left[\mathbf{P}^{1/2}_{k(-)} \ \ 0 \right] \tag{10.91}$$

where $\mathbf{P}^{1/2}_{k(-)}$ is triangular. The implementation in the m-file `housetri.m` on www.wiley.com/go/grewal/gnss produces an upper triangular result.

10.5.5.2 Carlson "Fast Triangular" Square Root Filter

This algorithmic implementation by Neal A. Carlson [36] maintains the square root matrix in triangular form, which reduces the computational requirements relative to that of the Potter implementation. It essentially modifies the Potter "rank-one modification" algorithm for the measurement update so that it maintains triangularity of the square root matrix. This cuts the computational requirements nearly in half. The MATLAB m-file `carlson.m` on www.wiley.com/go/grewal/gnss implements Carlson's measurement update. The time update is the same as Potter's.

10.5.5.3 Bierman–Thornton UD Filter

An alternative square root filter implementation that avoids the numerical calculation of square roots was developed by Gerald J. Bierman (1941–1987) and Catherine L. Thornton. It uses the alternative factorization

$$\mathbf{P} = \mathbf{U}\mathbf{D}\mathbf{U}^T \tag{10.92}$$

where the matrix $\mathbf{U}$ is *unit triangular* and $\mathbf{D}$ is a diagonal matrix with positive diagonal elements. Unit triangular matrices have ones along their main

9 Equation (10.89) is due to Schmidt [35].

diagonal and zeros on one side of it. In this case, the zeros are below the main diagonal and **U** is called *unit upper triangular*. The time update was derived by Thornton [34], based on what is called *modified Gram–Schmidt orthogonalization*. It is implemented in the m-file `thornton.m` on www.wiley .com/go/grewal/gnss. The measurement update is implemented in the m-file `bierman.m`. See [26] for a more detailed examination of its performance characteristics.

The m-file `udu.m` factors a symmetric positive definite matrix into its **U** and **D** factors. The m-file `UD_decomp.m` does this in-place, storing **D** on the main diagonal (which is otherwise known to contain only ones) and the other elements of **U** above the main diagonal.

Avoidance of scalar square roots can make a difference in implementations on processors without arithmetic square root instructions or microcoded square roots. The condition number of **U** in this decomposition tends to be relatively favorable for numerical stability, but that of **D** may scale up as that of **P**. This issue can be addressed in fixed-point implementations by rescaling.

10.5.5.4 Unscented Square Root Filter

The unscented Kalman filter (UKF) is a sampling-based approximation for nonlinear propagation means and square roots of covariance matrices ($\mathbf{P}^{1/2}$ or $\mathbf{R}^{1/2}$) using a set of sample values structured around the starting values of the estimate and its covariance. The square-root implementation of this filter approximates the square root of the nonlinearly transformed covariance using similar sampling strategies. This is discussed in Sections 10.6 and 10.6.5.3.

10.5.5.5 Square Root Information Filter (SRIF)

The *square root information filter* is usually abbreviated as *SRIF*. (The conventional square root filter is often abbreviated as *SRCF*, which stands for *square-root covariance filter*.) Like the SRCF, the SRIF is more robust against roundoff errors than the inverted covariance form of the filter.

Like most square root filters using triangularization, it admits more than one solution form. A complete formulation (i.e. including both updates) of the SRIF was developed by Dyer and McReynolds [37], using triangularization methods developed by Golub and Van Loan [20] and applied to sequential least-squares estimation by Lawson and Hanson [38]. The form developed by Dyer and McReynolds is shown in Table 10.3 [Ref. [18] in Chapter 4].

10.6 Nonlinear Approximations

For many of the estimation problems of practical interest, linearity tends to be more of an exception than the rule. However, this has not dissuaded

Table 10.3 Square-root information filter using triangularization. $\mathbf{\Omega}_{\Delta\text{meas}}$ and $\mathbf{\Omega}_{\Delta\text{temp}}$ are orthogonal matrices which lower triangularize the left-hand side matrices. Submatrices represented by "■" on the right-hand sides are extraneous.

Measurement update

$$\begin{bmatrix} \mathbf{Y}_{k(-)} & \mathbf{H}_k^T \mathbf{R}_k^{-1/2} \\ \hat{\mathbf{y}}_{k(-)}^T & \mathbf{z}_k^T \mathbf{R}_k^{-1/2} \end{bmatrix} \mathbf{\Omega}_{\Delta\text{meas}} = \begin{bmatrix} \mathbf{Y}_{k(+)} & 0 \\ \hat{\mathbf{y}}_{k(+)}^T & \blacksquare \end{bmatrix}$$

Time update

$$\begin{bmatrix} \mathbf{Q}_k^{-1/2} & -\Phi_k^{-T} \mathbf{Y}_{k(+)}^{1/2} \\ 0 & \Phi_k^{-T} \mathbf{Y}_{k(+)}^{1/2} \\ 0 & \hat{\mathbf{y}}_{k(+)}^T \end{bmatrix} \mathbf{\Omega}_{\Delta\text{temp}} = \begin{bmatrix} \blacksquare & 0 \\ \blacksquare & \mathbf{Y}_{k+1(-)}^{1/2} \\ \blacksquare & \hat{\mathbf{y}}_{k+1(-)}^T \end{bmatrix}$$

problem-solvers from using approximations to apply the more successful linear estimation methods to less-than-linear problems.

In the case of the Kalman filter, this generally involves transformations of the means and covariances of error distributions, the bread-and-butter of Kalman filtering. This approximation problem is "ill posed" in the mathematical sense. That is, there may be no unique solution. This is because nonlinear transformations of different PDFs with identical means and covariances can result in PDFs with distinctly different means and covariances. This issue of what might be called "ambiguity errors" is addressed in Appendix C (on www.wiley.com/go/grewal/gnss).

A method for assessing the magnitude of linearization errors is presented in the subsection 10.6.1, followed by some nonlinear implementations and approximation methods that have been used in Kalman filtering with relative success.

10.6.1 Linear Approximation Errors

The essential idea behind linear approximation of nonlinear PDF transformations is that, within reasonably expected variations of the state vector from its estimated value (as determined by the covariance of state estimation uncertainty), the mean-squared errors due to linearization should be dominated by the modeled uncertainties due to other sources. For measurement nonlinearities, the modeled uncertainties are characterized by the measurement noise covariance $\mathbf{R}$. For dynamic nonlinearities, the modeled uncertainties are partly characterized by the covariance matrix $\mathbf{P}$ of estimation errors and by the dynamic disturbance noise covariance $\mathbf{Q}$. The range of perturbations under which these conditions need to be met can be specified in terms of the expected magnitude of uncertainty in the estimate (e.g. within $\pm 3\sigma$).

The resulting statistical conditions for linearization can be stated in the following manner:

1. For the time state transition function $\boldsymbol{\phi}(\mathbf{x})$, the linear approximation error should be insignificant compared with $\mathbf{Q}$ when the state vector variations $\boldsymbol{\delta}_x$ of $\hat{\mathbf{x}}$ are statistically significant. This condition can be met if the values of $\boldsymbol{\delta}_x$ are smaller than the $N\sigma$-values of the estimated distribution of uncertainty in the estimate $\hat{\mathbf{x}}$, which is characterized by the covariance matrix $\mathbf{P}$, for $N \geq 3$. That is, for

$$(\boldsymbol{\delta}_x)^T \mathbf{P}^{-1}(\boldsymbol{\delta}_x) \leq N^2 \tag{10.93}$$

the linear approximation error

$$\varepsilon_{\phi}(\boldsymbol{\delta}_x) \stackrel{\text{def}}{=} \underbrace{\boldsymbol{\phi}(\hat{\mathbf{x}} + \boldsymbol{\delta}_x) - \left[\boldsymbol{\phi}(\hat{\mathbf{x}}) + \left.\frac{\partial \boldsymbol{\phi}}{\partial \mathbf{x}}\right|_{\hat{\mathbf{x}}} \boldsymbol{\delta}_x \right]}_{\text{approximation error}} \tag{10.94}$$

should be bounded by

$$\varepsilon_{\phi}^T(\boldsymbol{\delta}_x) \mathbf{Q}^{-1} \varepsilon_{\phi}(\boldsymbol{\delta}_x) \ll 1 \tag{10.95}$$

for $\boldsymbol{\delta}_x$ covering most of its expected range, in which case the nonlinear approximation errors should be dominated by modeled dynamic uncertainty.

2. *For the measurement/sensor transformation* $\boldsymbol{h}(\mathbf{x})$. for $N\sigma \geq 3\sigma$ perturbations of $\hat{\mathbf{x}}$, the linear approximation error is insignificant compared with $\mathbf{R}$. That is, for some $N \geq 3$, for all perturbations $\boldsymbol{\delta}_x$ of $\hat{\mathbf{x}}$ such that

$$(\boldsymbol{\delta}_x)^T \mathbf{P}^{-1}(\boldsymbol{\delta}_x) \leq N^2 \tag{10.96}$$

$$\varepsilon_{h}(\boldsymbol{\delta}_x) \stackrel{\text{def}}{=} \underbrace{h(\hat{\mathbf{x}} + \boldsymbol{\delta}_x) - \left[\boldsymbol{h}(\hat{\mathbf{x}}) + \left.\frac{\partial \boldsymbol{h}}{\partial \mathbf{x}}\right|_{\hat{\mathbf{x}}} \boldsymbol{\delta}_x \right]}_{\text{approximation error}} \tag{10.97}$$

$$\varepsilon_{h}^T(\boldsymbol{\delta}_x) \mathbf{R}^{-1} \varepsilon_{h}(\boldsymbol{\delta}_x) \ll 1 \tag{10.98}$$

for $\boldsymbol{\delta}_x$ covering most of its expected range, in which case the nonlinear approximation errors should be dominated by modeled measurement uncertainty.

The value of estimation uncertainty covariance $\mathbf{P}$ used in Eq. (10.96) would ordinarily be the a priori value, calculated before the measurement is used. If the measurement update uses what is called the *iterated extended Kalman filter* (IEKF), however, the a posteriori value can be used.

The range of variation $\boldsymbol{\delta}_x$ used in calculating the linear approximation error should encompass the "Nσ points" defined in terms of the eigenvalue–eigenvector decomposition of **P** as

$$\mathbf{P} = \sum_{i=1}^{n} \sigma_i^2 \mathbf{e}_i \mathbf{e}_i^T \tag{10.99}$$

$$\sigma_i^2 \stackrel{\text{def}}{=} i\text{th eigenvalue of } \mathbf{P}$$

$$\mathbf{e}_i \stackrel{\text{def}}{=} i\text{th eigenvector of } \mathbf{P}$$

$$\boldsymbol{\delta}_{xi} \stackrel{\text{def}}{=} \begin{cases} N\sigma_i \mathbf{e}_i, & 1 \leq i \leq n \\ 0, & i = 0 \\ -N\sigma_{-i}\mathbf{e}_{-i}, & -n \leq i \leq -1 \end{cases} \tag{10.100}$$

which would be located at the major axes on the manifold of the $N\sigma$ equiprobability hyperellipse if the PDF were Gaussian, but otherwise represent the principal $N\sigma$ values of $\boldsymbol{\delta}$ associated with **P**.

The mean-squared linearization errors calculated using the Nσ points could then be approximated as

$$\mathrm{E}\langle |\varepsilon_\phi|^2 \rangle \approx \frac{1}{2N+1} \sum_{i=-N}^{N} |\varepsilon_\phi(\boldsymbol{\delta}_{xi})|^2 \tag{10.101}$$

$$\mathrm{E}\langle |\varepsilon_h|^2 \rangle \approx \frac{1}{2N+1} \sum_{i=-N}^{N} |\varepsilon_h(\boldsymbol{\delta}_{xi})|^2 \tag{10.102}$$

Example 10.2 ***(Nonlinearity of satellite pseudorange measurements)***
The GNSS measurement sensitivity matrix is based on the partial derivatives of the measured pseudorange $\rho_\psi = |\boldsymbol{\psi}|$ with respect to the location of the receiver antenna, with the partial derivative evaluated at the estimated receiver antenna location.

One can use Eq. (10.97) to determine whether the pseudorange measurement is sufficiently linear, given the uncertainty in the receiver antenna position, the nonlinear pseudorange measurement model, and the uncertainty in the pseudorange measurement.

This was evaluated for satellite elevation angles of 0°, 30°, 60°, and 90° versus RMS position error on the three axes. The resulting 3σ nonlinearity errors for these perturbations are plotted versus position error in Figure 10.2 as a function of σ_{pos} for each of the four satellite elevation angles above the horizon. The four barely distinguishable solid diagonal lines in the plot are for these four different satellite elevation angles, which apparently have little influence on nonlinearity errors. The horizontal line represents the RMS pseudorange noise, indicating that nonlinear approximation errors are dominated by pseudorange noise for 3σ position uncertainties less than ~2 km.

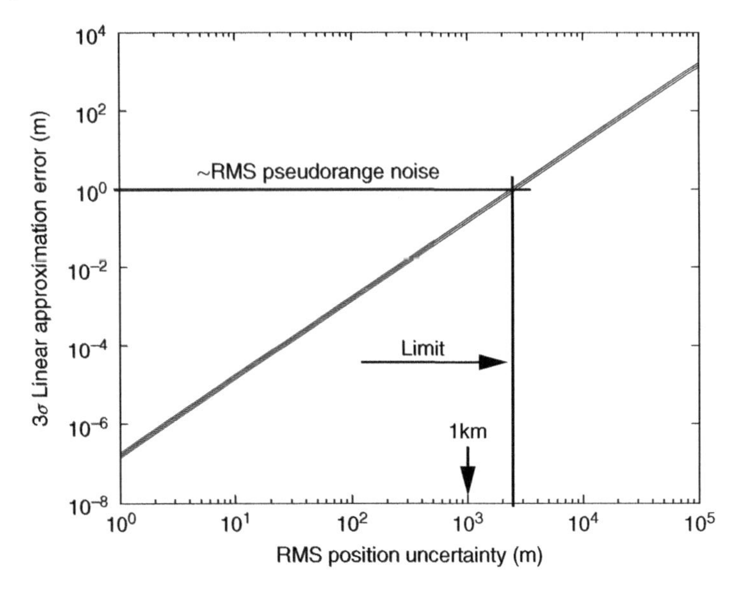

Figure 10.2 Linearization error analysis of pseudorange measurements.

10.6.2 Adaptive Kalman Filtering

This is a special class of nonlinear Kalman filtering problems in which some elements of the state vector are also parameters of the Kalman filter model. For example, making the scale factor of a sensor a part of the state vector makes the measurement a product of state vector components. This kind of nonlinearity, in particular, has sometimes been difficult to accommodate using extended Kalman filtering.

10.6.3 Taylor–Maclauren Series Approximations

Taylor–Maclauren series approximations for nonlinear Kalman filtering was suggested by Bucy [39] in 1965, after first-order (linear) approximations had been used in estimation for more than a century. Gauss first used his method of linear least squares in 1801 for solving a nonlinear problem: estimating the trajectory in space of the dwarf planet Ceres, and a first-order linearization of the Kalman filter had been introduced by Stanley F. Schmidt around 1960.

The Taylor–Maclauren series expansion of an infinitely continuously differentiable scalar function f of a scalar variable x in the vicinity of $x = x_0$ has the form

$$f(x) = \sum_{\ell=0}^{\infty} \frac{1}{\ell!} \left. \frac{\partial^{\ell} f}{\partial x^{\ell}} \right|_{x=x_0} (x - x_0)^{\ell} \tag{10.103}$$

However, when the variable and the function are n-vector-valued the equivalent mathematical representations for the ℓth-order derivatives become multidimensional structures with dimensions

$$\underbrace{n \times n \times n \times \cdots \times n}_{\ell+1 \text{ times}}$$

and the complexity of the resulting formulas soon become overwhelming. Therefore, the summed series is usually truncated after the first-order or second-order term.

10.6.3.1 First-Order: Extended Kalman Filter

Schmidt used first-order partial derivatives *evaluated at the estimated value of the state vector* for what is now called the *extended* Kalman filter. The possibly altered implementation equations are listed in Table 10.4. However, only the nonlinear parts need be altered. That is, if only the measurement is nonlinear, then only that part needs to be altered, and similarly for the time update.

Iterated Extended Kalman Filtering The measurement update can be iterated, using the partial derivative of $\mathbf{h}_k$ reevaluated at $\mathbf{x}_{k(+)}$ for a second guess at the approximation to $\mathbf{H}_k$.

10.6.3.2 Second-Order: Bass–Norum–Schwartz Filter

A second-order Taylor–Maclauren series approximation for nonlinear Kalman filtering was derived by Bass et al. [40, 41] in 1965. Its computational complexity and potential numerical instability have limited its acceptability, however [41].

10.6.4 Trajectory Perturbation Modeling

In the earliest application of the Kalman filter, Stanley F. Schmidt had linearized the space navigation problem by using Newton's laws to compute the first-order variations in the future position and velocity of a spacecraft due to variations

Table 10.4 Extended Kalman filter modifications

Extended Kalman filter: altered implementation equations	
Measurement update	Time update
$\nu_k \approx \mathbf{z}_k - \mathbf{h}_k(\mathbf{x}_{k(-)})$	$\hat{\mathbf{x}}_{k(-)} \approx \mathbf{f}_k(\hat{\mathbf{x}}_{k-1(+)})$
$\mathbf{H}_k \approx \left.\dfrac{\partial \mathbf{h}_k}{\partial \mathbf{x}}\right\|_{\mathbf{x}=\hat{\mathbf{x}}_{k(-)}}$	$\boldsymbol{\Phi}_k \approx \left.\dfrac{\partial \mathbf{f}_k}{\partial \mathbf{x}}\right\|_{\mathbf{x}=\hat{\mathbf{x}}_{k-1(+)}}$

in the position and velocity at the current time. This produced a 6×6 state transition matrix Φ for a Kalman filter model in which the state vector elements are the six perturbations of position and velocity from the reference trajectory.

In the case of orbital trajectories, his first-order linearized model could be simplified to the two-body models of Kepler, and the state transition matrices could then be approximated by analytical partial derivatives based on Kepler's equations. There is no equivalent analytical model for more than two bodies, however.

10.6.5 Structured Sampling Methods

These could be characterized as *extended Kalman filtering*, in that they extended the idea of nonlinear transformation of the estimate to include nonlinear transformation of the estimate and structured samples representing an approximation of its covariance matrix of estimation uncertainty.

Sampling-based methods got their start in 1946 when mathematician Stanislaw Ulam (1909–1984) was playing Canfield solitaire while recovering from viral encephalitis, and pondered what were his odds of winning. He soon found the problem too complicated to work out mathematically with pencil and paper, thought of the way gamblers determine odds by playing with many independently shuffled decks, imagined doing it faster using computers, and discovered what is now called Monte Carlo analysis [42]. This became a powerful tool for numerical approximation of nonlinear transformation of PDFs. It was originally based on using independent random samples from known probability distributions and used for estimating what happens to those probability distributions when the variates undergo nonlinear transformations.

Sampling-based estimation methods have since come to include the use of samples structured in such a way as to reduce the numbers of samples required, but these still generally depend on knowing more about the probability distributions involved than just their means and covariances. Nevertheless, some using only means and covariances have performed well compared with extended Kalman filtering in nonlinear applications.

The term "particle filter" generally refers to Kalman filters using structured sampling of some sort, with the terms "sigma point" or "unscented" referring to specific forms of sampling and processing. When used in Kalman filtering, the samples represent a PDF with mean equal to the estimate and covariance representing the second central moment about the mean. Therefore, they will generally be of the form

$$\sigma_i = \hat{\mathbf{x}} + \delta_i$$

where δ is a sample from a zero-mean PDF. Then the transformed mean is approximated as a weighted sum of the transformed samples and the transformed covariance is approximated by a weighted sum of the relevant outer products:

$$\mathbf{y} \stackrel{\text{def}}{=} \mathbf{f}(\mathbf{x}) + \mathbf{w} \tag{10.104}$$

$$\begin{aligned}\hat{\mathbf{y}} &\stackrel{\text{def}}{=} \mathrm{E}\langle \mathbf{y} \rangle \\ &\approx \sum_i \zeta_i \mathbf{f}(\boldsymbol{\sigma}_i)\end{aligned} \tag{10.105}$$

$$\begin{aligned}\mathbf{P}_{yy} &\stackrel{\text{def}}{=} \mathrm{E}\langle (\mathbf{y} - \hat{\mathbf{y}})(\mathbf{y} - \hat{\mathbf{y}})^T \rangle \\ &\approx \sum_i \zeta_i [\mathbf{f}(\boldsymbol{\sigma}_i) - \hat{\mathbf{y}}][\mathbf{f}(\boldsymbol{\sigma}_i) - \hat{\mathbf{y}}]^T + \mathrm{E}\langle \mathbf{w}\mathbf{w}^T \rangle\end{aligned} \tag{10.106}$$

where the ζ_i are positive weighting factors, generally constrained so that the correct linear results are obtained when $\mathbf{y}$ is linear.

These generally perform better than the extended Kalman filter. However, they do introduce errors that are not accounted for in the covariance of estimation uncertainty.

In practice, Kalman filtering tends to be conservative in its treatment of covariance matrices, biasing its estimates of estimation uncertainty to the high side just a bit to avoid filter divergence due to low Kalman gains.

10.6.5.1 Sigma-Point Filters

Sigma-points of Gaussian distributions are on the 1σ equiprobability ellipsoid at its principal axes, which are the eigenvectors of its covariance matrix – as illustrated in Figure 10.3 for a three-dimensional Gaussian distribution. In sigma-point filters, another sample is added at the center of the distribution, which is its mean.

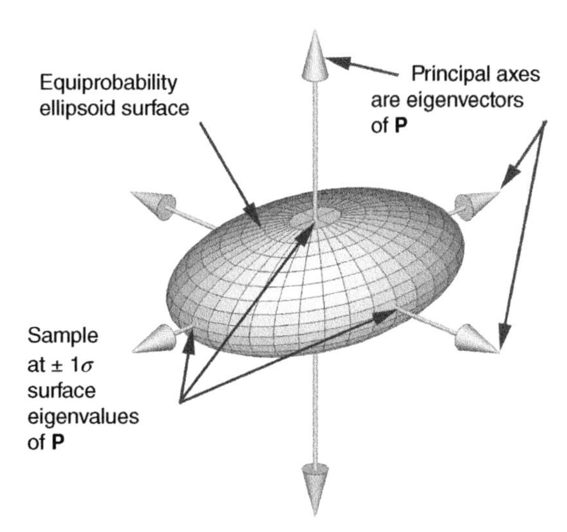

Figure 10.3 Sigma-points for $n = 3$.

The samples used for a PDF with mean $\hat{\mathbf{x}}_{k-1(+)}$ and $n \times n$ covariance matrix $\mathbf{P}_{k-1(+)}$ with eigenvalue–eigenvector decomposition

$$\mathbf{P}_{k-1(+)} = \sum_{i=1}^{n} \lambda_i \mathbf{e}_i \mathbf{e}_i^T \tag{10.107}$$

are at the $2n + 1$ vectors

$$\sigma_0 = \hat{\mathbf{x}}_{k-1(+)} \tag{10.108}$$

$$\sigma_i = \hat{\mathbf{x}}_{k-1(+)} + \sqrt{\lambda_i}\mathbf{e}_i \tag{10.109}$$

$$\sigma_{-i} = \hat{\mathbf{x}}_{k-1(+)} - \sqrt{\lambda_i}\mathbf{e}_i \tag{10.110}$$

for $1 \leq i \leq n$, where the λ_i are the eigenvalues of $\mathbf{P}_{k-1(+)}$ and the $\mathbf{e}_i$ are the corresponding unit eigenvectors.

Sigma-Point Transform For any transformation $\mathbf{f}(\mathbf{x})$ of the variate $\mathbf{x}$, the sigma-point approximated mean is

$$\hat{\mathbf{x}}_{k(-)} \approx \frac{1}{2n+1} \sum_{i=-n}^{+n} \mathbf{f}(\sigma_i) \tag{10.111}$$

and the approximated transformed covariance about the approximated mean is

$$\mathbf{P}_{k(-)} \approx \frac{1}{2n+1} \sum_{i=-n}^{+n} [\mathbf{f}(\sigma_i) - \hat{\mathbf{x}}_{k(-)}][\mathbf{f}(\sigma_i) - \hat{\mathbf{x}}_{k(-)}]^T \tag{10.112}$$

Note that Eqs. (10.111) and (10.112) produce the basic Kalman filter time update equations when $\mathbf{f}(\mathbf{x}) = \mathbf{\Phi}\mathbf{x}$ for $\mathbf{\Phi}$ an $n \times n$ matrix. That is, the sigma-point filter becomes the basic Kalman filter when the transformation is linear.

10.6.5.2 Particle Filters

There are several generalizations of the sigma-point filter, using different factorizations (e.g. Cholesky decomposition) of the covariance matrix and different weightings of the samples. Some are still called "sigma-point" filters, although they do not use conventional sigma points as samples.

10.6.5.3 The Unscented Kalman Filter

The term "unscented" was proposed by Jeffrey Uhlmann as an alternative to naming his approach the *Uhlman Kalman filter*. Uhlmann had begun development of the UKF in the 1990s by analyzing the performance of different sigma-point sampling and weighting strategies for large-scale nonlinear estimation problems. His analysis showed no significant variation in performance related to the particular square root factorization of the initial covariance matrix, and established some weighting strategies that can be "tuned" to improve performance for different applications. The results of this

Table 10.5 Unscented transform samples and weights.

Sampling strategy	Sample size[a)]	Sample values[b)]	Sample weights[c)]	Index values
Simplex	$n+2$	$\sigma_0 = \hat{\mathbf{x}}$	$0 \le \zeta_0 \le 1$(can be varied)	
		$\sigma_i = \sigma_0 + \gamma_i$	$\zeta_1 = 2^{-n}(1-\zeta_0)$	
		$\Gamma = \mathbf{C}\Psi$	$\zeta_2 = \zeta_1$	
		$\Psi = [\psi_1\ \psi_2\ \cdots\ \psi_{n+1}]$	$\zeta_i = 2^{i-1}\zeta_1$	$3 \le i \le n+1$
Symmetric	$2n$	$\sigma_i = \hat{\mathbf{x}} + \sqrt{n}\ \mathbf{c}_i$	$\zeta_i = 1/(2n)$	$1 \le i \le n$
		$\sigma_{i+n} = \hat{\mathbf{x}} - \sqrt{n}\ \mathbf{c}_i$	$\zeta_{i+n} = 1/(2n)$	$1 \le i \le n$
	$2n+1$	$\sigma_0 = \hat{\mathbf{x}}$	$\zeta_0 = \kappa/\sqrt{n+\kappa}$	
		$\sigma_i = \hat{\mathbf{x}} + \sqrt{n+\kappa}\ \mathbf{c}_i$	$\zeta_i = 1/[2(n+\kappa)]$	$1 \le i \le n$
		$\sigma_{i+n} = \hat{\mathbf{x}} - \sqrt{n+\kappa}\ \mathbf{c}_i$	$\zeta_{i+n} = 1/[2(n+\kappa)]$	$1 \le i \le n$
Scaled	$2n+1$	$\sigma_0 = \hat{\mathbf{x}}$	$\zeta_0 = \lambda/(n+\lambda)$	
		$\lambda = \alpha^2(n+\kappa) - n$	$\zeta_0^\star = \zeta_0 + (1-\alpha^2+\beta)$	
		$\sigma_i = \hat{\mathbf{x}} + \sqrt{n+\lambda}\ \mathbf{c}_i$	$\zeta_i = 1/[2(n+\kappa)]$	$1 \le i \le n$
		$\sigma_{n+i} = \hat{\mathbf{x}} - \sqrt{n+\lambda}\ \mathbf{c}_i$	$\zeta_{n+i} = 1/[2(n+\kappa)]$	$1 \le i \le n$

a) n = dimension of state space.
b) $\mathbf{c}_i$ = ith column of a Cholesky factor $\mathbf{C}$ of $\mathbf{P}$. γ_i is the ith column of Γ.
c) α, β, κ, and λ are "tuning" parameters, and $\zeta_0^\star$ is a separate weighting for covariance.

and collaborative studies is a suite of sample-based transforms, collectively called "unscented transforms," for approximating the means and covariances of PDFs after nonlinear transformations of the variate.

Examples of unscented transform strategies, in terms of the samples used and their associated weightings, are presented in Table 10.5. These are sample values σ_i and weightings ζ_i for estimating the nonlinearly transformed means and covariances using Eqs. (10.105) and (10.106) – with the exception that

$$\mathbf{P}_{yy} \approx \sum_{i \neq 0} \zeta_i [\mathbf{f}(\sigma_i) - \hat{\mathbf{y}}][\mathbf{f}(\sigma_i) - \hat{\mathbf{y}}]^T + \zeta_0^\star [\mathbf{f}(\sigma_0) - \hat{\mathbf{y}}][\mathbf{f}(\sigma_0) - \hat{\mathbf{y}}]^T + \mathrm{E}\langle \mathbf{w}\mathbf{w}^T \rangle \tag{10.113}$$

for the scaled unscented transform.

These methods have about the same computational complexity as the extended Kalman filter and have generally performed better for many nonlinear applications.

10.7 Diagnostics and Monitoring

10.7.1 Covariance Matrix Diagnostics

By definition, covariance matrices $\mathbf{P}$ are symmetric and positive semi-definite, meaning that $\mathbf{P} = \mathbf{P}^T$ and the eigenvalues of $\mathbf{P}$ are nonnegative. Neither of these is an issue with square root filtering, but they can be a problem in covariance or information filtering.

10.7.1.1 Symmetry Control

This is not a problem for square root filtering, although it was one of the symptoms of numerical instability leading to its discovery.

Verhaegen and Van Dooren [22] cite covariance matrix asymmetry as a destabilizing symptom, easily eliminated by computing only the upper or lower triangular submatrix of a covariance matrix, then forcing symmetry by storing results in the remaining symmetric positions. This also cuts the amount of computation nearly in half.

10.7.1.2 Eigenanalysis

Zero eigenvalues of a covariance matrix would imply perfect knowledge in some subspace of variates, which is practically impossible for Kalman filter applications. Therefore – if everything is working properly – $\mathbf{P}$ should have positive eigenvalues. The MATLAB function `eig` returns the eigenvalues and associated eigenvectors of a matrix. Eigenvectors corresponding to the largest eigenvalues are the directions in state space (i.e. combinations of state variables) with the greatest uncertainty.

10.7.1.3 Conditioning

The MATLAB function `cond` returns the ratio of the largest and smallest singular values of a matrix, which is a measure of its conditioning for inversion. The reserved variable `eps` is the smallest number ϵ for which $1 + \epsilon \neq 1$ in machine precision. The number `-log2(cond(P)*eps)` is about how many bits of precision will be useful in `inv(P)`. The only matrix inversion in Kalman filtering is $(\mathbf{R}_k + \mathbf{H}_k\mathbf{P}_{k(-)}\mathbf{H}_k^T)^{-1}$, but the conditioning of $\mathbf{P}$ may also be of interest in Kalman filter monitoring.

10.7.2 Innovations Monitoring

10.7.2.1 Kalman Filter Innovations

The intermediate ℓ_k-vector result

$$\boldsymbol{v}_k \stackrel{\text{def}}{=} \{\mathbf{z}_k - \mathbf{H}_k\hat{\mathbf{x}}_{k(-)}\} \tag{10.114}$$

from Eq. (10.45) is called "the kth innovation."[10] It is the difference between the current measurement $\mathbf{z}_k$ and the *expected measurement* $\mathbf{H}_k\hat{\mathbf{x}}_{k(-)}$ based on the a priori estimate $\hat{\mathbf{x}}_{k(-)}$. Its theoretical covariance

$$\mathrm{E}\langle \boldsymbol{v}_k\boldsymbol{v}_k^T\rangle = \mathbf{R}_k + \mathbf{H}_k\mathbf{P}_{k(-)}\mathbf{H}_k^T \tag{10.115}$$

10 This notation is from Thomas Kailath, who used the name *innovation* and gave it the Greek letter $\boldsymbol{v}$ (nu) because it represents "what is new" about the kth measurement.

includes the influence of measurement noise ($\mathbf{R}_k$) and prior estimation errors ($\mathbf{H}_k\mathbf{P}_{k(-)}\mathbf{H}_k^T$). Its inverse

$$\boldsymbol{\Upsilon}_k = (\mathbf{R}_k + \mathbf{H}_k\mathbf{P}_{k(-)}\mathbf{H}_k^T)^{-1} \tag{10.116}$$

is computed as a partial result in the computation of the Kalman gain in Eq. (10.44).

10.7.2.2 Information-Weighted Innovations Monitoring

This is a statistic used by Fred C. Schweppe [43] for signal detection using Kalman filter signal and noise models, and examined by Kailath and Poor [28] in a broader setting. Schweppe used the fact that these zero-mean innovations sequences, if they were Gaussian, would have the PDF

$$p(\mathbf{v}_k) = \frac{\det \boldsymbol{\Upsilon}_k}{\sqrt{(2\pi)^{\ell_k}}} \exp\left(-\frac{1}{2}\mathbf{v}_k^T\boldsymbol{\Upsilon}_k\mathbf{v}_k\right) \tag{10.117}$$

$$\log_e[p(\mathbf{v}_k)] = -\frac{1}{2}\mathbf{v}_k^T\boldsymbol{\Upsilon}_k\mathbf{v}_k + \log_e(\det \boldsymbol{\Upsilon}_k) - \frac{\ell_k}{2}\log_e(2\pi) \tag{10.118}$$

making $-\mathbf{v}_k^T\boldsymbol{\Upsilon}_k\mathbf{v}_k$ a likelihood function (or log-likelihood function) of $\mathbf{v}_k$.

However, this same statistic – without assuming gaussianity – can also be used to monitor the health of Kalman filters in terms of how well the model fits the data. It may be useful for monitoring overall Kalman filter goodness-of-fit, although it is generally not specific about likely sources of mis-modeling.

Given the innovations and their modeled information matrices, one can easily calculate the quadratic statistic

$$\chi_k^2 \stackrel{\text{def}}{=} \mathbf{v}_k^T\boldsymbol{\Upsilon}_k\mathbf{v}_k \tag{10.119}$$

which has the expected value

$$\mathrm{E}\langle\chi_k^2\rangle = \mathrm{E}\langle\mathbf{v}_k^T\boldsymbol{\Upsilon}_k\mathbf{v}_k\rangle \tag{10.120}$$

By using the properties of the matrix trace from Eq. (10.2), this can be maneuvered into the forms

$$\begin{aligned}
\mathrm{E}\langle\chi_k^2\rangle &= \mathrm{E}\langle\mathrm{Tr}(\mathbf{v}_k^T\boldsymbol{\Upsilon}_k\mathbf{v}_k)\rangle \\
&= \mathrm{E}\langle\mathrm{Tr}(\boldsymbol{\Upsilon}_k\mathbf{v}_k\mathbf{v}_k^T)\rangle \\
&= \mathrm{Tr}(\boldsymbol{\Upsilon}_k\mathrm{E}\langle\mathbf{v}_k\mathbf{v}_k^T\rangle) \\
&= \mathrm{Tr}(\boldsymbol{\Upsilon}_k\boldsymbol{\Upsilon}_k^{-1}) \\
&= \mathrm{Tr}(\mathbf{I}_{\ell_k}) \\
&= \ell_k
\end{aligned} \tag{10.121}$$

the dimension of the kth measurement vector. Even though we have not assumed gaussianity from Eq. (10.119) onward, this result is also the mean of a chi-squared distribution with ℓ_k degrees of freedom.

Example 10.3 ***(Simulated Innovations Monitoring)***
The m-file `ChiSqTest.m` on www.wiley.com/go/grewal/gnss calls the function `InnovChisqSim.m` (also on the website) to generate simulated innovations sequences from a Kalman filter model with

$$\mathbf{\Phi} = \begin{bmatrix} \exp(-1/3) & 0 & 0 \\ 0 & \exp(-1/9) & 0 \\ 0 & 0 & \exp(-1/27) \end{bmatrix} \quad (10.122)$$

$$\mathbf{Q} = (\mathbf{I} - \mathbf{\Phi}\mathbf{\Phi}^T) \quad (10.123)$$

$$\mathbf{H} = \begin{bmatrix} 0 & 1 & 1 \\ 1 & 0 & 1 \\ 1 & 1 & 0 \end{bmatrix} \quad (10.124)$$

$$\mathbf{R} = \mathbf{I} \quad (10.125)$$

This is a hypothetical stochastic model for demonstrating how underestimating noise covariance influences innovations. It models the state variable by three independent exponentially correlated processes with different correlation times, and noisy measurements of the sums of independent pairs, using the above Kalman filter model. The simulations use Gaussian pseudorandom noise with covariances equal to or greater than those assumed by the Kalman filter model (i.e. **Q** and **R**) to show the resulting relative shifts in the χ^2 statistic defined in Eq. (10.119) scaled relative to its expected value (ℓ_k, the dimension of the measurement vector). Results shown in Figure 10.4 include simulated values of χ_k^2/ℓ_k with all noise sources properly modeled (top plot), and the χ_k^2/ℓ_k values smoothed by a 100-sample moving average filter (bottom plot) for three cases:

1. All (simulated) noise sources modeled correctly.
2. Simulated dynamic disturbance noise doubled.
3. Simulated measurement noise doubled.

These results are intended to demonstrate that:

1. Raw values of χ_k^2/ℓ_k can be noisy (top plot).
2. Because χ_k^2/ℓ_k is theoretically a white noise process (i.e. uncorrelated) when the problem is properly modeled, averaged values are more steady (bottom plot).
3. Averaged values of χ_k^2/ℓ_k are close to 1 (one) if the problem is well-modeled (bottom plot).
4. Averaged values of χ_k^2/ℓ_k tend to increase if noise sources are underrated (bottom plot).

Innovations monitoring may offer a useful form of feedback for "tuning" Kalman filters by adjusting the model parameters.

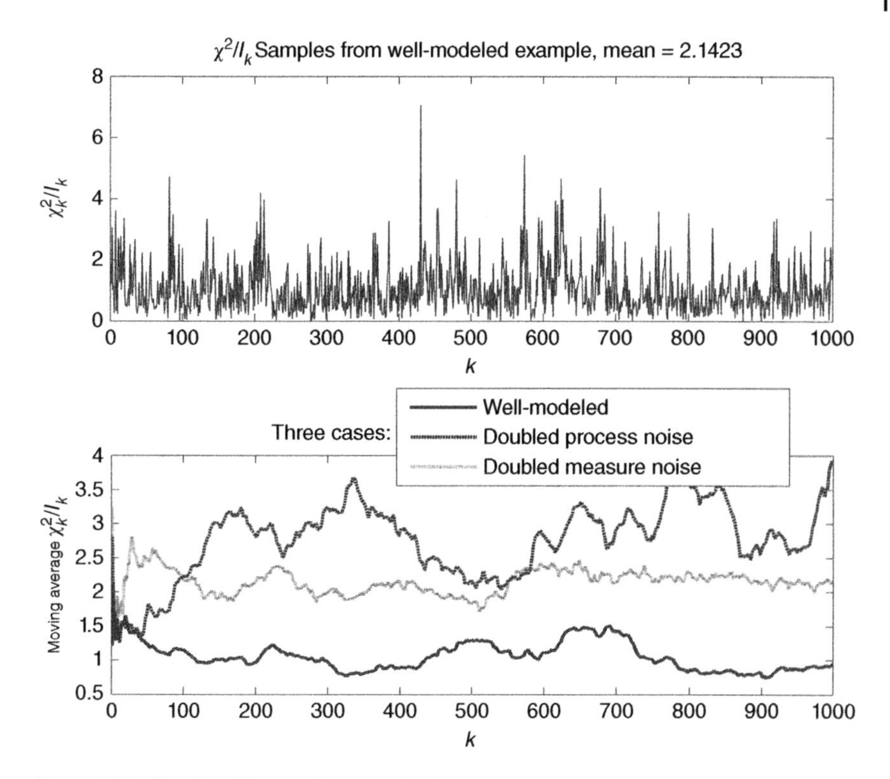

Figure 10.4 Simulated innovations monitoring.

10.8 GNSS-Only Navigation

Kalman filtering models required for any application include the following basic parameters:

Dynamic models defining the parameters
- **Φ**, the state transition matrix
- **Q**, the covariance of dynamic disturbance noise

Measurement models defining the parameters
- **H**, the measurement sensitivity matrix
- **R**, the covariance of measurement noise

This section is about the ways in which these Kalman filter parameters can be defined for representing the GNSS navigation problem in different applications. These model parameter distinctions are important enough that many handheld GNSS receivers include separate built-in Kalman filter dynamic model parameters for walking, bicycling, or driving.

10.8.1 GNSS Dynamic Models

10.8.1.1 Receiver Clock Bias Dynamics

GNSS is a time-based navigation system depending on the accuracy of clocks in the satellites and those in the receivers. Satellite clock synchronization is maintained at the system level, but individual receivers are responsible for synchronizing their internal clocks to GNSS time. The receiver clock synchronization problem is common to all GNSS navigation applications.

Clocks are frequency sources. Those on satellites are generally expensive atomic clocks with tight environmental controls, but those in receivers are mostly lower-cost crystal oscillators with less stringent environmental controls. Maintaining synchronization with the GNSS satellite clocks is part of the receiver's Kalman filter implementation. This generally adds a few state variables to the Kalman filter.

These clock state variables include time and frequency offsets from GNSS time, but may also include frequency drift rate, which tends to be more stable over short time periods. In the latter case, a potential dynamic model for the clock bias "state" $\varepsilon_{\text{clock}}$ would be of the form

$$\left.\begin{aligned} \varepsilon_{\text{clock}} &= \begin{bmatrix} \varepsilon_{\text{bias}} \\ \varepsilon_{\text{freq}} \\ \varepsilon_{\text{drift}} \end{bmatrix} \\ \frac{d}{dt}\varepsilon_{\text{clock}} &= \begin{bmatrix} 0 & 1 & 0 \\ 0 & 0 & 1 \\ 0 & 0 & -1/\tau_{\text{drift}} \end{bmatrix} \varepsilon_{\text{clock}} + \begin{bmatrix} 0 \\ 0 \\ w_{\text{drift}}(t) \end{bmatrix} \end{aligned}\right\} \tag{10.126}$$

where τ_{drift} is the exponential correlation time of frequency drift rate and $w_{\text{drift}}(t)$ is the drift rate dynamic disturbance noise, with

$$\left.\begin{aligned} \mathrm{E}\langle w_{\text{drift}}(t)\rangle &= 0 \\ \mathrm{E}\langle w_{\text{drift}}^2(t)\rangle &= q_{\text{drift}} \\ \sigma_{\text{drift}}^2 &\stackrel{\text{def}}{=} \lim_{t\to\infty} \mathrm{E}\langle \varepsilon_{\text{drift}}^2(t)\rangle \\ &= \frac{\tau_{\text{drift}} q_{\text{drift}}}{2} \\ \mathrm{E}\langle \varepsilon_{\text{drift}}(t)\varepsilon_{\text{drift}}(t+\Delta t)\rangle &= \sigma_{\text{drift}}^2 e^{-|\Delta t|/\tau_{\text{drift}}} \end{aligned}\right\} \tag{10.127}$$

This dynamic system is unstable in the variables $\varepsilon_{\text{bias}}$ and $\varepsilon_{\text{freq}}$, but stable in $\varepsilon_{\text{drift}}$. However, the Kalman filter estimate of all three variables can still be stable – depending on the measurements used.

The critical clock parameters for this part of the Kalman filter implementation are the mean-squared frequency drift rate σ_{drift}^2 and its correlation time τ_{drift}, which are generally determined by comparing test clocks to references standard clocks over long time periods.

10.8.1.2 Discrete Time Models

We have used the dynamic coefficient matrix $\mathbf{F}$ for continuous-time dynamics here because GNSS measurement times t_k are the GNSS signal time-mark arrival times, which may not be predictable and regularly spaced in time. The times between such events,

$$\Delta t_k = t_k - t_{k-1} \tag{10.128}$$

are not always predictable, and the equivalent state transition matrices Φ in discrete time,

$$\mathbf{\Phi}_k = \exp\left(\int_{t_{k-1}}^{t_k} \mathbf{F}(t)dt\right) \tag{10.129}$$

depend on Δt_k.

10.8.1.3 Exponentially Correlated Random Processes

Equations (10.127) define the linear time-invariant dynamic model for what is called an *exponentially correlated random process*, which is used extensively in Kalman filtering as a model for time-correlated zero-mean random processes with bounded mean-squared values. Its equivalent form in discrete time is

$$\left.\begin{aligned} \varepsilon_k &= \exp(-\Delta t/\tau)\varepsilon_{k-1} + w_k \\ \tau &> 0 \\ \mathrm{E}\langle w_k \rangle &= 0 \\ \mathrm{E}\langle w_k^2 \rangle &= q \\ \sigma_\infty^2 &\stackrel{\text{def}}{=} \lim_{k\to\infty} \mathrm{E}\langle \varepsilon_k^2 \rangle \\ &= \frac{q}{1-\exp(-2\Delta t/\tau)} \\ \mathrm{E}\langle \varepsilon_k \varepsilon_j \rangle &= \sigma_\infty^2 \exp(-|t_k - t_j|/\tau) \end{aligned}\right\} \tag{10.130}$$

where the essential design parameters are the correlation time-constant τ and σ_∞^2, the steady-state variance of error. The process noise variance q can be determined from τ and σ_∞^2, the steady-state variance of the exponentially correlated errors.

10.8.1.4 Host Vehicle Dynamics for Standalone GNSS Navigation

This is for navigation using only GNSS without an INS, in which case the dynamics of the host vehicle is defined only in terms of random process models.

Besides the clock correction variables, the other essential state variables for GNSS navigation are those determining the location of the receiver antenna relative to navigation coordinates – which are generally earth-fixed for terrestrial applications. The receiver antenna is the "holy point" for GNSS navigation. It is the point where all satellite signals are received, and its location relative to those of the satellite antennas determines the relative timing of the satellite signals at that point. Any signal delays between the antenna and the signal processor, plus processing delays, are common to all signals. These delays may not affect the navigation solution, but they do contribute to the solution time-lag.

For most applications, the dynamics of the antenna are determined by the dynamics of the host vehicle.[11] For applications in which the host vehicle has significant rotational freedom and the offsets between the antenna and the center of gravity of the host vehicle are large compared to measurement error, rotational dynamics of the host vehicle may also be part of the model. One of the advantages of GNSS/INS integration for this case is that an INS is at least an attitude sensor and – for strapdown INS – an attitude rate sensor.

10.8.1.5 Point Mass Dynamic Models

State vector dimension is determined by the degrees of freedom of movement of the host vehicle and by its mechanics of movement. GNSS measurements can only track the location of the receiver antenna, so the simplest models for host vehicle tracking are based on point-mass dynamics.

Velocity Random Walk Models The simplest usable tracking models are for point-mass dynamics including translational momentum mechanics, which adds one state variable (velocity) for each degree of freedom. In this case, the dynamic disturbance model has the form

$$\left.\begin{aligned} \mathbf{x}_{\text{dim}} &= \begin{bmatrix} x_{\text{pos}} \\ x_{\text{vel}} \end{bmatrix} \\ \frac{d}{dt}\mathbf{x}_{\text{dim}} &= \begin{bmatrix} 0 & 1 \\ 0 & 0 \end{bmatrix} \mathbf{x}_{\text{dim}} + \begin{bmatrix} 0 \\ w_{\text{acc}}(t) \end{bmatrix} \end{aligned}\right\} \tag{10.131}$$

for each dimension of travel.

Exponentially Correlated Velocity Models A point-mass stochastic model for random time-correlated host vehicle velocity with RMS value σ_{vel} and exponential correlation time τ has the form

$$\frac{d}{dt}\begin{bmatrix} x_{\text{pos}} \\ x_{\text{vel}} \end{bmatrix} = \begin{bmatrix} 0 & 1 \\ 0 & -1/\tau \end{bmatrix}\begin{bmatrix} x_{\text{pos}} \\ x_{\text{vel}} \end{bmatrix} + \begin{bmatrix} 0 \\ w_{\text{acc}} \end{bmatrix} \tag{10.132}$$

$$\mathrm{E}\langle w_{\text{acc}} \rangle = 0 \tag{10.133}$$

$$\mathrm{E}\langle w_{\text{acc}}^2 \rangle = \frac{2\sigma_{\text{vel}}^2}{\tau} \tag{10.134}$$

for each axis of freedom, but with potentially different values for τ and v_{rms} for different axes.

---

11 There are scientific applications in which the "host vehicle" is a bit of bedrock or glacier, in which case the state vector dynamic rates might be centimeters per year or meters per day.

Exponentially Correlated Acceleration Models These resemble the clock bias model of Eqs. (10.126) and (10.127) for each dimension of travel:

$$\left.\begin{array}{rcl} \mathbf{x}_{\text{dim}} & = & \begin{bmatrix} x_{\text{pos}} \\ x_{\text{vel}} \\ x_{\text{acc}} \end{bmatrix} \\ \frac{d}{dt}\mathbf{x}_{\text{dim}} & = & \begin{bmatrix} 0 & 1 & 0 \\ 0 & 0 & 1 \\ 0 & 0 & -1/\tau_{\text{acc}} \end{bmatrix} \mathbf{x}_{\text{dim}} + \begin{bmatrix} 0 \\ 0 \\ w_{\text{jerk}}(t) \end{bmatrix} \\ \mathrm{E}\langle w_{\text{jerk}}(t)\rangle & = & 0 \\ \mathrm{E}\langle w_{\text{jerk}}^2(t)\rangle & = & q_{\text{jerk}} \\ \sigma_{\text{acc}}^2 & \stackrel{\text{def}}{=} & \lim_{t\to\infty} \mathrm{E}\langle x_{\text{acc}}^2(t)\rangle \\ & = & \frac{\tau_{\text{acc}} q_{\text{jerk}}}{2} \\ \mathrm{E}\langle x_{\text{acc}}(t) x_{\text{acc}}(t+\Delta t)\rangle & = & \sigma_{\text{acc}}^2 \, e^{-|\Delta t|/\tau_{\text{acc}}} \end{array}\right\} \tag{10.135}$$

where the subscript "jerk" refers to the derivative of acceleration, which is called "jerk." This model may be more realistic for vehicles occupied by humans, who tend to become uncomfortable at acceleration levels greater than $3g$ ($\sim$30 m/s/s). It is also appropriate for wheeled vehicles, whose acceleration capabilities are limited by tire-to-surface contact stiction/friction to be generally $< 1g$. (An exception is in auto racing, where downward aerodynamic force is used to boost the tire limit to the driver's limit of $\sim 3G$.)

Models for Bounded RMS Velocity and Acceleration In this case, a linear drag coefficient d is used for bounding the RMS velocity in the above model, in which case the lower-right 2×2 submatrices of the steady-state covariance matrix $\mathbf{P}_{\infty}$ and its associated dynamic coefficient matrix

$$\mathbf{P}_{\infty\ 2\times2} = \begin{bmatrix} v_{\text{rms}}^2 & p_{1,2} \\ p_{1,2} & a_{\text{rms}}^2 \end{bmatrix} \tag{10.136}$$

$$\mathbf{F}_{2\times2} = \begin{bmatrix} -d & 1 \\ 0 & -\tau^{-1} \end{bmatrix} \tag{10.137}$$

satisfy the steady-state equation in continuous time:

$$\begin{aligned} 0 &= \frac{d}{dt}\mathbf{P}_{\infty\ 2\times2} \\ &= \mathbf{F}_{2\times2}\mathbf{P}_{\infty\ 2\times2} + \mathbf{P}_{\infty\ 2\times2}\mathbf{F}_{2\times2}^{T} + \begin{bmatrix} 0 & 0 \\ 0 & q \end{bmatrix} \end{aligned} \tag{10.138}$$

which can be solved for

$$q = \frac{2\, a_{\text{rms}}^2}{\tau} \tag{10.139}$$

$$d = \frac{\sqrt{4\,\tau^2 a_{\rm rms}^2 + v_{\rm rms}^2} - v_{\rm rms}}{2\,\tau\,v_{\rm rms}} \tag{10.140}$$

$$p_{1,2} = v_{\rm rms}^2\,d \tag{10.141}$$

where the q in Eq. (10.139) is for the dynamic model in continuous time. The equivalent $\mathbf{Q}_k$ in discrete time is given by Eq. (10.65).

This model has been converted to its equivalent implementation in discrete time in the MATLAB m-files `avRMS.m` and `avRMStestfunction.m`, a sample output from which is shown in Figure 10.5. This is from an example with 0.5g RMS acceleration with one minute correlation time, 8 m/s RMS velocity, one second discrete time steps, and 1000 simulated steps. Figure 10.5 shows the RMS position, velocity, and acceleration from the riccati equation solution, demonstrating steady-state behavior in velocity and acceleration, and unstable RMS uncertainty in position.

10.8.2 GNSS Measurement Models

10.8.2.1 Measurement Event Timing

GNSS receivers process satellite signals to compute the differences between the times the signals left their respective satellite antennas (encoded in each satellite signal) and the times when the same signals arrived at the receiver antenna

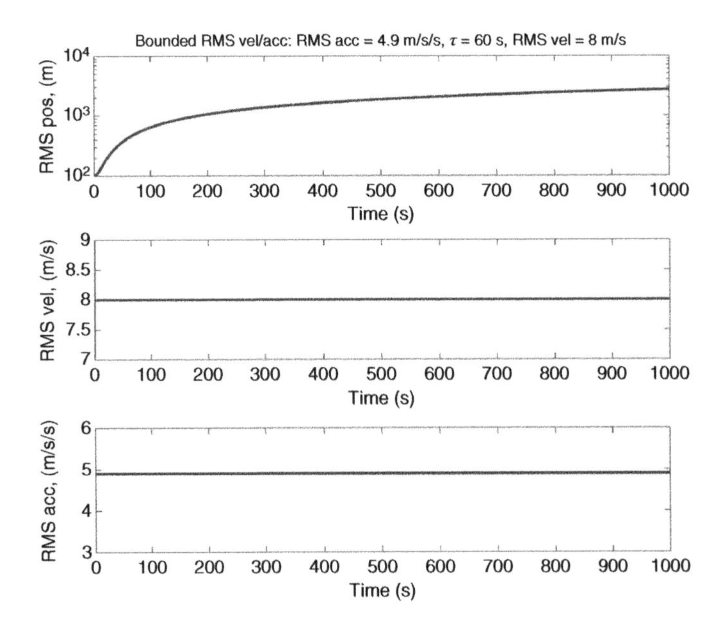

Figure 10.5 Sample 3 × 1 subplot output from `avRMStestfunction.m`, showing RMS uncertainties.

(determined by the receiver's clock). The effective measurement time t_k in this case is the time at which the signal timing mark arrived at the receiver antenna. The discrete time intervals between measurements are not necessarily uniform or predictable.

10.8.2.2 Pseudoranges

Those timing differences, multiplied by the propagation speed of the signal, are the *pseudoranges* between the satellite antennas at the transmission times and the receiver antenna at the reception times – as illustrated in Figure 10.6. For GPS, these total time differences are generally in the order of $\sim 75 \pm 10$ ms, with each nanosecond representing about 0.3 m of pseudorange distance.

10.8.2.3 Time and Distance Correlation

Some of the system error sources contributing to measurement error in GNSS pseudoranges are time-correlated, as well as distance correlated. However, we may not have good data on this for GPS III until after the next-generation upgrades are completed. Early on in the evolution of GPS, data collected worldwide showed a dominant exponentially time-correlation in the order of an hour and distance correlation in the order of 100 m [44]. We will not know any relevant correlation model parameters for upgraded and developing GNSS until some time after these upgrades are finished. The candidate dynamic model for GNSS navigation given below includes a single exponential time-correlation parameter, just in case it is still significant.

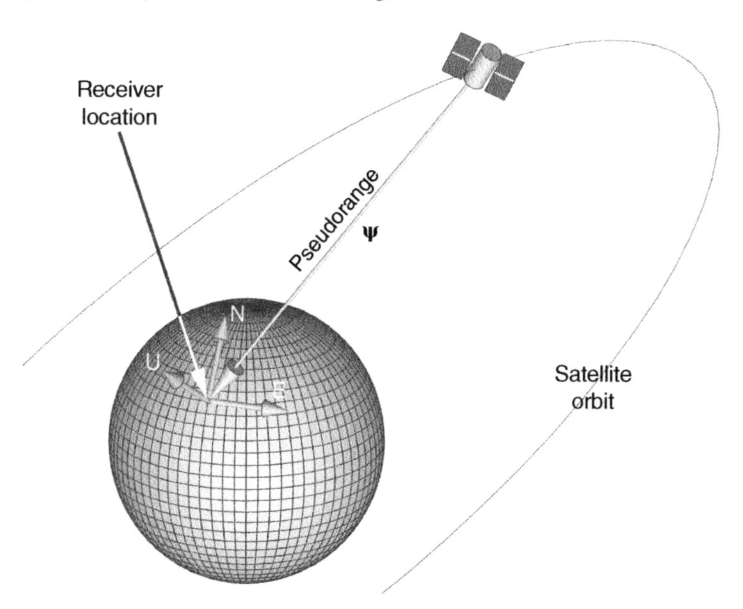

Figure 10.6 Pseudorange measurement geometry.

10.8.2.4 Measurement Sensitivity Matrix

Pseudoranges are scalar measurements, so the associated measurement sensitivity matrices **H** have but one row. At most four of the elements h_{1j} of **H** are nonzero, corresponding to

$\varepsilon_{\mathbf{bias}}$, the receiver clock time offset from GNSS time. This element of **H** will have the value $1/c$, where c is the signal propagation velocity.

$x_{\dim 1}$, the first dimension of receiver antenna location in navigation coordinates.

$x_{\dim 2}$, the second dimension of receiver antenna location in navigation coordinates.

$x_{\dim 3}$, the third dimension of receiver antenna location in navigation coordinates.

The last three of these elements of **H** will be the three components of the unit vector

$$\mathbf{u}_{\psi} = \frac{\boldsymbol{\psi}}{|\boldsymbol{\psi}|}$$

in the direction from the satellite antenna toward the receiver antenna, representing the sensitivity of calculated pseudorange to estimated receiver antenna location errors. This calculation requires that the satellite antenna position and receiver antenna position be represented in a common coordinate frame, which can be navigation coordinates (earth-centered earth-fixed).

10.8.2.5 Noise Model

The only noise statistics that matter in Kalman filtering are means and covariances of the Kalman model after time-correlated errors have been converted to white noise sources driving dynamic systems.

This can vary somewhat from one GNSS constellation to another, may be different from one satellite to another within the same GNSS constellation (as is the measurement sensitivity matrix **H**), and may include time-correlated and satellite-to-satellite correlated errors. The modeling of time-correlated measurement errors requires that at least one state variable be included for each satellite being used. The simplest time-correlated model for one satellite would be one in which there is a single exponentially time-correlated noise component ξ, with continuous-time and discrete-time models

$$\frac{d}{dt}\xi(t) = \frac{-1}{\tau}\xi + w(t) \tag{10.142}$$

$$\xi_{k(-)} = \phi_k \xi_{k-1(+)} + w_k \tag{10.143}$$

$$\phi_k = \exp\left(-\frac{t_k - t_{k-1}}{\tau}\right) \tag{10.144}$$

$$\mathrm{E}\langle w_k^2 \rangle = q_k \tag{10.145}$$

where τ is the correlation time-constant. In the time-invariant case in which $t_k - t_{k-1} = \Delta t$ is constant, the steady-state mean-squared noise will be the solution of the steady-state riccati equation

$$\begin{aligned}\sigma_\infty^2 &= e^{-2\,\Delta t/\tau}\sigma_\infty^2 + q \\ &= \frac{q}{1 - e^{-2\Delta t/\tau}}\end{aligned} \tag{10.146}$$

This might be a suitable model for system-level errors such as satellite clock errors or ephemeris errors, and it requires adding the variables ξ for each satellite to the state vector. There will still be uncorrelated measurement noise, as well, due to such sources as background noise, processing roundoff errors, or electromagnetic interference.

Example 10.4 ***(State Vector Example for Standalone GNSS Navigation)***
This is an example for what the Kalman filter state vector might look like for 3D GNSS navigation with a clock drift rate model, a host vehicle with jerk-noise dynamics and exponentially correlated pseudorange measurement noise on each of 30 satellites. Table 10.6 identifies the 42 elements of the resulting state vector in earth-fixed locally-level east-north-up coordinates. The first three state variables model receiver clock errors, the next nine

Table 10.6 Example state vector for standalone GNSS navigation.

Number	Symbol	Description
1	$x^\star_{\text{bias}}$	Receiver clock bias
2	x_{freq}	Receiver clock frequency
3	x_{drift}	Receiver clock drift rate
4	$x^\star_{\text{pos}\ E}$	Receiver antenna easting
5	$x_{\text{vel}\ E}$	Receiver antenna east velocity
6	$x_{\text{acc}\ E}$	Receiver antenna east acceleration
7	$x^\star_{\text{pos}\ N}$	Receiver antenna northing
8	$x_{\text{vel}\ N}$	Receiver antenna north velocity
9	$x_{\text{acc}\ N}$	Receiver antenna north acceleration
10	$x^\star_{\text{pos}\ U}$	Receiver antenna altitude
11	$x_{\text{vel}\ U}$	Receiver antenna vertical velocity
12	$x_{\text{acc}\ U}$	Receiver antenna vertical acceleration
13	$x^\star_{\text{sat}\ 1}$	Satellite #1 signal delay
14	$x^\star_{\text{sat}\ 2}$	Satellite #2 signal delay
⋮	⋮	⋮
42	$x^\star_{\text{sat}\ 30}$	Satellite #30 signal delay

model host vehicle dynamics, and the remaining 30 model time-correlated measurement noise. The starred variables (e.g. $x^{\star}_{\text{bias}}$) are the only components with nonzero entries at the corresponding positions in the associated 1×42 measurement sensitivity matrix **H**. This example gives some idea of what time-correlated measurement errors can do to the dimension of the state vector.

10.9 Summary

1. In navigation problems, the ***navigation solution*** must include values for all those variables needed for getting the host vehicle to its destination. These must include position of the vehicle being navigated but may also include its velocity, orientation, or any other attributes of interest.
2. Kalman filtering is used in navigation for estimating the current value of the navigation solution, given all measurements made up to the current time. It is also used for a host of other tracking problems related to navigation, including:
   (a) Keeping track of the precise orbits of all the GNSS satellites in a constellation, using measurements made from ground stations and between the satellites in orbit.
   (b) Using signals received from GNSS satellites to track the motions of the receiver antenna.
   (c) Using GNSS and INS together to obtain a better navigation solution and to attain some degraded level of navigational accuracy when GNSS signals are not available.
3. The Kalman filter itself is an algorithm based on:
   (a) The LLMSE of Carl Friedrich Gauss, which minimizes the mean squared error in the estimated navigation solution, using only
      (i) The linear relationships between measured variables and the navigation solution variables
      (ii) The means and variances (or covariances) of all error distributions, including
         A. Measurement noise and
         B. Estimation errors.
   (b) A recursive form of the linear least mean squared estimator, which
      (i) Allows the measurements to be processed one at a time, as they come in.
      (ii) Maintains the current value of the linear-least-mean-squares estimate, based on all measurements made up to the current time.
      (iii) Maintains the current value of estimation error covariance, which is needed for updating the estimate as measurements come in.

   (iv) Minimizes the mean-square estimation error, given all measurements up to the current time.
   (c) Linear stochastic process models (due to Andrey Andreyevich Markov) for representing
      (i) The known linear dynamic model for the host vehicle being navigated, plus
      (ii) Mean-squared contributions to navigation uncertainty due to random disturbances of the host vehicle.
   (d) Linear algebra (lots of it).
4. The Kalman filter does not depend on any attributes of probability distributions beyond their first and second moments. As a consequence, it cannot be used for calculating probabilities – unless one further assumes a particular parametric form of the PDF (e.g. Gaussian).
5. The basic implementation equations for the Kalman filter – as derived by Kalman [2] – are summarized in Table 10.1.
6. Those nonlinear matrix equations related to the propagation of the covariance matrix of estimation uncertainty are called *riccati equations*, named after the Venetian mathematician who first showed that they can be converted to linear equations by a change of variables. These can be solved independently from the estimation equations and used for predicting the expected estimation accuracy for a given measurement scenario.
7. Digital implementation of the Kalman filter in finite precision introduce roundoff errors that can degrade the accuracy of the solution to some degree, depending on the digital wordlength of the implementation and certain algebraic properties of the matrices involved – especially covariance matrices.
8. James E. Potter introduced the idea of using "square roots" of covariance matrices in alternative representations of Kalman filtering, which effectively cut digital wordlength requirements in half for achieving the same level of numerical accuracy.
9. Square root Kalman filter implementations also make use of orthogonal matrix transformations to improve robustness against roundoff errors, which is especially important for problems with large numbers of unknowns to be estimated. This includes Kalman filter implementations for keeping track of GNSS constellations.
10. Linearity of the dynamic equations and measurements is essential for Kalman filtering. However, it has been used with considerable success in quasi-linear applications, in which the dynamics and measurements are dominated by linearity within the essential range of the probability distributions involved. (See Appendix C for additional coverage.)
11. Kalman filters for GNSS navigation generally require stochastic models for random dynamics of the receiver antenna, the receiver clock error, and any time-correlated signal delay errors.

12. Monitoring the run-time health of a Kalman filter may include statistical assessment of measurement *innovations* (differences between predicted and actual measurements) and checking on algebraic properties of covariance matrices.

Problems

10.1 Can you name a PDF without a defined mean or variance? Why would this distinction be important in Kalman filtering?

10.2 What are the main reasons why the Kalman filter depends only on the means and covariances of PDFs?

10.3 Why are means and covariances of PDFs not sufficient when nonlinear transformations are used?

10.4 (Random Walk) Let the discrete stochastic process $\{x_k | k = 1,2,3, \dots\}$ be defined by

$$x_{k+1} = x_k + w_k, \quad \mathrm{E}\langle w_k \rangle = 0 \quad \text{for all} \quad k, \quad \mathrm{E}\langle w_k^2 \rangle = 1$$

(a) What is the equivalent state transition matrix for this system?
(b) If $\mathrm{E}\langle x_1 \rangle = 0$, what is $\mathrm{E}\langle x_k \rangle$ for $k = 2, 3, 4, 5$?
(c) If $\mathrm{E}\langle x_1^2 \rangle = 1$, what is $\mathrm{E}\langle x_k^2 \rangle$ for $k = 2, 3, 4, 5$?
(d) Are random walks stable?

10.5 (Random Walk with Measurements) In Problem 10.9.4 let there be measurements $z_k = h_k x_k + v_k$ with

$$Q = \mathrm{E}\langle w_k^2 \rangle = 1, \quad h = 1, \quad \mathrm{E}\langle v_k \rangle = 0, \quad \mathrm{E}\langle v_k^2 \rangle = 2 \quad \text{for all} \quad k$$

(a) If $P_{1(-)} = \mathrm{E}\langle x_1^2 \rangle = 2$, what will be the value of $P_{1(+)}$ after the measurement z_1 has been processed?
(b) What will be the value of $P_{2(-)}$ before the next measurement z_2 is processed?
(c) What will be the value of $P_{2(+)}$ after the measurement z_2 has been processed?
(d) What will be the values of $P_{k(-)}$ and $P_{k(+)}$ for $k = 1, 2, 3, 4$?
(e) In this example, is the estimation uncertainty of an (unstable) random walk stable?

10.6 Let

$$\mathbf{F} \stackrel{\text{def}}{=} \begin{bmatrix} 0 & 1 \\ 0 & 0 \end{bmatrix}$$

(a) What is

$$\mathbf{M}(t) = \int_0^t \mathbf{F}(s)ds?$$

(b) What is $\mathbf{M}^2(t)$?
(c) What is $\mathbf{M}^m(t)$ for $m > 1$?
(d) What is its matrix exponential

$$\exp(\mathbf{M}(t)) \stackrel{\text{def}}{=} \sum_{m=0}^{\infty} \frac{1}{m!}\mathbf{M}^m(t)$$

also known as a fundamental solution matrix for $\dot{\mathbf{x}} = \mathbf{F}\mathbf{x}$?

10.7 Prove that Eqs. (10.111) and (10.112) produce the basic Kalman filter temporal update equations when $\mathbf{f}(\mathbf{x}) = \mathbf{\Phi}\mathbf{x}$ for $\mathbf{\Phi}$ an $n \times n$ matrix.

10.8 Consider a scalar time-invariant dynamic system model or the sort

$$x_k = \phi x_{k-1} + w_{k-1}, \quad \text{where} \quad \mathrm{E}\langle w_m \rangle = 0$$
$$\text{and} \quad \mathrm{E}\langle w_m^2 \rangle = q > 0 \quad \text{for all} \quad m \tag{10.147}$$

Under what conditions on ϕ is this dynamic system stable or unstable? (See Section 10.4.5.)

10.9 Can the Kalman estimates of the state of an unstable dynamic system be stable (i.e. bounded)?

10.10 If $\phi = \frac{1}{2}$ and $q = \frac{1}{2}$ in Eq. (10.147), the initial variance of uncertainty

$$p_0 \stackrel{\text{def}}{=} \mathrm{E}\langle x_0^2 \rangle = 1$$

and there are no measurements, what is p_1?

10.11 Under the same conditions as in Problem 10.10, is there a steady-state solution for p_∞? If so, what is it?

10.12 Run the MATLAB function `avRMStestfunction.m` on www.wiley.com/go/grewal/gnss with the following parameter values:

Units	m/s/s	s	m/s	s	#	Possible
Inputs	`aRMS`	`tau`	`vRMS`	`Delta`	`Steps`	applications?
`avRMStestfunction(`	`2,`	`10,`	`10,`	`1,`	`1000)`	OFF-ROAD?
`avRMStestfunction(`	`4,`	`30,`	`20,`	`1,`	`1000)`	PAVEMENT-LIMITED?
`avRMStestfunction(`	`10,`	`30,`	`30,`	`1,`	`1000)`	HUMAN OCCUPANT?
`avRMStestfunction(`	`20,`	`10,`	`100,`	`1,`	`1000)`	ROCKET?

Observe the resulting multi-plots of RMS values and simulated examples for this model with bounded RMS acceleration and velocity. Note, however, that RMS position relative to the starting point is constantly increasing.
Here, the RMS (1σ) values can be used to make physical limits the $\sim 3\sigma$ values. For example,

- Off-road traction limits acceleration to be considerably less than $1g$ (9.8 m/s/s).
- Except for racing cars, pavement contact limits horizontal acceleration to about $1g$. Racing cars can utilize aerodynamic forces to boost that to the driver limit of about $3g$.
- Humans start to feel uncomfortable at around $3g$ for more than a few seconds.
- Unmanned hit-to-kill missile defense rockets can accelerate to dozens of *G*s and many times the speed of sound.

10.13 Run the m-file `ChiSqTest.m` on www.wiley.com/go/grewal/gnss to generate plots of the innovations statistic χ_k^2/ℓ_k for simulated measurement noise (Eq. (10.125)) and simulated dynamic disturbance noise (Eq. (10.123)) scaled up a factor of 2. The first of the three plots generated is for a well-modeled Kalman filter, and the second and third plots are for mis-scaled noise covariances. If everything is working properly, the outputs in all cases should be close to 1. Are these levels of mis-modeling apparent from the (simulated) values of χ_k^2/ℓ_k?

10.14 Run the m-file `GNSSonly.m` on the www.wiley.com/go/grewal/gnss. It simulates GNSS-only navigation of a vehicle traveling 100 kph on a 100 km figure-8 test track with a tunnel under the overpass where the tracks cross over one another, so that GNSS signals are lost briefly at that time. It also compares the RMS Monte Carlo simulation errors of the Kalman filter implementation with the riccati equation solution. The error buildup in the tunnel is largely determined by the stochastic model representing vehicle dynamics, which is rather simplistic in this case.

References

1 Gauss, C.F. (1995). In: *Theoria Combinationis Observationum Erribus Minimus Obnoxiae* (trans. G.W. Stewart). Philadelphia, PA: SIAM.

2 Kalman, R.E. (1960). A new approach to linear filtering and prediction problems. *ASME Transactions, Series D: Journal of Basic Engineering* 82: 35–45.

3 Levy, L.J. (1977). The Kalman filter: integrations's workhorse. *GPS World* (September 1977), pp. 65–71.
4 Gelb, A. (ed.) (1974). *Applied Optimal Estimation*. Cambridge, MA: MIT Press.
5 Antoulas, A.C (ed.) (1991). *Mathematical System Theory: The Influence of R.E. Kalman*. Berlin: Springer-Verlag.
6 Stigler, S.M. (1981). Gauss and the invention of least squares. *Annals of Statistics* 9 (3): 465–474.
7 Gauss, C.F. (1963). In: *Theoria Motus Corporum Coelestium* (trans. C.H. Davis). New York: Dover.
8 Legendre, A.M. (1805). *Nouvelles méthodes pour la détermination des orbites des comètes*. Paris: Didot.
9 Adrian, R. (1808). Research concerning the probabilities of the errors which happen in making observations, &c. *Analyst* 1 (4): 93–109.
10 Cooley, J.W. and Tukey, J.W. (1865). An algorithm for the machine calculation of complex Fourier series. *Mathematics of Computation* 19: 297–301.
11 Markov, A.A. (1898). The law of large numbers and the method of least squares. *Izvestia Fizika-Mathematika Obschestva Kazan University* 8 (2): 110–128 (in Russian).
12 Aitken, A.C. (1935). On least squares and linear combinations of observations. *Proceedings of the Royal Society of Edinburgh* 55: 42–48.
13 Duncan, W.J. (1944). Some devices for the solution of large sets of simultaneous linear equations. *The London, Edinburgh, and Dublin Philosophical Magazine and Journal of Science, Series 7* 35: 660–670.
14 Guttman, L. (1946). Enlargement methods for computing the inverse matrix. *Annals of Mathematical Statistics* 17 (3): 336–343.
15 Jarrow, R. and Protter, P. (2004). A short history of stochastic integration and mathematical finance: the early years, 1880–1970. *Institute of Mathematical Statistics, Monograph Series* 45: 75–91.
16 Itô, K. (1946). Stochastic integral. *Proceedings of the Imperial Academy Tokyo* 20 (1944): 519–524.
17 Jazwinski, A.H. (2009). *Stochastic Processes and Filtering Theory*. Dover Publications (reprint of 1970 edition by Academic Press).
18 Schmidt, S.F. (1981). The Kalman filter – its recognition and development for aerospace applications. *AIAA Journal of Guidance, Control, and Dynamics* 4 (1): 4–7.
19 Stewart, G.W. (1998). *Matrix Algorithms: Basic Decompositions*, vol. 1, 1e. Philadelphia, PA: SIAM.
20 Golub, G.H. and Van Loan, C.F. (2013). *Matrix Computations*, 4e. Baltimore, MD: Johns Hopkins University Press.
21 Kaminski, P.G. (1971). Square root filtering and smoothing for discrete processes. PhD thesis. Stanford, CA: Stanford University.

22 Verhaegen, M. and Van Dooren, P. (1986). Numerical aspects of different Kalman filter implementations. *IEEE Transactions on Automatic Control* AC-31: 907–917.

23 Fisher, R.A. (1925). Theory of statistical estimation. *Mathematical Proceedings of the Cambridge Philosophical Society* 22 (5): 700–725.

24 Grewal, M.S. and Kain, J. (2010). Kalman filter implementation with improved numerical properties. *IEEE Transactions on Automatic Control* 55 (9): 2058–2068.

25 Yu, W., Keegan, R.G., and Hatch, R.R. (2018). Satellite navigation receiver with fixed point sigma Rho filter. US Patent 20180313958, 1 November 2018.

26 Bierman, G.J. (2017). *Factorization Methods for Discrete Sequential Estimation*. New York: Dover.

27 Morf, M. and Kailath, T. (1975). Square root algorithms for least squares estimation. *IEEE Transactions on Automatic Control* AC-20: 487–497.

28 Kailath, T. and Poor, H.V. (1998). Detection of stochastic processes. *IEEE Transactions on Information Theory* 44 (6): 2230–2259.

29 Commendat Benoit (1924). Sur une méthode de résolution des équations normales provenant de l'application de la méthode des moindes carrés a un système d'équations linéaires en numbre inférieur a celui des inconnues—application de la m.éthode a la resolution d'un système defini d'équations linéaires (Procédé du Commandant Cholesky). *Bulletin Géodêsique et Géophysique Internationale* 2: 67–77.

30 Gram, J.P. (1883). Uber die Entwickelung reeler Funtionen in Reihen mittelst der Methode der kleinsten Quadrate. *Journal fur die reine und angewandte Mathematik* 94: 71–73.

31 Schmidt, E. (1907). Zur Theorie der linearen und nichtlinearen Integralgleichungen. I. Teil: Entwicklung willkulicher Funktionen nach Systemen vorgeschriebener. *Mathematische Annalen* 63: 433–476.

32 Givens, W. (1958). Computation of plane unitary rotations transforming a general matrix to triangular form. *Journal of the Society for Industrial and Applied Mathematics* 6 (1): 26–50.

33 Householder, A.S. (1958). Unitary triangularization of a nonsymmetric matrix. *Journal of the ACM* 5 (4): 339–342.

34 Thornton, C.L. (1976). Triangular covariance factorizations for Kalman filtering. PhD thesis. University of California at Los Angeles, School of Engineering.

35 Schmidt, S.F. (1970). Computational techniques in Kalman filtering. In: *Theory and Applications of Kalman Filtering*, AGARDograph 139. London: NATO Advisory Group for Aerospace Research and Development.

36 Carlson, N.A. (1973). Fast triangular formulation of the square root filter. *AIAA Journal* 11 (9): 1259–1265.

37 Dyer, P. and McReynolds, S. (1969). Extension of square-root filtering to include process noise. *Journal of Optimization Theory and Applications* 3: 444–458.
38 Lawson, C.L. and Hanson, R.J. (1974). *Solving Least Squares Problems*. Englewood Cliffs, NJ: Prentice-Hall.
39 Bucy, R.S. (1965). Nonlinear filtering theory. *IEEE Transactions on Automatic Control* 10 (2): 198.
40 Bass, R., Norum, V., and Schwartz, L. (1966). Optimal multichannel nonlinear filtering. *J. Math. Anal. Appl.* 16: 152–164.
41 Bucy, R.S. and Lo, J.T. (1998). Seminal contributions of Bob Bass to control. Proceedings of the 1998 American Control Conference. IEEE.
42 Ulam, S.M. (1976). *Adventures of a Mathematician*. New York: Charles Scribner's Sons.
43 Schweppe, F.C. (1965). Evaluation of likelihood functions for Gaussian signals. *IEEE Transactions on Information Theory* 11: 61–70.
44 Bierman, G.S. (1995). Error modeling for differential GPS. MS thesis. MIT.

11

Inertial Navigation Error Analysis

Football is a game of errors.
The team that makes the fewest errors in a game usually wins.
Paul Eugene Brown (1908–1991), Co-founder and Namesake, Cleveland Browns Professional Football Franchise

11.1 Chapter Focus

Design and development of inertial navigation systems is driven by error management, from the control of inertial sensor error characteristics during manufacture to the assessment of how the remaining errors contribute to essential navigation performance metrics during operation, and to what mitigation methods might improve performance, and by how much.

The purpose of this chapter is to show where mathematical models characterizing the dynamics of INS navigation errors come from, and how these models can be used for predicting INS performance. In the next chapter, some of these models will be augmented with error models for global navigation satellite systems (GNSS) for integrating GNSS with inertial navigation systems (INS).

Models for sensor error characteristics generally come from materials science and physics, as verified through testing. These generally fall into three types:

1. Unpredictable uncorrelated noise. Although the standard Kalman filter models for sensor noise are uncorrelated, the Kalman modeling formalism can also represent a wide variety of correlated noise models with parameters that can be estimated from test results and exploited to improve sensor performance. A model of this sort is commonly used for vertical stabilization of inertial navigation using a barometric altimeter.
2. Predictable errors such as those due to fixed variations in scale factors, biases, and input axis misalignments. The Kalman filter is a useful tool for identifying which parametric models best represent these errors, using

Global Navigation Satellite Systems, Inertial Navigation, and Integration,
Fourth Edition. Mohinder S. Grewal, Angus P. Andrews, and Chris G. Bartone.

Companion website: www.wiley.com/go/grewal/gnss

input–output pairs under controlled test conditions for estimating the parameters of those models and using the resulting parametric model to compensate for the modeled errors during operation. This is very common in inertial navigation and may include the use of thermometers for temperature compensation.
3. Correlated errors due to slowly changing values of the sensor error compensation parameters, which may be mitigated to some degree by using auxiliary navigation sensors such as altimeters (this chapter) or GNSS (next chapter).

We will demonstrate the general approach to sensor error modeling and compensation using hypothetical error models common in inertial navigation. Models for specific sensors may include additional mathematical forms, but the general approach would be the same.

Models for how sensor errors contribute to navigation errors in terrestrial environments have largely evolved since the introduction of Kalman filtering in 1960. These models can generally use linear approximations because sensor errors are so small relative to full-scale sensor outputs, and navigation errors are so small relative to the navigation solution. This modeling often ignores small effects that would be essential in the navigation solution but have insignificant effect on error propagation.

Navigation performance models generally depend on the trajectories of the host vehicle, as well as sensor errors. Therefore, trajectories used in performance analysis should represent the intended application(s) of the INS.

These models and methods have been used for integrating INS with auxiliary sensors from the beginning. The general subject has been called *inertial systems analysis*, and it has become a powerful tool for the development of INS and its integration with GNSS. One of the early textbooks on inertial navigation systems analysis was recently republished [1]. There is another classic treatise on the subject by Widnall and Grundy [2], but it is long out of print.

11.2 Errors in the Navigation Solution

The *navigation solution* for inertial navigation is what it takes to propagate the initial navigation solution forward in time, starting with the initial conditions of position, velocity, and orientation. These initial conditions are never known exactly, nor is their forward propagation model. As a result, there will always be errors in the navigation solution. It turns out that these errors also have distinctive ways of corrupting the calculations in the solution implementation, with the effect that these errors have their own dynamic evolution model.

This section is about those kinds of models. They are generally linear and depend on the error variables used in inertial systems analysis being sufficiently small compared with the navigation solution that linearity dominates their relationships. Modeled errors in the values of navigation variables are generally several orders of magnitude smaller than the navigation variables themselves.

11.2.1 Navigation Error Variables

For inertial navigation, navigation errors include errors in position, velocity, and orientation, each of which is three-dimensional. As described in Section 3.5, these are the same variables required for initializing the navigation solution, and they are used with the sensor outputs to carry the navigation solution forward in time. The resulting nine "core" inertial navigation error variables will also define how navigation errors propagate over time.

11.2.2 Coordinates Used for INS Error Analysis

Inertial navigation systems analysis can be complicated in any coordinate system. It was originally developed using locally level coordinates, because early systems were gimbaled with locally level inertial measurement unit (IMU) axes for navigation. This approach was natural in some ways, because INS performance requirements were specified in locally level coordinates. It is also natural because locally level coordinates represent the direction of gravity, which is very important for inertial navigation in the terrestrial environment. Locally level coordinates are used here for the same reasons. These could be either north–east–down (NED) or east–north–up (ENU). ENU coordinates are used here, although NED coordinates have been used elsewhere. Derivations in NED coordinates would follow similar lines. Derivations with respect to latitude and longitude are not that dissimilar.

11.2.3 Model Variables and Parameters

Table 11.1 is a list of symbols and definitions of the parameters and variables used in the modeling and derivations of navigation error dynamics. These include nine navigation error variables in ENU coordinates, and other variables and parameters used in the INS implementation. The "hatted" variables (e.g. $\hat{\phi}$) represent values used in the INS implementation.

All error variables are represented in ENU coordinates centered at the actual location of the inertial sensor assembly.

The state variable representations of the nine core INS errors are listed in Table 11.2.

11.2.3.1 INS Orientation Variables and Errors

INS orientation with respect to gravity and the earth rotation axis is typically represented in terms of the coordinate transformation between the INS sensor axes and locally level ENU coordinates. It will simplify the derivation somewhat if we assume that the INS sensor axes are nominally parallel to the ENU axes, except for small orientation errors. By "small," we mean that they can be represented by rotations in the order of a milliradian or less. At those levels, error

Table 11.1 List of symbols and approximations used.

			Earth model parameters[a]		
$\overline{R}_\oplus$	$\stackrel{\text{def}}{=}$	Mean radius			$\approx 0.637\ 100\ 9 \times 10^7$ m
$a_\oplus$	$\stackrel{\text{def}}{=}$	Equatorial radius			$\approx 0.637\ 813\ 7 \times 10^7$ m
$f_\oplus$	$\stackrel{\text{def}}{=}$	Flattening			$\approx 1/298.257\ 223\ 56$
					$\approx 0.0033\ 528\ 106\ 647\ 5$
$\text{GM}_\oplus$	$\stackrel{\text{def}}{=}$	Gravity constant			$\approx 0.398\ 600\ 4 \times 10^{15} \text{m}^3/\text{s}^2$
$\Omega_\oplus$	$\stackrel{\text{def}}{=}$	Rotation rate			$\approx 0.729\ 211\ 5 \times 10^{-4}$ rad/s
			INS navigation solution		
$\hat{\phi}$	$=$	ϕ	$+$	$\varepsilon_N/\overline{R}_\oplus$	Latitude (rad)
$\hat{\theta}$	$=$	θ	$+$	$\varepsilon_E/(\overline{R}_\oplus \cos \phi)$	Longitude (rad)
$\hat{E}$	$=$	E	$+$	ε_E	Easting with respect to INS (m)
$\hat{N}$	$=$	N	$+$	ε_N	Northing with respect to INS (m)
$\hat{h}$	$=$	h	$+$	ε_U	Altitude (m)
$\hat{v}_E$	$=$	v_E	$+$	$\dot{\varepsilon}_E$	East INS velocity (m/s)
$\hat{v}_N$	$=$	v_N	$+$	$\dot{\varepsilon}_N$	North INS velocity (m/s)
$\hat{v}_U$	$=$	v_U	$+$	$\dot{\varepsilon}_U$	Vertical INS velocity (m/s)
$\mathbf{C}^{\text{INS}}_{\text{ENU}}$	$\approx$	$\mathbf{I}$	$+$	$\boldsymbol{\rho} \otimes$	Coordinate transformation matrix, INS to ENU
ρ_E	$\stackrel{\text{def}}{=}$	INS misalignment about east axis (rad)			
ρ_N	$\stackrel{\text{def}}{=}$	INS misalignment about north axis (rad)			
ρ_U	$\stackrel{\text{def}}{=}$	INS misalignment about vertical axis (rad)			
			Miscellaneous variables and symbols		
ψ	$=$	Horizontal velocity direction, measured counter-clockwise from east (rad)			
$\boldsymbol{\omega}_\oplus$	$\stackrel{\text{def}}{=}$	Earthrate vector (rad/s)			
$\mathbf{x}$	$=$	Vector from the center of the Earth to the INS			
$\otimes$	$\stackrel{\text{def}}{=}$	Vector cross-product. As a suffix, it transforms a vector into its equivalent skew-symmetric matrix			

a) ⊕ is the astronomical symbol for Earth.

Table 11.2 State variables for the nine core INS errors.

		State vector	
ξ	$\stackrel{\text{def}}{=}$	$\begin{bmatrix} \varepsilon \\ \dot{\varepsilon} \\ \rho \end{bmatrix}$	INS navigation error
		Sub-vectors	
ε	$\stackrel{\text{def}}{=}$	$\begin{bmatrix} \varepsilon_E \\ \varepsilon_N \\ \varepsilon_U \end{bmatrix}$	INS position error
$\dot{\varepsilon}$	$\stackrel{\text{def}}{=}$	$\begin{bmatrix} \dot{\varepsilon}_E \\ \dot{\varepsilon}_N \\ \dot{\varepsilon}_U \end{bmatrix}$	INS velocity error
ρ	$\stackrel{\text{def}}{=}$	$\begin{bmatrix} \rho_E \\ \rho_N \\ \rho_U \end{bmatrix}$	INS orientation error

dynamics can be modeled in terms of first-order variations, and second-order effects can be ignored.

Misalignments and Tilts INS orientation errors are represented by misalignment variables, which are different enough from the other state variables that some clarification about what they mean and how they are used may be useful.

Misalignments represent the rotational difference between these locally level reference directions and what the navigation solution has estimated for them.

Errors in INS-calculated directions can be divided into three categories:

1. INS misalignments with respect to locally level coordinates at the actual position of the INS. These cause errors in the calculation of:
   a) Gravity, which is needed for navigation in an accelerating coordinate frame. These orientation errors are sometimes called "tilt errors," because they are equivalent to rotating the gravity vector about horizontal axes.
   b) The direction of the rotation axis of the Earth, which is needed for navigation in a rotating coordinate frame.
2. Errors in the INS estimate of its position. Errors in the estimated latitude, in particular, cause errors in the expected elevation angle of the Earth's rotation axis above or below the horizon. These are not true misalignments,

but their effects must be taken into account when determining the dynamic cross-coupling between different error types. Errors in longitude do not have much influence on dynamic coupling of navigation errors, although they do influence pointing accuracies with respect to objects in space.

Effect of INS Misalignments INS misalignments are represented in terms of a coordinate transformation between what the INS *believes* to be locally level ENU coordinates and the actual locally level ENU coordinates. If the misalignments are sufficiently "small" (in the order of milliradians or less) the coordinate transformation $\mathbf{C}_{\text{ENU}}^{\text{INS}}$ from INS coordinates to ENU coordinates can be approximated as

$$\mathbf{C}_{\text{ENU}}^{\text{INS}}(\boldsymbol{\rho}) \approx \mathbf{I} + \boldsymbol{\rho}\otimes = \begin{bmatrix} 1 & -\rho_U & \rho_N \\ \rho_U & 1 & -\rho_E \\ -\rho_N & \rho_E & 1 \end{bmatrix} \tag{11.1}$$

This approximation includes just the first two terms in the series expansion of the matrix exponential of the skew-symmetric matrix $\boldsymbol{\rho}\otimes$, which is the exact form of a coordinate transformation equivalent to a rotation. This small-angle approximation is used where necessary to transform INS variables to ENU coordinates.

Figure 11.1 shows a misalignment vector $\boldsymbol{\rho}$ representing a small-angle[1] rotation of the reference navigation coordinates used by the INS. True ENU coordinate axes at the location of the INS are labeled *E*, *N*, and *U*. However, the orientation error of the INS is such that its estimated coordinate directions are actually those labeled $\hat{E}$, $\hat{N}$, and $\hat{U}$ in the figure. This effective coordinate change is equivalent to rotation about the vector $\boldsymbol{\rho}$ by a small angle $|\boldsymbol{\rho}|$.

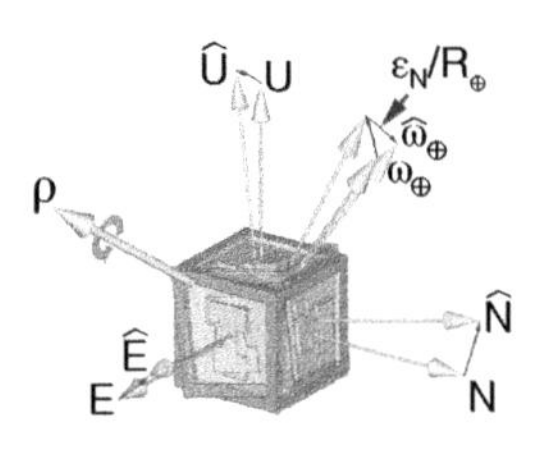

Figure 11.1 Misalignment of INS navigation coordinates.

Misalignments contribute to navigation errors in a number of ways. Figure 11.1 illustrates how misalignments cause miscalculation of the value of earth rotation used in the INS for maintaining its reference axes locally level, and how northing errors compound the error. The actual rotation rate vector in true ENU coordinates is labeled $\boldsymbol{\omega}_\oplus$ in the figure. However, due to misalignment error, the INS representation of this direction has been rotated twice: once by misalignments and again by miscalculation of latitude due to northing error ε_N divided by $\overline{R}_\oplus$ the mean radius of the Earth. As a consequence, in true ENU coordinates, the actual earth rotation rate vector (labeled $\boldsymbol{\omega}_\oplus$ in the figure) is moved to the direction labeled $\hat{\boldsymbol{\omega}}_\oplus$ in the figure by INS navigation errors.

1 For illustrative purposes only, the actual rotation angle in the figure has been made relatively large.

The resulting estimated value $\hat{\omega}_{\oplus}$ is used in the INS for maintaining its reference navigation coordinate aligned with what it believes to be true ENU coordinates and for compensating for Coriolis effect. The Coriolis implementation will have further compounding effects due to errors in estimated velocity. The result is a fairly complex structure of how navigation errors cause even more errors through how inertial navigation is implemented. This complexity is reduced somewhat by using only first-order error modeling.

In the first-order analysis model, what the small-angle approximation does to any vector $\boldsymbol{v}$ is approximated by

$$\begin{aligned} \hat{\boldsymbol{v}} &\approx [\mathbf{I} + \boldsymbol{\rho}\otimes]\boldsymbol{v} \\ &= \boldsymbol{v} + \boldsymbol{\rho} \otimes \boldsymbol{v} \\ &= \boldsymbol{v} - \boldsymbol{v} \otimes \boldsymbol{\rho} \\ &= \boldsymbol{v} + \begin{bmatrix} 0 & v_3 & -v_2 \\ -v_3 & 0 & v_1 \\ v_2 & -v_1 & 0 \end{bmatrix} \boldsymbol{\rho} \end{aligned} \tag{11.2}$$

These formulas are used repeatedly to represent how misalignments corrupt the estimated variables used in the INS implementation, and how this affects navigation errors.

Small-Angle Rotation Rate Approximation The differential equations

$$\begin{aligned} \frac{d}{dt}\mathbf{C}_{\text{TO}}^{\text{FROM}} &= \mathbf{C}_{\text{TO}}^{\text{FROM}}[\boldsymbol{\omega}_{\text{TO}}\otimes] \\ &= [\boldsymbol{\omega}_{\text{FROM}}\otimes]\mathbf{C}_{\text{TO}}^{\text{FROM}} \end{aligned} \tag{11.3}$$

model the time-rates-of-change of a coordinate transformation matrix $\mathbf{C}_{\text{TO}}^{\text{FROM}}$ from coordinate frame "FROM" to coordinate frame "TO" due to a rotation rate vector $\boldsymbol{\omega}_{\text{TO}}$ represented in the "TO" coordinate frame or $\boldsymbol{\omega}_{\text{FROM}}$ represented in the "FROM" coordinate frame. These formulas are generally not recommended for implementing strapdown rotation rate integration, but they are adequate for deriving first-order models for INS orientation error propagation. Also,

$$\text{if } \mathbf{C} \approx \mathbf{I} + \boldsymbol{\rho}\otimes, \text{ then } \frac{d}{dt}\mathbf{C} \approx \left[\frac{d}{dt}\boldsymbol{\rho}\right]\otimes$$

Effects of Position Errors Figure 11.2 is an illustration of the effective small-angle rotations associated with horizontal position errors. The north component of position error, ε_N, is equivalent to a small-angle rotation of

$$\Delta_{\text{Lat}} = \frac{\varepsilon_N}{R_{\oplus}} \tag{11.4}$$

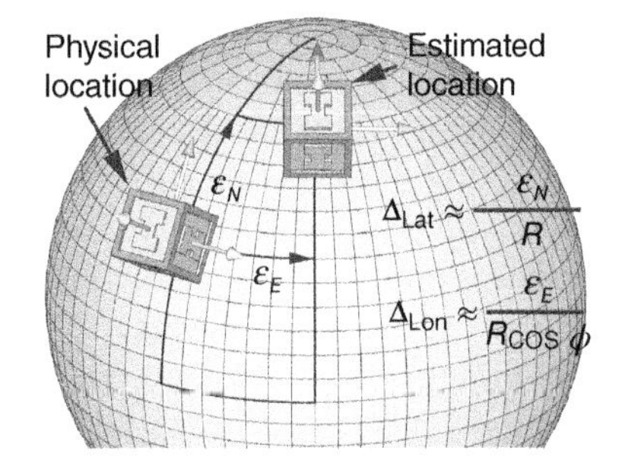

Figure 11.2 Effects of INS position errors.

about the negative east axis. Consequently, the equivalent small-angle rotation vector would be

$$\tilde{\rho}(\varepsilon_N) \approx \begin{bmatrix} -\frac{\varepsilon_N}{\overline{R}_\oplus} \\ 0 \\ 0 \end{bmatrix} \tag{11.5}$$

The primary effect of north position error is in the miscalculation of latitude, which has a direct effect on the calculation of earthrate. Fortunately, gyrocompass alignment of an INS ordinarily estimates latitude.

Because longitude is measured eastward, the east component of position error, ε_E, is equivalent to a rotation of

$$\Delta_{\text{Lon}} = \frac{\varepsilon_E}{\overline{R}_\oplus \cos\phi} \tag{11.6}$$

about the Earth's rotation axis (the polar axis), which has no effect on the estimated direction of the Earth's rotation axis in locally level ENU coordinates. This change in orientation is equivalent to the small-angle rotation vector

$$\tilde{\rho}(\varepsilon_E) \approx \begin{bmatrix} 0 \\ \frac{\varepsilon_E}{\overline{R}_\oplus} \\ \frac{\sin\phi\varepsilon_E}{\overline{R}_\oplus \cos\phi} \end{bmatrix} \tag{11.7}$$

where ϕ is the latitude of the INS position. *Initial gyrocompass alignment of an INS makes no determination of longitude.* It must be initialized by other means.

The net coordinate rotation effect from position errors can then be represented as

$$\tilde{\rho}(\varepsilon) \approx \begin{bmatrix} -\dfrac{\varepsilon_N}{\overline{R}_\oplus} \\ \dfrac{\varepsilon_E}{\overline{R}_\oplus} \\ \dfrac{\sin\phi\varepsilon_E}{\overline{R}_\oplus \cos\phi} \end{bmatrix} \tag{11.8}$$

although the last two components have little influence on the propagation of navigation errors.

11.2.4 Dynamic Coupling Mechanisms

Dynamic coupling Dynamic coupling of variables is the coupling of one variable into the derivative of another. This happens in inertial navigation because things that cannot be sensed have to be calculated using the estimated values of the navigation solution. This includes gravity, for example, which cannot be sensed by the accelerometers. It must then be estimated using the estimated values of INS position, and orientation, and added to the sensed accelerations to calculate INS acceleration relative to the Earth. It is through errors in calculating gravity that the other navigation errors become dynamically coupled to velocity errors.

For inertial navigation in locally level coordinates, there are other variables that cannot be sensed directly and must be calculated in the same manner. Figure 11.3 illustrates how the INS calculates and uses such variables. In this flowchart of the INS navigation implementation, the boxes numbered 1–6 represent six such calculations that use the estimated values of the navigation solution (enclosed in the dashed box) to estimate variables that cannot be sensed directly, but must be estimated from the navigation solution. A seventh dynamic coupling mechanism (labeled at the point where sensed accelerations are transformed to INS coordinates) comes from the effective misalignments of the accelerometer input axes due to INS orientation errors.

The numbered error coupling processes then include:

1. *Integration of velocities.* INS position is part of the navigation solution, although it cannot be measured directly. It must be inferred from the integrals of the velocities. However, any errors in estimated velocities are also integrated, adding to position errors. This will dynamically couple velocity errors into position errors.
2. *Estimating gravity.* Gravity cannot be sensed and it is far too big to be ignored. It must be calculated using the navigation solution and integrated along with the sensed accelerations to calculate velocity relative to the

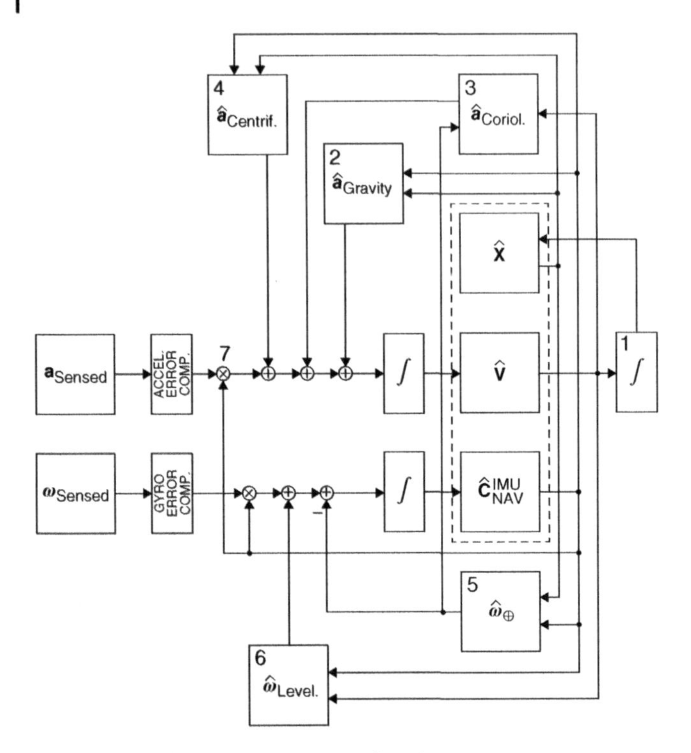

Figure 11.3 INS navigation solution flowchart.

Earth. Miscalculations due to navigation errors will dynamically couple them into velocity errors.

3. *Coriolis acceleration.* Coriolis accelerations in rotating coordinates cannot be sensed, either. They must be estimated from the navigation solution from the cross-product of velocity vector and the coordinate rotation rate vector. Miscalculation of the Coriolis effect dynamically couples navigation errors into velocity errors.
4. *Centrifugal acceleration.* This is another type of acceleration in rotating coordinates that cannot be sensed, so it must be estimated from the navigation solution using the estimated values of the earth rotation rate vector and the INS velocity vector in INS coordinates. Miscalculation of centrifugal acceleration due to navigation error will dynamically couple navigation errors into velocity errors.
5. *Earthrate compensation in leveling.* Leveling is the process of maintaining a locally level reference frame for terrestrial navigation. Due to the rotation of the Earth, the INS senses rotation rates even when it is stationary with respect to Earth's surface. For navigating in earth-fixed coordinates, the INS must estimate the contribution due to the earth rotation rate, based on its

navigation solution for latitude and orientation, and subtract it to maintain its locally level orientation. Errors in the navigation solution will corrupt this calculation, which will then dynamically couple position and orientation errors into orientation errors.

6. *Velocity leveling*. Leveling is also required for maintaining locally level reference directions while the INS moves over the curved surface of the Earth. The necessary coordinate rotation rates are calculated using the estimated values of velocity with respect to the Earth and estimated orientation. Miscalculation of this correction due to navigation errors will dynamically couple velocity and orientation errors into orientation errors.
7. *Acceleration and misalignment errors*. This causes errors in sensed accelerations due to inertial sensor assembly (ISA) misalignment errors. The resulting velocity errors are coupled into position errors by #1, above.

In the previously mentioned listing, note that:

1. Only the first of these (velocity integration) dynamically couples other navigation errors into position errors.
2. The next three (gravity, Coriolis, and centrifugal accelerations) dynamically couple navigation errors to velocity errors.
3. The fifth and sixth (earthrate compensation and leveling) dynamically couple navigation errors to orientation errors.
4. The seventh couples INS misalignment errors into velocity errors.

This partitioning will determine the data structure of the dynamic coefficient matrix in the model for propagating navigation errors. The resulting time-varying linear dynamic model for navigation errors will have the partitioned form:

$$\frac{d}{dt}\begin{bmatrix} \varepsilon \\ \dot{\varepsilon} \\ \rho \end{bmatrix} = \begin{bmatrix} \mathbf{F}_{11} & \mathbf{F}_{12} & \mathbf{F}_{13} \\ \mathbf{F}_{21} & \mathbf{F}_{22} & \mathbf{F}_{23} \\ \mathbf{F}_{31} & \mathbf{F}_{32} & \mathbf{F}_{33} \end{bmatrix}\begin{bmatrix} \varepsilon \\ \dot{\varepsilon} \\ \rho \end{bmatrix} \tag{11.9}$$

where

$\mathbf{F}_{11}$ represents the dynamic coupling of position errors into position errors
$\mathbf{F}_{12}$ represents the dynamic coupling of velocity errors into position errors
$\mathbf{F}_{13}$ represents the dynamic coupling of orientation errors into position errors
$\mathbf{F}_{21}$ represents the dynamic coupling of position errors into velocity errors
$\mathbf{F}_{22}$ represents the dynamic coupling of velocity errors into velocity errors
$\mathbf{F}_{23}$ represents the dynamic coupling of orientation errors into velocity errors
$\mathbf{F}_{31}$ represents the dynamic coupling of position errors into orientation errors
$\mathbf{F}_{32}$ represents the dynamic coupling of velocity errors into orientation errors
$\mathbf{F}_{33}$ represents the dynamic coupling of orientation errors into orientation errors

In the next section, mathematical models for each of the navigation calculations referenced to the numbered boxes in Figure 11.3 are used for deriving a model for the dynamics of the core navigation error variables, and the results are organized into formulas for the respective sub-matrices $\mathbf{F}_{ij[k]}$, where the $[k]$ denotes which of the six navigation calculations is responsible for the dynamic coupling contribution. The complete dynamic coefficient matrix will be the sum of the contributions from all six.

Evaluating the six 3×9 Jacobians requires taking $6 \times 3 \times 9 = 162$ partial derivatives, which were all evaluated in a symbolic mathematics programming environment.

11.3 Navigation Error Dynamics

For each of the six suspect calculations, the first-order contributions of the kth contributor to the 9×9 matrix of Eq. (11.9) are calculated as the Jacobian matrix of the respective function $\mathbf{f}_{[k]}$ it performs with respect to the navigation error variables, evaluated at $\xi = 0$. The resulting matrices are used for modeling first-order sensitivities of INS implementation error dynamics to navigation errors:

$$\mathbf{F}_{[k]} \approx \left. \frac{\partial \mathbf{f}_{[k]}}{\partial \xi} \right|_{\xi=0} \tag{11.10}$$

As mentioned in the paragraph following Eq. 11.9, each function $\mathbf{f}_{[k]}$ will contribute to only one of the three-row block rows of Eq. (11.9). Table 11.3 is a breakdown of that partitioning.

Table 11.3 Dynamic coefficient sub-matrix sources.

Corrupted INS implementation processes	Error coupling mechanisms: Position error	Velocity error	Tilt error
1. Velocity integ.	$\mathbf{F}_{1,1\ [1]}$	$\mathbf{F}_{1,2\ [1]}$	$\mathbf{F}_{1,3\ [1]}$
2. Gravity acc.	$\mathbf{F}_{2,1\ [2]}$	$\mathbf{F}_{2,2\ [2]}$	$\mathbf{F}_{2,3\ [2]}$
3. Coriolis acc.	$\mathbf{F}_{2,1\ [3]}$	$\mathbf{F}_{2,2\ [3]}$	$\mathbf{F}_{2,3\ [3]}$
4. Centrifugal acc.	$\mathbf{F}_{3,1\ [4]}$	$\mathbf{F}_{3,2\ [4]}$	$\mathbf{F}_{3,3\ [4]}$
5. Earthrate comp.	$\mathbf{F}_{3,1\ [5]}$	$\mathbf{F}_{3,2\ [5]}$	$\mathbf{F}_{3,3\ [5]}$
6. Leveling	$\mathbf{F}_{3,1\ [6]}$	$\mathbf{F}_{3,2\ [6]}$	$\mathbf{F}_{3,3\ [6]}$
7. Accel. misalign.	$\mathbf{F}_{2,1\ [7]}$	$\mathbf{F}_{2,2\ [7]}$	$\mathbf{F}_{2,3\ [7]}$

The final dynamic coefficient matrix of the navigation error model will be composed of the sums from each of the seven potentially contributing computational processes:

$$\mathbf{F}_{11} = \mathbf{F}_{11\ [1]} \tag{11.11}$$

$$\mathbf{F}_{12} = \mathbf{F}_{12\ [1]} \tag{11.12}$$

$$\mathbf{F}_{13} = \mathbf{F}_{13\ [1]} \tag{11.13}$$

$$\mathbf{F}_{21} = \mathbf{F}_{21\ [2]} + \mathbf{F}_{21\ [3]} + \mathbf{F}_{21\ [4]} + \mathbf{F}_{21\ [7]} \tag{11.14}$$

$$\mathbf{F}_{22} = \mathbf{F}_{22\ [2]} + \mathbf{F}_{22\ [3]} + \mathbf{F}_{22\ [4]} + \mathbf{F}_{22\ [7]} \tag{11.15}$$

$$\mathbf{F}_{23} = \mathbf{F}_{23\ [2]} + \mathbf{F}_{23\ [3]} + \mathbf{F}_{23\ [4]} + \mathbf{F}_{23\ [7]} \tag{11.16}$$

$$\mathbf{F}_{31} = \mathbf{F}_{31\ [5]} + \mathbf{F}_{31\ [6]} \tag{11.17}$$

$$\mathbf{F}_{32} = \mathbf{F}_{32\ [5]} + \mathbf{F}_{32\ [6]} \tag{11.18}$$

$$\mathbf{F}_{33} = \mathbf{F}_{33\ [5]} + \mathbf{F}_{33\ [6]} \tag{11.19}$$

This is the approach used for deriving the first-order error propagation model for the nine core navigation error variables. The following seven subsections contain the derivations of the component submatrices $\mathbf{F}_{ij[k]}$ listed earlier.

11.3.1 Error Dynamics Due to Velocity Integration

This is the most straightforward part of the model. The navigation error sub-vector $\dot{\varepsilon}$ is the error in the INS velocity estimate, and the only modeled error corruption due to velocity integration will be the accumulation of navigation position error ε through integration of velocity error $\dot{\varepsilon}$. That is,

$$\begin{aligned} \frac{d}{dt}\varepsilon &= \dot{\varepsilon} \\ &= \left[\mathbf{F}_{11\ [1]}\ \mathbf{F}_{12\ [1]}\ \mathbf{F}_{13\ [1]} \right] \xi \end{aligned} \tag{11.20}$$

$$\mathbf{F}_{11\ [1]} = 0 \text{ (}3 \times 3\text{ zero matrix)} \tag{11.21}$$

$$\mathbf{F}_{12\ [1]} = \mathbf{I} \text{ (}3 \times 3\text{ identity matrix)} \tag{11.22}$$

$$\mathbf{F}_{13\ [1]} = 0 \text{ (}3 \times 3\text{ zero matrix),} \tag{11.23}$$

where the “[1]” in the subscript refers to the first error coupling source listed in Table 11.3.

11.3.2 Error Dynamics Due to Gravity Miscalculations

11.3.2.1 INS Gravity Modeling

The model used for calculating gravity has to be rather accurate to attain reasonable inertial navigation performance [3]. Ellipsoidal earth models generally limit INS performance to $\sim 10^{-1}$ nautical mile per hour circular error probable (CEP) rate (defined in Section 3.7.3.1). The ellipsoidal gravity models will usually include a term of the sort

$$\hat{\mathbf{g}}_{\mathrm{INS}} \approx \begin{bmatrix} 0 \\ 0 \\ \frac{-\mathrm{GM}_{\oplus}}{[R_{\oplus}(\hat{\phi})+\hat{h}]^2} \end{bmatrix}$$

$$\hat{R}_{\oplus}(\hat{\phi}) \approx a_{\oplus}(1 - f_{\oplus}\sin^2 \hat{\phi})$$

where the variable $R_{\oplus}(\phi)$ is the reference ellipsoidal geoid surface radius, h is the INS height above that surface, $a_{\oplus}$ is the semi-major axis of the reference ellipsoid, $f_{\oplus}$ its flattening, and ϕ the geodetic latitude of the INS.

More general geoid models add spherical harmonics[2] to the ellipsoidal equipotential surface height.

11.3.2.2 Navigation Error Model for Gravity Calculations

The dominant term in the dynamic coefficient matrix is the one due to the vertical gradient of gravity, which is about -3×10^{-6} m/s^2 per meter of elevation increase at the surface of the Earth. It is the one that makes unaided inertial navigation unstable in the vertical direction.

The other major navigation error propagation effects of gravity miscalculation are caused by misalignments (ρ). To capture these effects, the gravity model can be stripped down to a simpler form:

$$\hat{\mathbf{g}}_{\mathrm{INS}} \approx \begin{bmatrix} 0 \\ 0 \\ \frac{-\mathrm{GM}_{\oplus}}{[R_{\oplus}+\varepsilon_U]^2} \end{bmatrix} \tag{11.24}$$

for which the equivalent value in ENU coordinates will be

$$\hat{\mathbf{g}}_{\mathrm{ENU}} \approx C^{\mathrm{INS}}_{\mathrm{ENU}} \hat{\mathbf{g}}_{\mathrm{INS}} \tag{11.25}$$

The navigation errors in this approximation will cause acceleration errors, which are the time-derivatives of $\dot{\varepsilon}$

$$\begin{aligned} \frac{d}{dt}\dot{\varepsilon} &\approx \left.\frac{\partial \hat{\mathbf{g}}_{\mathrm{ENU}}}{\partial \xi}\right|_{\xi=0} \xi \\ &= \left[\mathbf{F}_{21\,[2]} \ \mathbf{F}_{22\,[2]} \ \mathbf{F}_{23\,[2]} \right] \xi \end{aligned} \tag{11.26}$$

---

2 Up to order 60 in some models.

$$\mathbf{F}_{21\ [2]} = \begin{bmatrix} 0 & 0 & 0 \\ 0 & 0 & 0 \\ 0 & 0 & \frac{2\text{GM}}{\overline{R}_{\oplus}^{3}} \end{bmatrix} \tag{11.27}$$

$$\mathbf{F}_{22\ [2]} = \begin{bmatrix} 0 & 0 & 0 \\ 0 & 0 & 0 \\ 0 & 0 & 0 \end{bmatrix} \tag{11.28}$$

$$\mathbf{F}_{23\ [2]} = \begin{bmatrix} 0 & -\frac{\text{GM}}{\overline{R}_{\oplus}^{2}} & 0 \\ \frac{\text{GM}}{\overline{R}_{\oplus}^{2}} & 0 & 0 \\ 0 & 0 & 0 \end{bmatrix}, \tag{11.29}$$

where the "[2]" in the subscript refers to the second error coupling source listed in Table 11.3.

The lone nonzero element in $\mathbf{F}_{21\ [2]}$ is the cause of vertical error instability, which will be addressed in Section 11.3.9.

11.3.3 Error Dynamics Due to Coriolis Acceleration

The calculation for Coriolis correction is

$$\frac{d}{dt}\mathbf{v}_{\text{Coriolis}} = -2\omega_{\oplus} \otimes \mathbf{v} \tag{11.30}$$

However, navigation errors will corrupt its implementation as

$$\frac{d}{dt}\hat{\mathbf{v}}_{\text{Coriolis}} = -2\hat{\omega}_{\oplus} \otimes (\mathbf{v} + \dot{\varepsilon}) \tag{11.31}$$

$$\hat{\omega}_{\oplus} = \mathbf{C}_{\text{ENU}}^{\text{INS}} \begin{bmatrix} 0 \\ \cos\hat{\phi}\ \Omega_{\oplus} \\ \sin\hat{\phi}\ \Omega_{\oplus} \end{bmatrix} \tag{11.32}$$

$$\cos\hat{\phi} \approx \cos\phi - \sin\phi \frac{\varepsilon_N}{\overline{R}_{\oplus}} \tag{11.33}$$

$$\sin\hat{\phi} \approx \sin\phi + \cos\phi \frac{\varepsilon_N}{\overline{R}_{\oplus}} \tag{11.34}$$

the contribution of which to navigation error dynamics will be

$$\frac{d}{dt}\dot{\varepsilon} \approx \left.\frac{\partial \hat{\mathbf{v}}_{\text{Coriolis}}}{\partial \xi}\right|_{\xi=0} \xi$$
$$= \left[\mathbf{F}_{21\ [3]}\ \mathbf{F}_{22\ [3]}\ \mathbf{F}_{23\ [3]} \right] \xi \tag{11.35}$$

$$\mathbf{F}_{21\ [3]} = \begin{bmatrix} 0 & 2\,\dfrac{\Omega_\oplus \sin\phi\, v_U}{\overline{R}_\oplus} + 2\,\dfrac{\Omega_\oplus \cos\phi\, v_N}{\overline{R}_\oplus} & 0 \\ 0 & -2\,\dfrac{\Omega_\oplus \cos\phi\, v_E}{\overline{R}_\oplus} & 0 \\ 0 & -2\,\dfrac{\Omega_\oplus \sin\phi\, v_E}{\overline{R}_\oplus} & 0 \end{bmatrix} \tag{11.36}$$

$$\mathbf{F}_{22\ [3]} = \begin{bmatrix} 0 & 2\,\Omega_\oplus \sin\phi & -2\,\Omega_\oplus \cos\phi \\ -2\,\Omega_\oplus \sin\phi & 0 & 0 \\ 2\,\Omega_\oplus \cos\phi & 0 & 0 \end{bmatrix} \tag{11.37}$$

$$\mathbf{F}_{23\ [3]} = \begin{bmatrix} 2\,\Omega_\oplus \sin\phi\, v_U + 2\,\Omega_\oplus \cos\phi\, v_N & 0 & 0 \\ -2\,\Omega_\oplus \cos\phi\, v_E & 2\,\Omega_\oplus \sin\phi\, v_U & -2\,\Omega_\oplus \cos\phi\, v_U \\ -2\,\Omega_\oplus \sin\phi\, v_E & -2\,\Omega_\oplus \sin\phi\, v_N & 2\,\Omega_\oplus \cos\phi\, v_N \end{bmatrix} \tag{11.38}$$

where the "[3]" in the subscript refers to the third error coupling source listed in Table 11.3.

11.3.4 Error Dynamics Due to Centrifugal Acceleration

Centrifugal acceleration is built into the terrestrial gravity model at the surface of the reference geoid. It is the primary reason for its equatorial bulge. As a consequence, the vector sum of gravitational and centrifugal acceleration at sea level should be orthogonal to the surface.

The problem is that the respective gradients of centrifugal and gravitational accelerations at the surface are quite different. Gravity decreases as the inverse square of radius, but centrifugal acceleration increases linearly with radius. We will examine further in Section 11.3.9 how the gravity gradient influences the dynamics of inertial navigation errors, and we need to do the same for centrifugal acceleration. For that purpose, the formula for centrifugal acceleration is derived in section B.5 of Appendix B. It has the form

$$\hat{\mathbf{a}}_{\text{Centrifugal}} = -[\hat{\omega}_\oplus \otimes][\hat{\omega}_\oplus \otimes]\hat{\mathbf{x}} \tag{11.39}$$

$$\hat{\mathbf{x}} = \mathbf{x} + \varepsilon \tag{11.40}$$

where **x** is the vector from Earth's center to the INS, and $\hat{\boldsymbol{\omega}}_\oplus$ defined by Eq. (11.32). The relevant Jacobian

$$\left.\frac{\partial \hat{\mathbf{a}}_{\text{Centrifugal}}}{\partial \xi}\right|_{\xi=0} = \left[\,\mathbf{F}_{21\,[4]}\;\mathbf{F}_{22\,[4]}\;\mathbf{F}_{23\,[4]}\,\right] \tag{11.41}$$

$$\mathbf{F}_{21\,[4]} = \begin{bmatrix} 0 & 0 & 0 \\ 0 & \Omega_\oplus{}^2\sin^2 phi - \Omega_\oplus{}^2\cos^2 phi & -\Omega_\oplus{}^2 \sin\phi\cos\phi \\ 0 & -2\,\Omega_\oplus{}^2 \sin\phi\cos\phi & \Omega_\oplus{}^2\cos^2\phi \end{bmatrix} \tag{11.42}$$

$$\mathbf{F}_{22\,[4]} = \begin{bmatrix} 0 & 0 & 0 \\ 0 & 0 & 0 \\ 0 & 0 & 0 \end{bmatrix} \tag{11.43}$$

$$\mathbf{F}_{23\,[4]} = \begin{bmatrix} 0 & \Omega_\oplus{}^2\cos^2\phi\,\overline{R}_\oplus & \Omega_\oplus{}^2\sin\phi\cos\phi\,\overline{R}_\oplus \\ -\Omega_\oplus{}^2\cos^2\phi\,\overline{R}_\oplus & 0 & 0 \\ -\Omega_\oplus{}^2\sin\phi\cos\phi\,\overline{R}_\oplus & 0 & 0 \end{bmatrix} \tag{11.44}$$

11.3.5 Error Dynamics Due to Earthrate Leveling

Maintaining locally level reference directions on a rotating Earth requires that the estimated earth rotation rate be subtracted from the measured rotation rates. The formula for the error in the estimated earth rotation rate vector has already been used in the error analysis of the Coriolis correction, in Eq. (11.32). The Jacobian of the calculated negative earth rotation rate with respect to navigation errors is then

$$\left.\frac{\partial(-\hat{\boldsymbol{\omega}}_\oplus)}{\partial \xi}\right|_{\xi=0} = \left[\,\mathbf{F}_{31\,[5]}\;\mathbf{F}_{32\,[5]}\;\mathbf{F}_{33\,[5]}\,\right] \tag{11.45}$$

$$\mathbf{F}_{31\,[5]} = \begin{bmatrix} 0 & 0 & 0 \\ 0 & \frac{\Omega_\oplus\sin\phi}{\overline{R}_\oplus} & 0 \\ 0 & -\frac{\Omega_\oplus\cos\phi}{\overline{R}_\oplus} & 0 \end{bmatrix} \tag{11.46}$$

$$\mathbf{F}_{32\,[5]} = \begin{bmatrix} 0 & 0 & 0 \\ 0 & 0 & 0 \\ 0 & 0 & 0 \end{bmatrix} \tag{11.47}$$

$$\mathbf{F}_{33\,[5]} = \begin{bmatrix} 0 & -\Omega_\oplus\sin\phi & \Omega_\oplus\cos\phi \\ \Omega_\oplus\sin\phi & 0 & 0 \\ -\Omega_\oplus\cos\phi & 0 & 0 \end{bmatrix} \tag{11.48}$$

where the "[5]" in the subscript refers to the fourth error coupling source listed in Table 11.3.

11.3.6 Error Dynamics Due to Velocity Leveling

Locally level reference directions need to rotate at some vector rate $\boldsymbol{\omega}_v$ as the INS moves with velocity $\mathbf{v}$ relative to the curved surface of the Earth, as illustrated in Figure 11.4. The model for this in the navigation implementation can be rather sophisticated, because rotation rate depends on the radius of curvature of the reference geoid, and this can be different in different directions, and at different positions on the Earth. However, this level of rigor is not necessary for modeling the dynamics of navigation errors. A spherical earth model will suffice.

In that case, the operative formula for the locally level coordinate rotation rate as a function of position, velocity, and orientation will be

$$\hat{\boldsymbol{\omega}}_v = \frac{1}{\overline{R}_\oplus + \varepsilon_U}\hat{\mathbf{u}}_U \otimes \hat{\mathbf{v}} = \frac{1}{\overline{R}_\oplus + \varepsilon_U} C_{\text{ENU}}^{\text{INS}} \begin{bmatrix} 0 \\ 0 \\ 1 \end{bmatrix} \otimes (\mathbf{v} + \dot{\varepsilon}) \tag{11.49}$$

$$\left.\frac{\partial \hat{\boldsymbol{\omega}}_v}{\partial \xi}\right|_{\xi=0} = \left[\, \mathbf{F}_{31\,[6]}\ \mathbf{F}_{32\,[6]}\ \mathbf{F}_{33\,[6]} \,\right] \xi \tag{11.50}$$

$$\mathbf{F}_{31\,[6]} = \begin{bmatrix} 0 & 0 & \frac{v_N}{\overline{R}_\oplus^{\,2}} \\ 0 & 0 & -\frac{v_E}{\overline{R}_\oplus^{\,2}} \\ 0 & 0 & 0 \end{bmatrix} \tag{11.51}$$

$$\mathbf{F}_{32\,[6]} = \begin{bmatrix} 0 & -\overline{R}_\oplus^{\,-1} & 0 \\ \overline{R}_\oplus^{\,-1} & 0 & 0 \\ 0 & 0 & 0 \end{bmatrix} \tag{11.52}$$

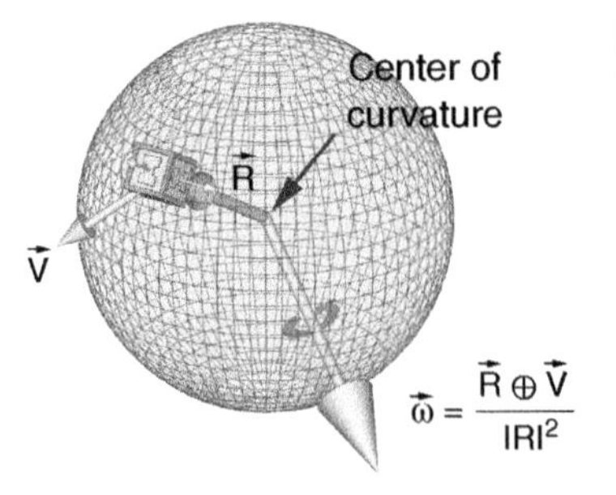

Figure 11.4 Velocity leveling for terrestrial navigation.

$$\mathbf{F}_{33\ [6]} = \begin{bmatrix} -\frac{v_U}{R_\oplus} & 0 & 0 \\ 0 & -\frac{v_U}{R_\oplus} & 0 \\ \frac{v_E}{R_\oplus} & \frac{v_N}{R_\oplus} & 0 \end{bmatrix}, \tag{11.53}$$

where the "[6]" in the subscript refers to the sixth error coupling source listed in Table 11.3.

11.3.7 Error Dynamics Due to Acceleration and IMU Alignment Errors

INS acceleration in INS coordinates is

$$\begin{aligned} \mathbf{a}_{\mathrm{INS}} &= \mathbf{C}_{\mathrm{INS}}^{\mathrm{ENU}} \mathbf{a}_{\mathrm{ENU}} \\ &= \{\mathbf{I} - \rho\otimes\}\mathbf{a}_{\mathrm{ENU}} \\ &= \mathbf{a}_{\mathrm{ENU}} - \rho \otimes \mathbf{a}_{\mathrm{ENU}} \\ &= \mathbf{a}_{\mathrm{ENU}} + \mathbf{a}_{\mathrm{ENU}} \otimes \rho \\ &= \mathbf{a}_{\mathrm{ENU}} + \begin{bmatrix} 0 & -a_U & a_N \\ a_U & 0 & -a_E \\ -a_N & a_E & 0 \end{bmatrix} \rho \end{aligned} \tag{11.54}$$

where $\mathbf{a}_{\mathrm{ENU}}$ is INS acceleration in ENU coordinates. The vector $\mathbf{a}_{\mathrm{ENU}}$ in this case is not sensed acceleration, but the physical acceleration of the INS in ENU coordinates.

The last term of the last equation above represents an acceleration error that will be integrated by the INS and added to velocity error. That is, the first-order contribution of this term to navigation error dynamics is a term coupling misalignments into the derivative of velocity error.

$$\mathbf{F}_{21\ [7]} = 0 \tag{11.55}$$

$$\mathbf{F}_{22\ [7]} = 0 \tag{11.56}$$

$$\mathbf{F}_{23\ [7]} = \begin{bmatrix} 0 & -a_U & a_N \\ a_U & 0 & -a_E \\ -a_N & a_E & 0 \end{bmatrix} \tag{11.57}$$

There is a corresponding term for rotation rates coupling misalignments into the derivative of misalignments. For gimbaled systems, those rotation rates in INS coordinates are those commanded by the INS to remain locally level, and these terms have already been factored into the model as leveling errors.

11.3.8 Composite Model from All Effects

Summing up the contributions from all six INS procedures according to Eqs. (11.11)–(11.66), the full dynamic coefficient matrix for the core variables will have the sub-matrices

$$\mathbf{F}_{11} = \mathbf{F}_{11\ [1]} = \begin{bmatrix} 0 & 0 & 0 \\ 0 & 0 & 0 \\ 0 & 0 & 0 \end{bmatrix} \tag{11.58}$$

$$\mathbf{F}_{12} = \mathbf{F}_{12\ [1]} = \begin{bmatrix} 1 & 0 & 0 \\ 0 & 1 & 0 \\ 0 & 0 & 1 \end{bmatrix} \tag{11.59}$$

$$\mathbf{F}_{13} = \mathbf{F}_{13\ [1]} = \begin{bmatrix} 0 & 0 & 0 \\ 0 & 0 & 0 \\ 0 & 0 & 0 \end{bmatrix} \tag{11.60}$$

$$\begin{aligned} \mathbf{F}_{21} &= \mathbf{F}_{21\ [2]} + \mathbf{F}_{21\ [3]} + \mathbf{F}_{21\ [4]} \\ &= \begin{bmatrix} 0 & 2\,\frac{\Omega_\oplus\ \sin\,\phi\ v_U}{\overline{R}_\oplus} + 2\,\frac{\Omega_\oplus\ \cos\,\phi\ v_N}{\overline{R}_\oplus} & 0 \\ 0 & -2\,\frac{\Omega_\oplus\ \cos\,\phi\ v_E}{\overline{R}_\oplus} + \Omega_\oplus{}^2 \sin^2\,\phi - \Omega_\oplus{}^2 \cos^2\,\phi & -\Omega_\oplus{}^2 \sin\,\phi\,\cos\,\phi \\ 0 & -2\,\frac{\Omega_\oplus\ \sin\,\phi\ v_E}{\overline{R}_\oplus} - 2\,\Omega_\oplus{}^2 \sin\,\phi\,\cos\,\phi & 2\,\frac{\mathrm{GM}}{\overline{R}_\oplus^3} + \Omega_\oplus{}^2 \cos^2\,\phi \end{bmatrix} \end{aligned} \tag{11.61}$$

$$\begin{aligned} \mathbf{F}_{22} &= \mathbf{F}_{22\ [2]} + \mathbf{F}_{22\ [3]} + \mathbf{F}_{22\ [4]} \\ &= \begin{bmatrix} 0 & 2\,\Omega_\oplus\ \sin\,\phi & -2\,\Omega_\oplus\ \cos\,\phi \\ -2\,\Omega_\oplus\ \sin\,\phi & 0 & 0 \\ 2\,\Omega_\oplus\ \cos\,\phi & 0 & 0 \end{bmatrix} \end{aligned} \tag{11.62}$$

$$\begin{aligned} \mathbf{F}_{23} &= \mathbf{F}_{23\ [2]} + \mathbf{F}_{23\ [3]} + \mathbf{F}_{23\ [3]} + \mathbf{F}_{23\ [7]} \\ &= \left[\begin{matrix} 2\,\Omega_\oplus\ \sin\,\phi\ v_U + 2\,\Omega_\oplus\ \cos\,\phi\ v_N & \cdots \\ a_U + \frac{\mathrm{GM}}{\overline{R}_\oplus^2} - 2\,\Omega_\oplus\ \cos\,\phi\ v_E - \Omega_\oplus{}^2 \cos^2\,\phi\ \overline{R}_\oplus & \cdots \\ -a_N - 2\,\Omega_\oplus\ \sin\,\phi\ v_E - \Omega_\oplus{}^2 \sin\,\phi\,\cos\,\phi\ \overline{R}_\oplus & \cdots \end{matrix}\right. \\ &\qquad \left.\begin{matrix} \cdots & -a_U - \frac{\mathrm{GM}}{\overline{R}_\oplus^2} + \Omega_\oplus{}^2 \cos^2\,\phi\ \overline{R}_\oplus & a_N + \Omega_\oplus{}^2 \sin\,\phi\,\cos\,\phi\ \overline{R}_\oplus \\ \cdots & 2\,\Omega_\oplus\ \sin\,\phi\ v_U & -a_E - 2\,\Omega_\oplus\ \cos\,\phi\ v_U \\ \cdots & a_E - 2\,\Omega_\oplus\ \sin\,\phi\ v_N & 2\,\Omega_\oplus\ \cos\,\phi\ v_N \end{matrix}\right] \end{aligned} \tag{11.63}$$

$$\mathbf{F}_{31} = \mathbf{F}_{31\ [5]} + \mathbf{F}_{31\ [6]} = \begin{bmatrix} 0 & 0 & \frac{v_N}{\overline{R}_\oplus^2} \\ 0 & \frac{\Omega_\oplus \sin\phi}{\overline{R}_\oplus} & -\frac{v_E}{\overline{R}_\oplus^2} \\ 0 & -\frac{\Omega_\oplus \cos\phi}{\overline{R}_\oplus} & 0 \end{bmatrix} \tag{11.64}$$

$$\mathbf{F}_{32} = \mathbf{F}_{32\ [5]} + \mathbf{F}_{32\ [6]} = \begin{bmatrix} 0 & -\overline{R}_\oplus^{-1} & 0 \\ \overline{R}_\oplus^{-1} & 0 & 0 \\ 0 & 0 & 0 \end{bmatrix} \tag{11.65}$$

$$\mathbf{F}_{33} = \mathbf{F}_{33\ [5]} + \mathbf{F}_{33\ [6]} = \begin{bmatrix} -\frac{v_U}{\overline{R}_\oplus} & -\Omega_\oplus \sin\phi & \Omega_\oplus \cos\phi \\ \Omega_\oplus \sin\phi & -\frac{v_U}{\overline{R}_\oplus} & 0 \\ -\Omega_\oplus \cos\phi + \frac{v_E}{\overline{R}_\oplus} & \frac{v_N}{\overline{R}_\oplus} & 0 \end{bmatrix} \tag{11.66}$$

and the corresponding equation numbers for the component sub-matrices $\mathbf{F}_{..\ [k]}$ are listed in Table 11.4.

The full 9×9 dynamic coefficient matrix for navigation errors, in 3×3 block form, is then

$$\mathbf{F}_{\text{core}} = \begin{bmatrix} \mathbf{F}_{11} & \mathbf{F}_{12} & \mathbf{F}_{13} \\ \mathbf{F}_{21} & \mathbf{F}_{22} & \mathbf{F}_{23} \\ \mathbf{F}_{31} & \mathbf{F}_{32} & \mathbf{F}_{33} \end{bmatrix} \tag{11.67}$$

where the corresponding values of the 3×3 sub-matrices are defined.

These equations are implemented to form $\mathbf{F}_{\text{core}}$ in the MATLAB® m-file `Fcore9.m` on www.wiley.com/go/grewal/gnss. The 7×7 dynamic coefficient matrix (i.e. without the unstable vertical channel described below) is computed by the MATLAB® m-file `Fcore7.m`.

11.3.9 Vertical Navigation Instability

The term "vertical channel" is sometimes used for the vertical components of position and velocity in inertial navigation. The vertical channel is the Dark Side of inertial navigation, because stand-alone inertial navigation errors in the vertical channel are naturally unstable due to the negative vertical gradient of gravitational acceleration (about -3×10^{-6} m/s^2 per meter of elevation increase).

Table 11.4 Equation references for dynamic coefficient sub-matrices.

	Error coupling mechanisms		
Corrupted INS implementation processes	**Position error**	**Velocity error**	**Tilt error**
1. Velocity integ.	$\mathbf{F}_{1,1\ [1]}$	$\mathbf{F}_{1,2\ [1]}$	$\mathbf{F}_{1,3\ [1]}$
(Equations)	(Eq. (11.21))	(Eq. (11.22))	(Eq. (11.23))
2. Gravity	$\mathbf{F}_{2,1\ [2]}$	$\mathbf{F}_{2,2\ [2]}$	$\mathbf{F}_{2,3\ [2]}$
(Equations)	(Eq. (11.27)	(Eq. (11.28))	(Eq. (11.29))
3. Coriolis	$\mathbf{F}_{2,1\ [3]}$	$\mathbf{F}_{2,2\ [3]}$	$\mathbf{F}_{2,3\ [3]}$
(Equations)	(Eq. (11.36))	(Eq. (11.37))	(Eq. (11.38))
4. Centrifugal	$\mathbf{F}_{2,1\ [4]}$	$\mathbf{F}_{2,2\ [4]}$	$\mathbf{F}_{2,4\ [3]}$
(Equations)	(Eq. (11.42))	(Eq. (11.43))	(Eq. (11.44))
5. Earthrate comp.	$\mathbf{F}_{3,1\ [5]}$	$\mathbf{F}_{3,2\ [5]}$	$\mathbf{F}_{3,3\ [5]}$
(Equations)	(Eq. (11.46))	(Eq. (11.47))	(Eq. (11.48))
6. Leveling	$\mathbf{F}_{3,1\ [6]}$	$\mathbf{F}_{3,2\ [6]}$	$\mathbf{F}_{3,3\ [6]}$
(Equations)	(Eq. (11.51))	(Eq. (11.52))	(Eq. (11.53))
6. Accel. misalign.	$\mathbf{F}_{2,1\ [7]}$	$\mathbf{F}_{2,2\ [7]}$	$\mathbf{F}_{2,3\ [7]}$
(Equations)	(Eq. (11.55))	(Eq. (11.56))	(Eq. (11.57))

The lower right matrix element in Eq. (11.27) is the source of vertical navigation instability. Its value is positive, meaning that upward acceleration error *increases* with upward position (altitude) error:

$$\frac{d}{dt}\dot{\varepsilon}_U = 2\,\frac{\mathrm{GM}_\oplus}{\overline{R}_\oplus^3}\varepsilon_U \tag{11.68}$$

or the equivalent state-space form for navigation errors in the vertical channel,

$$\frac{d}{dt}\begin{bmatrix}\varepsilon_U \\ \dot{\varepsilon}_U\end{bmatrix} = \begin{bmatrix}0 & 1 \\ \tau_U^{-2} & 0\end{bmatrix}\begin{bmatrix}\varepsilon_U \\ \dot{\varepsilon}_U\end{bmatrix} \tag{11.69}$$

$$\begin{aligned}\tau_U &= \sqrt{\frac{\overline{R}_\oplus^3}{2\,\mathrm{GM}_\oplus}} \\ &\approx 570 \text{ seconds} \\ &\approx 9.5 \text{ minutes}\end{aligned} \tag{11.70}$$

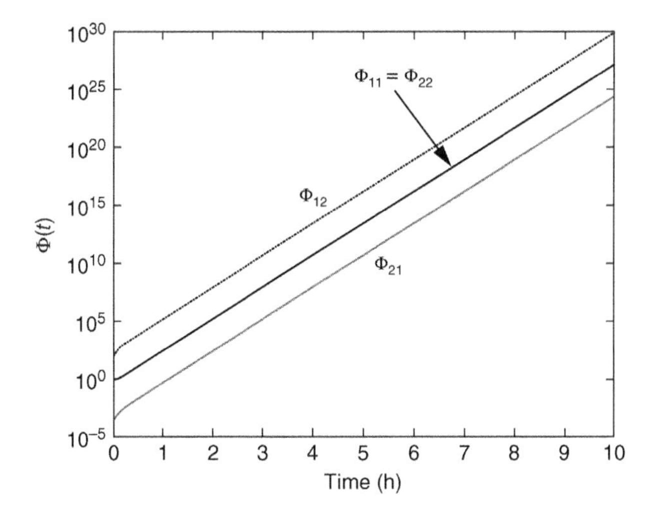

Figure 11.5 Growth of vertical channel errors over 10 hours.

The general solution to this linear time-invariant system, as a function of time $t > t_0$ and initial conditions at $t = t_0$, is

$$\begin{bmatrix} \varepsilon_U(t) \\ \dot{\varepsilon}_U(t) \end{bmatrix} = \underbrace{\begin{bmatrix} \cosh\left(\frac{t-t_0}{\tau_U}\right) & \tau_U \sinh\left(\frac{t-t_0}{\tau_U}\right) \\ \sinh\left(\frac{t-t_0}{\tau_U}\right)\tau_U{}^{-1} & \cosh\left(\frac{t-t_0}{\tau_U}\right) \end{bmatrix}}_{\mathbf{\Phi}_U} \begin{bmatrix} \varepsilon_U(t_0) \\ \dot{\varepsilon}_U(t_0) \end{bmatrix} \tag{11.71}$$

This solution diverges exponentially with time, as shown in Figure 11.5. This is a plot of the elements of the state transition matrix $\mathbf{\Phi}_U$ over a period of 10 hours, during which the solution grows by a factor approaching Avogadro's number ($\approx 6.022\,141\,5 \times 10^{23}$).

Inertial navigation for ballistic missiles must include vertical navigation, because that is the direction of most of the action. However, an INS for ballistic missiles typically has extremely small initial navigation errors, it spends long periods of time in self-calibration before it is launched, and its total period of navigation is in the order of several minutes. Fortunately, all these factors reduce the impact of vertical navigation instability on performance to levels that can be tolerated.

Vertical navigation instability is not a problem for surface ships at sea, either, because they do not need to navigate in the vertical direction.

It could have been a problem for aircraft navigation, except that aircraft already had reasonably reliable vertical information from altimeters. They would be able to navigate more like a ship but use their altimeters for vertical navigation. They could also use altimeters for "aiding" a full three-dimensional INS, in much the same way it is done in GNSS/INS integration. The latter approach led to augmentation of the navigation error model to include error characteristics of the altimeter, and eventually to methods for recalibrating critical sensors for even better performance.

11.3.9.1 Altimeter Aiding

This was originally done using barometric altimeters, which have biases with long-term drift due to ambient barometric pressure variations in the dynamic atmosphere. These variations have different statistics at different altitudes and different parts of the world, but they generally have correlation times in the order of a day for most of the continental United States [4]. The root mean squared (RMS) variations are in the order of a few millibars, roughly equivalent to 100 meters in altitude. The resulting altimeter bias is usually modeled as an exponentially correlated random process with a correlation time-constant of ~ 1 day and a steady-state RMS value in the order of 10^2 m. In that case, the dynamic coefficient matrix for the augmented system will have the block form

$$\mathbf{F}_{10} = \begin{bmatrix} \mathbf{F}_9 & 0 \\ 0 & -1/\tau_{\text{alt}} \end{bmatrix} \tag{11.72}$$

where τ_{alt} is the correlation time for altimeter bias errors.

A modeled demonstration of altimeter damping of an INS, using this model, is implemented in the MATLAB® m-file `F10CEPrate.m` on www.wiley.com/go/grewal/gnss. This script solves the riccati equation for this system with specified initial uncertainties and plots the resulting RMS uncertainties in all 10 state variables. It also plots the resulting CEP as a function of time and performs a least-squares fit to estimate CEP rate. The output for the latter is shown in Figure 11.6. The estimated CEP rate is ~ 0.07 nautical miles per hour. This may seem small, but it is for a system in which the only process noise is from the altimeter model, and the only source of sensor noise is on the altimeter output. The results do appear to indicate that the presence of even small altitude errors does corrupt system dynamics, even without error sources other than those due to the altimeter.

Example 11.1 ***(Three-state model for altimeter aiding)***

Exponentially correlated processes have dynamic models of the sort

$$\dot{\varepsilon}_{\text{alt}} = -\frac{\varepsilon_{\text{alt}}}{\tau_{\text{alt}}} + w_{\text{alt}}(t) \tag{11.73}$$

where τ_{alt} is the exponential correlation time of the altimeter errors, $w_{\text{alt}}(t)$ is a zero-mean white noise process with variance

$$q_{\text{alt}} = \frac{2\sigma_{\text{alt}}^2}{\tau_{\text{alt}}} \tag{11.74}$$

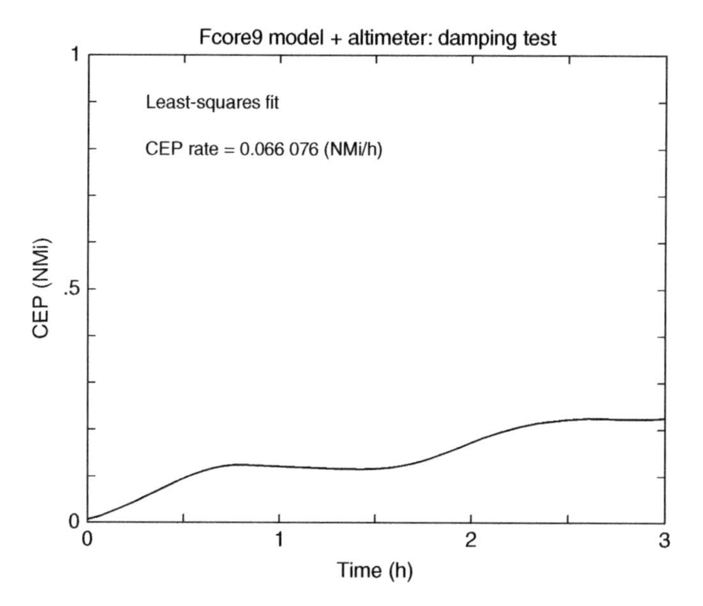

Figure 11.6 CEP plot from m-file F10CEPrate.m.

and σ_{alt}^2 is the mean-squared altimeter error.

If this altimeter were used as an auxiliary sensor for measuring altitude, then the three-state (altitude, altitude rate, and altimeter bias) dynamic model for just the vertical channel with altimeter aiding would be

$$\frac{d}{dt}\begin{bmatrix} \varepsilon_U \\ \dot{\varepsilon}_U \\ \varepsilon_{\text{alt}} \end{bmatrix} = \underbrace{\begin{bmatrix} 0 & 1 & 0 \\ \tau_U^{-2} & 0 & 0 \\ 0 & 0 & \frac{-1}{\tau_{\text{alt}}} \end{bmatrix}}_{\mathbf{F}_{\text{alt}}} \begin{bmatrix} \varepsilon_U \\ \dot{\varepsilon}_U \\ \varepsilon_{\text{alt}} \end{bmatrix} + \begin{bmatrix} 0 \\ w_U(t) \\ w_{\text{alt}}(t) \end{bmatrix} \tag{11.75}$$

where $\mathbf{F}_{\text{alt}}$ is the 3×3 dynamic coefficient matrix for this system.

The altimeter output is the sum of real altitude and the slowly varying altimeter bias, so the associated measurement sensitivity matrix for the altimeter output has the form

$$\mathbf{H}_{\text{alt}} = \begin{bmatrix} 1 & 0 & 1 \end{bmatrix} \tag{11.76}$$

Whether the covariance matrix $\mathbf{P}_\infty$ of mean-squared vertical channel uncertainties settles to a finite value with this setup is determined by whether or not the steady-state matrix riccati equation for this system

$$0 = \mathbf{F}_{\text{alt}}\mathbf{P}_\infty + \mathbf{P}_\infty\mathbf{F}_{\text{alt}}^T - \frac{\mathbf{P}_\infty\mathbf{H}_{\text{alt}}^T\mathbf{H}_{\text{alt}}\mathbf{P}_\infty}{R_{\text{alt}}} + \begin{bmatrix} 0 & 0 & 0 \\ 0 & q_U & 0 \\ 0 & 0 & q_{\text{alt}} \end{bmatrix} \tag{11.77}$$

has a solution. In this representation, R_{alt} is the mean-squared altimeter noise, q_U is the covariance of vertical accelerometer noise, and $\mathbf{P}_\infty$ is the steady-state covariance of state estimation uncertainty (if it exists). This riccati equation is equivalent to six independent quadratic equations constraining the six independent elements of $\mathbf{P}_\infty$. The solution can be determined numerically, however, and it is finite.

Example 11.2 (*Ten-state INS error model for altimeter aiding*)
The simulation result shown in Figure 11.6 was generated using a 10-state model with the altimeter bias as the 10th state variable. The measurement sensitivity matrix for the altimeter will be the 1×10 matrix

$$\mathbf{H}_{\text{alt}} = \begin{bmatrix} 0 & 0 & 1 & 0 & 0 & 0 & 0 & 0 & 0 & 1 \end{bmatrix} \tag{11.78}$$

modeling the fact that the altimeter measures the vertical position error, as well as its own bias. and some values for the altimeter bias time-constant and mean-squared altimeter error due to ambient atmospheric pressure variations. In this example,

$$\sigma_{\text{alt}} = 100 \text{ m, the effect of ambient pressure variation} \tag{11.79}$$

$$\tau_{\text{alt}} = 1 \text{ day correlation time for ambient atmospheric pressure variation} \tag{11.80}$$

with sampling every 10 seconds. The atmospheric pressure standard deviations and correlation time are from [4]. The values chosen are in the mid-range of Klein's hemisphere-wide values, but reasonable for the United States.

This test is only the beginning of the verification process for the **Fcore9** model, which requires some means of holding vertical channel errors in check while the rest of the INS error model is verified. This simple case includes no inertial sensor noise.

11.3.9.2 Using GNSS for Vertical Channel Stabilization

GNSS aiding can also be used to stabilize the INS vertical channel, but using GNSS only for this purpose carries the risk of navigation instability whenever GNSS signal availability is lost or compromised.

11.3.10 Schuler Oscillations

In 1906, the German gyrocompass inventor Hermann Anschütz-Kaempfe (1832–1931) invited his cousin, Maximilian Schuler (1882–1972), to look into why Anschütz-Kaempfe's gyrocompasses were not providing reliable bearing information on ships in high seas. In analyzing the physics of the situation, Schuler determined that the problem had to do with torques induced by lateral accelerations, and the best solution would be to design a pendulous suspension for the gyrocompass with a period of about 84 minutes, that being the period of an ideal pendulum (i.e. with massless support arm) with support arm length equal to the radius of the Earth, $R_\oplus$ [5]. The dynamic equation for small displacements δ of a pendulum with that support arm length in the near-earth gravitational field is

$$\begin{aligned}
\frac{d^2}{dt^2}\delta &\approx \frac{-g}{R_\oplus}\delta \\
&= -\frac{\mathrm{GM}}{R_\oplus^3}\delta \\
&= -\omega_{\text{Schuler}}^2\delta \\
\omega_{\text{Schuler}} &= \sqrt{\frac{\mathrm{GM}}{R_\oplus^3}} \\
&\approx \sqrt{\frac{0.398\,600\,4 \times 10^{15}}{(0.637\,100\,9 \times 10^7)^3}} \\
&\approx 0.001\,241\,528 \text{ rad/s} \\
&\approx 0.000\,197\,595\,3 \text{ Hz} \\
T_{\text{Schuler}} &= 1/f_{\text{Schuler}} \\
&\approx 1/0.000\,197\,595\,3 \\
&\approx 84 \text{ minutes}
\end{aligned}$$

called the *Schuler period.* It was rediscovered with the advent of inertial navigation, a consequence of gravity modeling on Earth with radius $R_\oplus$, just like the Schuler pendulum. As a consequence, due to the Coriolis effect, horizontal INS errors tend to behave like an ideal Schuler/Foucault pendulum.

11.3.10.1 Schuler Oscillations with Coriolis Coupling

Figure 11.7 shows a surface plot of the simulated trajectory of INS position errors, generated by the MATLAB® m-file `Fcore7Schuler.m` from www.wiley.com/go/grewal/gnss. This dynamic simulation uses the 7-state INS error model (without the unstable vertical channel) and begins with an initial north velocity error, exciting Schuler oscillations that – coupled with Coriolis accelerations – turn the plane of Schuler oscillation clockwise in the northern hemisphere at a rate that is latitude dependent. During the related model vetting process, this behavior must be compared with the behavior of the INS system under study under the same initial conditions.

11.3.11 Core Model Validation and Tuning

All models need to be verified by comparing modeled behavior with the behavior of the system they were designed to represent. It is often the case that the

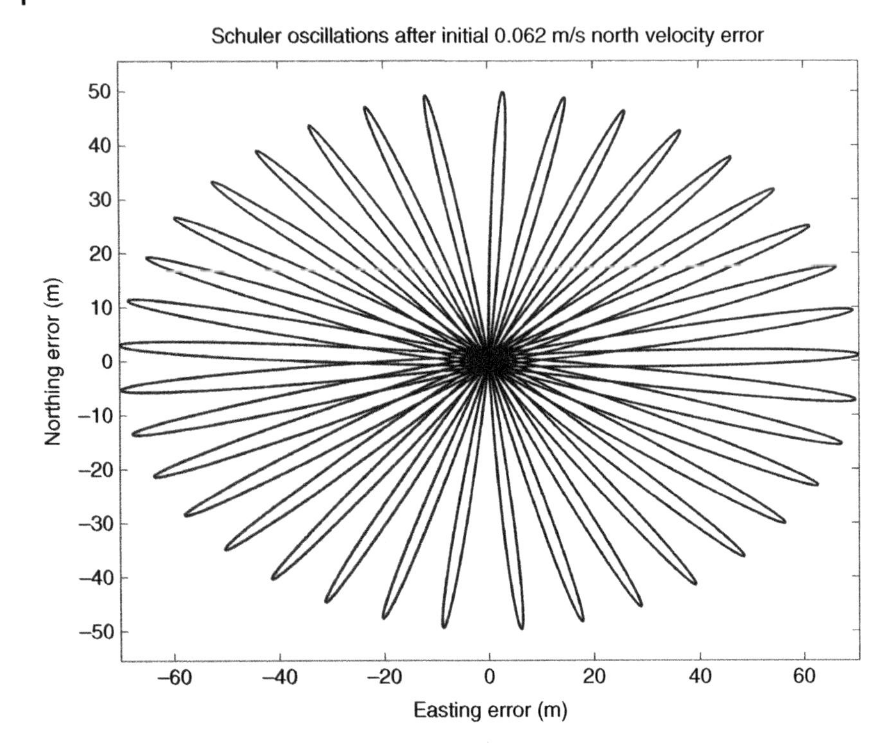

Figure 11.7 Schuler/Coriolis demonstration using seven-state horizontal error model.

INS and its model evolve together, so that each validates the other as development progresses.

After the INS and its model are ready to be tested, the first step in INS/model verification is usually to deliberately introduce navigation errors into the INS and its model and compare the respective responses. As a minimum, this involves nine tests for the INS and its model: one for each error variable.

Model "tuning" is the modification of model parameters to better represent the behavior of the modeled system. These modifications are usually related to sensor error characteristics, which have not yet been addressed.

The model derived earlier is for a hypothetical gimbaled INS that has not been vetted in this manner for any actual INS, but it can be used to demonstrate the verification procedure. As a rule, any model should be trusted only after verification with the INS it is designed to represent.

11.3.11.1 Horizontal Inertial Navigation Model

Ships afloat do not need to navigate vertically. They still need three gyroscopes, but for gimbaled navigation they can eliminate the vertical accelerometer and the calculations that would otherwise doubly integrate its outputs.

For that type of INS, one can also eliminate altitude and altitude rate from the core nine-state model, eliminating two state variables. The result is the dynamic coefficient matrix implemented in the MATLAB® m-file `Fcore7.m` on www.wiley.com/go/grewal/gnss. It was obtained by removing the third and sixth rows and columns of the dynamic coefficient matrix from the 10-state model and setting vertical velocity to zero. Seven-state models can be used for predicting INS response to deliberate initialization errors in each of the seven state variables.

11.4 Inertial Sensor Noise Propagation

A major source of dynamic process noise driving navigation error distributions is from noise in the gyroscopes and accelerometers, described in Section 3.3. How sensor noise is treated in Kalman filtering will depend on statistical properties determined from sensor outputs sampled with zero inputs. The mean output (bias) is then removed and the remaining zero-mean samples analyzed to establish their time-correlation characteristics from their autocovariance functions or their fourier transforms, their power spectral densities (PSDs).

11.4.1 $1/f$ Noise

PSD shapes of some common sensor noise models are plotted in Figure 11.8. The only one of these that is not readily modeled in Kalman filtering is the one labeled "1/f" noise (also called "one-upon-f" noise). It is not commonly associated with earlier inertial sensor designs, but it has been observed in "shot noise" from currents in electronic materials – including materials used in some micro-electrostatic gyroscope (MESG) sensors. It is not uncommonly modeled in Kalman filtering by a combination of white noise and exponentially correlated noise covering the frequency range of interest. All the other noise types in the figure are modeled by white noise put through what is called a *shaping filter* with a linear time-invariant dynamic system model.

11.4.2 White Noise

Zero-mean white sensor noise does not change the dynamic model structure of inertial navigation errors, except by adding a noise process $\mathbf{w}(t)$:

$$\frac{d}{dt}\xi = \mathbf{F}_{\text{core}}\xi + \mathbf{w}(t) \tag{11.81}$$

The white noise from the accelerometers is integrated into velocity error $\dot{\varepsilon}$, and white noise from the gyroscopes is integrated into orientation error. This results

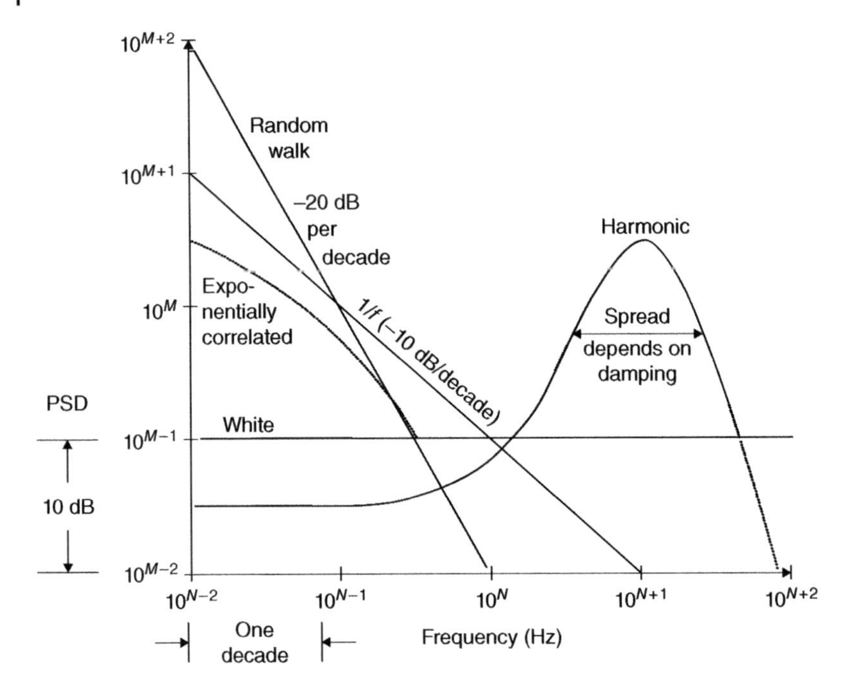

Figure 11.8 PSDs of common sensor noise models.

in a process noise covariance with covariance structure

$$\mathbf{Q}_{\text{sensor noise}} = \text{E}\langle \mathbf{w}(t)\mathbf{w}^T(t)\rangle = \begin{bmatrix} 0 & 0 & 0 \\ 0 & \mathbf{Q}_{\text{acc.}} & 0 \\ 0 & 0 & \mathbf{Q}_{\text{gyro}} \end{bmatrix} \tag{11.82}$$

where $\mathbf{Q}_{\text{acc.}}$ is the accelerometer noise covariance and $\mathbf{Q}_{\text{gyro}}$ is the gyro noise covariance. Unless there is data confirming noise correlation between sensors, these sub-matrices may be scalar matrices:

$$\mathbf{Q}_{\text{acc.}} = q_{\text{acc.}}\mathbf{I} \tag{11.83}$$

$$\mathbf{Q}_{\text{gyro}} = q_{\text{gyro}}\mathbf{I} \tag{11.84}$$

where the accelerometer noise variance $q_{\text{acc.}}$ is the same for all accelerometers, and gyro noise variance q_{gyro} is the same for all gyros. However, empirical values of $\mathbf{Q}_{\text{acc.}}$ and $\mathbf{Q}_{\text{gyro}}$ can be determined by computing the actual 3×3 covariances from all three accelerometers and from all three gyroscopes.

The propagation equation for the covariance matrix $\mathbf{P}_\xi$ of navigation errors in this case has either of the forms

$$\frac{d}{dt}\mathbf{P}_\xi = \mathbf{F}_{\text{core}}\mathbf{P}_\xi + \mathbf{P}_\xi\mathbf{F}_{\text{core}}^T + \mathbf{Q}_{\text{sensor noise}} \tag{11.85}$$

$$\mathbf{P}_{\xi[k]} \approx \mathbf{\Phi}_{\xi[k-1]}\mathbf{P}_{\xi[k-1]}\mathbf{\Phi}_{\xi[k-1]}^T + \Delta t\ \mathbf{Q}_{\text{sensor noise}} \tag{11.86}$$

$$\mathbf{\Phi}_{\xi[k-1]} = \exp[\mathbf{F}_{\text{core}}(\hat{\xi}(t_{k-1})\ \Delta t)] \tag{11.87}$$

in either continuous time or discrete time. These equations characterize the expected performance of an INS with given sensor noise, in terms of how fast the mean-squared navigation errors can be expected to deteriorate over time.

11.4.3 Horizontal CEP Rate Versus Sensor Noise

The m-file `CEPvsSensorNoise.m` on www.wiley.com/go/grewal/gnss uses Eq. (11.86) to obtain least-squares estimates of CEP rate over a range of gyro and accelerometer noise levels for horizontal inertial navigation, with only sensor noise and no sensor compensation errors. The results, plotted in Figure 11.9, would indicate that gyro noise dominates INS performance if RMS gyro noise (in radians per second per root-second) is greater than $\sim 10^{-4}$ times RMS accelerometer noise (in meters per second-squared per root-second),

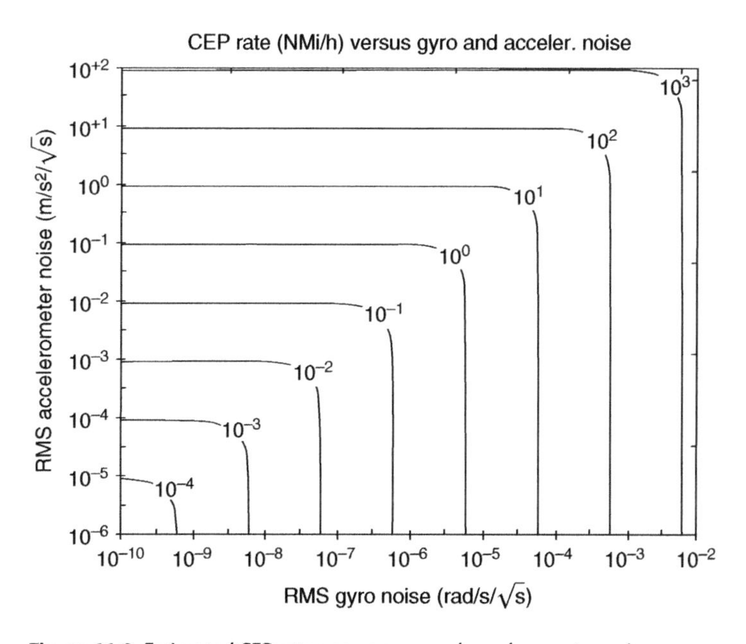

Figure 11.9 Estimated CEP rate versus gyro and accelerometer noise.

and accelerometer noise dominates INS performance if RMS accelerometer noise is greater than $\sim 10^4$ times RMS gyro noise. However, *these results do not include the effects of vertical navigation errors.*

11.5 Sensor Compensation Errors

Sensor compensation parameters

If the accuracy requirements for inertial sensors cannot be achieved through manufacturing precision, it can often be achieved using sensor calibration to characterize the residual error pattern, so that these errors can be adequately compensated during navigation. The models and methods used for this purpose are described in Section 3.3.3.

Parameter drift

It is not uncommon for the values of sensor compensation parameters to change noticeably between system turn-ons, and even to change during operation after turn-ons. The phenomenon has been reduced in some instances by design changes, but not totally eliminated. As an alternative, many schemes have been used for partial recalibration of the more suspect compensation parameters. These have generally relied on using additional sensors. With the arrival of GNSS, much attention has been focused on using essentially the same approach for INS recalibration using GNSS. The necessary models are developed in this chapter. The application off these models with GNSS models is the subject of the next chapter.

11.5.1 Sensor Compensation Error Models

The sensor compensation models used here are the same as those used in Chapter 3, except the compensation parameters are no longer necessarily constant. Attention is focused primarily on those compensation parameters deemed more likely to drift. These are primarily the zero-order (bias) and first-order (scale factor and input axis misalignment) compensation parameters. Input axis misalignments tend to be relatively stable, so the usual suspects are scale factors and biases.

The notation must change to reflect the fact that we are dealing with variables, not constants. The variations due to drift in the bias, scale factor, and input axis misalignment for the accelerometers and gyroscopes in the general case can then be modeled as

$$\boldsymbol{\delta}_a(t) = \begin{bmatrix} \delta_{abE} \\ \delta_{abN} \\ \delta_{abU} \end{bmatrix} + \mathbf{M}_{\mu a}\mathbf{a}_{\text{comp.0}} \tag{11.88}$$

$$\boldsymbol{\delta}_{\omega}(t) = \begin{bmatrix} \delta_{gbE} \\ \delta_{gbN} \\ \delta_{gbU} \end{bmatrix} + \mathbf{M}_{\mu g}\boldsymbol{\omega}_{\text{comp.0}} \tag{11.89}$$

$$\mathbf{M}_{a\mu} = \begin{bmatrix} \delta_{asE} & \delta_{a\mu EN} & \delta_{a\mu EU} \\ \delta_{a\mu NE} & \delta_{asN} & \delta_{a\mu NU} \\ \delta_{a\mu UE} & \delta_{a\mu UN} & \delta_{asU} \end{bmatrix} \tag{11.90}$$

$$\mathbf{M}_{g\mu} = \begin{bmatrix} \delta_{gsE} & \delta_{g\mu EN} & \delta_{g\mu EU} \\ \delta_{g\mu NE} & \delta_{gsN} & \delta_{g\mu NU} \\ \delta_{g\mu UE} & \delta_{g\mu UN} & \delta_{gsU} \end{bmatrix} \tag{11.91}$$

where

$\mathbf{a}_{\text{comp.0}}$ is the vector of accelerometer outputs compensated using the prior values of the compensation parameters
$\boldsymbol{\omega}_{\text{comp.0}}$ is the vector of gyro outputs compensated using the prior values of the compensation parameters
$\boldsymbol{\delta}_{a}(t)$ is the error in compensated sensed acceleration due to compensation parameter drift
$\boldsymbol{\delta}_{\omega}(t)$ is the error in compensated sensed rotation rate due to compensation parameter drift
$\delta_{ab\bigstar}$ is the accelerometer bias drift (with $\bigstar = E, N, \text{or } U$)
$\delta_{gb\bigstar}$ is the gyro bias drift (with $\bigstar = E, N, \text{or } U$)
$\mathbf{M}_{a\mu}$ is the drift in the accelerometer scale factor and input axis misalignment matrix
$\mathbf{M}_{g\mu}$ is the drift in the gyro scale factor and input axis misalignment matrix

With the assumption that the drift in input axis misalignments can be ignored, the model for each sensor type (accelerometer or gyroscope) has only six drifting parameters: three biases and three scale factors. The models for the errors in the sensed quantities then have only diagonal entries in the scale factor an input axis misalignment matrices:

$$\mathbf{M}_{a\mu} = \begin{bmatrix} \delta_{asE} & 0 & 0 \\ 0 & \delta_{asN} & 0 \\ 0 & 0 & \delta_{asU} \end{bmatrix} \tag{11.92}$$

$$\mathbf{M}_{g\mu} = \begin{bmatrix} \delta_{gsE} & 0 & 0 \\ 0 & \delta_{gsN} & 0 \\ 0 & 0 & \delta_{gsU} \end{bmatrix} \tag{11.93}$$

11.5.1.1 Exponentially Correlated Parameter Drift Models

Models for slowly varying sensor compensation errors are usually random walk models or exponentially correlated process models. The latter are preferred because they have finite bounds on their variances, and they allow the analyst to use model parameters (i.e. RMS error and error correlation time) based on test data.

The dynamic model for an independent exponentially correlated random process $\{\eta(t)\}$ has the general form

$$\frac{d}{dt}\eta = \frac{-1}{\tau}(\eta - \overline{\eta}) + w(t) \tag{11.94}$$

where $\overline{\eta}$ is the long-term mean of the process, τ is the process correlation time, and the zero-mean white process noise $w(t)$ has variance $2\,\sigma^2/\tau$, where $\sigma^2 = \mathrm{E}\langle(\eta - \overline{\eta})^2\rangle$ is the mean-squared variation in the sensor compensation parameter. In the case of parameter drift, the long-term average drift can usually be assumed to be zero, in which case $\overline{\delta} = 0$ and the model becomes

$$\frac{d}{dt}\delta = \frac{-1}{\tau}\delta + w(t) \tag{11.95}$$

with the same parameters.

Equation (11.95) is already in the form of linear stochastic equation. If 12 of these (for scale factor and bias drift of six sensors) were to be added to the dynamic system model for an inertial system, it would add a 12×12 scalar matrix to the dynamic coefficient matrix,

$$\mathbf{F}_{\mathrm{drift}} = -\tau^{-1}\mathbf{I} \tag{11.96}$$

assuming all parameters have the same correlation times. If not, each element on the diagonal of $\mathbf{F}_{\mathrm{drift}}$ can have a different negative value.

The corresponding process noise covariances may be different for accelerometers and gyroscopes, and different for bias and scale factor for each type of sensor. In general, the process noise covariance may have a 12×12 diagonal structure with 3×3 diagonal blocks:

$$\mathbf{Q}_{\mathrm{drift}} = \begin{bmatrix} \frac{2\,\sigma^2_{\mathrm{ab}}}{\tau_{\mathrm{ab}}}\mathbf{I} & 0 & 0 & 0 \\ 0 & \frac{2\,\sigma^2_{\mathrm{as}}}{\tau_{\mathrm{as}}}\mathbf{I} & 0 & 0 \\ 0 & 0 & \frac{2\sigma^2_{\mathrm{gb}}}{\tau_{\mathrm{gb}}}\mathbf{I} & 0 \\ 0 & 0 & 0 & \frac{2\sigma^2_{\mathrm{gs}}}{\tau_{\mathrm{gs}}}\mathbf{I} \end{bmatrix} \tag{11.97}$$

where, for the case in which all sensors of each type have the same parameter drift parameters but statistically independent distributions,

σ^2_{ab}, τ_{ab} are the variance and correlation time, respectively, of accelerometer biases

σ^2_{as}, τ_{as} are the variance and correlation time, respectively, of accelerometer scale factors
σ^2_{gb}, τ_{gb} are the variance and correlation time, respectively, of gyro biases
σ^2_{gs}, τ_{gs} are the variance and correlation time, respectively, of gyro scale factors
I is a 3×3 identity matrix

However, it is generally possible – and recommended – that the process noise parameters for compensation errors be determined from actual sensor error data under operating conditions. There is always the possibility that sensor compensation error drift is driven by common-mode random processes such as temperature or power-level variation, and that information can be exploited to make a more useful model.

11.5.1.2 Dynamic Coupling into Navigation Errors

There is no first-order dynamic coupling of navigation errors into sensor errors, but sensor errors do couple directly into the time-derivatives of navigation errors. The general formulas for navigation error dynamics due to compensation errors are

$$\frac{d}{dt}\begin{bmatrix} \dot{\varepsilon}_E \\ \dot{\varepsilon}_N \\ \dot{\varepsilon}_U \end{bmatrix} = \boldsymbol{\delta}_a(t) = \begin{bmatrix} \delta_{abE} \\ \delta_{abN} \\ \delta_{abU} \end{bmatrix} + \mathbf{M}_{\mu a} a_{\text{comp.0}} \tag{11.98}$$

$$\frac{d}{dt}\begin{bmatrix} \rho_E \\ \rho_N \\ \rho_U \end{bmatrix} = \boldsymbol{\delta}_\omega(t) = \begin{bmatrix} \delta_{gbE} \\ \delta_{gbN} \\ \delta_{gbU} \end{bmatrix} + \mathbf{M}_{\mu g} \omega_{\text{comp.0}} \tag{11.99}$$

where the six affected INS navigation errors are

$\dot{\varepsilon}_E$, $\dot{\varepsilon}_N$, and $\dot{\varepsilon}_U$, the INS velocity errors.
ρ_E, ρ_N, and ρ_U, the INS tilt and heading errors,

and the matrices $\mathbf{M}_{\mu a}$ and $\mathbf{M}_{\mu g}$ are defined by Eqs. (11.92) and (11.93).

11.5.1.3 Augmented Dynamic Coefficient Matrix

Let the order of the 12 state variables in the state vector for sensor compensation errors be

$$\boldsymbol{\delta} = \begin{bmatrix} \boldsymbol{\delta}_a \\ \boldsymbol{\delta}_\omega \end{bmatrix} \tag{11.100}$$

$$\boldsymbol{\delta}_a = \begin{bmatrix} \boldsymbol{\delta}_{ab} \\ \boldsymbol{\delta}_{as} \end{bmatrix}, \boldsymbol{\delta}_\omega = \begin{bmatrix} \boldsymbol{\delta}_{gb} \\ \boldsymbol{\delta}_{gs} \end{bmatrix}, \tag{11.101}$$

$$\boldsymbol{\delta}_{ab} = \begin{bmatrix} \delta_{abE} \\ \delta_{abN} \\ \delta_{abU} \end{bmatrix}, \quad \boldsymbol{\delta}_{as} = \begin{bmatrix} \delta_{asE} \\ \delta_{asN} \\ \delta_{asU} \end{bmatrix}, \boldsymbol{\delta}_{gb} = \begin{bmatrix} \delta_{gbE} \\ \delta_{gbN} \\ \delta_{gbU} \end{bmatrix}, \quad \boldsymbol{\delta}_{gs} = \begin{bmatrix} \delta_{gsE} \\ \delta_{gsN} \\ \delta_{gsU} \end{bmatrix} \tag{11.102}$$

then the augmented dynamic coefficient matrix for a system including an altimeter would have the block form

$$\mathbf{F}_{\text{aug}} = \begin{bmatrix} \mathbf{F}_{\text{core}} & 0 & \mathbf{F}_{\delta\to\xi} \\ 0 & -\tau_{\text{alt}}^{-1} & 0 \\ 0 & 0 & \mathbf{F}_{\text{drift}} \end{bmatrix} \tag{11.103}$$

where the 9×12 sub-matrix coupling compensation parameter errors into navigation errors is

$$\mathbf{F}_{\delta\to\xi} = \begin{bmatrix} 0 & 0 & 0 & 0 \\ \mathbf{I} & \mathbf{D}_a & 0 & 0 \\ 0 & 0 & \mathbf{I} & \mathbf{D}_\omega \end{bmatrix} \tag{11.104}$$

$$\mathbf{D}_a = \begin{bmatrix} a_E & 0 & 0 \\ 0 & a_N & 0 \\ 0 & 0 & a_U \end{bmatrix} \tag{11.105}$$

$$\mathbf{D}_\omega = \begin{bmatrix} \omega_E & 0 & 0 \\ 0 & \omega_N & 0 \\ 0 & 0 & \omega_U \end{bmatrix} \tag{11.106}$$

with diagonal entries equal to input acceleration and rotation rate components, respectively.

The corresponding process noise covariance for the resulting $9 + 1 + 12 = 22$-state INS error model would then be

$$\mathbf{Q}_{\text{INS}} = \begin{bmatrix} \mathbf{Q}_{\text{sensor noise}} & 0 & 0 \\ 0 & \mathbf{q}_{\text{alt}} & 0 \\ 0 & 0 & \mathbf{Q}_{\text{drift}} \end{bmatrix} \tag{11.107}$$

where $\mathbf{Q}_{\text{sensor noise}}$ is defined by Eq. (11.82), $\mathbf{q}_{\text{alt}}$ is defined by Eq. (11.74), and $\mathbf{Q}_{\text{drift}}$ is defined by Eq. (11.97).

At this point, $\mathbf{q}_{\text{alt}}$ is essentially a place-holder for what will become GNSS signal delay covariance in Chapter 12. The INS error dynamic model requires something to stabilize vertical navigation errors.

The alternative 20-state model for horizontal inertial navigation has the same structure, with the third and sixth rows and columns of **F** and **Q** deleted.

The MATLAB® m-file `CEPvsAccComp.m` on www.wiley.com/go/grewal/gnss analyzes the influence of accelerometer bias and scale factor drift on INS CEP rates, assuming no other error sources. CEP rate is evaluated for 2500 sets of sensor compensation drift rates, using six hours of simulation in each case. Compensation error is modeled as an exponentially correlated random process with a correlation time of one hour. The analysis uses simulated test track conditions to provide sensor inputs to drive the navigation errors. Otherwise, sensitivity to accelerometer scale factor errors would not be detectable. Using a closed figure-eight track for the test conditions makes the average acceleration equal to zero, which may cast some aspersions on the applicability of the results to other test conditions. Performance is evaluated using two different estimates of CEP rate: one simply making a straight-line fit through the origin, and the other computing independent slope and intercept values. Also, the simulation is for a gimbaled system, which is not sensitive to gyro scale factor errors. Results showing straight-line-fit CEP rates as a function of RMS accelerometer compensation errors are shown in Figure 11.10.

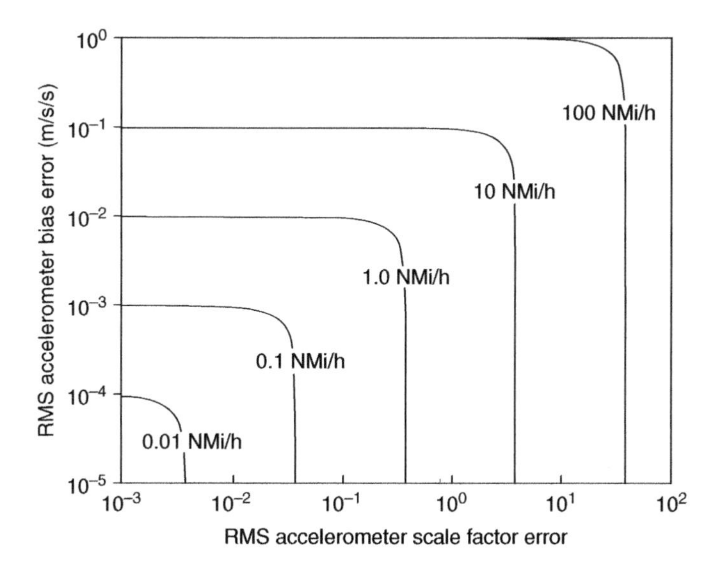

Figure 11.10 CEP slope versus accelerometer compensation drift parameters, based on test track simulation.

11.5.2 Carouseling and Indexing

As mentioned in Section 3.3.5, these methods have been used since the 1960s for mitigating the influence of inertial sensor errors on navigation errors, by rotating sensor input axes either continuously (carouseling) or in discrete steps (indexing). Both methods were originally introduced for gimbaled systems but have also been applied to strapdown systems by rotating the ISA about the vehicle yaw axis at about 1 rpm. They can reduce the effect of some sensor errors on navigation errors by an order of magnitude or more.

11.6 Chapter Summary

1. Inertial navigation systems measure three components of acceleration and three components of rotation, and estimate three components each representing the velocity, position, and orientation of the ISA. The primary navigational error propagation model for inertial navigation also has three components each, representing the velocity, position, and orientation of the ISA – nine state variables in total. Models for the dynamics of these state variables are derived.
2. The *navigation solution* for an INS includes estimates of position, velocity, and orientation. These estimates are propagated forward in time, using only the inputs from gyroscopes and accelerometers.
3. Inertial systems are ordinarily initialized with the starting values of position, velocity, and orientation, but – without the use of auxiliary sensors – there is no way of detecting errors in the navigation solution thereafter.
4. *Inertial systems analysis* is the study of how errors in the navigation solution behave over time, and how this behavior depends on the error characteristics of the sensors and the dynamics of the intended mission applications.
5. This kind of analysis is a fundamental part of the development cycle for inertial navigation systems. If the model is properly vetted with the hardware it represents, this provides a reliable model for judging how the selection of sensors influences system performance, for verifying INS software implementations, and for diagnosing test results.
6. Although modeling accuracy is extremely important in the implementation of inertial navigation, the same is not true for error analysis. Error variables typically represent relative errors in the order of parts per thousand or less. As a consequence, first-order approximations are generally adequate for INS error analysis. Although matrix operations in covariance analysis of inertial navigation systems may require good numerical precision to maintain stability of the results, the analysis results are not overly sensitive to the accuracy of the models used in the INS implementation.

7. However, the first-order-only modeling used here is not exact and is not intended to represent all inertial sensors or all INS implementations. Dynamic effects deemed insignificant for the theoretical sensor have been ignored. Before deciding what is insignificant, it is always necessary to validate the model by comparing its behavior to that of the INS it is intended to represent, using the dynamic conditions the INS is designed for. The models derived here have not been vetted in this manner for any hardware implementation.
8. Inertial navigation in the terrestrial environment exhibits horizontal position error oscillations with a period of about 84 minutes. This is called Schuler oscillation, and it is due to the shape of the gravitational field. This is not necessarily a bad thing, however. The change in the modeled inertial direction of gravitational acceleration with horizontal displacement results in an effective restoring force on the position error, which forces the position error back toward zero. If it were not for this effect, inertial navigation errors would be worse.
9. The same gravitational field shape causes instability of vertical errors in inertial navigation. Fortunately, an altimeter (or GNSS) solves the problem quite well at aircraft altitudes.
10. Navigation in earth-fixed coordinates also introduces error dynamics due to the Coriolis effect. That and Schuler oscillation can make horizontal position errors follow a trajectory similar to that of a Foucault pendulum.
11. The dynamic model for navigation errors can be augmented to include sensor noise, and this model can be used to predict how INS performance depends on sensor noise.
12. The result is a model for how sensor compensation errors influence navigation errors over time. These model derivations are limited to zeroth-order errors (sensor biases) and first-order errors (scale factors and input axis misalignments). The sensor models required for any particular INS application will be determined by the error characteristics of the sensors to be used and on the dynamic conditions they are likely to experience on the intended mission trajectories.
13. The influence on INS performance of "aiding" with other sensors is also a part of INS analysis. An analytical model was used for assessing the influence of an altimeter on vertical navigation. The next chapter will cover INS aiding using GNSS.

11.6.1 Further Reading

Additional background material on inertial navigation systems analysis can be found in the 1971 textbook by Britting [1], and – if you can find it – a 1973 Intermetics report by Widnall and Grundy [2]. However, even these models may not cover all sensors and all inertial navigation implementations. Strapdown modeling and analysis is covered in some detail by Titterton and Weston [6], and in even greater detail in the two-volume treatise by Paul Savage [7]. A more recent error analysis specific to MESG strapdown systems is presented by Ramalingam et al. [8].

Problems

11.1 Derive a calibration model for a sensor with third-order errors. That is, its output z_{out} has both affine (linear plus offset) and cubic dependency on its input z_{in}:

$$z_{\text{out}} = c_0 + c_1\, z_{\text{in}} + c_3\, z_{\text{in}}^3$$

Given n samples of input–output pairs $\{[z_{k,in}, z_{k,out}] \mid k = 1, 2, \ldots, n\}$, what is the least-squares estimate of the sensor bias c_0, scale factor c_1, and cubic coefficient c_3?

11.2 What is the minimum number of input–output pairs required for solving the problem earlier? Is it just the number of samples that determines observability? What, exactly, does determine observability for least-squares problems?

11.3 What is the compensation model corresponding to the error model

$$z_{\text{out}} = c_0 + c_1\, z_{\text{in}} + c_3\, z_{\text{in}}^3?$$

That is, what is the formula for z_{in}, given z_{out}, c_0, c_1, and c_3?

11.4 Derive the first-order compensation error model for the third-order model earlier, by taking the partial derivatives of the compensated sensor output with respect to the coefficients c_0 (bias), c_1 (scale factor), and c_3 (cubic error coefficient).

11.5 Replace the sensor noise model in the previous problem with a model for sensor compensation errors, so that sensor errors are ignored but replaced by sensor compensation errors. Use the equations in this chapter for the dynamic coefficient matrix (Eq. (11.103)) and process noise covariance matrix (Eq. (11.97)). Assume correlation times of an hour for all parameters, and find what combinations of compensation error covariances will yield CEP rates in the order of 1 nautical mile per hour. Assume that the corresponding process noise covariances for gyro parameters are 10^{-4} times the corresponding parameters for accelerometers, and vary the relative covariances of biases and scale factors to find the ratios at which dominance shifts between bias errors and scale factor errors.

11.6 Run the m-file `CEPvsAccComp.m`, which estimates CEP rates as a function of accelerometer compensation parameter drift. It also plots the radial error as a function of time for the case that no inertial navigation is used at all, with the estimated position remaining at the starting point. How would you compare the other results with the no-navigation case? What might this say about basing INS performance on one set of test conditions?

References

1 Britting, K.R. (2010). *Inertial Navigation Systems Analysis*. Artech House.
2 Widnall, W.S. and Grundy, P.A. (1973). Inertial Navigation System Error Model. Report TR-03-73. Intermetrics Inc.
3 Edwards, R.M. (1982). Gravity model performance in inertial navigation. *AIAA Journal of Guidance and Control* 5: 73–78.
4 Klein, W.H. (1951). A hemispheric study of daily pressure variability at sea level and aloft. *Journal of Meteorology* 8: 332–346.
5 Schuler, M. (1923). Die Störung von Pendul-und Kreiselapparaten durch die Beschleunigung der Fahrzeuges. *Physicalische Zeitschrift B* 24: 344–350.
6 Titterton, D.H. and Weston, J.L. (2005) *Strapdown Inertial Navigation Technology*. London, UK: IEE.
7 Savage, P.G. (1996). *Introduction to Strapdown Inertial Navigation Systems*, vols. 1 & 2. Maple Plain, MN: Strapdown Associates.
8 Ramalingam, R., Anitha, G., and Shanmugam, J. (2009) Microelectromechanical systems inertial measurement unit error modelling and error analysis for low-cost strapdown inertial navigation systems. *Defense Science Journal* 59 (6): 650–658.

12

GNSS/INS Integration

Nature laughs at the difficulties of integration.
Pierre Simon de Laplace (1749–1827)

12.1 Chapter Focus

Inertial navigation uses acceleration and rotation sensors and Newtonian mechanics to estimate the common trajectory of these sensors over time.

Global navigation satellite systems (GNSS) navigation use position measurements over time and Newtonian mechanics to estimate the trajectories of their receiver antennas.

This chapter is about how best to combine both measurement modes to obtain a better navigation solution with respect to practical measures of what is considered "better." This chapter covers many of the opportunities for improvement in this regard, developing mathematical methods for sample implementations. The focus will be on specific capabilities offered by integration, how to achieve them, and how to assess their performance.

We also address the issue of selecting measures of integrated system performance appropriate for different system-level applications. Depending on the intended application, one might assign different emphasis to different performance metrics. One of the goals of this chapter is to identify such metrics for different types of applications, and demonstrate how these metrics can be evaluated for different design choices. One of the results of this analysis is that – for some of these performance metrics – the trajectories of the host vehicle during operation also influence performance.

Most of the discussion is about tailoring the design of the inertial navigation system (INS) and the GNSS receiver, not the overall design of the GNSS system. GNSS system-level design generally involves satisfying different performance requirements across a spectrum of different applications, including integrated

Global Navigation Satellite Systems, Inertial Navigation, and Integration,
Fourth Edition. Mohinder S. Grewal, Angus P. Andrews, and Chris G. Bartone.

Companion website: www.wiley.com/go/grewal/gnss

GNSS/INS applications. To that purpose, the presentation includes different types of performance requirements and a general methodology for assessing whether different system-level designs can meet those requirements.

12.2 New Application Opportunities

Integrating satellite and inertial navigation not only offers some opportunities for improved navigational capabilities, including better navigational accuracy, but also some capabilities not available with either systems by itself – as described in the following subsections.

12.2.1 Integration Advantages

12.2.1.1 Exploiting Complementary Error Characteristics

Error characteristics of GNSS and INS navigation are complementary in a number of ways, including the following:

1. Uncertainty about the dynamics of the host vehicle had been a significant source of error for GNSS-only navigation, and a contributor to the filter lag. That source is eliminated when GNSS is integrated with another system that measures those dynamics directly.
2. By the same mechanism, signal phase tracking loops in GNSS receivers can potentially mitigate the filter lag error due to host vehicle dynamic uncertainty.
3. INS position information can be used to reduce the signal reacquisition search time after GNSS signal loss.
4. INS velocity estimates can be used to reduce Doppler uncertainty during GNSS signal acquisition and reacquisition.
5. The GNSS navigation solution and Doppler values can be used for aiding INS velocity estimates during INS initialization.
6. GNSS can eliminate the need for an altimeter to stabilize the INS vertical channel – so long as GNSS is available.
7. An INS cannot estimate its longitude and takes minutes to estimate latitude. GNSS provides near-instantaneous initialization of INS position.
8. Depending on the accuracy of its sensors, initial rotational alignment of the INS with respect to navigation coordinates can require several minutes or more in a quasi-static state before navigation can begin. GNSS allows navigation to begin immediately while the INS is aligned in motion, using the GNSS navigation solution to estimate INS alignment variables.
9. The INS error models from Chapter 11 can be used to keep track of how position errors are correlated with other error sources, and position measurements from GNSS can exploit these correlations to reduce estimation

uncertainties in the correlated variables. The variables being estimated in this way may include inertial sensor compensation parameters whose values may drift over time.

10. Improvements in INS navigational accuracy due to GNSS integration provide better INS-only performance while GNSS signals are not available.

12.2.1.2 Cancelling Vulnerabilities

The major vulnerability of inertial navigation is degradation of accuracy over time. GNSS provides better position measurements that do not degrade over time. These position measurements can be used to update the INS navigation solution, including velocities and (potentially) attitude.

GNSS depends on constant signal availability, which can be compromised by blockage, interference (including multipath), or jamming. The INS solution is immune from all these.

12.2.2 Enabling New Capabilities

12.2.2.1 Real-Time Inertial Sensor Error Compensation

Integration has the potential for exploiting sensor redundancies to self-correct some drifting error characteristics of inertial sensors, which would improve stand-alone INS performance when GNSS measurements are compromised. Models for implementing this were derived in Chapter 11.

12.2.2.2 INS Initialization on the Move

Integration also has the potential for using GNSS – which measures only position – to initialize the INS navigation solution, which requires sensor orientation information that can potentially be inferred from the position history and INS outputs.

INS Alignment Using GNSS The objective of INS alignment is to determine the directions of inertial sensor input axes with respect to navigation coordinates. Using GNSS measurements for rotational alignment of INS sensor axes with respect to navigation coordinates by Kalman filtering is problematic in that rotational mathematics is decidedly nonlinear. An intermediate objective is first to get the alignment error small enough for linear approximation. After that, the misalignment error would be updated and corrected by the integrating Kalman filter.

This nonlinear initialization of orientation is easier for host vehicles designed to move along a vehicle-fixed axis such as the roll axes of airplanes or wheeled surface vehicles, in which case that axis will be aligned to some degree with the GNSS-derived velocity. Roll can then be inferred from the measured acceleration required to counter gravity. This procedure may benefit from using stochastic vehicle dynamic models with acceleration states in the GNSS implementation, and from aiding with a three-axis magnetometer.

12.2.2.3 Antenna Switching

Surface-mounted GNSS receiver antennas generally offer hemispherical coverage, but highly maneuverable host vehicles may require additional antennas to maintain signal availability at all times. INS orientation and rotation rates can be used for predictive GNSS antenna-switching for such applications and for tailoring this to signals for specific satellites.

12.2.2.4 Antenna-INS Offsets

The GNSS navigation solution is for the location of its receiver antenna, and the INS navigation solution is for the location of its inertial sensors. When you are striving for relative navigational accuracy in the order of a meter or less, the relative offset of the receiver antenna from the INS matters. For rigid-body host vehicles, this can be specified in vehicle-fixed coordinates and converted to navigation coordinates using the INS attitude solution. It can be more complicated for non-rigid bodies. It also becomes more complicated with satellite-specific antenna-switching, with each pseudorange depending on which antenna is being used for which satellite.

12.2.3 Economic Factors

12.2.3.1 Economies of Scale

There are two interrelated economies of scale for navigation systems: one related to market size and another related to implementation scale.

Market Size This relates to how the size of the potential market influences user costs and how user costs influences the potential market.

The Transit GNSS system was initially funded and developed for military applications, but it was also being used to some degree for civilian navigation at sea before it was superseded by GPS in the 1990s. This commercial Transit navigation market was relatively small, the user costs were high, and the market was limited to high-value applications.

GPS was also developed for military uses, but with a degraded public channel. The original receiver costs per unit for military users were in the order of tens of thousands of dollars. The potential size of the combined military and civilian applications market soon drove the unit cost to military and civilian users down by orders of magnitude.

Manufacturing Technology This relates to how manufacturing technologies and implementation scale influence the user costs. GPS receiver implementations soon moved to semiconductor chip-level implementations, and MEMS technologies soon led to chip-level inertial sensors and systems. The combination of the two has made unit costs of components comparable to those for fast food.

12.2.3.2 Implementation Tradeoffs

These relate to how development and implementation costs and risks for integrated GNSS/INS navigation depend on the necessary changes required in more-or-less standard GNSS and INS implementations.

Satellite and inertial navigation may be complementary modes of navigation, but integrating these within a common Kalman filtering framework presents some daunting challenges.

The Curse of Dimensionality The computational complexity of a Kalman filter implementation grows as the cube of the dimension of the state vector, and the state vector generally includes every sensor input and the intermediate derivatives and/or integrals between them. This would include, as a bare minimum:

3 components of angular rotation rates, as inputs to rate gyros.

3 components representing the orientation of the inertial sensor assembly.

3 components of acceleration, as inputs to the accelerometers.

3 components of velocity.

3 components of position, as inputs to pseudorange measurements.

$\sim$30 time-correlated satellite signal delay biases.

$\sim$45 essential state vector components.

$\sim 10^4$ arithmetic operations per update

for just the bare-bones implementation.

But this does not include variables required for representing additional error compensation variables that may become observable with the expanded model. These may include:

6 time-correlated biases of the six inertial sensors

6 time-correlated scale factor errors of the six inertial sensors

12 time-correlated sensor input axis misalignments

– making the total closer to 3×10^5 arithmetic operations per Kalman filter update.

Sensor Output Rates The Kalman filter is updated in each sensor output cycle. Output rates of inertial sensors are in the order of 10^3 samples per second (higher for the Bortz implementation). Kalman filter updates at this rate would require something in the order of $10^3 \times 10^4 = 10^7$ arithmetic operations per second for the bare-bones implementation and 3×10^8 arithmetic operations per second for some INS sensor error correction. Furthermore, Kalman filter implementations with this many state variables generally require more precision. This has led developers to search for alternative implementations.

Early Implementation Alternatives The Transit GNSS was designed from the beginning for integration with the US Navy's Ships Inertial Navigation System, the software of which was modified for that purpose. Details of this have been classified, but it eventually evolved into something making use of Transit's position measurements to make adjustments in the INS navigation solution with Kalman filtering. Because Trident position measurement intervals were generally in the order of hours or days, the Kalman filter could use the INS to carry forward the velocity[1] and position estimates between measurement updates.

Although a common GNSS/INS navigation model was known at the beginning of GPS deployment, it did not start with anything like the full-blown model on the first attempt. One reason was to avoid the scheduling risks from making significant alterations to functioning stand-alone GPS receivers and inertial navigators. Another reason may have been to avoid development risks associated with restructuring the signal processing hardware and software. Also, available flight-qualified processors at the time had rather limited processing throughput and much slower data transfer rates than we have grown accustomed to. For the first proof-of-concept testing, a "looser" form of GPS/INS integration was chosen.

Figure 12.1 shows a ranking of approaches to GNSS/INS integration, based on how certain implementations functions are altered, and ranked according to the degree of coupling between the otherwise independent implementations of

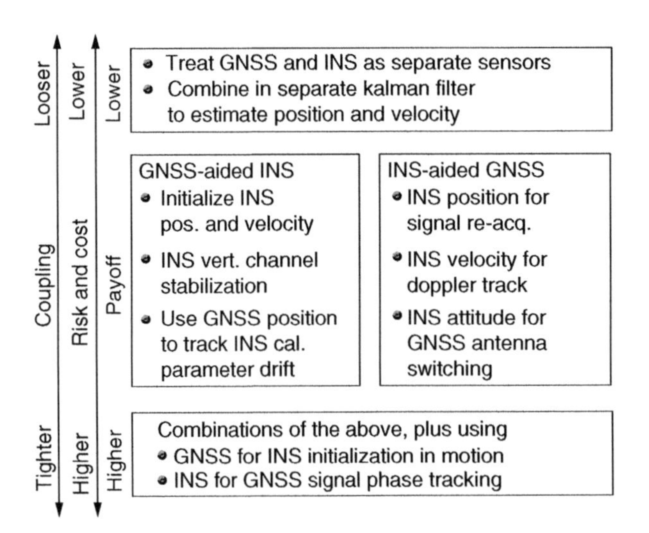

Figure 12.1 Loosely and tightly coupled implementations.

1 Submarines also use electromagnetic logs (water-speed sensors) as auxiliary velocity sensors, with a time-and-distance-correlated sensor bias error model to represent unknown drift currents.

GNSS receivers and inertial navigators. The rankings are loosely called "loose" and "tight," according to the degree to which the independent implementations are modified.

Loosely Coupled Implementations The most loosely coupled implementation uses only the standard navigation solutions from the GNSS receiver and INS as inputs to a filter (a Kalman filter, usually), the output of which is the combined estimate of the navigation solution. Although each subsystem (GNSS or INS) may already include its own Kalman filter, this "ultra-loosely coupled" integration architecture does not necessarily modify those.

The downside of this approach is that neither the GNSS solution nor the INS solution is made any better by it. As the INS estimate gets worse over time, its filter weighting gets smaller. Eventually, it is essentially ignored, and the integrated solution is the GNSS solution. In situations with inadequate GNSS signal coverage, the INS may be of limited utility.

More Tightly Coupled Implementations As a rule, the more tightly coupled implementations have greater impact on the internal implementations of the GNSS receiver and the INS and have better overall performance.

For example, the more tightly coupled implementations may use non-standard subsystem outputs. Raw pseudoranges from the GNSS receiver may be used independently of the GNSS navigation solution they produce. Similarly, raw accelerations from the inertial navigator may be used independently of the navigation solution they produce. These outputs generally require software changes within both the GNSS receiver and the INS, and it may even require hardware changes to make such data available for the combined solution.

In the more tightly coupled implementations, all data are available for a common navigation solution. In that case, the common filter model used for system integration must include variables such as GNSS signal propagation delays, or accelerometer bias and scale factor errors. The estimated values of these variables may be used in the internal implementations of the GNSS receiver and INS.

Ultra-tightly Coupled Integration This term is applied to GNSS/INS integration in which signal tracking loops in the GNSS receiver are augmented by using accelerations and rotations sensed by the INS. This approach can reduce phase-tracking filter lags and improve phase-locking during periods of high maneuvering, reduced signal strength, or signal jamming. It also improves GNSS signal acquisition times by reducing Doppler shift uncertainties. This effectively improves navigation accuracy by reducing these navigation error sources and provides additional operational margins for extreme dynamics, weak GNSS signals, or interference from other signals – including jamming. For a discussion of this approach using Kalman filtering, see [1].

Limitations This loose/tight ordering is not "complete," in the sense that it is not always possible to decide whether one implementation is strictly looser or tighter than another. There are just too many degrees of freedom in making implementation changes for them to be ordered in one dimension. However, the ranking does provide some notion of what is involved. Also, even the more tightly coupled implementations may not include rotation rates and accelerations in the Kalman filter state vector, but use the INS to implement the time-propagation of velocity, position, and orientation estimates between position fixes. This approach is used in the next section.

12.3 Integrated Navigation Models

12.3.1 Common Navigation Models

In this approach, the navigation solution is the one determined by the integrated GNSS/INS implementation. Key features of this approach are as follows:

1. The Kalman filter implementation may be non-standard, in that the inputs to the INS sensors are not necessarily treated as state variables. However, the output biases and scale factors of those same sensors may be treated as state variables for some implementations.
2. The INS itself replaces the forward state vector propagation model for the subset of navigation variables including INS position, velocity, and orientation. Propagation of the associated covariance matrix of state estimation uncertainty is implemented by the conventional matrix riccati equations.
3. The implementation uses a Kalman filter to combine all "measurements," some of which may be derived from INS sensor outputs. This could include all the navigation error variables of the INS error model, plus all the error variables of GNSS navigation.
4. The combined navigation solution is updated by measurements of all types (pseudoranges, variables derived from sensed accelerations, and inertial attitude rates).
5. In the short times between GNSS solutions, the combined navigation solution is updated from INS sensor measurements. These updates use only sensed accelerations and sensed attitude rates, as compensated by the current estimates of sensor compensation parameters. Updates of the covariance matrix of state estimation uncertainty during these periods reflect changes due to these measurements, as well as the effects of inertial sensor noise. Incremental changes in estimation uncertainties due to sensor noise are modeled by the formulas in Section 11.4. The measurement

update of the covariance matrix is the most computationally intensive part of a Kalman filter, but it does not necessarily have to be run each time the inertial sensor outputs are sampled. Inertial sensor sampling intervals generally range from milliseconds (for most strapdown systems) to tens of milliseconds (for most gimbaled systems).
6. Each GNSS pseudorange is used to update the navigation solution, as well as observable variables of the error model. These observable variables include the time-correlated satellite signal propagation delays and inertial sensor compensation parameters, which are not observable from the INS measurements alone. The intersample intervals for GNSS pseudoranges are typically in the order of seconds for each satellite.
7. The complete navigation solution includes the locations and velocities of the INS and of the GNSS receiver antenna, which cannot be co-located as a rule. The acceleration of the antenna can be determined from the measured INS rotation rates and accelerations, which could be used to reduce signal phase and frequency tracking error.
8. The solution may also include updates of inertial sensor error compensation parameters, and updates of location and velocity with estimated INS navigation errors removed.
9. Performance is determined by the solution of the riccati equations, which is an integral part of the Kalman filter implementation. However, the model used for solving the riccati equation does not necessarily need to be as precise as the models used for the rest of the navigation solution.

The approach uses a common model for all the dynamic variables required for GNSS/INS integration. It represents the best possible navigation solution from GNSS/INS integration, given all effective measurements and their associated uncertainties.

A common model for integrated GNSS/INS navigation would include everything that contributes to navigation error of the integrated system, including all the error sources and error dynamics, and especially any variables that can be estimated from the combined measurements of the GNSS receiver and the INS.

This would include, as a minimum, state variables for all nine navigation error variables (seven if for horizontal navigation only), all significant inertial sensor compensation errors, variables representing the receiver clock error (bias and drift, as a minimum), and variables representing unknown signal propagation delays due to atmospheric effects.

The common model includes terms from all sensor subsystems, including GNSS and INS. The model used for errors in the GNSS receiver clock is described next.

12.3.2 GNSS Error Models

12.3.2.1 GNSS Time Synchronization

UTC and GNSS time *Coordinated Universal Time* (UTC), also called *Greenwich Mean Time* or *Zulu Time*,[2] is an international time standard used worldwide for communications and aviation, among other things. It is a mean solar time referenced to the Royal Observatory at Greenwich, the inertial rotation rate of the Earth, the inertial direction of its rotation axis, and the orbital period of the Earth around the Sun. The Earth is not a perfect clock, however, because its rotation axis and rotation rate are also influenced by its moments of inertia and the net relative rotational inertia of its turbulent faction (e.g. oceans and atmosphere), none of which are exactly constant. UTC is then coordinated among several atomic clocks at different locations around the world and adjusted periodically by adding or subtracting "leap seconds" to keep it within $\pm\frac{1}{2}$ second of our earth clock model.

GNSS time, on the other hand, is synchronized among the transmitting satellites to time-tag their transmissions, and among its users to synchronize their clocks to a common GNSS time. It is not exactly synchronized with UTC time, but it does keep track of the relative time offset in leap-seconds. Satellites cannot use UTC time directly because a leap-second change would be equivalent to a pseudorange change of approximately 300 000 km. The relative offsets from UTC time can then be used to keep track of the relative GNSS time offsets between different GNSS systems. These relative offsets can then be used in the pseudorange calculations of multisystem GNSS receivers.

12.3.2.2 Receiver Clock Error Model

A GNSS receiver clock model must be included in the analysis, because errors in the receiver clock must also be taken into account when assessing performance.

GNSS receiver clocks generally keep the "GNSS time" for the particular GNSS system they are using. These are all referenced to UTC. Clock error models used in GNSS synchronization analysis generally use distance units for clock bias and velocity units for clock drift rates. This is because the clock is being used for timing wavefronts that are traveling at or near the speed of light, and distance traveled is the variable of interest. The equivalent variables and parameters in the GNSS receiver clock model are obtained by multiplying the usual time units by signal propagation speed.

The GNSS receiver clock model used in the analysis has two state variables:

$\delta_{\text{clock bias}}$, clock bias, in meters.

$\delta_{\text{clock drift}}$, clock drift rate in meters per second.

---

2 Military terminology using phoneticized letters to designate time zones.

Clock drift rate is assumed to be exponentially correlated. Its stochastic dynamic model has the form

$$\frac{d}{dt}\begin{bmatrix}\delta_{\text{clock bias}}\\ \delta_{\text{clock drift}}\end{bmatrix} = \begin{bmatrix}0 & 1\\ 0 & -1/\tau_{\text{clock drift}}\end{bmatrix}\begin{bmatrix}\delta_{\text{clock bias}}\\ \delta_{\text{clock drift}}\end{bmatrix} + \begin{bmatrix}0\\ w_{\text{clock drift}}(t)\end{bmatrix} \tag{12.1}$$

with noise variance

$$\begin{aligned} q_{\text{clock drift}} &\stackrel{\text{def}}{=} \mathrm{E}_t\langle w^2_{\text{clock drift}}(t)\rangle \\ &= \frac{2\sigma^2_{\text{clock drift}}}{\tau_{\text{clock drift}}} \end{aligned} \tag{12.2}$$

where $\sigma^2_{\text{clock drift}}$ is the steady-state variance of clock drift.

This model does not include short-term phase-flicker noise, which is assumed to be filtered out by the clock electronics.

Model Parameter Values Values used in this analysis will assume root mean square (RMS) drift rates of 10 m/s[3] and a correlation time of an hour, that is,

$$\sigma_{\text{clock drift}} \approx 10 \text{ m/s} \tag{12.3}$$

$$\tau_{\text{clock drift}} \approx 3600 \text{ s} \tag{12.4}$$

$$q_{\text{clock drift}} \approx 0.05 \text{ m}^2/\text{s}^3 \tag{12.5}$$

Cold-start Initialization In a "cold-start" simulation without prior satellite tracking, the initial value for RMS drift rate uncertainty would be its steady-state value,

$$\begin{aligned} \sigma_{\text{clock drift}}(t_0) &= \sigma_{\text{clock drift}}(\infty) \\ &= 10 \text{ m/s} \end{aligned} \tag{12.6}$$

$$\sigma_{\text{clock bias}}(\infty) = \infty \tag{12.7}$$

That is, there is no corresponding finite steady-state value for clock bias uncertainty with this model. Fortunately, bootstrap satellite signal acquisition requires significantly less timing accuracy than navigation. Before satellite signal acquisition, the receive clock time is used primarily for selecting candidate satellites from stored ephemerides, which does not require accurate timing. Signal acquisition requires less clock frequency accuracy than navigation, and the clock bias correction is usually initialized as part of acquisition. In navigation analysis, its RMS initial uncertainty is usually set closer to its steady-state tracking value, which is in the order of a few meters, equivalent to an RMS receiver timing error of $\sim 10^{-8}$ seconds.

3 This is comparable to the performance of the quartz oscillator in a Timex wristwatch with ambient indoor temperatures [2].

The MATLAB® function `ClockModel2` on www.wiley.com/go/grewal/gnss calculates the values of the dynamic coefficient matrix, process noise covariance matrix, pseudorange sensitivity matrix, and initial covariance matrix value as a function of $\sigma_{\text{clock drift}}$ and $\tau_{\text{clock drift}}$.

12.3.2.3 Propagation Delay

Dual-frequency GNSS The world's first satellite, *Sputnik 1*, broadcast at 20 and 40 MHz to measure differential propagation delay, and the world's first GNSS, *Transit*, used two frequencies to compensate for variable ionospheric propagation delays. GPS was originally conceived as a dual-frequency system encrypted for military use only, but – following the 1983 Korean Air Lines Flight 007 disaster – then-President Ronald Reagan signed an executive order to include an unencrypted civilian channel. Because that happened during the cold war, the accuracy of the GPS civilian channel was doubly degraded to discourage its use by military adversaries by (i) making it single-frequency[4] and (ii) corrupting the signal information further by something called "Selective Availability."[5] Selective Availability was removed after the cold war had ended, and a second GPS civilian frequency has been added. The remaining propagation delay errors are due to a number of factors, including small-scale free-electron density inhomogeneities in the ionosphere, which tend to be time- and distance-correlated.

Propagation delay errors are modeled in distance units, in much the same way as clock errors. These are modeled as independent, exponentially correlated errors. In which case, the associated dynamic coefficient matrix and process noise covariance matrix are diagonal matrices of the same dimension:

$$\mathbf{F}_{\text{sat. del.}} = \frac{-1}{\tau_{\text{sat. del.}}} \begin{bmatrix} 1 & 0 & 0 & \cdots & 0 \\ 0 & 1 & 0 & \cdots & 0 \\ 0 & 0 & 1 & \cdots & 0 \\ \vdots & \vdots & \vdots & \ddots & \vdots \\ 0 & 0 & 0 & \cdots & 1 \end{bmatrix} \tag{12.8}$$

$$\mathbf{Q}_{\text{sat. del.}} = q_{\text{sat. del.}} \begin{bmatrix} 1 & 0 & 0 & \cdots & 0 \\ 0 & 1 & 0 & \cdots & 0 \\ 0 & 0 & 1 & \cdots & 0 \\ \vdots & \vdots & \vdots & \ddots & \vdots \\ 0 & 0 & 0 & \cdots & 1 \end{bmatrix} \tag{12.9}$$

4 This was mitigated to some degree by broadcasting parameters for a theoretical model for ionospheric delay [3].

5 This was largely compensated when the US Coast Guard began broadcasting locally determined delay corrections.

Steady-state Delay Covariance The steady-state equation for propagation delay covariance $\mathbf{P}_{\text{sat. del.}}$ is then a diagonal matrix satisfying

$$0 = \mathbf{F}_{\text{sat. del.}}\mathbf{P}_{\text{sat. del.}} + \mathbf{P}_{\text{sat. del. }\infty}\mathbf{F}^{T}_{\text{sat. del.}} + \mathbf{Q}_{\text{sat. del.}} \tag{12.10}$$

$$0 = \frac{-2}{\tau_{\text{sat. del.}}} p_{\text{sat. del. }\infty\text{ ii}} + q_{\text{sat. del.}} \tag{12.11}$$

$$p_{\text{sat. del. }\infty\text{ ii}} = \frac{1}{2}\tau_{\text{sat. del.}} q_{\text{sat. del.}} \tag{12.12}$$

12.3.2.4 Pseudorange Measurement Noise

The measurements are pseudoranges, but the measured time delay values will generally contain some short-term additive noise in the nanosecond range, due to electronic noise and signal processing noise. This is modeled as a measurement noise covariance matrix. This will be a diagonal matrix, unless there are significant common-mode error sources such as additive power supply noise or grounding noise.

12.3.3 INS Error Models

12.3.3.1 Navigation Error Model

The dynamic coefficient matrix has already been derived in Eq. (11.67).

12.3.3.2 Sensor Compensation Errors

Changes in input axis misalignments may not be significant enough to be included in the model. That eliminates twelve variables. Also, because the model is for a gimbaled system, errors in gyro scale factors can be ignored. The resulting model has only nine error state variables. Because compensation parameter drift is modeled as an exponentially correlated process, the resulting dynamic coefficient matrix and process noise covariance matrix are diagonal.

The dynamic coupling of compensation errors into navigation errors, given in Eq. (11.104), will be missing its final three rows and columns:

$$\mathbf{F}_{\delta\to\xi} = \begin{bmatrix} 0 & 0 & 0 \\ \mathbf{I} & \mathbf{D}_a & 0 \\ 0 & 0 & \mathbf{I} \end{bmatrix} \tag{12.13}$$

$$\mathbf{D}_a = \begin{bmatrix} a_E & 0 & 0 \\ 0 & a_N & 0 \\ 0 & 0 & a_U \end{bmatrix} \tag{12.14}$$

12.3.4 GNSS/INS Error Model

12.3.4.1 State Variables

So far, the error models for GNSS and INS include the following variables that can potentially be estimated as part of the generalized navigation solution:

1. GNSS error sources
   (a) Receiver clock error (2 variables)
   (b) Satellite signal delay errors (= number of satellites[6])
2. INS error sources
   (a) Errors in the navigation solution
      i. Position errors (3 components)
      ii. Velocity errors (3 components)
      iii. Orientation errors (3 components)
   (b) Inertial sensor errors
      i. Accelerometer errors
         A. Output bias errors (3 components, total)
         B. Scale factor errors (3 components, total)
         C. Input axis misalignment errors (6 components, total)
      ii. Gyroscope errors
         A. Output bias errors (3 components, total)
         B. Scale factor errors (3 components, total)
         C. Input axis misalignment errors (6 components, total)

The grand total, if all of these were in the model, would be 66 state variables in an integrated GNSS/INS error model. However, there are reasons for leaving some of these out of the model.

12.3.4.2 Numbers of State Variables

There are, in total, 9 navigation error variable, 24 first-order sensor compensation parameters, and 2 clock model variables. Assuming around a dozen satellites may be used at any one time, the potential number of state variables for an integrated system would be around $9 + 24 + 2 + 12 = 47$. However, not all of these variables need to be updated in real time. Input axis misalignments, for example, tend to be relatively stable. If just the inertial sensor biases and scale factors (12 in all) will require updating, the total number would decrease to $9 + 12 + 2 + 12 = 35$.

The model that is used in the rest of this chapter is for a gimbaled system, for which gyro scale factors are but a factor in the feedback gains used in stabilizing the platform, and do not contribute significantly to navigation error. That would leave only 31 state variables. However, to simplify the state variable switching required as satellites come into use and fall out of use, this model includes the signal delays of all 31 satellites from the GPS constellation of 7 July 2012, rather

6 Thirty-one in the ephemerides used in the simulations.

than the 12-or-so used at any one time. That brings the total number of state variables to $9 + 9 + 2 + 31 = 51$.

12.3.4.3 Dynamic Coefficient Matrix

The resulting dynamic coefficient matrix for just those 51 state variables will have the structure shown in Figure 12.2, where

$\mathbf{F}_{\text{core}}$ is the 9×9 dynamic coefficient matrix for the core navigation errors – including vertical navigation errors.

$\mathbf{F}_{\delta}$ is the 9×9 dynamic coefficient matrix for the nine sensor compensation parameters used. These include three each of accelerometer bias, accelerometer scale factor, and gyro bias.

$\mathbf{F}_{\delta \to \varepsilon}$ is the dynamic coupling matrix of sensor compensation errors into navigation errors.

$\mathbf{F}_{\text{clock}}$ is the dynamic coefficient matrix for the receiver clock model. In this case, the model has only clock bias, but it can easily be expanded to include drift.

$\mathbf{F}_{\text{sig.del.}}$ is the dynamic coefficient matrix for satellite signal delay error due to unmodeled atmospheric effects.

12.3.4.4 Process Noise Covariance

The common model has process noise sources for the following:

1. Inertial sensor noise.
2. Drifting inertial sensor compensation parameters.
3. Receiver clock error.
4. Atmospheric propagation delays.

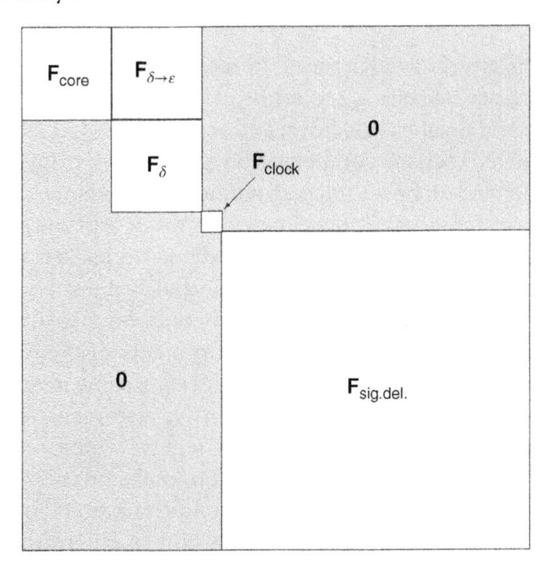

Figure 12.2 Block structure of dynamic coefficient matrix.

12.3.4.5 Measurement Sensitivities

The sensitivity of pseudorange measurements to position is represented in terms of a unit vector in the direction from the satellite antenna to the receiver antenna. That is the essential "input axis" for the receiver as a pseudorange sensor. The longer the pseudorange, the more the estimated location should be adjusted in that direction. Each row of the measurement sensitivity matrix, corresponding to the pseudorange measurement from the ith satellite, would have the structure

$$\mathbf{H}_i = \begin{bmatrix} u_{iE} & u_{iN} & u_{iU} & 0 & \cdots & 0 & 1 & 0 & \cdots & 0 & 1 & 0 & \cdots \end{bmatrix} \tag{12.15}$$

where

$$\mathbf{u}_i = \begin{bmatrix} u_{iE} \\ u_{iN} \\ u_{iU} \end{bmatrix} \tag{12.16}$$

is the ith pseudorange measurement input axis (a unit vector in the direction from the satellite to the antenna location), the first "1" after that is in the 19th column (clock bias), and the final "1" is in the $(20 + i)$th column (propagation delay for ith satellite).

Sensitivities of all pseudorange measurements to clock error are all equal to +1 if the clock error is in distance units and positive clock errors mean the clock is "fast."

12.4 Performance Analysis

12.4.1 The Influence of Trajectories

Host vehicle dynamics influence integrated GNSS/INS navigation performance because some terms in the self-calibration model depend on attitudes, attitude rates, velocities, and accelerations. As a consequence, the observability of INS sensor compensation parameters from GNSS receiver pseudoranges depend on host vehicle dynamics. For example, the scale factors of sensors with zero input have no influence on the sensor outputs, nor on navigation errors. This phenomenon was well understood half a century ago, when methods were developed for alignment of inertial systems aboard aircraft on carrier decks. The roll and pitch of the ship could be measured by the INS on the aircraft and compared with the measurements made in the ship's own INS. However, this approach depends on their being some identifiable pitching and rolling. A similar situation arises in "transfer alignment" of inertial systems in stand-off weapons attached to aircraft wings, which may require specially designed maneuvers by the host aircraft to make this alignment more observable.

The same is true of test conditions used for performance assessment of GNSS/INS systems. It is important that performance prediction be evaluated on representative trajectories from the intended missions. Systems intended for commercial aircraft should be evaluated on representative routes, and systems for military applications need to be evaluated under expected mission conditions. Each also needs to be evaluated for representative satellite geometries. Trans-polar flights, for example, will have distinctively different satellite geometries.

The combined influence of host vehicle dynamics on integrated GNSS/INS performance can usually be characterized by a representative set of host vehicle trajectories, from which we can estimate the expected performance of the integrated GNSS/INS system over an ensemble of mission applications.

For example, a representative set of trajectories for a regional passenger jet might include several trajectories from gate to gate, including taxiing, takeoff, climb-out, cruise, midcourse heading and altitude changes, approach, and landing. Trajectories with different total distances and headings should represent the expected range of applications. These can even be weighted according to the expected frequency of use. With such a set of trajectories, one can assess expected performances with different INS error characteristics and different satellite and pseudolite geometries.

Similarly, for a standoff air-to-ground weapon, an ensemble of trajectories with different approach and launch geometries and different target impact constraints can be used to evaluate RMS miss distances with GNSS jamming at different ranges from the target.

We demonstrated integrated GNSS/INS performance using a simple trajectory simulator, just to show the benefits of GNSS/INS integration. In order to quantify the expected performance for a specific application, however, you must use your own representative set of trajectories.

12.4.2 Performance Metrics

12.4.2.1 Application-Dependent Performance Metrics

These are generally applications driven and related to specific product uses, such as:

1. Precision of positioning control for such applications as autonomous grading, plowing, or harvesting. In this case, inertial sensors are also needed for vehicle steering control.
2. Precise location of racecars on a track, used for generating visual cues on track videos to identify individual cars.
3. Determining which track a train is on, in which case the possible trajectories are essentially one-dimensional, but there is more than one possible trajectory. Because the cost of failure in this application can be extremely high, this requires very high confidence in the implementation method used.

4. Guidance and control of drones, using the inertial sensors for control as well as for navigation. The INS rotation sensors may also be used for control of cameras in some applications.
5. INS initialization in motion, using GNSS. In this case, observability of the INS navigation solution is generally influenced by the trajectory of the host vehicle during initialization, and whether host vehicle maneuvering capabilities are confined to moving along its roll axis.
6. Using GNSS to correct drifting inertial sensor error compensation parameters.
7. Sustaining a specified level of navigational accuracy for a specified time after loss of GNSS signals, which can be important in applications such as delivery of precision munitions in GNSS jamming environments or guiding first-responders in indoor environments with or without other navigation aids.

12.4.2.2 General-Purpose Metrics

These are used in design of integrated GNSS/INS systems for predicting performance and for verification of predicted performance from data collected during tests and trials.

Predictive design of sensor suites. This is often an iterative process using mathematical models to compare relative expected performance.

Error budgeting. This involves allocation of accuracy requirements to subsystems for achieving a specified level of accuracy at the system level. Budgeting allows designers to identify critical accuracy requirements and possible error re-allocation strategies for lowering risk.

12.4.2.3 Mean Squared Error Metrics

These are based on the mathematical systems analysis of Kalman [4], which is based, in part, on the least mean squares estimator of Gauss.

GNSS/INS integration is but another example of how profoundly Kalman filtering has transformed so many technologies involving measurement and estimation. It has become the *lingua franca* for solving tracking problems, especially those with only stochastically predictable dynamics (as in navigation). It solves not only the software implementation problem, but also the problem of tailoring the system design to meet application-specific performance goals. The same riccati equations used in the Kalman filter for propagating the covariance matrices of estimation uncertainties can be solved all by themselves – independent of the estimation solution – for determining how mission trajectories and sensor characteristics influence applications-specific metrics of performance.

12.4.2.4 Probabilistic Metrics

CEP The *circle of equal probability CEP*[7] for a navigation system is defined as the *median radial horizontal navigation error*, usually determined empirically from the results of a specified set of trials. It has become a standard unit of accuracy for military weapon systems of all sorts, from artillery to ICBMs, and it was adopted early on for navigation, as well. It is easily determinable from the results of testing under controlled representative conditions of use, and it is easy to understand and apply as a measure of military efficacy.

It still is a common metric for navigational accuracy, even though it is not analytically determinable within the linear stochastic formalism of navigation systems analysis as introduced by Kalman. The problem is that the Kalman filter propagates only the means and covariances of navigation errors, not their medians.

Medians of PDFs The *median* of a univariate probability density function (PDF) is defined as that value in its domain at which half the distribution is greater than that value and half is less than that value. There is no standard notation for the median of a PDF $p(x)$, but if we let it be denoted by $\tilde{x}$, then

$$\int_{x=-\infty}^{x=\tilde{x}} p(x)\,dx = \int_{x=\tilde{x}}^{x=+\infty} p(x)\,dx = \frac{1}{2} \tag{12.17}$$

Although radial horizontal navigation error will always have a univariate distribution, the median of that distribution is not uniquely determinable from the means and covariances propagated by the Kalman filter. The Kalman filter depends only on means and covariances, but those statistics alone do not uniquely determine the median radial horizontal error. One must either know or assume the actual error PDF to do that. Gaussianity is a common assumption used for transforming covariances into CEPs,

12.4.3 Dynamic Simulation Model

12.4.3.1 State Transition Matrices

The dynamic coefficient matrix for navigation and sensor compensation errors has terms that depend on velocities and accelerations, so it is a time-varying system. In that case, the dynamic coefficient matrix must be evaluated at each simulation time-step. If it were time-invariant, it would need to be computed only once. The lower 33×33 diagonal sub-matrix, representing clock and propagation delay errors, is time-invariant. In this case, the block-diagonal structure of the dynamic coefficient can be exploited to make the computation involved a little easier.

---

7 See Section 3.7.3.1.

For any block-diagonal matrix

$$\mathbf{F} = \begin{bmatrix} \mathbf{F}_{11} & 0 & \cdots & 0 \\ 0 & \mathbf{F}_{22} & \cdots & 0 \\ \vdots & \vdots & \ddots & \vdots \\ 0 & 0 & \cdots & \mathbf{F}_{kk} \end{bmatrix}, \tag{12.18}$$

its equivalent state transition matrix for the discrete time-step Δt is the matrix exponential

$$\begin{aligned} \mathbf{\Phi} &= \exp(\Delta t\ \mathbf{F}) \\ &= \begin{bmatrix} \exp(\Delta t\ \mathbf{F}_{11}) & 0 & \cdots & 0 \\ 0 & \exp(\Delta t\ \mathbf{F}_{22}) & \cdots & 0 \\ \vdots & \vdots & \ddots & \vdots \\ 0 & 0 & \cdots & \exp(\Delta t\ \mathbf{F}_{kk}) \end{bmatrix}. \end{aligned} \tag{12.19}$$

Consequently, if any of the diagonal blocks is time-invariant, it only needs to be computed once. This is true for the dynamic coefficient sub-matrices for clock errors and signal propagation delay errors. The remaining time-varying block

$$\begin{bmatrix} \mathbf{F}_{\varepsilon}(v_{\text{ENU}}(t)) & \mathbf{F}_{\delta\to\varepsilon}(a_{\text{ENU}}(t)) \\ 0 & \mathbf{D}_{-1/\tau} \end{bmatrix}$$

has a lower-right time-invariant part, but the off-diagonal block $\mathbf{F}_{\delta\to\varepsilon}$ is time-varying. Therefore, the upper-left 18×18 diagonal block of the dynamic coefficient matrix must be evaluated, multiplied by the time-step, and transformed into its matrix exponential. Because the computational complexity of taking matrix exponentials increases as the cube of the dimension, this trick can save a lot of unnecessary computing.

12.4.3.2 Dynamic Simulation

The MATLAB® function `Big8TrackSimENU.m` on www.wiley.com/go/grewal/gnss simulates dynamic conditions for traveling at 100 kph on a 100-km figure-eight test track. The simulation has the same average speed of 100 kph, so the lap time is one hour. Dynamic test conditions are plotted in Figure 12.3, including position, velocity, acceleration, and inertial rotation rates. The rotation rates assume near-critical banking in the turns, as describe in the simulation m-file `Big8SimENU`. However, these rates have little effect on gimbaled systems.

This same track simulation was used for navigation simulations using GNSS only, inertial navigation only, and integrated GNSS/INS.

12.4.4 Sample Results

12.4.4.1 Stand-Alone GNSS Performance

Simulating Satellite Selection GNSS receivers have their own built-in routines for selecting and using GNSS signals. Error simulation of GNSS navigation

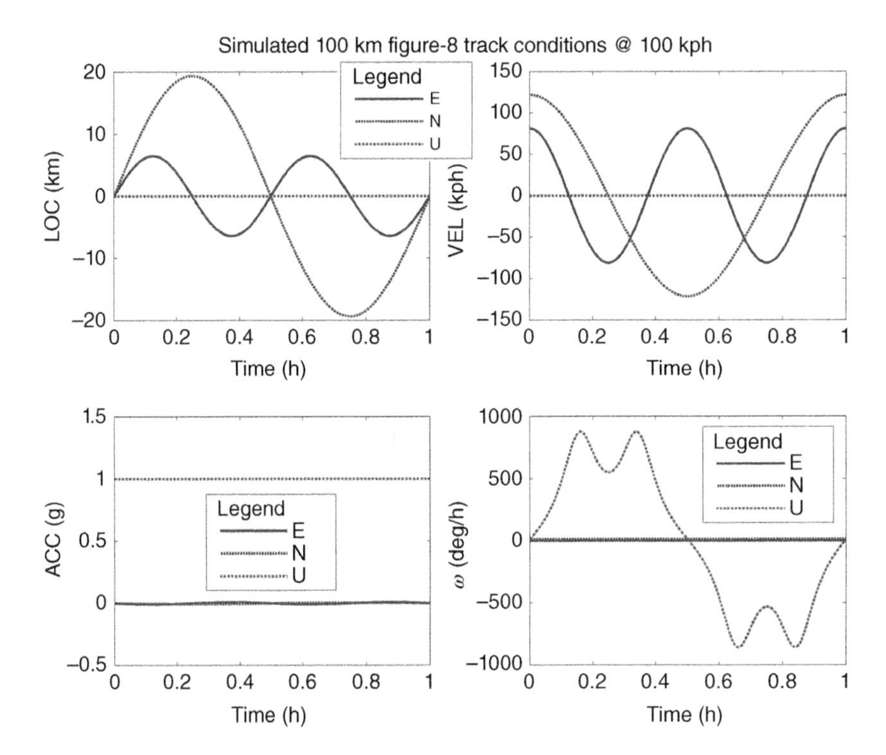

Figure 12.3 Dynamic conditions on 100-km figure-8 test track.

needs to simulate the receiver methods. Doing so in locally level coordinates uses three different coordinate systems, as illustrated in Figure 12.4.

1. Satellite ephemerides are in earth-centered inertial (ECI) coordinates.
2. The location of the GNSS receiver antenna is known from its own navigation solution, usually specified in earth-centered earth-fixed (ECEF) coordinates in terms of latitude, longitude, and altitude.
3. The horizon mask at the receiver (for avoiding surface multipath reflections and rejecting satellites too close to the horizon) is specified with respect to locally level coordinates. In Figure 12.4, lines coming from the antenna (shown atop an INS, represented as a block in the figure) identify those satellites above a 15° horizon mask.

GNSS-Only MATLAB® Implementation The MATLAB® m-file `GNSShalftime.m` on www.wiley.com/go/grewal/gnss runs a six-hour simulation using only GNSS with the a vehicle tracking filter model with parameters set by the m-file `avRMS.m`, described in Chapter 10. This filter includes estimates for the position, velocity, and acceleration of the receiver antenna, and the

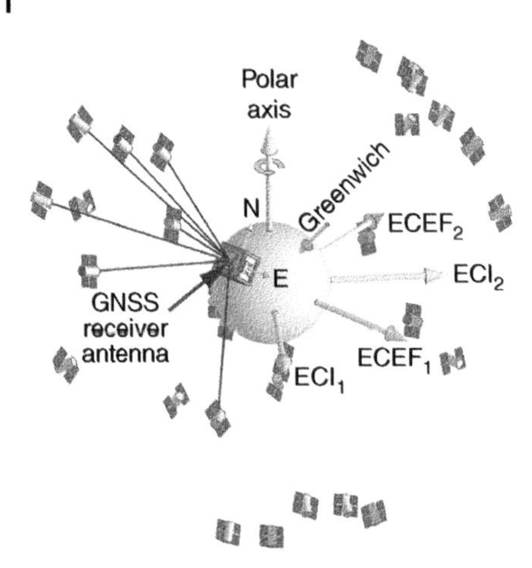

Figure 12.4 GNSS navigation geometry.

tracking filter bounds the RMS dynamic conditions during signal outages. The filter model is statistically adjusted to track dynamics, using the track statistics determined by the m-file `Big8TrackStats.m`, the results of which are also listed as comments in `GNSShalftime.m`. The empirically determined state dynamics are different for the east, north, and vertical components of position, velocity, and acceleration, and this information is used in tuning the filter to the application. The simulation includes a Kalman filter estimating the vehicle state. The program compares the simulated mean-squared position estimation error to the values determined from the riccati equations in the Kalman filter. The agreement is remarkably close when the GNSS receiver has satellite signal access but diverges significantly after signal outage.

The simulation was run with the satellite signals alternately available for an hour and unavailable for an hour, to show the tracking filter response to signal outage. Figure 12.5 is one of the plots generated by `GNSShalftime.m`, showing the history of CEP based on the covariance of horizontal position error in the Kalman filter used for determining the navigation solution. Also shown is the estimate of CEP from simulating the actual navigation solution with simulated Gaussian pseudorandom noise sources.

12.4.4.2 INS-Only Performance

The MATLAB® file `INS7only0.m` on www.wiley.com/go/grewal/gnss simulates an INS under the same dynamic conditions as those in Figure 12.5, with a set of sensor error parameters consistent with performance in the order of a nautical mile per hour. Figure 12.6 is one of its outputs. In order to avoid the problem of vertical channel instability, the 14-state error model includes only horizontal dynamics and sensor compensation errors for two horizontal accelerometers and three gyroscopes. In this case, GNSS signal outages have no effect on INS performance.

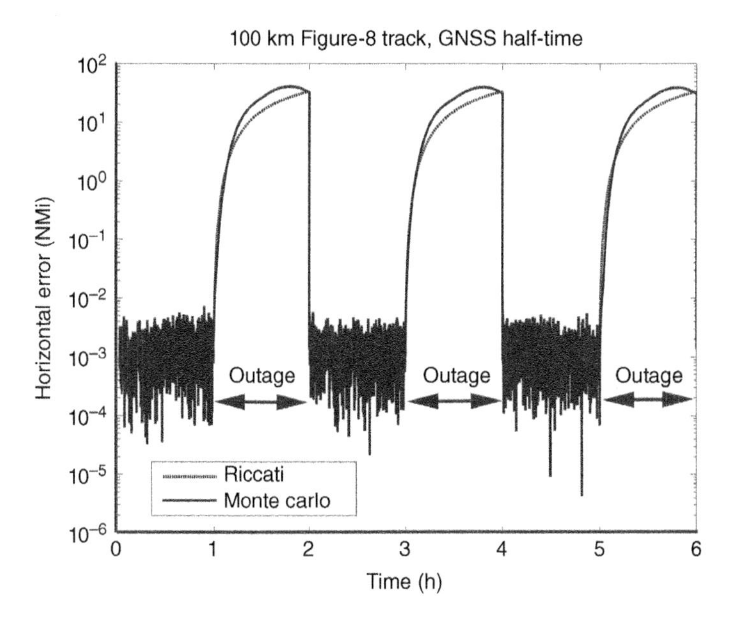

Figure 12.5 GNSS-only navigation simulation results.

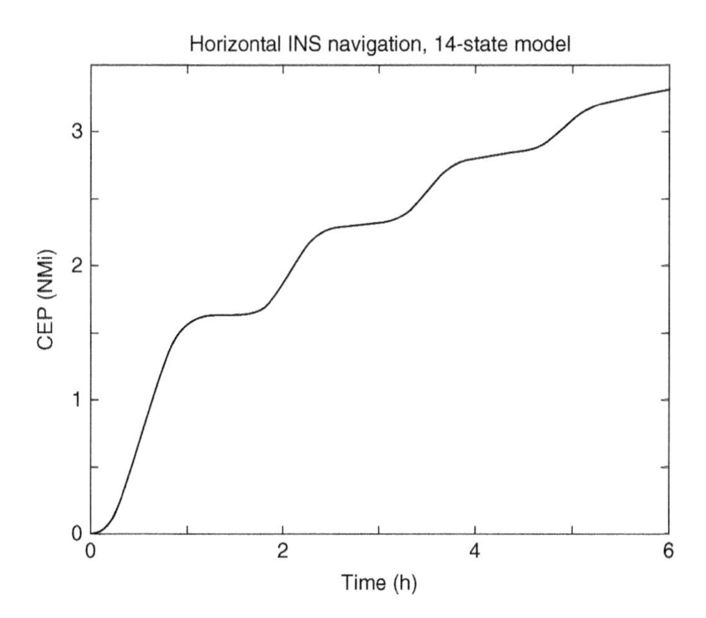

Figure 12.6 INS-only navigation simulation results.

12.4.4.3 Integrated GNSS/INS Performance

Figure 12.7 is a plot of the predicted performance of the same two systems from Figures 12.5 and 12.6, but now integrated using a common navigation model with nine navigation variables, nine sensor compensation variables, two GNSS receiver clock variables, and 31[8] GPS satellites.

Figures 12.5–12.7 capture the essential lesson here: integrated GNSS/INS systems significantly outperform either of their independent subsystems.

Note that performance of the integrated system is superior to that of either subsystem when GNSS signals are available and even performs better than either when signals are unavailable. In this example, integrated performance during a one-hour signal outage is about 80 times better than from using GNSS alone. The secret to this is that the integrated system self-calibrates the inertial sensors when it has signal coverage, so that performance is much better when signals are lost.

The plot in Figure 12.7 – and many more – are generated by the MATLAB® m-file `GNSSINSInt0.m` on www.wiley.com/go/grewal/gnss. The other plots show what is happening to the uncertainties of the various inertial sensor compensation variables when there is good GNSS signal coverage, and how slowly things deteriorate when the signals are unavailable.

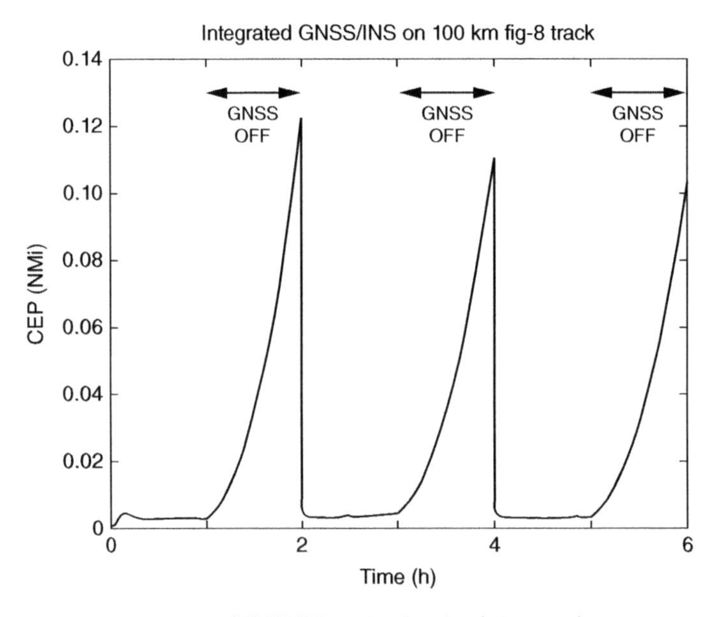

Figure 12.7 Integrated GNSS/INS navigation simulation results.

8 GPS configuration of 7 July 2012, used in the simulations.

Those sensor performance parameters that determine integrated system performance are variables that can be altered in the m-file scripts to simulate a range of inertial navigators.

Keep in mind that these results are for a hypothetical gimbaled INS, and are intended to demonstrate the potential benefits of GNSS/INS integration with INS error compensation updates.

12.5 Summary

1. GPS and INS were both developed for worldwide navigation capability, and together they have taken that capability to new levels of performance that neither approach could have achieved on its own.
2. The payoff in military costs and capabilities had originally driven development of GPS by the Department of Defense of the United States of America during the cold war between the Soviet Bloc of nations and the NATO allies.
3. Some of its proponents and developers had foreseen many of the markets for GNSS/INS integration, including such applications as automating field equipment operations for farming and grading.
4. This insight came from using mathematical systems analysis of the integrated navigator to demonstrate its superior performance, just as we have done in this chapter.
5. The integrated system uses a common navigation model for both types of sensors.
6. This model may also include parameters used for compensating inertial sensor output errors.
7. Sensor compensation parameters then become part of the navigation solution. That is, they are estimated right along with position and velocity, using the GNSS receiver pseudoranges and the inertial sensor outputs.
8. As a consequence, whenever GNSS signals are unavailable, improved accuracy of the sensor compensation parameters results in improved INS stand-alone navigation until GNSS signals become available again.
9. This has led to new automated control applications for integrated GNSS/INS systems, which require inertial sensors for precise and reliable operation under dynamic conditions. Integration with GNSS has brought the costs and capabilities of the resulting systems to very practical levels. The results of integrating inertial systems with GPS has made enormous improvements in achievable operational speed and efficiency.
10. GNSS systems architectures continue to change with the addition of more systems, more satellites, more signal channels, and more aiding systems, and integration-compatible inertial systems have continued to improve as the market expands, driving hardware costs further downward. It is a

part of the Silicon Revolution, harnessing the enormous power and low cost of electronic systems to make our lives more enjoyable and efficient. As costs continue downward, the potential applications market continues to expand.

11. The MATLAB® m-files on www.wiley.com/go/grewal/gnss can be modified by altering sensor performance parameters for exploring different INS design alternatives. They can also be altered to represent new GNSS constellations and signal characteristics, such as two-frequency methods for reducing atmospheric propagation delay errors.
12. This book is not intended to give you the answer to GNSS/INS integration. GNSS systems are changing constantly and every INS is unique in some way, so that answer is always changing. The purpose of this book is to show you how to find that answer yourself.

Problems

12.1 Why might you want to include a chip-level barometric altimeter in an integrated GNSS/INS navigation system?

12.2 What might be gained by including a three-axis magnetometer, as well?

12.3 The m-file `Problem12pt3.m` on www.wiley.com/go/grewal/gnss is a copy of `GNSSINSInt0.m`, which performs simulated performance analysis of an integrated GNSS/INS navigation system. Modify it to increase and decrease each of the following parameters by a factor of 10:

(a) RMS accelerometer noise.
(b) RMS gyro noise.
(c) RMS accelerometer bias.
(d) RMS accelerometer scale factor.
(e) RMS gyro bias.

Record the peak CEP during satellite signal outage for each case.
Questions:

(f) Which has the greatest influence on integrated GNSS/INS navigation performance?
(g) How linear is the relationship? That is, does RMS navigation performance go down by a factor of 10 when any of these is increased by a factor of 10?
(h) What does this say about which sensor performance specifications might be relaxed?

12.4 Run the MATLAB® m-file `Problem12pt4.m` on www.wiley.com/go/grewal/gnss to see how the RMS noise on the vertical accelerometer of an INS influences the accuracy of the GNSS-aided INS in altitude.

References

1 Babu, R. and Wang, J. (2005). Ultra-tight GPS/INS/PL integration: Kalman filter performance analysis. GNSS, Hong Kong, pp. 8–10.
2 Lombardi, M.A. (2008). The accuracy and stability of quartz watches. *Horological Journal* 150 (2): 57–59.
3 Klobuchar, J.A. (1976). Ionospheric time delay corrections for advanced satellite ranging systems. NATO AGARD Conference Proceedings No. 209, in Propagation Limitations of Navigation and Positioning Systems, NATO AGARD, Paris, France.
4 Antoulas, A.C. (ed.) (1991). *Mathematical System Theory: The Influence of R.E. Kalman*. Berlin: Springer-Verlag.

Appendix A

Software

> *Software is like entropy. It is difficult to grasp, weighs nothing, and obeys the second law of thermodynamics; i.e. it always increases.*
>
> Norman R. Augustine
>
> Law Number XVII, p. 114, *Augustine's Laws*, AIAA, 1997.

A.1 Software Sources

The MATLAB® m-files on the companion Wiley website at www.wiley.com/go/grewal/gnss are intended to demonstrate to the reader how the methods described in this book actually work.

This is not "commercial grade" software, and it is not intended to be used as part of any commercial design process or product implementation software. The authors and publisher do not claim that this software meets any standards of merchantability, and we cannot assume any responsibility for the results if they are used for such purposes.

There is better, more reliable commercial software available for global navigation satellite system (GNSS) and inertial navigation system (INS) analysis, implementation, and integration. There are commercial products available for these purposes. Many of the providers of such software maintain internet websites describing their products and services, and the interested user is encouraged to search the internet for suitable sources. The Mathworks File Exchange at www.mathworks.com/matlabcentral/fileexchange also has some useful user-generated utilities (for Kalman filtering, in particular).

Most of the MATLAB routines referred to in the text are self-documented by comments. The following sections contain short descriptions of those mentioned in the different chapters. Routines they call are also self-documented.

Global Navigation Satellite Systems, Inertial Navigation, and Integration,
Fourth Edition. Mohinder S. Grewal, Angus P. Andrews, and Chris G. Bartone.

Companion website: www.wiley.com/go/grewal/gnss

A.2 Software for Chapter 2

GPS_perf.m performs analysis of expected performance of a GPS receiver.

init_var.m initializes parameters and variables for GPS_perf.m.

calcH.m calculates **H** matrix for GPS_perf.m.

gdop.m calculates geometric dilution of precision (GDOP) for chosen constellation for GPS_perf.m.

covar.m solves the Riccati equation for GPS_perf.m.

plot_covar.m plots results from GPS_perf.m.

A.3 Software for Chapter 3

fBortz.m computes the Bortz "noncommutative rate vector" as a function of measured body rates (ω) and cumulative rotation vector (ρ).

A.4 Software for Chapter 4

The MATLAB script ephemeris.m calculates a GPS satellite position in Earth-centered Earth-fixed (ECEF) coordinates from its ephemeris parameters. The ephemeris parameters comprise a set of Keplerian orbital parameters and describe the satellite orbit during a particular time interval. From these parameters, ECEF coordinates are calculated using the equations from the text. Note that time t is the GPS time at transmission and t_k(tk) in the script is the total time difference between time t and the epoch time t_{oe}(toe). Kepler's equation for eccentric anomaly is nonlinear in E_k(Ek) and is solved numerically using the Newton–Raphson method.

The following MATLAB scripts calculate satellite position for 24 hours using almanac data:

ephemeris.m computes the satellite position using ephemeris data.

GPS_position(PRN#) plots satellite position using pseudorandom noise (PRN) one at a time for all satellites using almanac data.

GPS_position_3D plots satellite position for all PRN in three dimensions. Use rotate option in MATLAB to see the satellite positions from the equator, north pole, south pole, and so on, using almanac data.

GPS_el_az (PRN#, 33.8825, −117.8833) plots satellite trajectory for a PRN from Fullerton, California (GPS laboratory located at California State University, Fullerton) using almanac data.

GPS_el_az_all (PRN#, 33.8825, −117.8833) plots for all PRN visible from Fullerton, California (GPS laboratory located at California State University, Fullerton) using almanac data.

GPS_el_az_one_time (PRN#, 14.00, 33.8825, −117.8833) plots at 14:00 time all visible PRN from Fullerton, California (GPS laboratory located at California State University, Fullerton) using almanac data.

A.5 Software for Chapter 7

The following MATLAB scripts compute and plot ionospheric delays using Klobuchar models:

Klobuchar_fix plots the ionospheric delays for geostationary earth orbit (GEO) stationary satellites for 24 hours, such as PanAmSat, GalaxyXV.

Klobuchar (PRN#) plots the ionospheric delays for a satellite specified by the argument PRN, when that satellite is visible.

Iono_delay (PRN#) plots the ionospheric delays for a PRN using dual-frequency data, when a satellite is visible. It uses the pseudorange carrier phase data for L1 and L2 signals. Plots are overlaid for comparison.

init _var initializes parameters and variables for GPS_perf.m.

GPS_perf.m performs covariance analysis of expected performance of a GPS receiver using a Kalman filter. GPS_perm.m evaluates the nonobservability (when three satellites are available, as compared to four satellites) by covariance analysis.

A.6 Software for Chapter 10

VanLoan.m converts the stochastic model coefficients (F, $Q(t)$) of the linear time-invariant matrix differential riccati equation to those of its discrete-time equivalent (Φ, Q_k).

modchol.m computes modified cholesky factors U, D of a symmetric positive definite matrix P such that U is unit upper triangular, D is diagonal, and $P = U^*D^*U^T$.

utchol.m computes the upper triangular cholesky factor of a symmetric positive-definite matrix.

housetri.m performs Householder upper triangularization of a matrix.

potter.m implements the measurement update of the Potter square-root filter.

carlson.m implements the measurement update of the Carlson square-root filter.

thornton.m implements the time update of the Bierman–Thornton square-root filter.

bierman.m implements the measurement update of the Bierman–Thornton square root filter.

udu.m performs modified cholesky decomposition of a symmetric positive-definite matrix.

UD_decomp.m performs in-place modified cholesky decomposition of a symmetric positive-definite matrix.

ChiSqTest.m demonstrates innovations monitoring of simulated Kalman filter implementations.

InnovChisqSim.m called by ChiSqTest.m to simulate innovations monitoring.

avRMS.m returns the parameters for a linear stochastic system with specified root mean square (RMS) velocity, acceleration, and acceleration correlation time constant.

avRMStestfunction.m uses pseudorandom simulation to verify avRMS.m.

GNSSonly1.m simulates GNSS tracking of a vehicle traveling 100 km/h on a 100 km-long figure-8 test track with GNSS signal outage for one minute of every hour, using the host vehicle stochastic model derived in Eqs. (10.153)–(10.159).

GNSSonly2.m simulates the same tracking trajectory as GNSSonly1.m, but using a host vehicle stochastic model with RMS position as an additional statistic.

avRMS.m returns the alternative host vehicle tracking model parameters used in GNSSonly2.m, based on the same track simulation statistics.

YUMAdata.m parses online GPS satellite ephemerides for simulation period.

ClockModel2.m generates receiver clock model matrix parameters.

Big8TrackSimENU.m simulates test track dynamics, called by GNSSonly.

Big8TrackStats.m calculates test track dynamic statistics used by Big8Track SimENU.

HSatENU.m generates GNSS measurement sensitivity matrix for all satellites in view.

A.7 Software for Chapter 11

Fcore7.m calculates the 7×7 dynamic coefficient matrix for the 7 core horizontal navigation error variables, excluding vertical velocity and altitude.

Fcore7Schuler.m simulates Schuler oscillations after an initial north velocity error of 0.062 m/s.

Fcore9.m calculates the 9×9 dynamic coefficient matrix for the 9 "core" navigation error variables, including vertical velocity and altitude.

Fcore10.m calculates the 10×10 dynamic coefficient matrix for the 9 "core" navigation error variables (including vertical velocity and altitude) plus the bias state of a barometric altimeter for stabilizing the vertical channel.

F10CEPrate.m demonstrates INS vertical channel stabilization using an altimeter.

CEPvsSensorNoise.m computes representative contour plots of INS circular error probable (CEP) as a function of accelerometer and gyroscope noise levels, with no other error sources.

CEPvsAccComp.m computes representative contour plots of INS CEP rates versus accelerometer error compensation errors, using figure-8 racetrack trajectories to represent typical usage and a gimbaled system as the INS model. Because each set of compensation errors requires a six hour simulation to compute CEP rates, it runs a very long time.

A.8 Software for Chapter 12

avRMS.m solves a transcendental equation for parameters of a host vehicle tracking filter for given RMS acceleration and velocity.

Big8TrackSimENU.m simulates dynamic conditions for a host vehicle traveling 100 km/h on a 100 km figure-8 test track.

Big8TrackStats.m generates RMS dynamic conditions for a host vehicle traveling 100 km/h on a 100 km figure-8 test track. Used for specifying tracking filter parameters.

ClockModel2.m calculates the stochastic model parameters for a clock in a GNSS receiver.

DAMP3Params.m solves a transcendental equation for parameters of a host vehicle tracking filter for given RMS acceleration and velocity.

ECEF2ENU.m returns the coordinate transformation matrix from Earth-centered Earth-fixed coordinates to east-north-up coordinates.

ECI2ECEF.m returns the coordinate transformation matrix from Earth-centered inertial coordinates to Earth-centered Earth-fixed coordinates.

GNSShalftime.m simulates GNSS racking of a test vehicle on a figure-8 track with GNSS availability only every other hour.

GNSSINSInt0.m simulates integrated GNSS/INS navigation of a test vehicle on a figure-8 track with GNSS availability only every other hour.

HSatENU.m returns measurement sensitivity matrix for satellite configuration at specified time, using loaded almanac data.

INS7only0.m simulates INS navigation of a test vehicle on a figure-8 track.

Problem12pt3.m m-file for Problem 12.3.

Problem12pt4.m m-file for Problem 12.4.

YUMAdata.m loads the global arrays RA (right ascension) and PA (perigee and mean anomaly) from GPS almanac data downloaded from online almanac data.

A.9 Software for Appendix B

Euler2CTMat.m computes the coordinate transformation matrix represented by Euler angles.

CTMat2Euler.m computes the Euler angles equivalent to a given coordinate transformation matrix.

RotVec2CTMat.m computes the coordinate transformation matrix represented by a rotation vector.

CTMat2RotVec.m computes the rotation vector represented by a coordinate transformation matrix.

Quat2RotVec.m computes the rotation vector represented by a quaternion.

RotVec2Quat.m computes the quaternion represented by a rotation vector.

Quat2CTMat.m computes the coordinate transformation matrix represented by a quaternion.

CTMat2Quat.m computes the quaternion represented by a coordinate transformation matrix.

A.10 Software for Appendix C

MSErrVsEst.m demonstrates that Eq. (10.20) (Eq. (C.5) in Appendix C) depends only on the means and variances of the probability density function involved, by computing and plotting it for 10 different probability density functions (PDFs).

PDFAmbiguityErr.m computes and plots the variations in nonlinearly propagated means and variances due to remaining ambiguities about the PDFs beyond means and variances. This illustrates a generally overlooked error source in nonlinear approximations for Kalman filtering.

pUniform.m computes the uniform probability density function as a function of the variate.

pTriangle.m computes the triangle probability density function as a function of the variate.

pNormal.m computes the normal (Gaussian) probability density function as a function of the variate.

pLogNormal.m computes the uniform probability density function as a function of the variate.

pQuadratic.m computes the quadratic probability density function as a function of the variate.

pCosine.m computes the cosine probability density function as a function of the variate.

pHalfCosine.m computes the half-cosine probability density function as a function of the variate.

pU.m computes the U probability density function as a function of the variate.

pChiSquared.m computes the chi-squared probability density function as a function of the variate.

pRayleigh.m computes the uniform probability density function as a function of the variate.

A.11 GPS Almanac/Ephemeris Data Sources

Several ephemeris sources are used for generating satellite positions as a function of time. Some are m-files with right ascensions, arguments of perigee, and satellite angles stored in matrices. These were downloaded from government sources listed in the comments. For example,

YUMAdata.m converts data from https://cddis.nasa.gov/Data_and_Derived_Products/GNSS/broadcast_ephemeris_data.html, sorts it into arrays of right ascension and phase angles.

Prior downloads from the US Coast Guard website have been downloaded in ASCII, converted to MATLAB arrays of right ascension and the sum of argument of perigee and satellite angle, and converted to ".dat" files using the MATLAB save command. Two of these are named RA.dat (right ascensions) and PA.dat (perigee angle plus satellite angle). The sum of perigee angle and satellite angle represents the satellite location with respect to the south-to north equatorial plane crossing at the time the prime meridian is at the vernal equinox.

The MATLAB m-file FetchYUMAdata.m converts ASCII files download from the US Coast Guard website, which you can use to obtain more recent ephemerides. The ASCII file YUMAdata.txt was downloaded from this website on 7 July 2012, and used for creating the data files RA.dat and PA.dat. See the file YUMAdata.txt for an example of the data downloaded. Instructions for navigating the Coast Guard website to download the current GPS Almanac are given in the comments at the end of YUMAdata.txt.

Appendix B

Coordinate Systems and Transformations

Navigation makes use of coordinates that are natural to the problem at hand: inertial coordinates for inertial navigation, orbital coordinates for global navigation satellite system (GNSS) navigation, and Earth-fixed coordinates for representing locations on the Earth.

The principal coordinate systems used, and the transformations between these different coordinate systems, are summarized in this appendix. These are primarily Cartesian (orthogonal) coordinates, and the transformations between them can be represented by orthogonal matrices.[1] However, the coordinate transformations can also be represented by rotation vectors or quaternions, and all representations are used in the derivations and implementation of GNSS/inertial navigation system (INS) integration.

B.1 Coordinate Transformation Matrices

B.1.1 Notation

We use the notation $\mathbf{C}_{\text{TO}}^{\text{FROM}}$ to denote a coordinate transformation matrix from one coordinate frame (designated by "FROM") to another coordinated frame (designated by "TO"). For example,

$\mathbf{C}_{\text{ENU}}^{\text{ECI}}$ denotes the coordinate transformation matrix from Earth-centered inertial (ECI) coordinates (Figure B.1) to Earth-fixed, locally-level, east–north–up (ENU) coordinates (Figure B.7).

$\mathbf{C}_{\text{NED}}^{\text{RPY}}$ denotes the coordinate transformation matrix from vehicle body-fixed roll–pitch–yaw (RPY) coordinates (Figure B.9) to Earth-fixed north–east–down (NED) coordinates (Section B.3.7.4).

1 Readers who may need to refresh their knowledge of matrix theory and notation may want to prepare by reading a current textbook on that subject.

Global Navigation Satellite Systems, Inertial Navigation, and Integration,
Fourth Edition. Mohinder S. Grewal, Angus P. Andrews, and Chris G. Bartone.

Companion website: www.wiley.com/go/grewal/gnss

B.1.2 Definitions

What we mean by a coordinate transformation matrix is that if a vector $\mathbf{v}$ has the representation

$$\mathbf{v} = \begin{bmatrix} v_x \\ v_y \\ v_z \end{bmatrix} \tag{B.1}$$

in XYZ coordinates and the same vector $\mathbf{v}$ has the alternative representation

$$\mathbf{v} = \begin{bmatrix} v_u \\ v_v \\ v_w \end{bmatrix} \tag{B.2}$$

in UVW coordinates, then

$$\begin{bmatrix} v_x \\ v_y \\ v_z \end{bmatrix} = \mathbf{C}^{UVW}_{XYZ} \begin{bmatrix} v_u \\ v_v \\ v_w \end{bmatrix} \tag{B.3}$$

where "XYZ" and "UVW" stand for any two Cartesian coordinate systems in three-dimensional space.

B.1.3 Unit Coordinate Vectors

The components of a vector in either coordinate system can be expressed in terms of the vector components along unit vectors parallel to the respective coordinate axes. For example, if one set of coordinate axes is labeled X, Y, and Z, and the other set of coordinate axes are labeled U, V, and W, then the same vector $\mathbf{v}$ can be expressed in either coordinate frame as

$$\begin{aligned} \mathbf{v} &= v_x \vec{\mathbf{1}}_x + v_y \vec{\mathbf{1}}_y + v_z \vec{\mathbf{1}}_z \\ &= v_u \vec{\mathbf{1}}_u + v_v \vec{\mathbf{1}}_v + v_w \vec{\mathbf{1}}_w \end{aligned} \tag{B.4}$$

where

- The unit vectors $\vec{\mathbf{1}}_x$, $\vec{\mathbf{1}}_y$, and $\vec{\mathbf{1}}_z$ are along the XYZ axes.
- The scalars v_x, v_y, and v_z are the respective components of $\mathbf{v}$ along the XYZ axes.
- The unit vectors $\vec{\mathbf{1}}_u$, $\vec{\mathbf{1}}_v$, and $\vec{\mathbf{1}}_w$ are along the UVW axes.
- The scalars v_u, v_v, and v_w are the respective components of $\mathbf{v}$ along the UVW axes.

B.1.4 Direction Cosines

The respective components can also be represented in terms of dot products of **v** with the various unit vectors,

$$v_x = \vec{\mathbf{1}}_x^T \mathbf{v} = v_u \vec{\mathbf{1}}_x^T \vec{\mathbf{1}}_u + v_v \vec{\mathbf{1}}_x^T \vec{\mathbf{1}}_v + v_w \vec{\mathbf{1}}_x^T \vec{\mathbf{1}}_w \tag{B.5}$$

$$v_y = \vec{\mathbf{1}}_y^T \mathbf{v} = v_u \vec{\mathbf{1}}_y^T \vec{\mathbf{1}}_u + v_v \vec{\mathbf{1}}_y^T \vec{\mathbf{1}}_v + v_w \vec{\mathbf{1}}_y^T \vec{\mathbf{1}}_w \tag{B.6}$$

$$v_z = \vec{\mathbf{1}}_z^T \mathbf{v} = v_u \vec{\mathbf{1}}_z^T \vec{\mathbf{1}}_u + v_v \vec{\mathbf{1}}_z^T \vec{\mathbf{1}}_v + v_w \vec{\mathbf{1}}_z^T \vec{\mathbf{1}}_w \tag{B.7}$$

which can be represented in matrix form as

$$\begin{bmatrix} v_x \\ v_y \\ v_z \end{bmatrix} = \begin{bmatrix} \vec{\mathbf{1}}_x^T \vec{\mathbf{1}}_u & \vec{\mathbf{1}}_x^T \vec{\mathbf{1}}_v & \vec{\mathbf{1}}_x^T \vec{\mathbf{1}}_w \\ \vec{\mathbf{1}}_y^T \vec{\mathbf{1}}_u & \vec{\mathbf{1}}_y^T \vec{\mathbf{1}}_v & \vec{\mathbf{1}}_y^T \vec{\mathbf{1}}_w \\ \vec{\mathbf{1}}_z^T \vec{\mathbf{1}}_u & \vec{\mathbf{1}}_z^T \vec{\mathbf{1}}_v & \vec{\mathbf{1}}_z^T \vec{\mathbf{1}}_w \end{bmatrix} \begin{bmatrix} v_u \\ v_v \\ v_w \end{bmatrix}$$

$$\stackrel{\text{def}}{=} \mathbf{C}_{XYZ}^{UVW} \begin{bmatrix} v_u \\ v_v \\ v_w \end{bmatrix} \tag{B.8}$$

which defines the coordinate transformation matrix $\mathbf{C}_{XYZ}^{UVW}$ from UVW to XYZ coordinates in terms of the dot products of unit vectors. However, dot products of unit vectors also satisfy the cosine rule:

$$\mathbf{v}^T \mathbf{w} = |\mathbf{v}||\mathbf{w}| \ \cos(\theta_{vw}) \tag{B.9}$$

where θ_{vw} is the angle between the vectors **v** and **w**. For unit vectors $\vec{\mathbf{1}}_a$ and $\vec{\mathbf{1}}_b$, this has the form

$$\vec{\mathbf{1}}_a^T \vec{\mathbf{1}}_b = \cos(\theta_{ab}) \tag{B.10}$$

where θ_{ab} is the angle between $\vec{\mathbf{1}}_a$ and $\vec{\mathbf{1}}_b$. As a consequence, the coordinate transformation matrix can also be written in the form

$$\mathbf{C}_{XYZ}^{UVW} = \begin{bmatrix} \cos(\theta_{xu}) & \cos(\theta_{xv}) & \cos(\theta_{xw}) \\ \cos(\theta_{yu}) & \cos(\theta_{yv}) & \cos(\theta_{yw}) \\ \cos(\theta_{zu}) & \cos(\theta_{zv}) & \cos(\theta_{zw}) \end{bmatrix} \tag{B.11}$$

which is why coordinate transformation matrices are also called **direction cosines matrices.**

B.1.5 Composition of Coordinate Transformations

Coordinate transformation matrices satisfy the composition rule

$$\mathbf{C}_C^B \mathbf{C}_B^A = \mathbf{C}_C^A \tag{B.12}$$

where A, B, and C represent different coordinate frames.

B.2 Inertial Reference Directions

The word *inertia* comes from the Latin stem *iners-*, meaning *immobile*,[2] and is used to describe coordinate systems in which Newton's laws of motion apply. Such coordinate systems are neither accelerating nor rotating.

We can use directions with respect to the "fixed" star background to define such inertial coordinate directions. The following are some in common use.

B.2.1 Earth's Polar Axis and the Equatorial Plane

In the northern hemisphere, there is a point in the heavens about which the stars appear to rotate. Thanks to Copernicus and others, we were eventually able to understand that it is the Earth that is rotating, and that the direction to that point defines the direction of the axis of rotation of Earth (also called its *North Pole*). Newton's law of rotational inertia states that – in the absence of other disturbing torques – the direction of that rotation axis remains constant in inertial coordinates. That direction provides one inertial coordinate axis.

The *equatorial plane* (the plane containing the equator) is orthogonal to the polar axis, so it would suffice to find some inertial coordinates in the equatorial plane to complete our coordinate system.

B.2.2 The Ecliptic and the Vernal Equinox

In inertial coordinates, the solution to Newton's equations for the motions of two bodies under the influence of gravity only (the so-called "two-body problem" in celestial mechanics) has the centers of mass of the two bodies orbiting about their common center of mass. In the case that the two bodies in question are our Earth and our Sun, that plane is called the *ecliptic* – because it is also the plane in which lunar and solar *eclipses* occur.

Some ancients were clever enough to determine that the direction from Earth to the Sun in the ecliptic plane is approximately toward the same point in the stellar background when spring starts and the length of darkness equals the length of sunlight here on Earth. That direction is called the *vernal (springtime)*

---

2 Originally it applied, as well, to social immobility due to lack of skill or motivation.

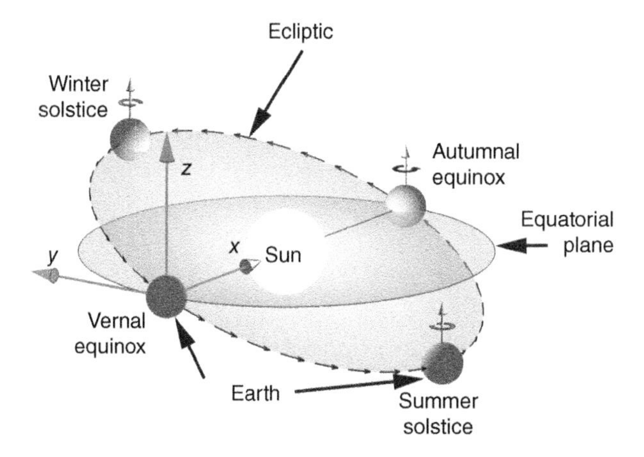

Figure B.1 Earth-centered inertial (ECI) coordinates.

equinox (equal night and day). At the time of the equinox, the Sun is also in the equatorial plane of Earth as it passes from the southern hemisphere to the northern hemisphere.

A third orthogonal coordinate axis can be defined by the cross-product of the first two, such that the three axes define a right-handed orthogonal coordinate system. A heliocentric view of these directions is illustrated in Figure B.1.

B.2.3 Earth-Centered Inertial (ECI) Coordinates

If, as illustrated in Figure B.1, the origin of this coordinate system is at the center of mass of the Earth, the coordinates are called Earth-centered inertial (ECI). ECI coordinates have their first (x) axis toward the vernal equinox, the third (z) axis parallel to the Earth's polar axis, and a second (y) axis to complete a right-handed coordinate system.

B.3 Application-dependent Coordinate Systems

Although we are concerned exclusively with coordinate systems in the three dimensions of the observable world, there are many ways of representing a location in that world by a set of coordinates. The coordinates presented here are those used in navigation with GNSS and/or INS.

B.3.1 Cartesian and Polar Coordinates

René Descartes (1596–1650) introduced the idea of representing points in three-dimensional space by a triplet of coordinates, called "Cartesian

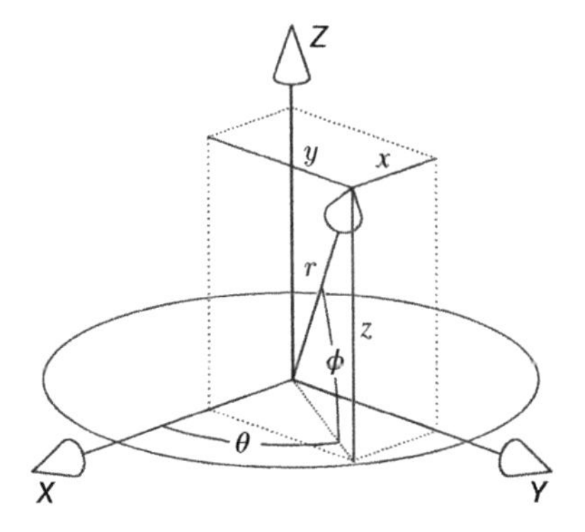

Figure B.2 Cartesian & polar coordinates.

coordinates" in his honor. They are also called "Euclidean coordinates," even though Euclid's *geometry* (from "land measurement" in Greek) is confined to two dimensions. The Cartesian coordinates (x, y, z) and polar coordinates (θ, ϕ, r) of a common reference point, as illustrated in Figure B.2, are related by the equations

$$x = r\cos(\theta)\cos(\phi) \tag{B.13}$$

$$y = r\sin(\theta)\cos(\phi) \tag{B.14}$$

$$z = r\sin(\phi) \tag{B.15}$$

$$r = \sqrt{x^2 + y^2 + z^2} \tag{B.16}$$

$$\phi = \arcsin\left(\frac{z}{r}\right) \quad (-\pi/2 \leq \phi \leq +\pi/2) \tag{B.17}$$

$$\theta = \arctan\left(\frac{y}{x}\right) \quad (-\pi < \theta \leq +\pi) \tag{B.18}$$

with the angle θ (in radians) undefined if $\phi = \pm\pi/2$.

B.3.2 Celestial Coordinates

Similar to ECI coordinates, the "celestial sphere" is a quasi-inertial polar coordinate system referenced to the polar axis of the Earth and the vernal equinox. The prime meridian of the celestial sphere is fixed to the vernal equinox. Polar celestial coordinates are **right ascension** (the celestial analog of longitude, measured eastward from the vernal equinox) and **declination** (the celestial analog of latitude), as illustrated in Figure B.3. Because the celestial sphere is used primarily as a reference for direction, no origin need be specified.

Right ascension is zero at the vernal equinox and increases eastward (in the direction the Earth turns). The units of right ascension (RA) can be radians, degrees, or hours (with 15 deg/h as the conversion factor).

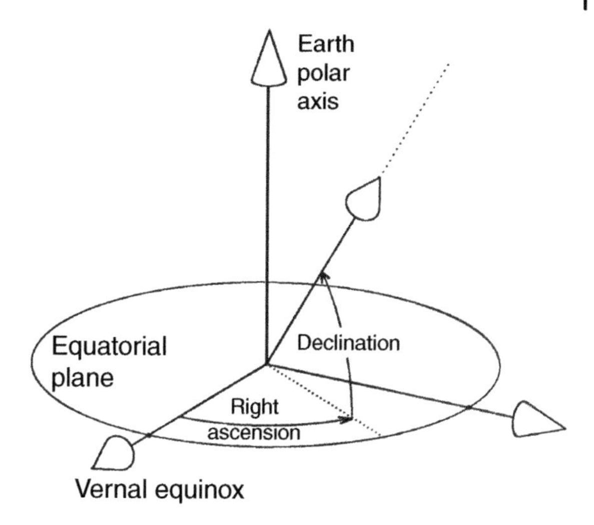

Figure B.3 Celestial coordinates.

By convention, declination is zero in the equatorial plane and increases toward the north pole, with the result that celestial objects in the northern hemisphere have positive declinations. Its units can be degrees or radians.

B.3.3 Satellite Orbit Coordinates

Johannes Kepler (1571–1630) discovered (with the aid of data collected with Tycho Brahe) the geometrical shapes of the orbits of planets, and the minimum number of parameters necessary to specify an orbit (called "Keplerian" parameters). Keplerian parameters used to specify GNSS satellite orbits in terms of their orientations relative to the equatorial plane and the vernal equinox (defined in Section B.2.2 and illustrated in Figure B.1) include

- Right ascension of the ascending node and orbit inclination, specifying the orientation of the orbital plane with respect to the vernal equinox and equatorial plane, as illustrated in Figure B.4.
  - Right ascension is defined in section B.3.2 and shown in Figure B.3.
  - The intersection of the orbital plane of a satellite with the equatorial plane is called its "line of nodes," where the "nodes" are the two intersections of the satellite orbit with this line. The two nodes are dubbed "ascending[3]" (i.e. ascending from the southern hemisphere to the northern hemisphere) and "descending." The right ascension of the ascending node (RAAN) is the angle in the equatorial plane from the vernal equinox to the ascending node, measured counterclockwise as seen looking down from the north pole direction.

3 The astronomical symbol for the ascending node is ☊, often read as "earphones."

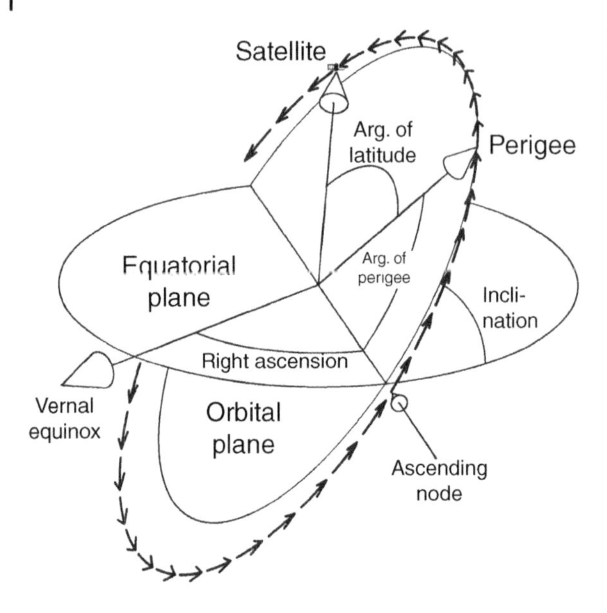

Figure B.4 Keplerian Parameters for Satellite Orbit

  - Orbital inclination is the dihedral angle between the orbital plane and the equatorial plane. It ranges from 0° (orbit in equatorial plane) to 180°.
- Semimajor axis a and semiminor axis b (defined on Section B.3.5.5 and illustrated in Figure B.6), specifying the size and shape of the elliptical orbit within the orbital plane.
- Orientation of the ellipse within its orbital plane, specified in terms of the "argument of perigee," the angle between the ascending node and the perigee of the orbit (closest approach to Earth), as illustrated in Figure B.4.
- Position of the satellite relative to perigee of the elliptical orbit, specified in terms of the angle from perigee – called the "argument of latitude" or "true anomaly," as illustrated in Figure B.4

For computer simulation demonstrations, GPS satellite orbits can usually be assumed to be circular with radius $a = b = R = 26\ 560$ km and inclined at 55° to the equatorial plane. This eliminates the need to specify the orientation of the elliptical orbit within the orbital plane. (The argument of perigee becomes overly sensitive to orbit perturbations when eccentricity is close to zero.)

B.3.4 Earth-Centered Inertial (ECI) Coordinates

The ECI coordinates illustrated in Figure B.1 are the favored *inertial* coordinates in the near-Earth environment. The origin of ECI coordinates is at the center of gravity of the Earth, with axis directions

x, in the direction of the vernal equinox.
z, parallel to the rotation axis (north polar axis) of the Earth.
y, an additional axis to make this a right-handed orthogonal coordinate system,

as illustrated in Figure B.1.

The equatorial plane of the Earth is also the equatorial plane of ECI coordinates, but the Earth itself is rotating relative to the vernal equinox at its sidereal rotation rate of about $7\ 292\ 115\ 167 \times 10^{-14}$ rad/s, or about 15.041 09 deg/h, as illustrated in Figure B.5.

B.3.5 Earth-Centered, Earth-Fixed (ECEF) Coordinates

Earth-centered, Earth-fixed (ECEF) coordinates have the same origin (Earth center) and third (polar) axis as ECI coordinates, but rotate with the Earth – as shown in Figure B.5. As a consequence, ECI and ECEF longitudes differ only by a linear function of time.

B.3.5.1 Longitude in ECEF Coordinates

Longitude in ECEF coordinates are measured east (+) and west (−) from the prime meridian passing through the principal transit instrument at the observatory at Greenwich, UK, a convention adopted by 41 representatives of 25 nations at the International Meridian Conference, held in Washington, DC, in October 1884.

B.3.5.2 Latitudes in ECEF Coordinates

Latitudes are measured with respect to the equatorial plane, but there is more than one kind of "latitude" on the planet.

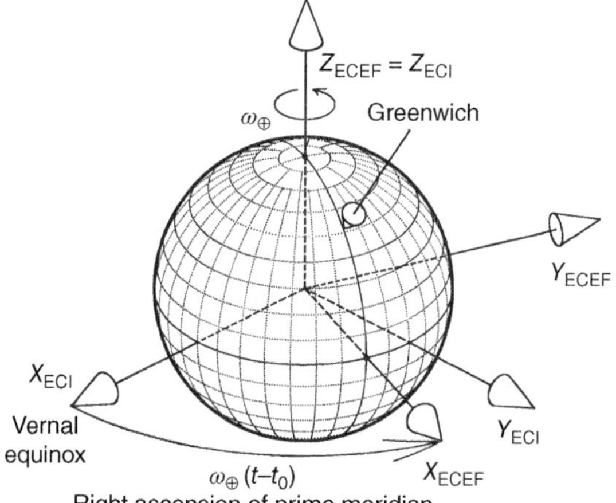

Figure B.5 ECI and ECEF coordinates.

Geocentric latitude would be measured as the angle between the equatorial plane and a line from the reference point to the center of the Earth, but this angle could not be determined accurately (before GNSS) without running a transit survey over vast distances.

The angle between the pole star and the local gravitational vertical direction can be measured more readily, and that angle is more closely approximated as *geodetic latitude*. There is yet a third latitude (parametric latitude) that is useful in analysis. These alternative latitudes are defined in the following subsections.

B.3.5.3 Latitude on an Ellipsoidal Earth

Geodesy is the study of the size and shape of the Earth, and the establishment of physical control points defining the origin and orientation of coordinate systems for mapping the Earth. Earth shape models are very important for navigation using either GNSS or INS, or both. INS alignment is with respect to the local vertical, which does not generally pass through the center of the Earth. That is because the Earth is not spherical.

At different times in history, the Earth has been regarded as being flat (first-order approximation), spherical (second-order), and ellipsoidal (third-order). The third-order model is an ellipsoid of revolution, with its shorter radius at the poles and its longer radius at the equator.

B.3.5.4 Parametric Latitude

For geoids based on ellipsoids of revolution, every meridian is an ellipse with equatorial radius a (also called "semi-major axis") and polar radius b (also called "semi-minor axis"). If we let z be the Cartesian coordinate in the polar direction and $x_{\text{MERIDIONAL}}$ be the equatorial coordinate in the meridional plane, as illustrated in Figure B.6, then the equation for this ellipse will be

$$\begin{aligned}\frac{x^2_{\text{MERIDIONAL}}}{a^2} + \frac{z^2}{b^2} &= 1\\ &= \cos^2(\phi_{\text{PARAMETRIC}}) + \sin^2(\phi_{\text{PARAMETRIC}})\\ &= \frac{a^2\cos^2(\phi_{\text{PARAMETRIC}})}{a^2} + \frac{b^2\sin^2(\phi_{\text{PARAMETRIC}})}{b^2}\\ &= \frac{\{a\cos(\phi_{\text{PARAMETRIC}})\}^2}{a^2} + \frac{\{b\sin(\phi_{\text{PARAMETRIC}})\}^2}{b^2}\end{aligned} \tag{B.19}$$

That is, a parametric solution for the ellipse is

$$x_{\text{MERIDIONAL}} = a\cos(\phi_{\text{PARAMETRIC}}) \tag{B.20}$$

$$z = b\sin(\phi_{\text{PARAMETRIC}}) \tag{B.21}$$

as illustrated in Figure B.6. Although the parametric latitude $\phi_{\text{PARAMETRIC}}$ has no physical significance, it is quite useful for relating geocentric and geodetic latitude, which do have physical significance.

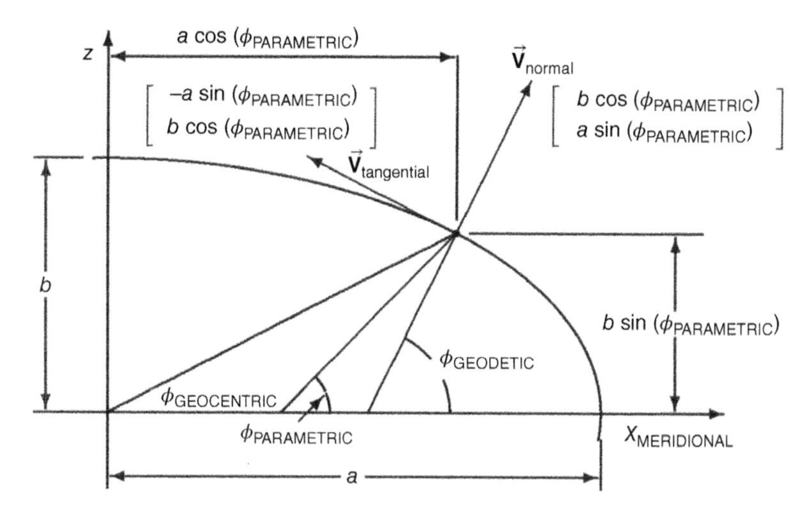

Figure B.6 Geocentric, parametric, and geodetic latitudes in meridional plane.

B.3.5.5 Geodetic Latitude

Geodetic latitude is defined as the elevation angle above (+) or below (−) the equatorial plane of the normal to the ellipsoidal surface. This direction can be defined in terms of the parametric latitude, because it is orthogonal to the meridional tangential direction.

The vector tangential to the meridian will be in the direction of the derivative to the elliptical equation solution with respect to parametric latitude:

$$\begin{aligned}\vec{\mathbf{v}}_{\text{tangential}} &\propto \frac{\partial}{\partial\phi_{\text{PARAMETRIC}}}\begin{bmatrix} a\cos(\phi_{\text{PARAMETRIC}}) \\ b\sin(\phi_{\text{PARAMETRIC}}) \end{bmatrix} \\ &= \begin{bmatrix} -a\sin(\phi_{\text{PARAMETRIC}}) \\ b\cos(\phi_{\text{PARAMETRIC}}) \end{bmatrix}\end{aligned} \tag{B.22}$$

and the meridional normal direction will be orthogonal to it, or

$$\vec{\mathbf{v}}_{\text{normal}} \propto \begin{bmatrix} b\cos(\phi_{\text{PARAMETRIC}}) \\ a\sin(\phi_{\text{PARAMETRIC}}) \end{bmatrix} \tag{B.23}$$

as illustrated in Figure B.6.

The tangent of geodetic latitude is then the ratio of the z- and x-components of the surface normal vector, or

$$\begin{aligned}\tan(\phi_{\text{GEODETIC}}) &= \frac{a\sin(\phi_{\text{PARAMETRIC}})}{b\cos(\phi_{\text{PARAMETRIC}})} \\ &= \frac{a}{b}\tan(\phi_{\text{PARAMETRIC}})\end{aligned} \tag{B.24}$$

from which, using some standard trigonometric identities,

$$\begin{aligned}\sin(\phi_{\text{GEODETIC}}) &= \frac{\tan(\phi_{\text{GEODETIC}})}{\sqrt{1+\tan^2(\phi_{\text{GEODETIC}})}} \\ &= \frac{a\sin(\phi_{\text{PARAMETRIC}})}{\sqrt{a^2\sin^2(\phi_{\text{PARAMETRIC}})+b^2\cos^2(\phi_{\text{PARAMETRIC}})}}\end{aligned} \tag{B.25}$$

$$\begin{aligned}\cos(\phi_{\text{GEODETIC}}) &= \frac{1}{\sqrt{1+\tan^2(\phi_{\text{GEODETIC}})}} \\ &= \frac{b\cos(\phi_{\text{PARAMETRIC}})}{\sqrt{a^2\sin^2(\phi_{\text{PARAMETRIC}})+b^2\cos^2(\phi_{\text{PARAMETRIC}})}}\end{aligned} \tag{B.26}$$

The inverse relationship is

$$\tan(\phi_{\text{PARAMETRIC}}) = \frac{b}{a}\tan(\phi_{\text{GEODETIC}}) \tag{B.27}$$

from which, using the same trigonometric identities as before,

$$\begin{aligned}\sin(\phi_{\text{PARAMETRIC}}) &= \frac{\tan(\phi_{\text{PARAMETRIC}})}{\sqrt{1+\tan^2(\phi_{\text{PARAMETRIC}})}} \\ &= \frac{b\sin(\phi_{\text{GEODETIC}})}{\sqrt{a^2\cos^2(\phi_{\text{GEODETIC}})+b^2\sin^2(\phi_{\text{GEODETIC}})}}\end{aligned} \tag{B.28}$$

$$\begin{aligned}\cos(\phi_{\text{PARAMETRIC}}) &= \frac{1}{\sqrt{1+\tan^2(\phi_{\text{PARAMETRIC}})}} \\ &= \frac{a\cos(\phi_{\text{GEODETIC}})}{\sqrt{a^2\cos^2(\phi_{\text{GEODETIC}})+b^2\sin^2(\phi_{\text{GEODETIC}})}}\end{aligned} \tag{B.29}$$

and the 2D X–Z Cartesian coordinates in the meridional plane of a point on the geoid surface will

$$\begin{aligned}x_{\text{MERIDIONAL}} &= a\cos(\phi_{\text{PARAMETRIC}}) \\ &= \frac{a^2\cos(\phi_{\text{GEODETIC}})}{\sqrt{a^2\cos^2(\phi_{\text{GEODETIC}})+b^2\sin^2(\phi_{\text{GEODETIC}})}}\end{aligned} \tag{B.30}$$

$$\begin{aligned}z &= b\sin(\phi_{\text{PARAMETRIC}}) \\ &= \frac{b^2\sin(\phi_{\text{GEODETIC}})}{\sqrt{a^2\cos^2(\phi_{\text{GEODETIC}})+b^2\sin^2(\phi_{\text{GEODETIC}})}}\end{aligned} \tag{B.31}$$

in terms of geodetic latitude.

Equations (B.30) and (B.31) apply only to points on the geoid surface. Orthometric height h above (+) or below (−) the geoid surface is measured along the surface normal, so that the X–Z coordinates for a point with altitude h will be

$$x_{\text{MERIDIONAL}} = \cos(\phi_{\text{GEODETIC}}) \times \left\{ h + \frac{a^2}{\sqrt{a^2\cos^2(\phi_{\text{GEODETIC}}) + b^2\sin^2(\phi_{\text{GEODETIC}})}} \right\} \tag{B.32}$$

$$z = \sin(\phi_{\text{GEODETIC}}) \times \left\{ h + \frac{b^2}{\sqrt{a^2\cos^2(\phi_{\text{GEODETIC}}) + b^2\sin^2(\phi_{\text{GEODETIC}})}} \right\} \tag{B.33}$$

In 3D ECEF coordinates, with X-axis passing through the equator at the prime meridian (at which longitude $\theta = 0$),

$$\begin{aligned} x_{\text{ECEF}} &= \cos(\theta)x_{\text{MERIDIONAL}} \\ &= \cos(\theta)\cos(\phi_{\text{GEODETIC}}) \\ &\quad\times \left\{ h + \frac{a^2}{\sqrt{a^2\cos^2(\phi_{\text{GEODETIC}}) + b^2\sin^2(\phi_{\text{GEODETIC}})}} \right\} \end{aligned} \tag{B.34}$$

$$\begin{aligned} y_{\text{ECEF}} &= \sin(\theta)x_{\text{MERIDIONAL}} \\ &= \sin(\theta)\cos(\phi_{\text{GEODETIC}}) \\ &\quad\times \left\{ h + \frac{a^2}{\sqrt{a^2\cos^2(\phi_{\text{GEODETIC}}) + b^2\sin^2(\phi_{\text{GEODETIC}})}} \right\} \end{aligned} \tag{B.35}$$

$$\begin{aligned} z_{\text{ECEF}} &= \sin(\phi_{\text{GEODETIC}}) \\ &\quad\times \left\{ h + \frac{b^2}{\sqrt{a^2\cos^2(\phi_{\text{GEODETIC}}) + b^2\sin^2(\phi_{\text{GEODETIC}})}} \right\} \end{aligned} \tag{B.36}$$

in terms of geodetic latitude ϕ_{GEODETIC}, longitude θ, and orthometric altitude h with respect to the reference geoid.

The inverse transformation, from ECEF XYZ to geodetic longitude–latitude–altitude coordinates, is

$$\theta = \text{atan2}(y_{\text{ECEF}}, x_{\text{ECEF}}) \tag{B.37}$$

$$\phi_{\text{GEODETIC}} = \text{atan2}(z_{\text{ECEF}} + e^2a^2\sin^3(\zeta)/b, \xi - e^2a\cos^3(\zeta)) \tag{B.38}$$

$$h = \frac{\xi}{\cos(\phi)} - R_T \tag{B.39}$$

where

$$\zeta = \text{atan2}(a\ z_{\text{GECEF}}, b\xi) \tag{B.40}$$

$$\xi = \sqrt{x_{\text{ECEF}}^2 + y_{\text{ECEF}}^2} \tag{B.41}$$

$$R_T = \frac{a}{\sqrt{1 - e^2\sin^2(\phi)}} \tag{B.42}$$

and R_T is the transverse radius of curvature on the ellipsoid, a is the equatorial radius, b is the polar radius, and e is elliptical eccentricity.

B.3.5.6 WGS84 Reference Geoid Parameters

There are several reference geoids used throughout the world, each with its own set of parameters. The one commonly used throughout this book has the following parameters:

Semi-major axis (equatorial radius)	6378137.0	m
Semi-minor axis (polar radius)	6356752.3142	m
Flattening	0.0033528106718309896	Unitless
Inverse flattening	298.2572229328697	Unitless
First eccentricity	0.08181919092890624	Unitless
First eccentricity squared	0.006694380004260827	Unitless
Second eccentricity	0.08209443803685366	Unitless
Second eccentricity squared	0.006739496756586903	Unitless

B.3.5.7 Geocentric Latitude

For points on the geoid surface, the tangent of geocentric latitude is the ratio of distance above (+) or below (−) the equator ($z = b\sin(\phi_{\text{PARAMETRIC}})$) to the distance from the polar axis ($x_{\text{MERIDIONAL}} = a\cos(\phi_{\text{PARAMETRIC}})$), or

$$\begin{aligned}\tan(\phi_{\text{GEOCENTRIC}}) &= \frac{b\sin(\phi_{\text{PARAMETRIC}})}{a\cos(\phi_{\text{PARAMETRIC}})} \\ &= \frac{b}{a}\tan(\phi_{\text{PARAMETRIC}}) \\ &= \frac{b^2}{a^2}\tan(\phi_{\text{GEODETIC}})\end{aligned} \tag{B.43}$$

from which, using the same trigonometric identities as were used for geodetic latitude,

$$\begin{aligned}\sin(\phi_{\text{GEOCENTRIC}}) &= \frac{\tan(\phi_{\text{GEOCENTRIC}})}{\sqrt{1+\tan^2(\phi_{\text{GEOCENTRIC}})}}\\ &= \frac{b\sin(\phi_{\text{PARAMETRIC}})}{\sqrt{a^2\cos^2(\phi_{\text{PARAMETRIC}})+b^2\sin^2(\phi_{\text{PARAMETRIC}})}}\\ &= \frac{b^2\sin(\phi_{\text{GEODETIC}})}{\sqrt{a^4\cos^2(\phi_{\text{GEODETIC}})+b^4\sin^2(\phi_{\text{GEODETIC}})}}\end{aligned} \tag{B.44}$$

$$\begin{aligned}\cos(\phi_{\text{GEOCENTRIC}}) &= \frac{1}{\sqrt{1+\tan^2(\phi_{\text{GEOCENTRIC}})}}\\ &= \frac{a\cos(\phi_{\text{PARAMETRIC}})}{\sqrt{a^2\cos^2(\phi_{\text{PARAMETRIC}})+b^2\sin^2(\phi_{\text{PARAMETRIC}})}}\\ &= \frac{a^2\cos(\phi_{\text{GEODETIC}})}{\sqrt{a^4\cos^2(\phi_{\text{GEODETIC}})+b^4\sin^2(\phi_{\text{GEODETIC}})}}\end{aligned} \tag{B.45}$$

The inverse relationships are

$$\tan(\phi_{\text{PARAMETRIC}}) = \frac{a}{b}\tan(\phi_{\text{GEOCENTRIC}}) \tag{B.46}$$

$$\tan(\phi_{\text{GEODETIC}}) = \frac{a^2}{b^2}\tan(\phi_{\text{GEOCENTRIC}}) \tag{B.47}$$

from which, using the same trigonometric identities again,

$$\begin{aligned}\sin(\phi_{\text{PARAMETRIC}}) &= \frac{\tan(\phi_{\text{PARAMETRIC}})}{\sqrt{1+\tan^2(\phi_{\text{PARAMETRIC}})}}\\ &= \frac{a\sin(\phi_{\text{GEOCENTRIC}})}{\sqrt{a^2\sin^2(\phi_{\text{GEOCENTRIC}})+b^2\cos^2(\phi_{\text{GEOCENTRIC}})}}\end{aligned} \tag{B.48}$$

$$\sin(\phi_{\text{GEODETIC}}) = \frac{a^2\sin(\phi_{\text{GEOCENTRIC}})}{\sqrt{a^4\sin^2(\phi_{\text{GEOCENTRIC}})+b^4\cos^2(\phi_{\text{GEOCENTRIC}})}} \tag{B.49}$$

$$\begin{aligned}\cos(\phi_{\text{PARAMETRIC}}) &= \frac{1}{\sqrt{1+\tan^2(\phi_{\text{PARAMETRIC}})}}\\ &= \frac{b\cos(\phi_{\text{GEOCENTRIC}})}{\sqrt{a^2\sin^2(\phi_{\text{GEOCENTRIC}})+b^2\cos^2(\phi_{\text{GEOCENTRIC}})}}\end{aligned} \tag{B.50}$$

$$\cos(\phi_{\text{GEODETIC}}) = \frac{b^2 \cos(\phi_{\text{GEOCENTRIC}})}{\sqrt{a^4 \sin^2(\phi_{\text{GEOCENTRIC}}) + b^4 \cos^2(\phi_{\text{GEOCENTRIC}})}} \tag{B.51}$$

B.3.5.8 Geocentric Radius

Geocentric radius $R_{\text{GEOCENTRIC}}$ is the distance to the center of the Earth. As a function of geodetic latitude ϕ_{GEODETIC},

$$R_{\text{GEOCENTRIC}}(\phi_{\text{GEODETIC}}) = \sqrt{\frac{(a^2 \cos \phi_{\text{GEODETIC}})^2 + (b^2 \sin \phi_{\text{GEODETIC}})^2}{(a \cos \phi_{\text{GEODETIC}})^2 + (b \sin \phi_{\text{GEODETIC}})^2}} \tag{B.52}$$

B.3.6 Ellipsoidal Radius of Curvature

The radius of curvature on the reference ellipsoidal surface is what determines how geodetic longitude and latitude change with distance traveled over the surface. It is generally different in different directions, except at the poles. At other places, it is specified by two different values, corresponding to two directions.

Meridional radius of curvature is measured in the north–south direction.[4]

$$R_M = \frac{(ab)^2}{[(a \cos \phi_{\text{GEODETIC}})^2 + (b \sin \phi_{\text{GEODETIC}}]^2)^{3/2}} \tag{B.53}$$

Transverse radius of curvature is measured in the east-west direction. It is defined by the angular rate of the local vertical about the north direction as a function of east velocity.

$$\omega_N = \frac{v_E}{R_T} \tag{B.54}$$

$$R_T = \frac{a^2}{\sqrt{(a \cos \phi_{\text{GEODETIC}})^2 + (b \sin \phi_{\text{GEODETIC}})^2}} \tag{B.55}$$

4 The direction measured ~240 BCE by the Greek polymath Eratosthenes (~276–194 BCE) to arrive at a value of about 250 000 Greek *stadia* (a somewhat ambiguous unit of measure) for the circumference of Earth – which (we think) is close to 40 000 km. French measurements of meridional arcs were later used in defining the meter so as to make the meridional circumferences exactly 40 000 km.

B.3.7 Local Tangent Plane (LTP) Coordinates

Local tangent plane (LTP) coordinates, also called "locally level coordinates," are a return to the first-order model of the Earth as being flat, where they serve as local reference directions for representing vehicle attitude and velocity for operation on or near the surface of the Earth. A common orientation for LTP coordinates has one horizontal axis (the north axis) in the direction of increasing latitude and the other horizontal axis (the east axis) in the direction of increasing longitude – as illustrated in Figure B.7.

Horizontal location components in this local coordinate frame are called "relative northing" and "relative easting."

B.3.7.1 Alpha Wander Coordinates

Maintaining east–north orientation was a problem for some inertial navigation systems at the poles, where north and east directions change by 180°. Early gimbaled inertial systems could not slew the platform axes fast enough for near-polar operation. This problem was solved by letting the platform axes "wander" from north, but keeping track of the angle α between north and a reference platform axis, as shown in Figure B.8. This LTP orientation came to be called "alpha wander."

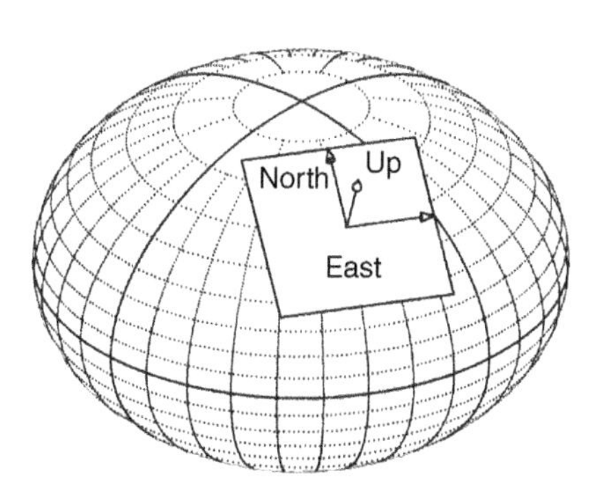

Figure B.7 ENU coordinates.

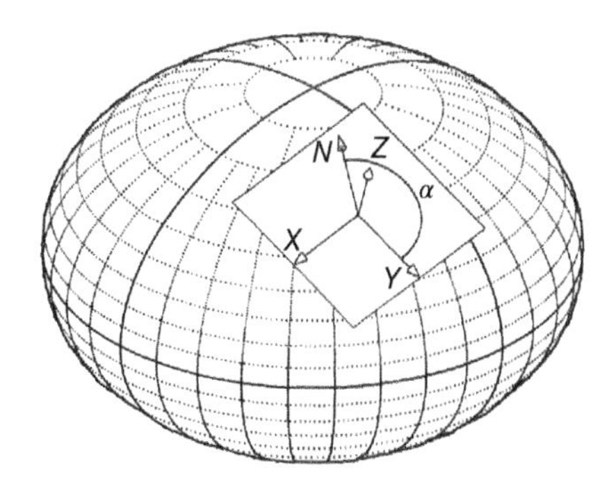

Figure B.8 Alpha wander.

B.3.7.2 ENU/NED Coordinates

East–north–up (ENU) and north–east–down (NED) are two common right-handed LTP coordinate systems. ENU coordinates may be preferred to NED coordinates because altitude increases in the upward direction. But NED coordinates may also be preferred over ENU coordinates because the direction of a right (clockwise) turn is in the positive direction with respect to a downward axis, and NED coordinate axes coincide with vehicle-fixed roll–pitch–yaw (RPY) coordinates (Figure B.9) when the vehicle is level and headed north.

The coordinate transformation matrix $\mathbf{C}_{\text{NED}}^{\text{ENU}}$ from ENU to NED coordinates and the transformation matrix $\mathbf{C}_{\text{ENU}}^{\text{NED}}$ from NED to ENU coordinates are one and the same:

$$\mathbf{C}_{\text{NED}}^{\text{ENU}} = \mathbf{C}_{\text{ENU}}^{\text{NED}} = \begin{bmatrix} 0 & 1 & 0 \\ 1 & 0 & 0 \\ 0 & 0 & -1 \end{bmatrix} \tag{B.56}$$

B.3.7.3 ENU/ECEF Coordinates

The unit vectors in local *east, north,* and *up* directions, as expressed in ECEF Cartesian coordinates, will be

$$\vec{\mathbf{1}}_E = \begin{bmatrix} -\sin(\theta) \\ \cos(\theta) \\ 0 \end{bmatrix} \tag{B.57}$$

$$\vec{\mathbf{1}}_N = \begin{bmatrix} -\cos(\theta)\sin(\phi_{\text{GEODETIC}}) \\ -\sin(\theta)\sin(\phi_{\text{GEODETIC}}) \\ \cos(\phi_{\text{GEODETIC}}) \end{bmatrix} \tag{B.58}$$

$$\vec{\mathbf{1}}_U = \begin{bmatrix} \cos(\theta)\cos(\phi_{\text{GEODETIC}}) \\ \sin(\theta)\cos(\phi_{\text{GEODETIC}}) \\ \sin(\phi_{\text{GEODETIC}}) \end{bmatrix} \tag{B.59}$$

and the unit vectors in the ECEF X, Y, and Z directions, as expressed in ENU coordinates, will be

$$\vec{\mathbf{1}}_X = \begin{bmatrix} -\sin(\theta) \\ -\cos(\theta)\sin(\phi_{\text{GEODETIC}}) \\ \cos(\theta)\cos(\phi_{\text{GEODETIC}}) \end{bmatrix} \tag{B.60}$$

$$\vec{\mathbf{1}}_Y = \begin{bmatrix} \cos(\theta) \\ -\sin(\theta)\sin(\phi_{\text{GEODETIC}}) \\ \sin(\theta)\cos(\phi_{\text{GEODETIC}}) \end{bmatrix} \tag{B.61}$$

$$\vec{\mathbf{1}}_Z = \begin{bmatrix} 0 \\ \cos(\phi_{\text{GEODETIC}}) \\ \sin(\phi_{\text{GEODETIC}}) \end{bmatrix} \tag{B.62}$$

B.3.7.4 NED/ECEF Coordinates

It is more natural in some applications to use NED directions for locally level coordinates. This coordinate system coincides with vehicle-body-fixed RPY coordinates (shown in Figure B.9) when the vehicle is level headed north. The unit vectors in local *north, east,* and *down* directions, as expressed in ECEF Cartesian coordinates, will be

$$\vec{\mathbf{1}}_N = \begin{bmatrix} -\cos(\theta)\sin(\phi_{\text{GEODETIC}}) \\ -\sin(\theta)\sin(\phi_{\text{GEODETIC}}) \\ \cos(\phi_{\text{GEODETIC}}) \end{bmatrix} \tag{B.63}$$

$$\vec{\mathbf{1}}_E = \begin{bmatrix} -\sin(\theta) \\ \cos(\theta) \\ 0 \end{bmatrix} \tag{B.64}$$

$$\vec{\mathbf{1}}_D = \begin{bmatrix} -\cos(\theta)\cos(\phi_{\text{GEODETIC}}) \\ -\sin(\theta)\cos(\phi_{\text{GEODETIC}}) \\ -\sin(\phi_{\text{GEODETIC}}) \end{bmatrix} \tag{B.65}$$

and the unit vectors in the ECEF X, Y, and Z directions, as expressed in NED coordinates, will be

$$\vec{\mathbf{1}}_X = \begin{bmatrix} -\cos(\theta)\sin(\phi_{\text{GEODETIC}}) \\ -\sin(\theta) \\ -\cos(\theta)\cos(\phi_{\text{GEODETIC}}) \end{bmatrix} \tag{B.66}$$

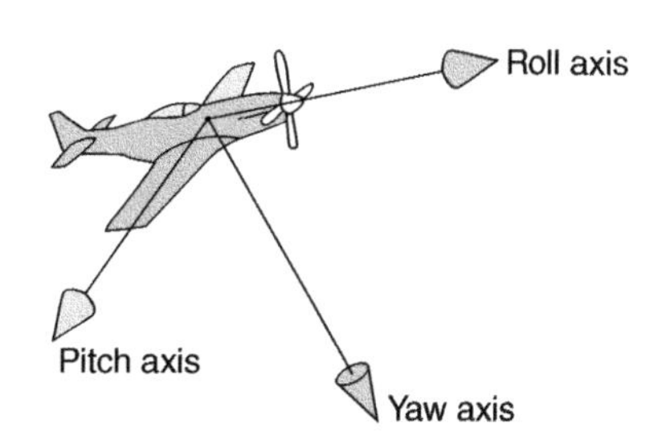

Figure B.9 Roll–pitch–yaw axes.

$$\vec{1}_Y = \begin{bmatrix} -\sin(\theta)\sin(\phi_{\text{GEODETIC}}) \\ \cos(\theta) \\ -\sin(\theta)\cos(\phi_{\text{GEODETIC}}) \end{bmatrix} \tag{B.67}$$

$$\vec{1}_Z = \begin{bmatrix} \cos(\phi_{\text{GEODETIC}}) \\ 0 \\ -\sin(\phi_{\text{GEODETIC}}) \end{bmatrix} \tag{B.68}$$

B.3.8 Roll–Pitch–Yaw (RPY) Coordinates

RPY coordinates are vehicle-fixed, with the roll axis in the nominal direction of motion of the vehicle, the pitch axis out the right-hand side, and the yaw axis such that turning to the right is positive, as illustrated in Figure B.9. The same orientations of vehicle-fixed coordinates are used for surface ships and ground vehicles. They are also called "SAE coordinates," because they are the standard body-fixed coordinates used by the Society of Automotive Engineers.

For rocket boosters with their roll axes vertical at lift-off, the pitch axis is typically defined to be orthogonal to the plane of the boost trajectory (also called the "pitch plane" or "ascent plane").

B.3.9 Vehicle Attitude Euler Angles

The attitude of the vehicle body with respect to local coordinates can be specified in terms of rotations about the vehicle roll, pitch and yaw axes, starting with these axes aligned with NED coordinates. The angles of rotation about each of these axes are called *Euler angles*, named for the Swiss mathematician Leonard Euler (1707–1783). It is always necessary to specify the order of rotations when specifying Euler (rhymes with "oiler") angles.

A fairly common convention for vehicle attitude Euler angles is illustrated in Figure B.10, where, starting with the vehicle level with roll axis pointed north.

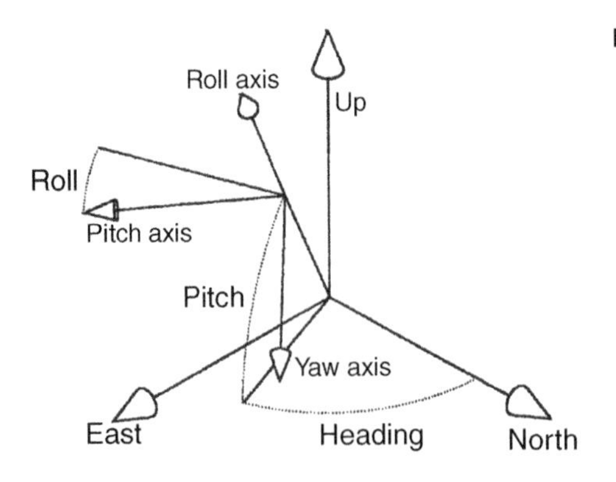

Figure B.10 Vehicle Euler angles.

1. *Yaw/heading*. Rotate through the yaw angle (Y) about the vehicle yaw axis to the intended azimuth (heading) of the vehicle roll axis. Azimuth is measured clockwise (east) from north.
2. *Pitch*. Rotate through the pitch angle (P) about the vehicle pitch axis to bring the vehicle roll axis to its intended elevation. Elevation is measured positive upward from the local horizontal plane.
3. *Roll*. Rotate through the roll angle (R) about the vehicle roll axis to bring the vehicle attitude to the specified orientation.

Euler angles are redundant for vehicle attitudes with 90° pitch, in which case the roll axis is vertical. In that attitude, heading changes also rotate the vehicle about the roll axis. This is the attitude of most rocket boosters at lift-off. Some boosters can be seen making a roll maneuver immediately after lift-off to align their yaw axes with the launch azimuth in the ascent plane. This maneuver may be required to correct for launch delays on missions for which launch azimuth is a function of launch time.

B.3.9.1 RPY/ENU Coordinates

With vehicle attitude specified by yaw angle (Y), pitch angle (P) and roll angle (R) as specified earlier, the resulting unit vectors of the roll, pitch and yaw axes in ENU coordinates will be:

$$\vec{\mathbf{1}}_R = \begin{bmatrix} \sin(Y)\cos(P) \\ \cos(Y)\cos(P) \\ \sin(P) \end{bmatrix} \tag{B.69}$$

$$\vec{\mathbf{1}}_P = \begin{bmatrix} \cos(R)\cos(Y) + \sin(R)\sin(Y)\sin(P) \\ -\cos(R)\sin(Y) + \sin(R)\cos(Y)\sin(P) \\ -\sin(R)\cos(P) \end{bmatrix} \tag{B.70}$$

$$\vec{\mathbf{1}}_Y = \begin{bmatrix} -\sin(R)\cos(Y) + \cos(R)\sin(Y)\sin(P) \\ \sin(R)\sin(Y) + \cos(R)\cos(Y)\sin(P) \\ -\cos(R)\cos(P) \end{bmatrix} \tag{B.71}$$

the unit vectors of the east, north, and up axes in RPY coordinates will be

$$\vec{\mathbf{1}}_E = \begin{bmatrix} \sin(Y)\cos(P) \\ \cos(R)\cos(Y) + \sin(R)\sin(Y)\sin(P) \\ -\sin(R)\cos(Y) + \cos(R)\sin(Y)\sin(P) \end{bmatrix} \tag{B.72}$$

$$\vec{\mathbf{1}}_N = \begin{bmatrix} \cos(Y)\cos(P) \\ -\cos(R)\sin(Y) + \sin(R)\cos(Y)\sin(P) \\ \sin(R)\sin(Y) + \cos(R)\cos(Y)\sin(P) \end{bmatrix} \tag{B.73}$$

$$\vec{\mathbf{1}}_U = \begin{bmatrix} \sin(P) \\ -\sin(R)\cos(P) \\ -\cos(R)\cos(P) \end{bmatrix} \tag{B.74}$$

and the coordinate transformation matrix from RPY coordinates to ENU coordinates will be

$$\mathbf{C}_{\text{ENU}}^{\text{RPY}} = \left[\, \vec{\mathbf{1}}_R \; \vec{\mathbf{1}}_P \; \vec{\mathbf{1}}_Y \,\right] = \begin{bmatrix} \vec{\mathbf{1}}_E^T \\ \vec{\mathbf{1}}_N^T \\ \vec{\mathbf{1}}_U^T \end{bmatrix}$$

$$= \begin{bmatrix} S_Y C_P & C_R C_Y + S_R S_Y S_P & -S_R C_Y + C_R S_Y S_P \\ C_Y C_P & -C_R S_Y + S_R C_Y S_P & S_R S_Y + C_R C_Y S_P \\ S_P & -S_R C_P & -C_R C_P \end{bmatrix} \tag{B.75}$$

where

$$S_R = \sin(R) \tag{B.76}$$

$$C_R = \cos(R) \tag{B.77}$$

$$S_P = \sin(P) \tag{B.78}$$

$$C_P = \cos(P) \tag{B.79}$$

$$S_Y = \sin(Y) \tag{B.80}$$

$$C_Y = \cos(Y) \tag{B.81}$$

B.3.10 GPS Coordinates

The parameter Ω in Figure B.12 is the *right ascension of the ascending node* (RAAN), which is the ECI longitude where the orbital plane intersects the equatorial plane as the satellite crosses from the southern hemisphere to the northern hemisphere. The orbital plane is specified by Ω and α, the inclination of the orbit plane with respect to the equatorial plane ($\alpha \approx 55^\circ$ for GPS satellite orbits). The θ parameter represents the location of the satellite within the orbit plane, as the angular phase in the circular orbit with respect to the ascending node. R is the radius of the circular orbit.

For GPS satellite orbits, the angle θ changes at a nearly constant rate of about 1.4584×10^{-4} rad/s and a period of about 43 082 seconds (half-a-day).

The nominal satellite position in ECEF coordinates is then

$$x = R[\cos\theta\cos\Omega - \sin\theta\sin\Omega\cos\alpha] \tag{B.82}$$

$$y = R[\cos\theta\sin\Omega + \sin\theta\cos\Omega\cos\alpha] \tag{B.83}$$

$$z = R \sin \theta \sin 55^\circ \tag{B.84}$$

$$\theta = \theta_0 + (t - t_0)\frac{360^\circ}{43\ 082} \tag{B.85}$$

$$\Omega = \Omega_0 - (t - t_0)\frac{360^\circ}{86\ 164} \tag{B.86}$$

$$R \approx 26\ 559\ 800 \text{ m} \tag{B.87}$$

GPS satellite positions in the transmitted navigation message are specified in ECEF coordinate system of WGS 84. A locally level x^1–y^1–z^1 reference coordinate system (described in Section B.3.7) is used by an observer location on Earth, where the x^1–y^1 plane is tangential to the surface of the Earth, x^1 pointing east, y^1 pointing north, and z^1 normal to the plane. See Figures B.11 and B.12.

$$X_{\text{ENU}} = \mathbf{C}_{\text{ENU}}^{\text{ECEF}}\ X_{\text{ECEF}} + S \tag{B.88}$$

$$\mathbf{C}_{\text{ENU}}^{\text{ECEF}} = \text{Coordinate transformation matrix from ECEF to ENU} \tag{B.89}$$

$$S = \text{Coordinate origin shift vector from ECEF to local reference} \tag{B.90}$$

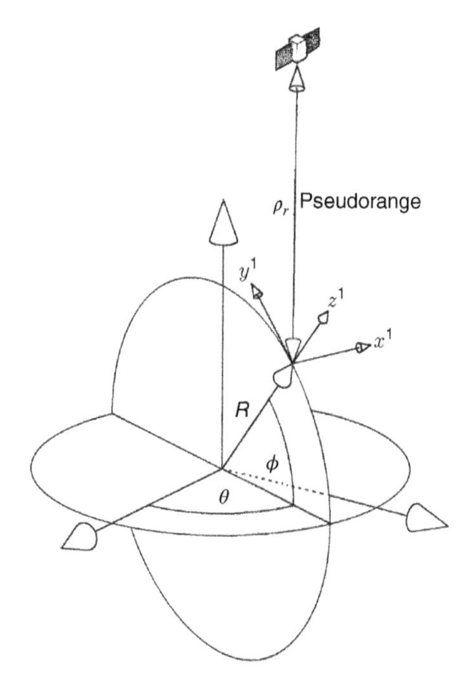

Figure B.11 Pseudorange.

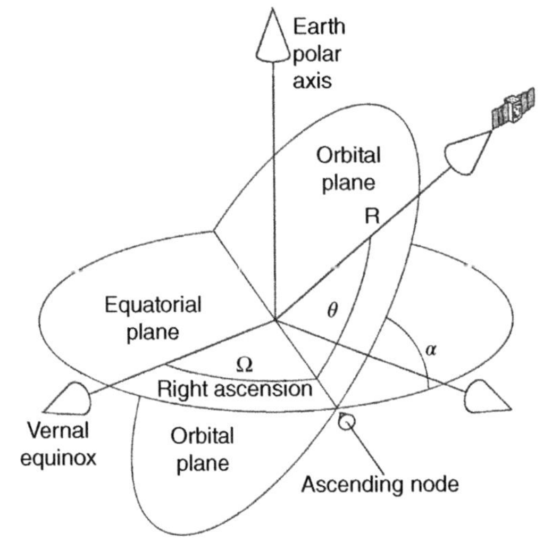

Figure B.12 Satellite coordinates.

$$\mathbf{C}_{\mathrm{ENU}}^{\mathrm{ECEF}} = \begin{bmatrix} -\sin\theta & \cos\theta & 0 \\ -\sin\phi\cos\theta & -\sin\phi\sin\theta & \cos\phi \\ \cos\phi\cos\theta & \cos\phi\sin\theta & \sin\theta \end{bmatrix} \tag{B.91}$$

$$S = \begin{bmatrix} X_U\sin\theta - Y_U\cos\theta \\ X_U\sin\phi\cos\theta + Y_U\sin\phi\sin\theta - Z_U\cos\phi \\ -X_U\cos\phi\cos\theta - Y_U\cos\phi\sin\theta - Z_U\sin\phi \end{bmatrix} \tag{B.92}$$

$$X_U,\ Y_U,\ Z_U = \text{User's position} \tag{B.93}$$

$$\theta = \text{Local reference longitude} \tag{B.94}$$

$$\phi = \text{Local geocentric latitude} \tag{B.95}$$

B.4 Coordinate Transformation Models

Coordinate transformations are methods for transforming a vector represented in one coordinate system into the appropriate representation in another coordinate system. These coordinate transformations can be represented in a number of different ways, each with its advantages and disadvantages.

These transformations generally involve translations (for coordinate systems with different origins) and rotations (for Cartesian coordinate systems with different axis directions) or transcendental transformations (between Cartesian and polar or geodetic coordinates). The transformations between Cartesian and polar coordinates have already been discussed in Section B.3.1 and translations are rather obvious, so we will concentrate on the rotations.

B.4.1 Euler Angles

Euler (rhymes with "oiler") angles were used for defining vehicle attitude in Section B.3.9, and vehicle attitude representation is a common use of Euler angles in navigation.

Euler angles are used to define a coordinate transformation in terms of a set of three angular rotations, performed in a specified sequence about three specified orthogonal axes, to bring one coordinate frame to coincide with another. The coordinate transformation from RPY coordinates to NED coordinates, for example, can be composed from three Euler rotation matrices

$$\mathbf{C}_{\text{NED}}^{\text{RPY}} = \overbrace{\begin{bmatrix} C_Y & -S_Y & 0 \\ S_Y & C_Y & 0 \\ 0 & 0 & 1 \end{bmatrix}}^{\text{Yaw}} \overbrace{\begin{bmatrix} C_P & 0 & S_P \\ 0 & 1 & 0 \\ -S_P & 0 & C_P \end{bmatrix}}^{\text{Pitch}} \overbrace{\begin{bmatrix} 1 & 0 & 0 \\ 0 & C_R & -S_R \\ 0 & S_R & C_R \end{bmatrix}}^{\text{Roll}}$$

$$= \underbrace{\begin{bmatrix} C_Y C_P & -S_Y C_R + C_Y S_P S_R & S_Y S_R + C_Y S_P C_R \\ S_Y C_P & C_Y C_R + S_Y S_P S_R & -C_Y S_R + S_Y S_P C_R \\ -S_P & C_P S_R & C_P C_R \\ \text{(Roll Axis)} & \text{(Pitch Axis)} & \text{(Yaw Axis)} \end{bmatrix}}_{\text{in NED coordinates,}} \tag{B.96}$$

where the matrix elements are defined in Eqs. (B.76)–(B.81). This matrix also rotates the NED coordinate axes to coincide with RPY coordinate axes. (Compare this with the transformation from RPY to ENU coordinates in Eq. (B.75).)

For example, the coordinate transformation for nominal booster rocket launch attitude (roll axis straight up) would be given by Eq. (B.96) with pitch angle $P = \pi/2$ ($C_P = 0,\ S_P = 1$), which becomes

$$\mathbf{C}_{\text{NED}}^{\text{RPY}} = \begin{bmatrix} 0 & \sin(R-Y) & \cos(R-Y) \\ 0 & \cos(R-Y) & -\sin(R-Y) \\ 1 & 0 & 0 \end{bmatrix} \tag{B.97}$$

That is, the coordinate transformation in this attitude depends only on the difference between roll angle (R) and yaw angle (Y). Euler angles are a concise representation for vehicle attitude. They are handy for driving cockpit displays such as compass cards (using Y) and artificial horizon indicators (using R and P), but they are not particularly handy for representing vehicle attitude dynamics. The reasons for the latter include

- Euler angles have discontinuities analogous to "gimbal lock" when the vehicle roll axis is pointed upward, as it is for launch of many rockets. In that orientation, tiny changes in vehicle pitch or yaw cause $\pm 180^\circ$ changes in heading angle. For aircraft, this creates a slewing rate problem for electromechanical compass card displays.
- The relationships between sensed body rates and Euler angle rates are mathematically complicated.

B.4.2 Rotation Vectors

All right-handed orthogonal coordinate systems with the same origins in three dimensions can be transformed one onto another by single rotations about fixed axes. The corresponding **rotation vectors** relating two coordinate systems are defined by the direction (rotation axis) and magnitude (rotation angle) of that transformation.

For example, the rotation vector for rotating ENU coordinates to NED coordinates (and *vice versa*) is

$$\vec{\rho}_{\text{NED}}^{\text{ENU}} = \begin{bmatrix} \pi/\sqrt{2} \\ \pi/\sqrt{2} \\ 0 \end{bmatrix} \tag{B.98}$$

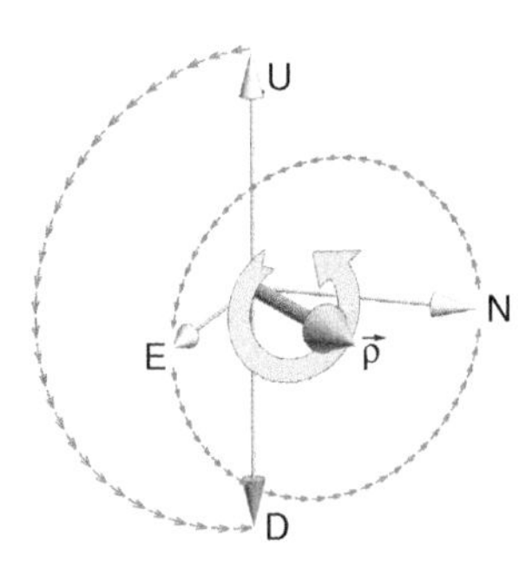

Figure B.13 Rotation from ENU to NED coordinates.

which has magnitude $|\vec{\rho}_{\text{NED}}^{\text{ENU}}| = \pi$ (180°) and direction north-east, as illustrated in Figure B.13. In transforming between ENU coordinates to NED coordinates, note that the transformation represents a change in coordinates, not a physical rotation of anything else. North is still north, but changes between being the second component to the first component. Up is still up, it is just that down is now the direction of the third component.

The rotation vector is another minimal representation of a coordinate transformation, along with Euler angles. Like Euler angles, rotation vectors are concise, but also have some drawbacks:

1. It is not a unique representation, in that adding multiples of $\pm 2\pi$ to the magnitude of a rotation vector has no effect on the transformation it represents.

2. It is a nonlinear and rather complicated representation, in that the result of one rotation followed by another is a third rotation, the rotation vector for which is a fairly complicated function of the first two rotation vectors.

But, unlike Euler angles, rotation vector models do not exhibit "gimbal lock."

B.4.2.1 Rotation Vector to Matrix

The rotation represented by a rotation vector

$$\vec{\rho} = \begin{bmatrix} \rho_1 \\ \rho_2 \\ \rho_3 \end{bmatrix} \tag{B.99}$$

can be implemented as multiplication by the matrix

$$\begin{aligned} \mathbf{C}(\vec{\rho}) &\stackrel{\text{def}}{=} \exp(\vec{\rho}\otimes) \\ &\stackrel{\text{def}}{=} \exp\left(\begin{bmatrix} 0 & -\rho_3 & \rho_2 \\ \rho_3 & 0 & -\rho_1 \\ -\rho_2 & \rho_1 & 0 \end{bmatrix}\right) \\ &= \cos(|\vec{\rho}|)\mathbf{I}_3 + \frac{1-\cos(|\vec{\rho}|)}{|\vec{\rho}|^2}\vec{\rho}\vec{\rho}^T + \frac{\sin(|\vec{\rho}|)}{|\vec{\rho}|}\begin{bmatrix} 0 & -\rho_3 & \rho_2 \\ \rho_3 & 0 & -\rho_1 \\ -\rho_2 & \rho_1 & 0 \end{bmatrix} \end{aligned} \tag{B.100a}$$

$$= \cos(\theta)\mathbf{I}_3 + (1-\cos(\theta))\vec{\mathbf{1}}_\rho\vec{\mathbf{1}}_\rho^T + \sin(\theta)\begin{bmatrix} 0 & -u_3 & u_2 \\ u_3 & 0 & -u_1 \\ -u_2 & u_1 & 0 \end{bmatrix} \tag{B.100b}$$

$$\theta \stackrel{\text{def}}{=} |\vec{\rho}| \tag{B.101}$$

$$\begin{bmatrix} u_1 \\ u_2 \\ u_3 \end{bmatrix} \stackrel{\text{def}}{=} \frac{\vec{\rho}}{|\vec{\rho}|} \tag{B.102}$$

That is, for any three-rowed column vector $\mathbf{v}$, $\mathbf{C}(\vec{\rho})\,\mathbf{v}$ rotates it through an angle of $|\vec{\rho}|$ radians about the vector $\vec{\rho}$.

The form of the matrix in Eq. (B.100b)[5] is better suited for computation when $\theta \approx 0$, but the form of the matrix in Eq. (B.100a) is useful for computing sensitivities using partial derivatives.

---

5 Linear combinations of the sort $a_1\mathbf{I}_{3\times3} + a_2[\vec{\mathbf{1}}_\rho\otimes] + a_3\vec{\mathbf{1}}_\rho\vec{\mathbf{1}}_\rho^T$, where $\mathbf{u}$ is a unit vector, form a subalgebra of 3×3 matrices with relatively simple rules for multiplication, inversion, etc.

For example, the rotation vector $\vec{\boldsymbol{\rho}}_{\text{NED}}^{\text{ENU}}$ in Eq. (B.98) transforming between ENU and NED has magnitude and direction

$$\theta = \pi(\sin(\theta) = 0,\ \cos(\theta) = -1) \tag{B.103}$$

$$\begin{bmatrix} u_1 \\ u_2 \\ u_3 \end{bmatrix} = \begin{bmatrix} 1/\sqrt{2} \\ 1/\sqrt{2} \\ 0 \end{bmatrix} \tag{B.104}$$

respectively, and the corresponding rotation matrix

$$\begin{aligned} \mathbf{C}_{\text{NED}}^{\text{ENU}} &= \cos(\pi)\mathbf{I}_3 + (1-\cos(\pi))\begin{bmatrix} u_1 \\ u_2 \\ u_3 \end{bmatrix}\begin{bmatrix} u_1 \\ u_2 \\ u_3 \end{bmatrix}^T + \sin(\pi)\begin{bmatrix} 0 & -u_3 & u_2 \\ u_3 & 0 & -u_1 \\ -u_2 & u_1 & 0 \end{bmatrix} \\ &= -\mathbf{I}_3 + 2\begin{bmatrix} u_1 \\ u_2 \\ u_3 \end{bmatrix}\begin{bmatrix} u_1 \\ u_2 \\ u_3 \end{bmatrix}^T + 0 \\ &= \begin{bmatrix} -1 & 0 & 0 \\ 0 & -1 & 0 \\ 0 & 0 & -1 \end{bmatrix} + \begin{bmatrix} 1 & 1 & 0 \\ 1 & 1 & 0 \\ 0 & 0 & 0 \end{bmatrix} \\ &= \begin{bmatrix} 0 & 1 & 0 \\ 1 & 0 & 0 \\ 0 & 0 & -1 \end{bmatrix} \end{aligned} \tag{B.105}$$

transforms from ENU to NED coordinates. (Compare this result to Eq. (B.56).) Because coordinate transformation matrices are orthogonal matrices and the matrix $\mathbf{C}_{\text{NED}}^{\text{ENU}}$ is also symmetric, $\mathbf{C}_{\text{NED}}^{\text{ENU}}$ is its own inverse. That is,

$$\mathbf{C}_{\text{NED}}^{\text{ENU}} = \mathbf{C}_{\text{ENU}}^{\text{NED}} \tag{B.106}$$

B.4.2.2 Matrix to Rotation Vector

Although there is a unique coordinate transformation matrix for each rotation vector, the converse is not true. Adding multiples of 2π to the magnitude of a rotation vector has no effect on the resulting coordinate transformation matrix. The following approach yields a unique rotation vector with magnitude $|\vec{\rho}| \leq \pi$.

The trace tr(**C**) of a square matrix **M** is the sum of its diagonal values. For the coordinate transformation matrix of Eq. (B.100a),

$$\mathrm{tr}(\mathbf{C}(\vec{\rho})) = 1 + 2\ \cos(\theta) \tag{B.107}$$

from which the rotation angle

$$\begin{aligned} |\vec{\rho}| &= \theta \\ &= \arccos\left(\frac{\mathrm{tr}(\mathbf{C}(\vec{\rho})) - 1}{2}\right) \end{aligned} \tag{B.108}$$

a formula that will yield a result in the range $0 < \theta < \pi$, but with poor fidelity near where the derivative of the cosine equals zero at $\theta = 0$ and $\theta = \pi$.

The values of θ near $\theta = 0$ and $\theta = \pi$ can be better estimated using the sine of θ, which can be recovered using the antisymmetric part of $\mathbf{C}(\vec{\rho})$,

$$\begin{aligned} \mathbf{A} &= \begin{bmatrix} 0 & -a_{21} & a_{13} \\ a_{21} & 0 & -a_{32} \\ -a_{13} & a_{32} & 0 \end{bmatrix} \\ &\stackrel{\text{def}}{=} \frac{1}{2}[\mathbf{C}(\vec{\rho}) - \mathbf{C}^T(\vec{\rho})] \\ &= \frac{\sin(\theta)}{\theta}\begin{bmatrix} 0 & -\rho_3 & \rho_2 \\ \rho_3 & 0 & -\rho_1 \\ -\rho_2 & \rho_1 & 0 \end{bmatrix} \end{aligned} \tag{B.109}$$

from which the vector

$$\begin{bmatrix} a_{32} \\ a_{13} \\ a_{21} \end{bmatrix} = \sin(\theta)\frac{1}{|\vec{\rho}|}\vec{\rho} \tag{B.110}$$

will have magnitude

$$\sqrt{a_{32}^2 + a_{13}^2 + a_{21}^2} = \sin(\theta) \tag{B.111}$$

and the same direction as $\vec{\rho}$. As a consequence, one can recover the magnitude θ of $\vec{\rho}$ from

$$\theta = \mathtt{atan2}\left(\sqrt{a_{32}^2 + a_{13}^2 + a_{21}^2}, \frac{\mathrm{tr}(\mathbf{C}(\vec{\rho})) - 1}{2}\right) \tag{B.112}$$

using the MATLAB® function `atan2`, and then the rotation vector $\vec{\rho}$ as

$$\vec{\rho} = \frac{\theta}{\sin(\theta)}\begin{bmatrix} a_{32} \\ a_{13} \\ a_{21} \end{bmatrix} \tag{B.113}$$

when $0 < \theta < \pi$.

B.4.2.3 Special Cases for $\sin(\theta) \approx 0$

For $\theta \approx 0$, $\vec{\rho} \approx 0$, although Eq. (B.113) may still work adequately for $\theta > 10^{-6}$, say.

For $\theta \approx \pi$, the symmetric part of $\mathbf{C}(\vec{\rho})$,

$$\begin{aligned} \mathbf{S} &= \begin{bmatrix} s_{11} & s_{12} & s_{13} \\ s_{12} & s_{22} & s_{23} \\ s_{13} & s_{23} & s_{33} \end{bmatrix} \\ &\stackrel{\text{def}}{=} \frac{1}{2}[\mathbf{C}(\vec{\rho}) + \mathbf{C}^T(\vec{\rho})] \\ &= \cos(\theta)\mathbf{I}_3 + \frac{1-\cos(\theta)}{\theta^2}\vec{\rho}\vec{\rho}^T \\ &\approx -\mathbf{I}_3 + \frac{2}{\theta^2}\vec{\rho}\vec{\rho}^T \end{aligned} \tag{B.114}$$

and the unit vector

$$\begin{bmatrix} u_1 \\ u_2 \\ u_3 \end{bmatrix} \stackrel{\text{def}}{=} \frac{1}{\theta}\vec{\rho} \tag{B.115}$$

satisfies

$$\mathbf{S} \approx \begin{bmatrix} 2u_1^2 - 1 & 2u_1u_2 & 2u_1u_3 \\ 2u_1u_2 & 2u_2^2 - 1 & 2u_2u_3 \\ 2u_1u_3 & 2u_2u_3 & 2u_3^2 - 1 \end{bmatrix} \tag{B.116}$$

which can be solved for a unique $\mathbf{u}$ by assigning $u_k > 0$ for

$$k = \text{argmax}\left(\begin{bmatrix} s_{11} \\ s_{22} \\ s_{33} \end{bmatrix}\right) \tag{B.117}$$

$$u_k = \sqrt{\frac{s_{kk} + 1}{2}} \tag{B.118}$$

then, depending on whether $k = 1, k = 2$, or $k = 3$,

$$\left.\begin{array}{cccc} & k=1 & k=2 & k=3 \\ u_1 \approx & \sqrt{\frac{s_{11}+1}{2}} & s_{12}/2u_2 & s_{13}/2u_3 \\ u_2 \approx & s_{12}/2u_1 & \sqrt{\frac{s_{22}+1}{2}} & s_{23}/2u_2 \\ u_3 \approx & s_{13}/2u_1 & s_{23}/2u_2 & \sqrt{\frac{s_{11}+1}{2}} \end{array}\right\} \tag{B.119}$$

and

$$\vec{\rho} = \theta \begin{bmatrix} u_1 \\ u_2 \\ u_3 \end{bmatrix} \tag{B.120}$$

B.4.2.4 MATLAB® Implementations

MATLAB® implementations of the transformations between different coordinate representations are labeled in Figure B.14 and located on www.wiley.com/go/grewal/gnss.

B.4.2.5 Time Derivatives of Rotation Vectors

The mathematical relationships between rotation rates ω_k and the time derivatives of the corresponding rotation vector $\vec{\rho}$ are fairly complicated, but they can be derived from Eq. (B.228) for the dynamics of coordinate transformation matrices.

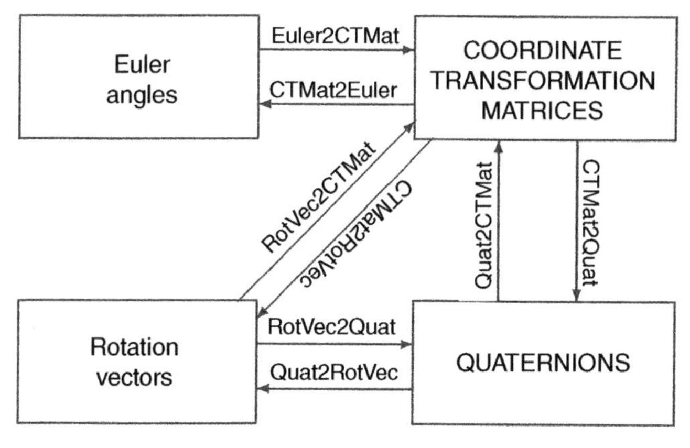

Figure B.14 MATLAB® transformations of coordinate representations.

Let $\vec{\rho}_{\text{ENU}}$ be the rotation vector represented in Earth-fixed ENU coordinates that rotates Earth-fixed ENU coordinate axes into vehicle body-fixed RPY axes, and let $\mathbf{C}(\vec{\rho})$ be the corresponding rotation matrix, so that, in ENU coordinates

$$\begin{aligned}
\vec{\mathbf{1}}_E &= \left[\,1\ 0\ 0\,\right]^T \\
\vec{\mathbf{1}}_N &= \left[\,0\ 1\ 0\,\right]^T \\
\vec{\mathbf{1}}_U &= \left[\,0\ 0\ 1\,\right]^T \\
\mathbf{C}(\vec{\rho}_{\text{ENU}})\vec{\mathbf{1}}_E &= \vec{\mathbf{1}}_R \\
\mathbf{C}(\vec{\rho}_{\text{ENU}})\vec{\mathbf{1}}_N &= \vec{\mathbf{1}}_P \\
\mathbf{C}(\vec{\rho}_{\text{ENU}})\vec{\mathbf{1}}_U &= \vec{\mathbf{1}}_Y \\
\mathbf{C}_{\text{ENU}}^{\text{RPY}} &= \left[\,\vec{\mathbf{1}}_R\ \vec{\mathbf{1}}_P\ \vec{\mathbf{1}}_Y\,\right] \\
&= \left[\,\mathbf{C}(\vec{\rho}_{\text{ENU}})\vec{\mathbf{1}}_E\ \mathbf{C}(\vec{\rho}_{\text{ENU}})\vec{\mathbf{1}}_N\ \mathbf{C}(\vec{\rho}_{\text{ENU}})\vec{\mathbf{1}}_U\,\right] \\
&= \mathbf{C}(\vec{\rho}_{\text{ENU}})\left[\,\vec{\mathbf{1}}_E\ \vec{\mathbf{1}}_N\ \vec{\mathbf{1}}_U\,\right] \\
&= \mathbf{C}(\vec{\rho}_{\text{ENU}})\begin{bmatrix} 1 & 0 & 0 \\ 0 & 1 & 0 \\ 0 & 0 & 1 \end{bmatrix}
\end{aligned} \tag{B.121}$$

$$\mathbf{C}_{\text{ENU}}^{\text{RPY}} = \mathbf{C}(\vec{\rho}_{\text{ENU}}) \tag{B.122}$$

That is, $\mathbf{C}(\vec{\rho}_{\text{ENU}})$ is the coordinate transformation matrix from RPY coordinates to ENU coordinates. As a consequence, from Eq. (B.228),

$$\begin{aligned}
\frac{d}{dt}\mathbf{C}(\vec{\rho}_{\text{ENU}}) &= \frac{d}{dt}\mathbf{C}_{\text{ENU}}^{\text{RPY}} \\
&= \begin{bmatrix} 0 & \omega_U & -\omega_N \\ -\omega_U & 0 & \omega_E \\ \omega_N & -\omega_E & 0 \end{bmatrix}\mathbf{C}_{\text{ENU}}^{\text{RPY}} \\
&\quad + \mathbf{C}_{\text{ENU}}^{\text{RPY}}\begin{bmatrix} 0 & -\omega_Y & \omega_P \\ \omega_Y & 0 & -\omega_R \\ -\omega_P & \omega_R & 0 \end{bmatrix}.
\end{aligned} \tag{B.123}$$

$$\begin{aligned}
\frac{d}{dt}\mathbf{C}(\vec{\rho}_{\text{ENU}}) &= \begin{bmatrix} 0 & \omega_U & -\omega_N \\ -\omega_U & 0 & \omega_E \\ \omega_N & -\omega_E & 0 \end{bmatrix}\mathbf{C}(\vec{\rho}_{\text{ENU}}) \\
&\quad + \mathbf{C}(\vec{\rho}_{\text{ENU}})\begin{bmatrix} 0 & -\omega_Y & \omega_P \\ \omega_Y & 0 & -\omega_R \\ -\omega_P & \omega_R & 0 \end{bmatrix}
\end{aligned} \tag{B.124}$$

where

$$\vec{\boldsymbol{\omega}}_{\text{RPY}} = \begin{bmatrix} \omega_R \\ \omega_P \\ \omega_Y \end{bmatrix} \tag{B.125}$$

is the vector of inertial rotation rates of the vehicle body, expressed in RPY coordinates, and

$$\vec{\boldsymbol{\omega}}_{\text{ENU}} = \begin{bmatrix} \omega_E \\ \omega_N \\ \omega_U \end{bmatrix} \tag{B.126}$$

is the vector of inertial rotation rates of the ENU coordinate frame, expressed in ENU coordinates.

The 3×3 matrix equation (Equation (B.124)) is equivalent to nine scalar equations:

$$\frac{\partial c_{11}}{\partial \rho_E}\dot{\rho}_E + \frac{\partial c_{11}}{\partial \rho_N}\dot{\rho}_N + \frac{\partial c_{11}}{\partial \rho_U}\dot{\rho}_U = -c_{1,3}\omega_P + c_{1,2}\omega_Y - c_{3,1}\omega_N + c_{2,1}\omega_U \tag{B.127}$$

$$\frac{\partial c_{12}}{\partial \rho_E}\dot{\rho}_E + \frac{\partial c_{12}}{\partial \rho_N}\dot{\rho}_N + \frac{\partial c_{12}}{\partial \rho_U}\dot{\rho}_U = c_{1,3}\omega_R - c_{1,1}\omega_Y - c_{3,2}\omega_N + c_{2,2}\omega_U \tag{B.128}$$

$$\frac{\partial c_{13}}{\partial \rho_E}\dot{\rho}_E + \frac{\partial c_{13}}{\partial \rho_N}\dot{\rho}_N + \frac{\partial c_{13}}{\partial \rho_U}\dot{\rho}_U = -c_{1,2}\omega_R + c_{1,1}\omega_P - c_{3,3}\omega_N + c_{2,3}\omega_U \tag{B.129}$$

$$\frac{\partial c_{21}}{\partial \rho_E}\dot{\rho}_E + \frac{\partial c_{21}}{\partial \rho_N}\dot{\rho}_N + \frac{\partial c_{21}}{\partial \rho_U}\dot{\rho}_U = -c_{2,3}\omega_P + c_{2,2}\omega_Y + c_{3,1}\omega_E - c_{1,1}\omega_U \tag{B.130}$$

$$\frac{\partial c_{22}}{\partial \rho_E}\dot{\rho}_E + \frac{\partial c_{22}}{\partial \rho_N}\dot{\rho}_N + \frac{\partial c_{22}}{\partial \rho_U}\dot{\rho}_U = c_{2,3}\omega_R - c_{2,1}\omega_Y + c_{3,2}\omega_E - c_{1,2}\omega_U \tag{B.131}$$

$$\frac{\partial c_{23}}{\partial \rho_E}\dot{\rho}_E + \frac{\partial c_{23}}{\partial \rho_N}\dot{\rho}_N + \frac{\partial c_{23}}{\partial \rho_U}\dot{\rho}_U = -c_{2,2}\omega_R + c_{2,1}\omega_P + c_{3,3}\omega_E - c_{1,3}\omega_U \tag{B.132}$$

$$\frac{\partial c_{31}}{\partial \rho_E}\dot{\rho}_E + \frac{\partial c_{31}}{\partial \rho_N}\dot{\rho}_N + \frac{\partial c_{31}}{\partial \rho_U}\dot{\rho}_U = -c_{3,3}\omega_P + c_{3,2}\omega_Y - c_{2,1}\omega_E + c_{1,1}\omega_N \tag{B.133}$$

$$\frac{\partial c_{32}}{\partial \rho_E}\dot{\rho}_E + \frac{\partial c_{32}}{\partial \rho_N}\dot{\rho}_N + \frac{\partial c_{32}}{\partial \rho_U}\dot{\rho}_U = c_{3,3}\omega_R - c_{3,1}\omega_Y - c_{2,2}\omega_E + c_{1,2}\omega_N \tag{B.134}$$

$$\frac{\partial c_{33}}{\partial \rho_E}\dot{\rho}_E + \frac{\partial c_{33}}{\partial \rho_N}\dot{\rho}_N + \frac{\partial c_{33}}{\partial \rho_U}\dot{\rho}_U = -c_{3,2}\omega_R + c_{3,1}\omega_P - c_{2,3}\omega_E + c_{1,3}\omega_N \tag{B.135}$$

where

$$\begin{bmatrix} c_{11} & c_{12} & c_{13} \\ c_{21} & c_{22} & c_{23} \\ c_{31} & c_{32} & c_{33} \end{bmatrix} \stackrel{\text{def}}{=} \mathbf{C}(\vec{\rho}_{\text{ENU}}) \tag{B.136}$$

and the partial derivatives

$$\frac{\partial c_{11}}{\partial \rho_E} = \frac{u_E(1-u_E^2)[2\ (1-\cos(\theta)) - \theta\ \sin(\theta)]}{\theta} \tag{B.137}$$

$$\frac{\partial c_{11}}{\partial \rho_N} = \frac{u_N[-2\ u_E^2(1-\cos(\theta)) - \theta\ \sin(\theta)(1-u_E^2)]}{\theta} \tag{B.138}$$

$$\frac{\partial c_{11}}{\partial \rho_U} = \frac{u_U[-2\ u_E^2(1-\cos(\theta)) - \theta\ \sin(\theta)(1-u_E^2)]}{\theta} \tag{B.139}$$

$$\frac{\partial c_{12}}{\partial \rho_E} = \frac{u_N(1-2\ u_E^2)(1-\cos(\theta)) + u_E u_U \sin(\theta) - \theta\ u_E u_U \cos(\theta) + \theta\ u_N u_E^2 \sin(\theta)}{\theta} \tag{B.140}$$

$$\frac{\partial c_{12}}{\partial \rho_N} = \frac{u_E(1-2\ u_N^2)(1-\cos(\theta)) + u_U u_N \sin(\theta) - \theta\ u_N u_U \cos(\theta) + \theta\ u_E u_N^2 \sin(\theta)}{\theta} \tag{B.141}$$

$$\frac{\partial c_{12}}{\partial \rho_U} = \frac{-2\ u_E u_N u_U(1-\cos(\theta)) - (1-u_U^2)\sin(\theta) - \theta\ u_U^2 \cos(\theta) + \theta\ u_U u_N u_E \sin(\theta)}{\theta} \tag{B.142}$$

$$\frac{\partial c_{13}}{\partial \rho_E} = \frac{u_U(1-2\ u_E^2)(1-\cos(\theta)) - u_E u_N \sin(\theta) + \theta\ u_E u_N \cos(\theta) + \theta\ u_U u_E^2 \sin(\theta)}{\theta} \tag{B.143}$$

$$\frac{\partial c_{13}}{\partial \rho_N} = \frac{-2\ u_E u_N u_U(1-\cos(\theta)) + (1-u_N^2)\sin(\theta) + \theta\ u_N^2 \cos(\theta) + \theta\ u_U u_N u_E \sin(\theta)}{\theta} \tag{B.144}$$

$$\frac{\partial c_{13}}{\partial \rho_U} = \frac{u_E(1-2\ u_U^2)(1-\cos(\theta)) - u_U u_N \sin(\theta) + \theta\ u_N u_U \cos(\theta) + \theta\ u_E u_U^2 \sin(\theta)}{\theta} \tag{B.145}$$

$$\frac{\partial c_{21}}{\partial \rho_E} = \frac{u_N(1-2\ u_E^2)(1-\cos(\theta)) - u_E u_U \sin(\theta) + \theta\ u_E u_U \cos(\theta) + \theta\ u_N u_E^2 \sin(\theta)}{\theta} \tag{B.146}$$

$$\frac{\partial c_{21}}{\partial \rho_N} = \frac{u_E(1-2\ u_N^2)(1-\cos(\theta)) - u_U u_N \sin(\theta) + \theta\ u_N u_U \cos(\theta) + \theta\ u_E u_N^2 \sin(\theta)}{\theta} \tag{B.147}$$

$$\frac{\partial c_{21}}{\partial \rho_U} = \frac{-2\ u_E u_N u_U(1-\cos(\theta)) + \sin(\theta)(1-u_U^2) + \theta\ u_U^2 \cos(\theta) + \theta\ u_U u_N u_E \sin(\theta)}{\theta} \tag{B.148}$$

$$\frac{\partial c_{22}}{\partial \rho_E} = \frac{u_E[-2\ u_N^2(1-\cos(\theta)) - \theta\ (1-u_N^2)\sin(\theta)]}{\theta} \tag{B.149}$$

$$\frac{\partial c_{22}}{\partial \rho_N} = \frac{u_N[1-u_N^2][2(1-\cos(\theta)) - \theta\ \sin(\theta)]}{\theta} \tag{B.150}$$

$$\frac{\partial c_{22}}{\partial \rho_U} = \frac{u_U[-2\ u_N^2(1-\cos(\theta)) - \theta\ (1-u_N^2)\sin(\theta)]}{\theta} \tag{B.151}$$

$$\frac{\partial c_{23}}{\partial \rho_E} = \frac{-2\ u_E u_N u_U(1-\cos(\theta)) - (1-u_E^2)\sin(\theta) - \theta\ u_E^2\cos(\theta) + \theta\ u_E u_N u_U \sin(\theta)}{\theta} \tag{B.152}$$

$$\frac{\partial c_{23}}{\partial \rho_N} = \frac{u_U(1-2\ u_N^2)(1-\cos(\theta)) + u_E u_N \sin(\theta) - \theta\ u_E u_N \cos(\theta) + \theta\ u_N^2 u_U \sin(\theta)}{\theta} \tag{B.153}$$

$$\frac{\partial c_{23}}{\partial \rho_U} = \frac{u_N(1-2\ u_U^2)(1-\cos(\theta)) + u_E u_U \sin(\theta) - \theta\ u_E u_U \cos(\theta) + \theta\ u_U^2 u_N \sin(\theta)}{\theta} \tag{B.154}$$

$$\frac{\partial c_{31}}{\partial \rho_E} = \frac{u_U(1-2\ u_E^2)(1-\cos(\theta)) + u_E u_N \sin(\theta) - \theta\ u_E u_N \cos(\theta) + \theta\ u_U u_E^2 \sin(\theta)}{\theta} \tag{B.155}$$

$$\frac{\partial c_{31}}{\partial \rho_N} = \frac{-2\ u_E u_N u_U(1-\cos(\theta)) - (1-u_N^2)\sin(\theta) - \theta\ u_N^2\cos(\theta) + \theta\ u_U u_N u_E \sin(\theta)}{\theta} \tag{B.156}$$

$$\frac{\partial c_{31}}{\partial \rho_U} = \frac{u_E(1-2\ u_U^2)(1-\cos(\theta)) + u_U u_N \sin(\theta) - \theta\ u_N u_U \cos(\theta) + \theta\ u_E u_U^2 \sin(\theta)}{\theta} \tag{B.157}$$

$$\frac{\partial c_{32}}{\partial \rho_E} = \frac{-2\ u_E u_N u_U(1-\cos(\theta)) + (1-u_E^2)\sin(\theta) + \theta\ u_E^2\cos(\theta) + \theta\ u_U u_N u_E \sin(\theta)}{\theta} \tag{B.158}$$

$$\frac{\partial c_{32}}{\partial \rho_N} = \frac{u_U(1-2\ u_N^2)(1-\cos(\theta)) - u_E u_N \sin(\theta) + \theta\ u_E u_N \cos(\theta) + \theta\ u_N^2 u_U \sin(\theta)}{\theta} \tag{B.159}$$

$$\frac{\partial c_{32}}{\partial \rho_U} = \frac{u_N(1-2\ u_U^2)(1-\cos(\theta)) - u_E u_U \sin(\theta) + \theta\ u_E u_U \cos(\theta) + \theta\ u_U^2 u_N \sin(\theta)}{\theta} \tag{B.160}$$

$$\frac{\partial c_{33}}{\partial \rho_E} = \frac{u_E[-2\ u_U^2(1-\cos(\theta)) - \theta\ \sin(\theta)(1+u_U^2)]}{\theta} \tag{B.161}$$

$$\frac{\partial c_{33}}{\partial \rho_N} = \frac{u_N[-2\ u_U^2(1-\cos(\theta)) - \theta\ \sin(\theta)(1+u_U^2)]}{\theta} \tag{B.162}$$

$$\frac{\partial c_{33}}{\partial \rho_U} = \frac{u_U[1-u_U^2][2\ (1-\cos(\theta)) - \theta\ \sin(\theta)]}{\theta} \tag{B.163}$$

for

$$\theta \stackrel{\text{def}}{=} |\vec{\rho}_{\text{ENU}}| \tag{B.164}$$

$$u_E \stackrel{\text{def}}{=} \rho_E/\theta \tag{B.165}$$

$$u_N \stackrel{\text{def}}{=} \rho_N/\theta \tag{B.166}$$

$$u_U \stackrel{\text{def}}{=} \rho_U/\theta \tag{B.167}$$

These nine scalar linear equations can be put into matrix form and solved in least squares fashion as

$$\mathbf{L}\begin{bmatrix}\dot{\rho}_E\\ \dot{\rho}_N\\ \dot{\rho}_U\end{bmatrix} = \mathbf{R}\begin{bmatrix}\omega_R\\ \omega_P\\ \omega_Y\\ \omega_E\\ \omega_N\\ \omega_U\end{bmatrix} \tag{B.168}$$

$$\begin{bmatrix}\dot{\rho}_E\\ \dot{\rho}_N\\ \dot{\rho}_U\end{bmatrix} = \underbrace{[\mathbf{L}^T\mathbf{L}]\backslash[\mathbf{L}^T\mathbf{R}]}_{\frac{\partial \dot{\rho}}{\partial \vec{\omega}}}\begin{bmatrix}\vec{\omega}_{\text{RPY}}\\ \vec{\omega}_{\text{ENU}}\end{bmatrix} \tag{B.169}$$

The matrix product $\mathbf{L}^T\mathbf{L}$ will always be invertible because its determinant

$$\det[\mathbf{L}^T\mathbf{L}] = 32\,\frac{(1-\cos(\theta))^2}{\theta^4} \tag{B.170}$$

$$\lim_{\theta\to 0}\det[\mathbf{L}^T\mathbf{L}] = 8 \tag{B.171}$$

and the resulting equation for $\dot{\rho}_{\text{ENU}}$ can be put into the form

$$\dot{\rho}_{\text{ENU}} = \left[\frac{\partial \dot{\rho}}{\partial \vec{\omega}}\right]\begin{bmatrix}\vec{\omega}_{\text{RPY}}\\ \vec{\omega}_{\text{ENU}}\end{bmatrix} \tag{B.172}$$

The 3×6 matrix $\frac{\partial \dot{\rho}}{\partial \vec{\omega}}$ can be partitioned as

$$\left[\frac{\partial \dot{\rho}}{\partial \vec{\omega}}\right] = \left[\frac{\partial \dot{\rho}}{\partial \vec{\omega}_{\text{RPY}}} \quad \frac{\partial \dot{\rho}}{\partial \vec{\omega}_{\text{ENU}}}\right] \tag{B.173}$$

with 3×3 submatrices

$$\begin{aligned}\frac{\partial \dot{\rho}}{\partial \vec{\omega}_{\text{RPY}}} &= \left[\frac{1}{|\vec{\rho}|^2} - \frac{\sin(|\vec{\rho}|)}{2\,|\vec{\rho}|[1-\cos(|\vec{\rho}|)]}\right]\vec{\rho}\vec{\rho}^T \\ &\quad + \frac{|\vec{\rho}|\sin(|\vec{\rho}|)}{2\,[1-\cos(|\vec{\rho}|)]}\mathbf{I} + \frac{1}{2}[\vec{\rho}\otimes] \\ &= \begin{bmatrix}u_1\\ u_2\\ u_3\end{bmatrix}\begin{bmatrix}u_1\\ u_2\\ u_3\end{bmatrix}^T + \frac{\theta\sin(\theta)}{2[1-\cos(\theta)]}\left[\mathbf{I} - \begin{bmatrix}u_1\\ u_2\\ u_3\end{bmatrix}\begin{bmatrix}u_1\\ u_2\\ u_3\end{bmatrix}^T\right] + \frac{\theta}{2}\left[\begin{bmatrix}u_1\\ u_2\\ u_3\end{bmatrix}\otimes\right]\end{aligned} \tag{B.174}$$

$$\lim_{|\vec{\rho}|\to 0} \frac{\partial \dot{\rho}}{\partial \vec{\omega}_{\mathrm{RPY}}} = \mathbf{I} \tag{B.175}$$

$$\begin{aligned}\frac{\partial \dot{\rho}}{\partial \vec{\omega}_{\mathrm{ENU}}} &= -\left[\frac{1}{|\vec{\rho}|^2} - \frac{\sin(|\vec{\rho}|)}{2\,|\vec{\rho}|[1-\cos(|\vec{\rho}|)]}\right]\vec{\rho}\vec{\rho}^T \\ &\quad - \frac{|\vec{\rho}|\sin(|\vec{\rho}|)}{2\,[1-\cos(|\vec{\rho}|)]}\mathbf{I} + \frac{1}{2}[\vec{\rho}\otimes] \\ &= -\begin{bmatrix}u_1\\u_2\\u_3\end{bmatrix}\begin{bmatrix}u_1\\u_2\\u_3\end{bmatrix}^T - \frac{\theta\sin(\theta)}{2[1-\cos(\theta)]}\left[\mathbf{I} - \begin{bmatrix}u_1\\u_2\\u_3\end{bmatrix}\begin{bmatrix}u_1\\u_2\\u_3\end{bmatrix}^T\right] + \frac{\theta}{2}\left[\begin{bmatrix}u_1\\u_2\\u_3\end{bmatrix}\otimes\right]\end{aligned} \tag{B.176}$$

$$\lim_{|\vec{\rho}|\to 0} \frac{\partial \dot{\rho}}{\partial \vec{\omega}_{\mathrm{ENU}}} = -\mathbf{I} \tag{B.177}$$

For locally leveled gimbaled systems, $\vec{\omega}_{RPH} = \vec{0}$. That is, the gimbals normally keep the accelerometer axes aligned to the ENU or NED coordinate axes, a process that is modeled by $\vec{\omega}_{\mathrm{ENU}}$ alone.

B.4.2.6 Time Derivatives of Matrix Expressions

The Kalman filter implementation for integrating GNSS with a strapdown INS will require derivatives with respect to time of the matrices

$$\frac{\partial \dot{\rho}_{\mathrm{ENU}}}{\partial \vec{\omega}_{RPH}} \text{ (Eq. (B.174)) and } \frac{\partial \dot{\rho}_{\mathrm{ENU}}}{\partial \vec{\omega}_{\mathrm{ENU}}} \text{ (Eq. (B.176)).}$$

We derive here a general-purpose formula for taking such derivatives, and then apply it to these two cases.

General Formulas There is a general-purpose formula for taking the time-derivatives $\frac{d}{dt}\mathbf{M}(\vec{\rho})$ of matrix expressions of the sort

$$\begin{aligned}\mathbf{M}(\vec{\rho}) &= \mathbf{M}(s_1(\vec{\rho}),\ s_2(\vec{\rho}),\ s_3(\vec{\rho})) \\ &= s_1(\vec{\rho})\,\mathbf{I}_3 + s_2(\vec{\rho})\,[\vec{\rho}\otimes] + s_3(\vec{\rho})\vec{\rho}\vec{\rho}^T\end{aligned} \tag{B.178}$$

i.e. as linear combinations of $\mathbf{I}_3$, $\vec{\rho}\otimes$ and $\vec{\rho}\vec{\rho}^T$ with scalar functions of $\vec{\rho}$ as the coefficients.

The derivation uses the time derivatives of the basis matrices,

$$\frac{d}{dt}\mathbf{I}_3 = \mathbf{0}_3 \tag{B.179}$$

$$\frac{d}{dt}[\vec{\rho}\otimes] = [\dot{\rho}\otimes] \tag{B.180}$$

$$\frac{d}{dt}\vec{\rho}\vec{\rho}^T = \dot{\rho}\vec{\rho}^T + \vec{\rho}\dot{\rho}^T \tag{B.181}$$

where the vector

$$\dot{\rho} = \frac{d}{dt}\vec{\rho} \tag{B.182}$$

and then uses the chain rule for differentiation to obtain the general formula:

$$\begin{aligned}\frac{d}{dt}\mathbf{M}(\vec{\rho}) = {} & \frac{\partial s_1(\vec{\rho})}{\partial \vec{\rho}}\dot{\rho}\mathbf{I}_3 + \frac{\partial s_2(\vec{\rho})}{\partial \vec{\rho}}\dot{\rho}[\vec{\rho}\otimes] + s_2(\vec{\rho})[\dot{\rho}\otimes] \\ & + \frac{\partial s_3(\vec{\rho})}{\partial \vec{\rho}}\dot{\rho}[\vec{\rho}\vec{\rho}^T] + s_3(\vec{\rho})[\dot{\rho}\vec{\rho}^T + \vec{\rho}\dot{\rho}^T]\end{aligned} \tag{B.183}$$

where the gradients $\frac{\partial s_i(\vec{\rho})}{\partial \vec{\rho}}$ are to be computed as row vectors and the inner products $\frac{\partial s_i(\vec{\rho})}{\partial \vec{\rho}}\dot{\rho}$ will be scalars.

Equation (B.183) is the general-purpose formula for the matrix forms of interest, which differ only in their scalar functions $s_i(\vec{\rho})$. These scalar functions $s_i(\vec{\rho})$ are generally rational functions of the following scalar functions (shown in terms of their gradients):

$$\frac{\partial}{\partial \vec{\rho}}|\vec{\rho}|^p = p|\vec{\rho}|^{p-2}\vec{\rho}^T \tag{B.184}$$

$$\frac{\partial}{\partial \vec{\rho}}\sin(|\vec{\rho}|) = \cos(|\vec{\rho}|)|\vec{\rho}|^{-1}\vec{\rho}^T \tag{B.185}$$

$$\frac{\partial}{\partial \vec{\rho}}\cos(|\vec{\rho}|) = -\sin(|\vec{\rho}|)|\vec{\rho}|^{-1}\vec{\rho}^T \tag{B.186}$$

Time Derivative of $\frac{\partial \dot{\rho}_{ENU}}{\partial \vec{\omega}_{RPY}}$ In this case (Eq. (B.174)),

$$s_1(\vec{\rho}) = \frac{|\vec{\rho}|\sin(|\vec{\rho}|)}{2\,[1-\cos(|\vec{\rho}|)]} \tag{B.187}$$

$$\frac{\partial s_1(\vec{\rho})}{\partial \vec{\rho}} = -\frac{1-|\vec{\rho}|^{-1}\sin(|\vec{\rho}|)}{2\,[1-\cos(|\vec{\rho}|)]}\vec{\rho}^T \tag{B.188}$$

$$s_2(\vec{\rho}) = \frac{1}{2} \tag{B.189}$$

$$\frac{\partial s_2}{\partial \vec{\rho}} = \mathbf{0}_{1\times 3} \tag{B.190}$$

$$s_3(\vec{\rho}) = \left[\frac{1}{|\vec{\rho}|^2} - \frac{\sin(|\vec{\rho}|)}{2\,|\vec{\rho}|[1-\cos(|\vec{\rho}|)]}\right] \tag{B.191}$$

$$\frac{\partial s_3(\vec{\rho})}{\partial \vec{\rho}} = \frac{1+|\vec{\rho}|^{-1}\sin(|\vec{\rho}|) - 4\,|\vec{\rho}|^{-2}[1-\cos(|\vec{\rho}|)]}{2\,|\vec{\rho}|^2[1-\cos(|\vec{\rho}|)]}\vec{\rho}^T \tag{B.192}$$

$$\begin{aligned}
\frac{d}{dt}\frac{\partial \dot{\rho}_{\mathrm{ENU}}}{\partial \vec{\omega}_{\mathrm{RPY}}} &= \frac{\partial s_1(\vec{\rho})}{\partial \vec{\rho}}\dot{\rho}\mathbf{I}_3 + \frac{\partial s_2(\vec{\rho})}{\partial \vec{\rho}}\dot{\rho}[\vec{\rho}\otimes] + s_2(\vec{\rho})[\dot{\rho}\otimes] \\
&\quad + \frac{\partial s_3(\vec{\rho})}{\partial \vec{\rho}}\dot{\rho}[\vec{\rho}\vec{\rho}^T] + s_3(\vec{\rho})[\dot{\rho}\vec{\rho}^T + \vec{\rho}\dot{\rho}^T] \qquad \text{(B.193)} \\
&= -\left\{\frac{1-|\vec{\rho}|^{-1}\sin(|\vec{\rho}|)}{2\,[1-\cos(|\vec{\rho}|)]}\vec{\rho}^T\right\}(\vec{\rho}^T\dot{\rho})\mathbf{I}_3 \\
&\quad + \frac{1}{2}(\vec{\rho})[\dot{\rho}\otimes] \\
&\quad + \left\{\frac{1+|\vec{\rho}|^{-1}\sin(|\vec{\rho}|) - 4\,|\vec{\rho}|^{-2}[1-\cos(|\vec{\rho}|)]}{2\,|\vec{\rho}|^2[1-\cos(|\vec{\rho}|)]}\right\} \\
&\quad \times(\vec{\rho}^T\dot{\rho})[\vec{\rho}\vec{\rho}^T] \\
&\quad + \left[\frac{1}{|\vec{\rho}|^2} - \frac{\sin(|\vec{\rho}|)}{2\,|\vec{\rho}|[1-\cos(|\vec{\rho}|)]}\right][\dot{\rho}\vec{\rho}^T + \vec{\rho}\dot{\rho}^T] \qquad \text{(B.194)}
\end{aligned}$$

Time Derivative of $\frac{\partial \dot{\rho}_{ENU}}{\partial \vec{\omega}_{ENU}}$ In this case (Eq. (B.176)),

$$s_1(\vec{\rho}) = -\frac{|\vec{\rho}|\sin(|\vec{\rho}|)}{2\,[1-\cos(|\vec{\rho}|)]} \qquad \text{(B.195)}$$

$$\frac{\partial s_1(\vec{\rho})}{\partial \vec{\rho}} = \frac{1-|\vec{\rho}|^{-1}\sin(|\vec{\rho}|)}{2\,[1-\cos(|\vec{\rho}|)]}\vec{\rho}^T \qquad \text{(B.196)}$$

$$s_2(\vec{\rho}) = \frac{1}{2} \qquad \text{(B.197)}$$

$$\frac{\partial s_2}{\partial \vec{\rho}} = \mathbf{0}_{1\times 3} \qquad \text{(B.198)}$$

$$s_3(\vec{\rho}) = -\left[\frac{1}{|\vec{\rho}|^2} - \frac{\sin(|\vec{\rho}|)}{2\,|\vec{\rho}|[1-\cos(|\vec{\rho}|)]}\right] \qquad \text{(B.199)}$$

$$\frac{\partial s_3(\vec{\rho})}{\partial \vec{\rho}} = -\frac{1+|\vec{\rho}|^{-1}\sin(|\vec{\rho}|) - 4\,|\vec{\rho}|^{-2}[1-\cos(|\vec{\rho}|)]}{2\,|\vec{\rho}|^2[1-\cos(|\vec{\rho}|)]}\vec{\rho}^T \qquad \text{(B.200)}$$

$$\begin{aligned}
\frac{d}{dt}\frac{\partial \dot{\rho}_{\mathrm{ENU}}}{\partial \vec{\omega}_{\mathrm{ENU}}} &= \frac{\partial s_1(\vec{\rho})}{\partial \vec{\rho}}\dot{\rho}\mathbf{I}_3 + \frac{\partial s_2(\vec{\rho})}{\partial \vec{\rho}}\dot{\rho}[\vec{\rho}\otimes] + s_2(\vec{\rho})[\dot{\rho}\otimes] \\
&\quad + \frac{\partial s_3(\vec{\rho})}{\partial \vec{\rho}}\dot{\rho}[\vec{\rho}\vec{\rho}^T] + s_3(\vec{\rho})[\dot{\rho}\vec{\rho}^T + \vec{\rho}\dot{\rho}^T] \\
&= \left\{\frac{1-|\vec{\rho}|^{-1}\sin(|\vec{\rho}|)}{2\,[1-\cos(|\vec{\rho}|)]}\vec{\rho}^T\right\}(\vec{\rho}^T\dot{\rho})\mathbf{I}_3
\end{aligned}$$

$$
\begin{aligned}
&+\frac{1}{2}(\vec{\rho})[\dot{\rho}\otimes] \\
&-\left\{\frac{1+|\vec{\rho}|^{-1}\sin(|\vec{\rho}|)-4\,|\vec{\rho}|^{-2}[1-\cos(|\vec{\rho}|)]}{2\,|\vec{\rho}|^{2}[1-\cos(|\vec{\rho}|)]}\vec{\rho}^{T}\right\} \\
&\times(\vec{\rho}^{T}\dot{\rho})[\vec{\rho}\vec{\rho}^{T}] \\
&-\left[\frac{1}{|\vec{\rho}|^{2}}-\frac{\sin(|\vec{\rho}|)}{2\,|\vec{\rho}|[1-\cos(|\vec{\rho}|)]}\right](\vec{\rho})[\dot{\rho}\vec{\rho}^{T}+\vec{\rho}\dot{\rho}^{T}]
\end{aligned}
\tag{B.201}
$$

B.4.2.7 Partial Derivatives with Respect to Rotation Vectors

Calculation of the dynamic coefficient matrices **F** and measurement sensitivity matrices **H** in linearized or extended Kalman filtering with rotation vectors $\vec{\rho}_{\mathrm{ENU}}$ as part of the system model state vector requires taking derivatives with respect to $\vec{\rho}_{\mathrm{ENU}}$ of associated vector-valued **f**- or **h**-functions, as

$$
\mathbf{F}=\frac{\partial\mathbf{f}(\vec{\rho}_{\mathrm{ENU}},\mathbf{v})}{\partial\vec{\rho}_{\mathrm{ENU}}}
\tag{B.202}
$$

$$
\mathbf{H}=\frac{\partial\mathbf{h}(\vec{\rho}_{\mathrm{ENU}},\mathbf{v})}{\partial\vec{\rho}_{\mathrm{ENU}}}
\tag{B.203}
$$

where the vector-valued functions will have the general form

$$
\begin{aligned}
&\mathbf{f}(\vec{\rho}_{\mathrm{ENU}},\mathbf{v})\ \text{or}\ \mathbf{h}(\vec{\rho}_{\mathrm{ENU}},\mathbf{v}) \\
&\quad=\{s_0(\vec{\rho}_{\mathrm{ENU}})\mathbf{I}_3+s_1(\vec{\rho}_{\mathrm{ENU}})[\vec{\rho}_{\mathrm{ENU}}\otimes]+s_2(\vec{\rho}_{\mathrm{ENU}})\vec{\rho}_{\mathrm{ENU}}\vec{\rho}_{\mathrm{ENU}}^{T}\}\mathbf{v}
\end{aligned}
\tag{B.204}
$$

and

s_0, s_1, and s_2 are scalar-valued functions of $\vec{\rho}_{\mathrm{ENU}}$.
v is a vector that does not depend on $\vec{\rho}_{\mathrm{ENU}}$.

We will derive here the general formulas that can be used for taking the partial derivatives

$$
\frac{\partial\mathbf{f}(\vec{\rho}_{\mathrm{ENU}},\mathbf{v})}{\partial\vec{\rho}_{\mathrm{ENU}}}\ \text{or}\ \frac{\partial\mathbf{h}(\vec{\rho}_{\mathrm{ENU}},\mathbf{v})}{\partial\vec{\rho}_{\mathrm{ENU}}}.
$$

These formulas can all be derived by calculating the derivatives of the different factors in the functional forms, and then using the chain rule for differentiation to obtain the final result.

Derivatives of Scalars The derivatives of the scalar factors s_0, s_1, and s_2 will

$$
\frac{\partial}{\partial\vec{\rho}_{\mathrm{ENU}}}s_i(\vec{\rho}_{\mathrm{ENU}})=\left[\frac{\partial s_i(\vec{\rho}_{\mathrm{ENU}})}{\partial\rho_E}\ \frac{\partial s_i(\vec{\rho}_{\mathrm{ENU}})}{\partial\rho_N}\ \frac{\partial s_i(\vec{\rho}_{\mathrm{ENU}})}{\partial\rho_U}\right]
\tag{B.205}
$$

a row-vector. Consequently, for any vector-valued function $\mathbf{g}(\vec{\rho}_{\text{ENU}})$, by the chain rule, the derivatives of the vector-valued product $s_i(\vec{\rho}_{\text{ENU}})\,\mathbf{g}(\vec{\rho}_{\text{ENU}})$ will be

$$\frac{\partial\{s_i(\vec{\rho}_{\text{ENU}})\,\mathbf{g}(\vec{\rho}_{\text{ENU}})\}}{\partial\vec{\rho}_{\text{ENU}}} = \underbrace{\mathbf{g}(\vec{\rho}_{\text{ENU}})\frac{\partial s_i(\vec{\rho}_{\text{ENU}})}{\partial\vec{\rho}_{\text{ENU}}}}_{3\times3\text{matrix}} + s_i(\vec{\rho}_{\text{ENU}})\underbrace{\frac{\partial\mathbf{g}(\vec{\rho}_{\text{ENU}})}{\partial\vec{\rho}_{\text{ENU}}}}_{3\times3\text{matrix}}, \tag{B.206}$$

the result of which will be the 3×3 Jacobian matrix of that sub-expression in $\mathbf{f}$ or $\mathbf{h}$.

Derivatives of Vectors The three potential forms of the vector-valued function $\mathbf{g}$ in Eq. (B.206) are

$$\mathbf{g}(\vec{\rho}_{\text{ENU}}) = \begin{cases} \mathbf{I}\mathbf{v} = \mathbf{v} \\ \vec{\rho}_{\text{ENU}} \otimes \mathbf{v} \\ \vec{\rho}_{\text{ENU}}\vec{\rho}_{\text{ENU}}^{\,T}\mathbf{v}, \end{cases} \tag{B.207}$$

each of which is considered independently:

$$\frac{\partial\mathbf{v}}{\partial\vec{\rho}_{\text{ENU}}} = 0_{3\times3} \tag{B.208}$$

$$\begin{aligned}\frac{\partial\vec{\rho}_{\text{ENU}} \otimes \mathbf{v}}{\partial\vec{\rho}_{\text{ENU}}} &= \frac{\partial[-\mathbf{v} \otimes \vec{\rho}_{\text{ENU}}]}{\partial\vec{\rho}_{\text{ENU}}} \\ &= -[\mathbf{v}\otimes] \\ &= -\begin{bmatrix} 0 & -v_3 & v_2 \\ v_3 & 0 & -v_1 \\ -v_2 & v_1 & 0 \end{bmatrix}\end{aligned} \tag{B.209}$$

$$\begin{aligned}\frac{\partial\vec{\rho}_{\text{ENU}}\vec{\rho}_{\text{ENU}}^{\,T}\mathbf{v}}{\partial\vec{\rho}_{\text{ENU}}} &= (\vec{\rho}_{\text{ENU}}^{\,T}\mathbf{v})\frac{\partial\vec{\rho}_{\text{ENU}}}{\partial\vec{\rho}_{\text{ENU}}} + \vec{\rho}_{\text{ENU}}\frac{\partial\vec{\rho}_{\text{ENU}}^{\,T}\mathbf{v}}{\partial\vec{\rho}_{\text{ENU}}} \\ &= (\vec{\rho}_{\text{ENU}}^{\,T}\mathbf{v})\mathbf{I}_{3\times3} + \vec{\rho}_{\text{ENU}}\mathbf{v}^T\end{aligned} \tag{B.210}$$

General Formula Combining the aforementioned formulas for the different parts, one can obtain the following general-purpose formula

$$
\begin{aligned}
&\frac{\partial}{\partial \vec{\rho}_{\mathrm{ENU}}}\{s_0(\vec{\rho}_{\mathrm{ENU}})\mathbf{I}_3 + s_1(\vec{\rho}_{\mathrm{ENU}})[\vec{\rho}_{\mathrm{ENU}}\otimes] + s_2(\vec{\rho}_{\mathrm{ENU}})\vec{\rho}_{\mathrm{ENU}}\vec{\rho}_{\mathrm{ENU}}^T\}\mathbf{v} \\
&\quad = \mathbf{v}\left[\begin{array}{ccc} \frac{\partial s_0(\vec{\rho}_{\mathrm{ENU}})}{\partial \rho_E} & \frac{\partial s_0(\vec{\rho}_{\mathrm{ENU}})}{\partial \rho_N} & \frac{\partial s_0(\vec{\rho}_{\mathrm{ENU}})}{\partial \rho_U} \end{array}\right] \\
&\quad + [\vec{\rho}_{\mathrm{ENU}} \otimes \mathbf{v}]\left[\begin{array}{ccc} \frac{\partial s_1(\vec{\rho}_{\mathrm{ENU}})}{\partial \rho_E} & \frac{\partial s_1(\vec{\rho}_{\mathrm{ENU}})}{\partial \rho_N} & \frac{\partial s_1(\vec{\rho}_{\mathrm{ENU}})}{\partial \rho_U} \end{array}\right] \\
&\quad - s_1(\vec{\rho}_{\mathrm{ENU}})[\mathbf{v}\otimes] \\
&\quad + (\vec{\rho}_{\mathrm{ENU}}^T\mathbf{v})\vec{\rho}_{\mathrm{ENU}}\left[\begin{array}{ccc} \frac{\partial s_2(\vec{\rho}_{\mathrm{ENU}})}{\partial \rho_E} & \frac{\partial s_2(\vec{\rho}_{\mathrm{ENU}})}{\partial \rho_N} & \frac{\partial s_2(\vec{\rho}_{\mathrm{ENU}})}{\partial \rho_U} \end{array}\right] \\
&\quad + s_2(\vec{\rho}_{\mathrm{ENU}})[(\vec{\rho}_{\mathrm{ENU}}^T\mathbf{v})\mathbf{I}_{3\times 3} + \vec{\rho}_{\mathrm{ENU}}\mathbf{v}^T]
\end{aligned}
\tag{B.211}
$$

applicable for any differentiable scalar functions s_0, s_1, s_2.

B.4.3 Direction Cosines Matrix

We have demonstrated in Eq. (B.11) that the coordinate transformation matrix between one orthogonal coordinate system and another is a matrix of direction cosines between the unit axis vectors of the two coordinate systems,

$$
\mathbf{C}_{XYZ}^{UVW} = \begin{bmatrix} \cos(\theta_{XU}) & \cos(\theta_{XV}) & \cos(\theta_{XW}) \\ \cos(\theta_{YU}) & \cos(\theta_{YV}) & \cos(\theta_{YW}) \\ \cos(\theta_{ZU}) & \cos(\theta_{ZV}) & \cos(\theta_{ZW}) \end{bmatrix} \tag{B.212}
$$

Because the angles do not depend on the order of the direction vectors (i.e. $\theta_{ab} = \theta_{ba}$) the inverse transformation matrix

$$
\begin{aligned}
\mathbf{C}_{UVW}^{XYZ} &= \begin{bmatrix} \cos(\theta_{UX}) & \cos(\theta_{UY}) & \cos(\theta_{UZ}) \\ \cos(\theta_{VX}) & \cos(\theta_{VY}) & \cos(\theta_{VZ}) \\ \cos(\theta_{WX}) & \cos(\theta_{WY}) & \cos(\theta_{WX}) \end{bmatrix} \\
&= \begin{bmatrix} \cos(\theta_{XU}) & \cos(\theta_{XV}) & \cos(\theta_{XW}) \\ \cos(\theta_{YU}) & \cos(\theta_{YV}) & \cos(\theta_{YW}) \\ \cos(\theta_{ZU}) & \cos(\theta_{ZV}) & \cos(\theta_{ZW}) \end{bmatrix}^T \\
&= (\mathbf{C}_{XYZ}^{UVW})^T
\end{aligned}
\tag{B.213}
$$

That is, the inverse coordinate transformation matrix is the transpose of the forward coordinate transformation matrix. This implies that the coordinate transformation matrices are orthogonal matrices.

B.4.3.1 Rotating Coordinates

Let "ROT" denote a set of rotating coordinates, with axes X_{ROT}, Y_{ROT}, Z_{ROT}, and let "NON" represent a set of non-rotating (i.e. inertial) coordinates, with axes X_{NON}, Y_{NON}, Z_{NON}, as illustrated in Figure B.15.

Any vector $\mathbf{v}_{\text{ROT}}$ in rotating coordinates can be represented in terms of its non-rotating components and unit vectors parallel to the non-rotating axes, as

$$\begin{aligned}
\mathbf{v}_{\text{ROT}} &= v_{x\ \text{NON}}\vec{\mathbf{1}}_{x\ \text{NON}} + v_{y\ \text{NON}}\vec{\mathbf{1}}_{y\ \text{NON}} + v_{z\ \text{NON}}\vec{\mathbf{1}}_{z\ \text{NON}} \\
&= \begin{bmatrix} \vec{\mathbf{1}}_{x\text{NON}} & \vec{\mathbf{1}}_{y\text{NON}} & \vec{\mathbf{1}}_{z\text{NON}} \end{bmatrix} \begin{bmatrix} v_{x\ \text{NON}} \\ v_{y\ \text{NON}} \\ v_{z\ \text{NON}} \end{bmatrix} \\
&= \mathbf{C}_{\text{ROT}}^{\text{NON}} \mathbf{v}_{\text{NON}}
\end{aligned} \tag{B.214}$$

where

$v_{x\ \text{NON}}$, $v_{y\ \text{NON}}$, $v_{z\ \text{NON}}$ are the non-rotating components of the vector,

$\vec{\mathbf{1}}_{x\ \text{NON}}$, $\vec{\mathbf{1}}_{y\ \text{NON}}$, $\vec{\mathbf{1}}_{z\ \text{NON}}$ are unit vectors along the X_{NON}, Y_{NON}, and Z_{NON} axes, as expressed in rotating coordinates,

$\mathbf{v}_{\text{ROT}}$ is the vector $\mathbf{v}$ expressed in RPY coordinates,

$\mathbf{v}_{\text{NON}}$ is the vector $\mathbf{v}$ expressed in ECI coordinates,

$\mathbf{C}_{\text{ROT}}^{\text{NON}}$ is the coordinate transformation matrix from non-rotating coordinates to rotating coordinates,
and

$$\mathbf{C}_{\text{ROT}}^{\text{NON}} = \begin{bmatrix} \vec{\mathbf{1}}_{x\ \text{NON}} & \vec{\mathbf{1}}_{y\ \text{NON}} & \vec{\mathbf{1}}_{z\ \text{NON}} \end{bmatrix} \tag{B.215}$$

The time derivative of $\mathbf{C}_{\text{ROT}}^{\text{NON}}$, as viewed from the non-rotating coordinate frame, can be derived in terms of the dynamics of the unit vectors $\vec{\mathbf{1}}_{x\ \text{NON}}$, $\vec{\mathbf{1}}_{y\ \text{NON}}$ and $\vec{\mathbf{1}}_{z\ \text{NON}}$ in rotating coordinates.

As seen by an observer fixed with respect to the non-rotating coordinates, the non-rotating coordinate directions will appear to remain fixed, but the external inertial reference directions will appear to be changing, as illustrated in Figure B.15. Gyroscopes fixed in the rotating coordinates would measure three

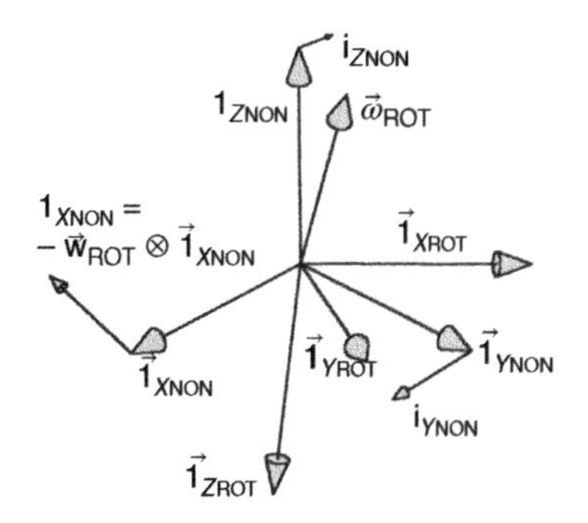

Figure B.15 Rotating coordinates.

components of the inertial rotation rate vector

$$\vec{\omega}_{\text{ROT}} = \begin{bmatrix} \omega_{x\ \text{ROT}} \\ \omega_{y\ \text{ROT}} \\ \omega_{z\ \text{ROT}} \end{bmatrix} \tag{B.216}$$

in rotating coordinates, but the non-rotating unit vectors, as viewed in rotating coordinates, appear to be changing in the opposite sense, as

$$\frac{d}{dt}\vec{\mathbf{1}}_{x\ \text{NON}} = -\vec{\omega}_{\text{ROT}} \otimes \vec{\mathbf{1}}_{x\ \text{NON}} \tag{B.217}$$

$$\frac{d}{dt}\vec{\mathbf{1}}_{y\ \text{NON}} = -\vec{\omega}_{\text{ROT}} \otimes \vec{\mathbf{1}}_{y\ \text{NON}} \tag{B.218}$$

$$\frac{d}{dt}\vec{\mathbf{1}}_{z\ \text{NON}} = -\vec{\omega}_{\text{ROT}} \otimes \vec{\mathbf{1}}_{z\ \text{NON}} \tag{B.219}$$

as illustrated in Figure B.15. The time-derivative of the coordinate transformation represented in Eq. (B.215) will then be

$$\begin{aligned} \frac{d}{dt}\mathbf{C}_{\text{ROT}}^{\text{NON}} &= \left[\frac{d}{dt}\vec{\mathbf{1}}_{x\ \text{NON}} \ \frac{d}{dt}\vec{\mathbf{1}}_{y\ \text{NON}} \ \frac{d}{dt}\vec{\mathbf{1}}_{z\ \text{NON}} \right] \\ &= \left[-\vec{\omega}_{\text{ROT}} \otimes \vec{\mathbf{1}}_{x\ \text{NON}} \ -\vec{\omega}_{\text{ROT}} \otimes \vec{\mathbf{1}}_{y\ \text{NON}} \ -\vec{\omega}_{\text{ROT}} \otimes \vec{\mathbf{1}}_{z\ \text{NON}} \right] \\ &= -[\vec{\omega}_{\text{ROT}} \otimes] \left[\vec{\mathbf{1}}_{x\ \text{NON}} \ \vec{\mathbf{1}}_{y\ \text{NON}} \ \vec{\mathbf{1}}_{z\ \text{NON}} \right] \\ &= -[\vec{\omega}_{\text{ROT}} \otimes]\mathbf{C}_{\text{ROT}}^{\text{NON}} \end{aligned} \tag{B.220}$$

$$[\vec{\omega}_{\text{ROT}} \otimes] \stackrel{\text{def}}{=} \begin{bmatrix} 0 & -\omega_{z\ \text{ROT}} & \omega_{y\ \text{ROT}} \\ \omega_{z\ \text{ROT}} & 0 & -\omega_{x\ \text{ROT}} \\ -\omega_{y\ \text{ROT}} & \omega_{x\ \text{ROT}} & 0 \end{bmatrix} \tag{B.221}$$

The inverse coordinate transformation

$$\begin{aligned} \mathbf{C}_{\text{NON}}^{\text{ROT}} &= (\mathbf{C}_{\text{ROT}}^{\text{NON}})^{-1} \\ &= (\mathbf{C}_{\text{ROT}}^{\text{NON}})^{T} \end{aligned} \tag{B.222}$$

the transpose of $\mathbf{C}_{\mathrm{ROT}}^{\mathrm{NON}}$, and its derivative

$$\begin{aligned}
\frac{d}{dt}\mathbf{C}_{\mathrm{NON}}^{\mathrm{ROT}} &= \frac{d}{dt}(\mathbf{C}_{\mathrm{ROT}}^{\mathrm{NON}})^T \\
&= \left(\frac{d}{dt}\mathbf{C}_{\mathrm{ROT}}^{\mathrm{NON}}\right)^T \\
&= (-[\vec{\boldsymbol{\omega}}_{\mathrm{ROT}}\otimes]\mathbf{C}_{\mathrm{ROT}}^{\mathrm{NON}})^T \\
&= -(\mathbf{C}_{\mathrm{ROT}}^{\mathrm{NON}})^T[\vec{\boldsymbol{\omega}}_{\mathrm{ROT}}\otimes]^T \\
&= \mathbf{C}_{\mathrm{NON}}^{\mathrm{ROT}}[\vec{\boldsymbol{\omega}}_{\mathrm{ROT}}\otimes]
\end{aligned} \tag{B.223}$$

In the case that "ROT" is "RPY" (roll–pitch–yaw coordinates) and "NON" is "*ECI*" (Earth-centered inertial coordinates), Eq. (B.223) becomes

$$\frac{d}{dt}\mathbf{C}_{ECI}^{\mathrm{RPY}} = \mathbf{C}_{ECI}^{\mathrm{RPY}}[\vec{\boldsymbol{\omega}}_{\mathrm{RPY}}\otimes], \tag{B.224}$$

and in the case that "ROT" is "ENU" (east–north–up coordinates) and "NON" is "*ECI*" (Earth-centered inertial coordinates), Eq. (B.220) becomes

$$\frac{d}{dt}\mathbf{C}_{\mathrm{ENU}}^{ECI} = -[\vec{\boldsymbol{\omega}}_{\mathrm{ENU}}\otimes]\mathbf{C}_{\mathrm{ENU}}^{ECI} \tag{B.225}$$

and the derivative of their product

$$\mathbf{C}_{\mathrm{ENU}}^{\mathrm{RPY}} = \mathbf{C}_{\mathrm{ENU}}^{ECI}\,\mathbf{C}_{ECI}^{\mathrm{RPY}} \tag{B.226}$$

$$\begin{aligned}
\frac{d}{dt}\mathbf{C}_{\mathrm{ENU}}^{\mathrm{RPY}} &= \left[\frac{d}{dt}\mathbf{C}_{\mathrm{ENU}}^{ECI}\right]\mathbf{C}_{ECI}^{\mathrm{RPY}} + \mathbf{C}_{\mathrm{ENU}}^{ECI}\left[\frac{d}{dt}\mathbf{C}_{ECI}^{\mathrm{RPY}}\right] \\
&= [[-\vec{\boldsymbol{\omega}}_{\mathrm{ENU}}\otimes]\mathbf{C}_{\mathrm{ENU}}^{ECI}]\,\mathbf{C}_{ECI}^{\mathrm{RPY}} + \mathbf{C}_{\mathrm{ENU}}^{ECI}\,[\mathbf{C}_{ECI}^{\mathrm{RPY}}[\vec{\boldsymbol{\omega}}_{\mathrm{RPY}}\otimes]] \\
&= [-\vec{\boldsymbol{\omega}}_{\mathrm{ENU}}\otimes]\underbrace{\mathbf{C}_{\mathrm{ENU}}^{ECI}\,\mathbf{C}_{ECI}^{\mathrm{RPY}}}_{\mathbf{C}_{\mathrm{ENU}}^{\mathrm{RPY}}} + \underbrace{\mathbf{C}_{\mathrm{ENU}}^{ECI}\,\mathbf{C}_{ECI}^{\mathrm{RPY}}}_{\mathbf{C}_{\mathrm{ENU}}^{\mathrm{RPY}}}[\vec{\boldsymbol{\omega}}_{\mathrm{RPY}}\otimes]
\end{aligned} \tag{B.227}$$

$$\frac{d}{dt}\mathbf{C}_{\mathrm{ENU}}^{\mathrm{RPY}} = -[\vec{\boldsymbol{\omega}}_{\mathrm{ENU}}\otimes]\mathbf{C}_{\mathrm{ENU}}^{\mathrm{RPY}} + \mathbf{C}_{\mathrm{ENU}}^{\mathrm{RPY}}[\vec{\boldsymbol{\omega}}_{\mathrm{RPY}}\otimes] \tag{B.228}$$

Equation (B.228) was originally used for maintaining vehicle attitude information in strapdown INS implementations, where the variables

$$\begin{aligned}
\vec{\boldsymbol{\omega}}_{\mathrm{RPY}} &= \text{vector of inertial rates measured by the gyroscopes} \\
\vec{\boldsymbol{\omega}}_{\mathrm{ENU}} &= \vec{\boldsymbol{\omega}}_{\mathrm{earthrate}} + \vec{\boldsymbol{\omega}}_{v_E} + \vec{\boldsymbol{\omega}}_{v_N}
\end{aligned} \tag{B.229}$$

$$\vec{\boldsymbol{\omega}}_{\oplus} = \omega_{\oplus}\begin{bmatrix} 0 \\ \cos(\phi_{\mathrm{GEODETIC}}) \\ \sin(\phi_{\mathrm{GEODETIC}}) \end{bmatrix} \tag{B.230}$$

$$\vec{\omega}_{v_E} = \frac{v_E}{r_T + h} \begin{bmatrix} 0 \\ 1 \\ 0 \end{bmatrix} \tag{B.231}$$

$$\vec{\omega}_{v_N} = \frac{v_N}{r_M + h} \begin{bmatrix} -1 \\ 0 \\ 0 \end{bmatrix} \tag{B.232}$$

and

$\omega_{\oplus}$ is Earth rotation rate.

ϕ_{GEODETIC} is geodetic latitude.

v_E is the east component of velocity with respect to the surface of the Earth.

r_T is the transverse radius of curvature of the ellipsoid.

v_N is the north component of velocity with respect to the surface of the Earth.

r_M is the meridional radius of curvature of the ellipsoid (i.e. radius of the osculating circle tangent to the meridian in the meridional plane).

h is altitude above (+) or below (−) the reference ellipsoid surface (≈ mean sea level).

Unfortunately, Eq. (B.228) was found to be not particularly well-suited for accurate integration in finite-precision arithmetic. This integration problem was eventually solved using quaternions.

B.4.4 Quaternions

The term *quaternions* is used in several contexts to refer to sets of four. In mathematics, it refers to an algebra in four dimensions discovered by the Irish physicist and mathematician Sir William Rowan Hamilton (1805–1865). The utility of quaternions for representing rotations (as points on a sphere in four dimensions) was known before strapdown systems, they soon became the standard representation of coordinate transforms in strapdown systems, and they have since been applied to computer animation.

B.4.4.1 Quaternion Matrices

For people already familiar with matrix algebra, the algebra of quaternions can be defined by using an isomorphism between 4×1 **quaternion vectors q** and real 4×4 **quaternion matrices Q**:

$$\mathbf{q} = \begin{bmatrix} q_1 \\ q_2 \\ q_3 \\ q_4 \end{bmatrix} \leftrightarrow \mathbf{Q} = \begin{bmatrix} q_1 & -q_2 & -q_3 & -q_4 \\ q_2 & q_1 & -q_4 & q_3 \\ q_3 & q_4 & q_1 & -q_2 \\ q_4 & -q_3 & q_2 & q_1 \end{bmatrix}$$

$$= q_1\mathcal{Q}_1 + q_2\mathcal{Q}_2 + q_3\mathcal{Q}_3 + q_4\mathcal{Q}_4 \tag{B.233}$$

$$\mathcal{Q}_1 \stackrel{\text{def}}{=} \begin{bmatrix} 1 & 0 & 0 & 0 \\ 0 & 1 & 0 & 0 \\ 0 & 0 & 1 & 0 \\ 0 & 0 & 0 & 1 \end{bmatrix} \tag{B.234}$$

$$\mathcal{Q}_2 \stackrel{\text{def}}{=} \begin{bmatrix} 0 & -1 & 0 & 0 \\ 1 & 0 & 0 & 0 \\ 0 & 0 & 0 & -1 \\ 0 & 0 & 1 & 0 \end{bmatrix} \tag{B.235}$$

$$\mathcal{Q}_3 \stackrel{\text{def}}{=} \begin{bmatrix} 0 & 0 & -1 & 0 \\ 0 & 0 & 0 & 1 \\ 1 & 0 & 0 & 0 \\ 0 & -1 & 0 & 0 \end{bmatrix} \tag{B.236}$$

$$\mathcal{Q}_4 \stackrel{\text{def}}{=} \begin{bmatrix} 0 & 0 & 0 & -1 \\ 0 & 0 & -1 & 0 \\ 0 & 1 & 0 & 0 \\ 1 & 0 & 0 & 0 \end{bmatrix} \tag{B.237}$$

in terms of four 4×4 **quaternion basis matrices,** $\mathcal{Q}_1, \mathcal{Q}_2, \mathcal{Q}_3, \mathcal{Q}_4$, the first of which is an identity matrix and the rest of which are antisymmetric.

B.4.4.2 Addition and Multiplication

Addition of quaternion vectors is the same as that for ordinary vectors. Multiplication is defined by the usual rules for matrix multiplication applied to the

Table B.1 Multiplication of quaternion basis matrices.

First Factor	Second Factor $\mathcal{Q}_1$	$\mathcal{Q}_2$	$\mathcal{Q}_3$	$\mathcal{Q}_4$
$\mathcal{Q}_1$	$\mathcal{Q}_1$	$\mathcal{Q}_2$	$\mathcal{Q}_3$	$\mathcal{Q}_4$
$\mathcal{Q}_2$	$\mathcal{Q}_2$	$-\mathcal{Q}_1$	$\mathcal{Q}_4$	$-\mathcal{Q}_3$
$\mathcal{Q}_3$	$\mathcal{Q}_3$	$-\mathcal{Q}_4$	$-\mathcal{Q}_1$	$\mathcal{Q}_2$
$\mathcal{Q}_4$	$\mathcal{Q}_4$	$\mathcal{Q}_3$	$-\mathcal{Q}_2$	$-\mathcal{Q}_1$

four quaternion basis matrices, the multiplication table for which is given in Table B.1. Note that, like matrix multiplication, **quaternion multiplication is non-commutative**. That is, the result depends on the order of multiplication.

Using the quaternion basis matrix multiplication table (B.1), the ordered product **AB** of two quaternion matrices

$$\mathbf{A} = a_1\mathcal{Q}_1 + a_2\mathcal{Q}_2 + a_3\mathcal{Q}_3 + a_4\mathcal{Q}_4 \tag{B.238}$$

$$\mathbf{B} = b_1\mathcal{Q}_1 + b_2\mathcal{Q}_2 + b_3\mathcal{Q}_3 + b_4\mathcal{Q}_4 \tag{B.239}$$

can be shown to be

$$\begin{aligned}\mathbf{AB} = & (a_1b_1 - a_2b_2 - a_3b_3 - a_4b_4)\mathcal{Q}_1 \\ & + (a_2b_1 + a_1b_2 - a_4b_3 + a_3b_4)\mathcal{Q}_2 \\ & + (a_3b_1 + a_4b_2 + a_1b_3 - a_2b_4)\mathcal{Q}_3 \\ & + (a_4b_1 - a_3b_2 + a_2b_3 + a_1b_4)\mathcal{Q}_4\end{aligned} \tag{B.240}$$

in terms of the coefficients a_k, b_k, and the quaternion basis matrices.

B.4.4.3 Conjugation

Conjugation of quaternions is a unary operation analogous to conjugation of complex numbers, in that the real part (the first component of a quaternion) is unchanged and the other parts change sign. For quaternions, this is equivalent to transposition of the associated quaternion matrix

$$\mathbf{Q} = q_1\mathcal{Q}_1 + q_2\mathcal{Q}_2 + q_3\mathcal{Q}_3 + q_4\mathcal{Q}_4 \tag{B.241}$$

so that

$$\begin{aligned}\mathbf{Q}^T &= q_1\mathcal{Q}_1 - q_2\mathcal{Q}_2 - q_3\mathcal{Q}_3 - q_4\mathcal{Q}_4 \\ &\leftrightarrow \mathbf{q}^\star\end{aligned} \tag{B.242}$$

$$\begin{aligned}\mathbf{Q}^T\mathbf{Q} &= (q_1^2 + q_2^2 + q_3^2 + q_4^2)\mathcal{Q}_1 \\ &\leftrightarrow \mathbf{q}^\star\mathbf{q} = |\mathbf{q}|^2\end{aligned} \tag{B.243}$$

B.4.4.4 Representing Rotations

The problem with rotation vectors as representations for rotations is that the rotation vector representing successive rotations $\vec{\rho}_1, \vec{\rho}_2, \vec{\rho}_3, \dots, \vec{\rho}_n$ is not a simple function of the respective rotation vectors.

This representation problem is solved rather elegantly using quaternions, such that the quaternion representation of the successive rotations is represented by the quaternion product $\mathbf{q}_n \times \mathbf{q}_{n-1} \times \cdots \times \mathbf{q}_3 \times \mathbf{q}_2 \times \mathbf{q}_1$. That is, each successive rotation can be implemented by a single quaternion product.

The quaternion equivalent of the rotation vector $\vec{\rho}$ with $|\vec{\rho}| = \theta$,

$$\vec{\rho} \stackrel{\text{def}}{=} \begin{bmatrix} \rho_1 \\ \rho_2 \\ \rho_3 \end{bmatrix} \stackrel{\text{def}}{=} \theta \begin{bmatrix} u_1 \\ u_2 \\ u_3 \end{bmatrix} \tag{B.244}$$

(i.e. where $\mathbf{u}$ is a unit vector) is

$$\mathbf{q}(\vec{\rho}) \stackrel{\text{def}}{=} \begin{bmatrix} \cos\left(\frac{\theta}{2}\right) \\ \rho_1 \sin\left(\frac{\theta}{2}\right)/\theta \\ \rho_2 \sin\left(\frac{\theta}{2}\right)/\theta \\ \rho_3 \sin\left(\frac{\theta}{2}\right)/\theta \end{bmatrix} = \begin{bmatrix} \cos\left(\frac{\theta}{2}\right) \\ u_1 \sin\left(\frac{\theta}{2}\right) \\ u_2 \sin\left(\frac{\theta}{2}\right) \\ u_3 \sin\left(\frac{\theta}{2}\right) \end{bmatrix} \tag{B.245}$$

and the vector $\mathbf{w}$ resulting from the rotation of any three-dimensional vector

$$\mathbf{v} \stackrel{\text{def}}{=} \begin{bmatrix} v_1 \\ v_2 \\ v_3 \end{bmatrix} \tag{B.246}$$

through the angle θ about the unit vector $\mathbf{u}$ is implemented by the quaternion product

$$\begin{aligned} \mathbf{q}(\mathbf{w}) &\stackrel{\text{def}}{=} \mathbf{q}(\vec{\rho})\mathbf{q}(\mathbf{v})\mathbf{q}^{\star}(\vec{\rho}) \\ &\stackrel{\text{def}}{=} \begin{bmatrix} \cos\left(\frac{\theta}{2}\right) \\ u_1 \sin\left(\frac{\theta}{2}\right) \\ u_2 \sin\left(\frac{\theta}{2}\right) \\ u_3 \sin\left(\frac{\theta}{2}\right) \end{bmatrix} \times \begin{bmatrix} 0 \\ v_1 \\ v_2 \\ v_3 \end{bmatrix} \times \begin{bmatrix} \cos\left(\frac{\theta}{2}\right) \\ -u_1 \sin\left(\frac{\theta}{2}\right) \\ -u_2 \sin\left(\frac{\theta}{2}\right) \\ -u_3 \sin\left(\frac{\theta}{2}\right) \end{bmatrix} \\ &= \begin{bmatrix} 0 \\ w_1 \\ w_2 \\ w_3 \end{bmatrix} \end{aligned} \tag{B.247}$$

$$w_1 = \cos(\theta)v_1 + [1 - \cos(\theta)][u_1(u_1v_1 + u_2v_2 + u_3uv_3)] + \sin(\theta)[u_2v_3 - u_3v_2] \tag{B.248}$$

$$w_2 = \cos(\theta)v_2 + [1 - \cos(\theta)][u_2(u_1v_1 + u_2v_2 + u_3uv_3)] + \sin(\theta)[u_3v_1 - u_1v_3] \tag{B.249}$$

$$w_3 = \cos(\theta)v_3 + [1 - \cos(\theta)][u_3(u_1v_1 + u_2v_2 + u_3uv_3)] + \sin(\theta)[u_1v_2 - u_2v_1] \tag{B.250}$$

or

$$\begin{bmatrix} w_1 \\ w_2 \\ w_3 \end{bmatrix} = \mathbf{C}(\vec{\rho}) \begin{bmatrix} v_1 \\ v_2 \\ v_3 \end{bmatrix} \tag{B.251}$$

where the rotation matrix $\mathbf{C}(\vec{\rho})$ is defined in Eq. (B.100b) and Eq. (B.247) implements the same rotation of $\mathbf{v}$ as the matrix product $\mathbf{C}(\vec{\rho})\mathbf{v}$. Moreover, if

$$\mathbf{q}(\mathbf{w}_0) \stackrel{\text{def}}{=} \mathbf{v} \tag{B.252}$$

and

$$\mathbf{q}(\mathbf{w}_k) \stackrel{\text{def}}{=} \mathbf{q}(\vec{\rho}_k)\mathbf{q}(\mathbf{w}_{k-1})\mathbf{q}^\star(\vec{\rho}_k) \tag{B.253}$$

for $k = 1, 2, 3, \ldots, n$, then the nested quaternion product

$$\mathbf{q}(\mathbf{w}_n) = \mathbf{q}(\vec{\rho}_n) \cdots \mathbf{q}(\vec{\rho}_2)\mathbf{q}(\vec{\rho}_1)\mathbf{q}(\mathbf{v})\mathbf{q}^\star(\vec{\rho}_1)\mathbf{q}^\star(\vec{\rho}_2) \cdots \mathbf{q}^\star(\vec{\rho}_n) \tag{B.254}$$

implements the succession of rotations represented by the rotation vectors $\vec{\rho}_1$, $\vec{\rho}_2, \vec{\rho}_3, \ldots, \vec{\rho}_n$, and the single quaternion

$$\begin{aligned} \mathbf{q}_{[n]} &\stackrel{\text{def}}{=} \mathbf{q}(\vec{\rho}_n)\mathbf{q}(\vec{\rho}_{n-1}) \cdots \mathbf{q}(\vec{\rho}_3)\mathbf{q}(\vec{\rho}_2)\mathbf{q}(\vec{\rho}_1) \\ &= \mathbf{q}(\vec{\rho}_n)\mathbf{q}_{[n-1]} \end{aligned} \tag{B.255}$$

then represents the net effect of the successive rotations as

$$\mathbf{q}(\mathbf{w}_n) = \mathbf{q}_{[n]}\mathbf{q}(\mathbf{w}_0)\mathbf{q}^\star_{[n]} \tag{B.256}$$

The initial value $\mathbf{q}_{[0]}$ for the rotation quaternion will depend upon the initial orientation of the two coordinate systems. The initial value

$$\mathbf{q}_{[0]} \stackrel{\text{def}}{=} \begin{bmatrix} 1 \\ 0 \\ 0 \\ 0 \end{bmatrix} \tag{B.257}$$

applies to the case that the two coordinate systems are aligned. In strapdown system applications, the initial value $\mathbf{q}_{[0]}$ is determined during the INS alignment procedure.

Equation (B.255) is the much-used quaternion representation for successive rotations, and Eq. (B.256) is how it is used to perform coordinate transformations of any vector $\mathbf{w}_0$.

This representation uses the four components of a unit quaternion to maintain the transformation from one coordinate frame to another through a succession of rotations. In practice, computer roundoff may tend to alter the magnitude of the allegedly unit quaternion, but it can easily be rescaled to a unit quaternion by dividing by its magnitude.

B.5 Newtonian Mechanics in Rotating Coordinates

Using Eqs. (B.100a) and (B.2) derived earlier, one can easily derive formulas for the corrections to Newtonian mechanics in rotating coordinate systems. For that purpose, we substitute the rotation vector

$$\rho = -\omega t \tag{B.258}$$

where t is time and $\boldsymbol{\omega}$ is the rotation rate vector of the rotating coordinate system with respect to the non-rotating coordinate system. There is a negative sign on the right-hand side of Eq. (B.258) because the rotation matrix of Eqs. (B.100a) and (B.100b) represents the physical rotation of a coordinate system, whereas the application here represents the coordinate transformation from the un-rotated (i.e. inertial) coordinate system to the rotated coordinate system.

B.5.1 Rotating Coordinates

Let $x_{\text{ROT}}, \dot{x}_{\text{ROT}}, \ddot{x}_{\text{ROT}}, \ldots$ represent the position, velocity, acceleration, etc. of a point mass in the rotating (but non-accelerating) coordinate system, rotating at angular rate $|\boldsymbol{\omega}|$ about a vector $\boldsymbol{\omega}$. That is, the components of $\boldsymbol{\omega}$ are the components of rotation rate about the coordinate axes, and the axis of rotation passes through the origin of the rotating coordinate system.

In order to relate these dynamic variables to Newtonian mechanics, let $x_{\text{NON}}, \dot{x}_{\text{NON}}, \ddot{x}_{\text{NON}}, \ldots$ represent the position, velocity, acceleration, etc. of the same point mass in an inertial (i.e. non-rotating) coordinate system coincident with the rotating system at some arbitrary reference time t_0.

Then the coordinate transform from inertial to rotating coordinates can be represented in terms of a rotation matrix:

$$x_{\text{ROT}}(t) = C_{\text{ROT}}^{\text{NON}}(\boldsymbol{\omega}, t) x_{\text{NON}}(t) \tag{B.259}$$

$$\begin{aligned} C_{\mathrm{ROT}}^{\mathrm{NON}}(\boldsymbol{\omega},\ t) &= \frac{\boldsymbol{\omega}\boldsymbol{\omega}^T}{|\boldsymbol{\omega}|^2} - \frac{\cos(|\boldsymbol{\omega}|(t-t_0))}{|\boldsymbol{\omega}|^2}[\boldsymbol{\omega}\otimes][\boldsymbol{\omega}\otimes] - \frac{\sin(|\boldsymbol{\omega}|(t-t_0))}{|\boldsymbol{\omega}|}[\boldsymbol{\omega}\otimes] \\ &= \cos(|\boldsymbol{\omega}|(t-t_0))I_3 + \frac{[1-\cos(|\boldsymbol{\omega}|(t-t_0))]}{|\boldsymbol{\omega}|^2}\boldsymbol{\omega}\boldsymbol{\omega}^T \\ &\quad - \frac{\sin(|\boldsymbol{\omega}|(t-t_0))}{|\boldsymbol{\omega}|}[\boldsymbol{\omega}\otimes] \end{aligned} \tag{B.260}$$

B.5.2 Time Derivatives of Matrix Products

Leibniz rule for derivatives of products applies to vector–matrix products, as well. Using this rule, the time-derivatives of $x_{\mathrm{ROT}}(t)$ can be expressed in terms of the time-derivatives of $C_{\mathrm{ROT}}^{\mathrm{NON}}(\boldsymbol{\omega},\ t)$ and $x_{\mathrm{ROT}}(t)$ as:

$$\dot{x}_{\mathrm{ROT}}(t) = \dot{C}_{\mathrm{ROT}}^{\mathrm{NON}}(\boldsymbol{\omega},\ t)x_{\mathrm{NON}}(t) + C_{\mathrm{NON}}^{\mathrm{NON}}(\boldsymbol{\omega},\ t)\dot{x}_{\mathrm{ROT}}(t) \tag{B.261}$$

$$\begin{aligned} \ddot{x}_{\mathrm{ROT}}(t) &= \ddot{C}_{\mathrm{ROT}}^{\mathrm{NON}}(\boldsymbol{\omega},\ t)x_{\mathrm{NON}}(t) + 2\ \dot{C}_{\mathrm{ROT}}^{\mathrm{NON}}(\boldsymbol{\omega},\ t)\dot{x}_{\mathrm{NON}}(t) \\ &\quad + C_{\mathrm{ROT}}^{\mathrm{NON}}(\boldsymbol{\omega},\ t)\ddot{x}_{\mathrm{NON}}(t) \end{aligned} \tag{B.262}$$

The respective time-derivatives of $C_{\mathrm{ROT}}^{\mathrm{NON}}(\boldsymbol{\omega},\ t)$ can be derived by straightforward differentiation:

$$\dot{C}_{\mathrm{ROT}}^{\mathrm{NON}}(\boldsymbol{\omega},\ t) = \frac{\sin(|\boldsymbol{\omega}|(t-t_0))}{|\boldsymbol{\omega}|}[\boldsymbol{\omega}\otimes][\boldsymbol{\omega}\otimes] - \cos(|\boldsymbol{\omega}|(t-t_0))[\boldsymbol{\omega}\otimes] \tag{B.263}$$

$$\ddot{C}_{\mathrm{ROT}}^{\mathrm{NON}}(\boldsymbol{\omega},\ t) = \cos(|\boldsymbol{\omega}|(t-t_0))[\boldsymbol{\omega}\otimes][\boldsymbol{\omega}\otimes] + |\boldsymbol{\omega}|\sin(|\boldsymbol{\omega}|(t-t_0))[\boldsymbol{\omega}\otimes] \tag{B.264}$$

B.5.3 Solving for Centrifugal and Coriolis Accelerations

Evaluated at $t = t_0$, these become

$$C_{\mathrm{ROT}}^{\mathrm{NON}}(\boldsymbol{\omega},\ t_0) = I_3 \tag{B.265}$$

$$\dot{C}_{\mathrm{ROT}}^{\mathrm{NON}}(\boldsymbol{\omega},\ t_0) = -[\boldsymbol{\omega}\otimes] \tag{B.266}$$

$$\ddot{C}_{\mathrm{ROT}}^{\mathrm{NON}}(\boldsymbol{\omega},\ t_0) = [\boldsymbol{\omega}\otimes][\boldsymbol{\omega}\otimes] \tag{B.267}$$

and the corresponding time-derivatives at $t = t_0$ are:

$$x_{\mathrm{ROT}}(t_0) = x_{\mathrm{NON}}(t_0) \tag{B.268}$$

$$\begin{aligned} \dot{x}_{\mathrm{ROT}}(t_0) &= -[\boldsymbol{\omega}\otimes]x_{\mathrm{NON}}(t_0) + \dot{x}_{\mathrm{NON}}(t_0) \\ &= \underbrace{-[\boldsymbol{\omega}\otimes]x_{\mathrm{ROT}}(t_0)}_{\text{TANGENTIAL}} + \underbrace{\dot{x}_{\mathrm{NON}}(t_0)}_{\text{INERTIAL}} \end{aligned} \tag{B.269}$$

$$\dot{x}_{\mathrm{NON}}(t_0) = \dot{x}_{\mathrm{ROT}}(t_0) + [\boldsymbol{\omega}\otimes]x_{\mathrm{ROT}}(t_0) \tag{B.270}$$

$$\begin{aligned}\ddot{x}_{\mathrm{ROT}}(t_0) &= [\boldsymbol{\omega}\otimes][\boldsymbol{\omega}\otimes]x_{\mathrm{ROT}}(t_0) - 2\,[\boldsymbol{\omega}\otimes]\{\dot{x}_{\mathrm{ROT}}(t_0) + [\boldsymbol{\omega}\otimes]x_{\mathrm{ROT}}(t_0)\} + \ddot{x}_{\mathrm{NON}}(t_0) \\ &= \underbrace{-[\boldsymbol{\omega}\otimes][\boldsymbol{\omega}\otimes]x_{\mathrm{ROT}}(t_0)}_{\text{CENTRIFUGAL}} \underbrace{- 2\,[\boldsymbol{\omega}\otimes]\dot{x}_{\mathrm{ROT}}(t_0)}_{\text{CORIOLIS}} + \underbrace{\ddot{x}_{\mathrm{NON}}(t_0)}_{\text{SPECIFIC FORCE}}\end{aligned} \tag{B.271}$$

where the labels under the various terms in the formula for acceleration observed in rotating coordinates identify the *centrifugal, Coriolis*, and *specific-force* accelerations in rotating coordinates. The *specific-force* acceleration is that due to inertial forces applied to the point mass. The others are artifacts due to rotation.

In similar fashion, the velocity term in rotating coordinates labeled "INERTIAL" is the point mass velocity in inertial coordinates, and that labeled "TANGENTIAL" is an artifact due to rotation.

Because the reference time t_0 was chosen arbitrarily, these formulas apply for all times.

Appendix C

PDF Ambiguity Errors in Nonlinear Kalman Filtering

... nonlinear interactions almost always make the behavior of the aggregate more complicated than would be predicted by summing or averaging.

John Henry Holland (1929–2015) Hidden Order: How Adaptation Builds Complexity, Addison-Wesley, New York, 1995.

C.1 Objective

It has long been known that the problem of nonlinear transformation of means and variances of probability distributions is ill-posed, in the sense that there may be no unique solution, given just the starting means and variances. There may be a range of possible solutions, depending on the initial probability density functions (PDFs) and the nonlinear transformation(s) involved.

This appendix presents a preliminary look at the magnitude of the range of possible solutions for a sampling of potential PDFs and a parameterized range of nonlinear transformations [1]. These PDFs all have the same starting means and variances, but their nonlinearly transformed PDFs have distinctly different means and variances. The resulting "PDF ambiguity errors" are due to underrepresentation of the starting PDFs. The results show significant relative PDF ambiguity errors at relatively low levels of nonlinearity.

The question this poses for nonlinear Kalman filtering is whether PDF ambiguity errors can somehow be taken into account within the Kalman filter implementation. The Riccati equations in the Kalman filter implementations are a form of "statistical accounting" for keeping track of uncertainties used for minimizing mean squared estimation error. This has been used for accommodating random disturbances in vehicle dynamics, and for random noise in sensor outputs. The question then is whether the same mechanisms might be exploited for accommodating nonlinear dynamic uncertainties due to approximation errors. In essence, this appendix poses a problem without providing a solution.

Global Navigation Satellite Systems, Inertial Navigation, and Integration,
Fourth Edition. Mohinder S. Grewal, Angus P. Andrews, and Chris G. Bartone.

Companion website: www.wiley.com/go/grewal/gnss

There has been some success in making better use of the mean and covariance data for reducing nonlinear approximation errors, but the remaining approximation errors and PDF ambiguity errors are not otherwise taken into account in Kalman filter implementations. We know that underrepresentation of some uncertainties in Kalman filtering tends to reduce the Kalman gain, which can lead to poor filter convergence. Traditionally, this risk has been mitigated in practice by a deliberate bias toward overrepresentation of uncertainty. This has not been done with nonlinear approximation errors and PDF ambiguity errors.

The main objective of the study was to quantify the magnitude of PDF ambiguity errors for limited sets of nonlinear transformations and PDFs, using a somewhat efficient methodology for nonlinear transformations of PDFs. In practice, this approach would need to be specialized to specific applications to get the necessary statistics to quantify the relevant errors.

C.2 Methodology

Gauss was able to demonstrate characteristics of his linear least mean square estimator (LLMSE) with different error PDFs by using the calculus to perform integrations. The same is done here using a computer and MATLAB, and some simplifying shortcuts.

The approach has been simplified to univariate PDFs, in which case the majority of the computation can be done using matrix products.

C.2.1 Computing Expected Values

All PDFs are assumed to have a common domain on the real line as functions of the variate x and to have identical means and variances. This is essentially the model for LLMSE and Kalman filtering, both of which are insensitive to attributes of PDFs beyond means and variances.

The problem then is to compute the means and variances after various nonlinear transformations of x, which are expected values of functions of x. Means are the expected values of the transformed variate, and variances are the expected values of the squared difference between the transformed variate and its computed mean. It therefore suffices to use a general method for computing expected values.

The essential idea is to vectorize the variate x as a monotonically increasing series of samples x_k with uniform spacing δ_x, then vectorize each PDF as a column vector $\mathbf{p}_h$ of the values $p_h(x_k)$. The same approach can be used to vectorize any other function $f(x)$ of x. Then the expected value of $f(x)$ with respect to the PDF $p_h(x)$ can be calculated as the dot product

$$\underset{h}{\mathrm{E}}\langle f(x)\rangle \stackrel{\text{def}}{=} \int f(x)\, p_h(x)\, dx$$

$$\approx \delta_x \sum_{k=1}^{L} f(x_k)\, p_h(x_k)$$

$$= \delta_x\, \mathbf{f}^T \mathbf{p}_h \tag{C.1}$$

of the vectorized functions

$$\mathbf{f} \stackrel{\text{def}}{=} \begin{bmatrix} f(x_1) \\ f(x_2) \\ f(x_3) \\ \vdots \\ f(x_L) \end{bmatrix} \text{ and } \mathbf{p}_h \stackrel{\text{def}}{=} \begin{bmatrix} p_h(x_1) \\ p_h(x_2) \\ p_h(x_3) \\ \vdots \\ p_h(x_L) \end{bmatrix}$$

Furthermore, if both the functions and PDFs are matricized as

$$\mathbf{F}_f \stackrel{\text{def}}{=} \left[\, \mathbf{f}_1 \;\, \mathbf{f}_2 \;\, \mathbf{f}_3 \;\cdots\; \mathbf{f}_N \,\right] \tag{C.2}$$

$$\mathbf{P}_{\text{PDF}} \stackrel{\text{def}}{=} \left[\, \mathbf{p}_1 \;\, \mathbf{p}_2 \;\, \mathbf{p}_3 \;\cdots\; \mathbf{p}_M \,\right] \tag{C.3}$$

then the scaled matrix product

$$\mathbf{M} = \delta_x \mathbf{F}_f^T \mathbf{P}_{\text{PDF}} \tag{C.4}$$

is an $N \times M$ array with rows representing nonlinear transformations (of means or variances) and columns representing PDFs.

C.2.2 Representative Sample of PDFs

No "PDF of PDFs" is assumed for the ensemble of representative PDFs used in the analysis. The ensemble was chosen to represent possible PDFs, with no assigned application-dependent likelihoods.

The effects of linear and nonlinear transformations on PDFs are demonstrated using 10 different univariate PDFs $p_h(x)$ with identical means μ_h and variances (for univariate distributions) σ_h^2 about the mean. These are plotted in Figure C.1.

These all have the same mean ($\mu_h = 2$) and variance about the mean ($\sigma_h^2 = 4$). The first eight of these distributions are used for representing measurement errors in metrology analysis (from [2] and [3]), and the first three of these were used by Gauss in his analysis of linear least mean squares estimation [4]. The last two (chi-squared and Rayleigh) were included to extend the range of *skewness*

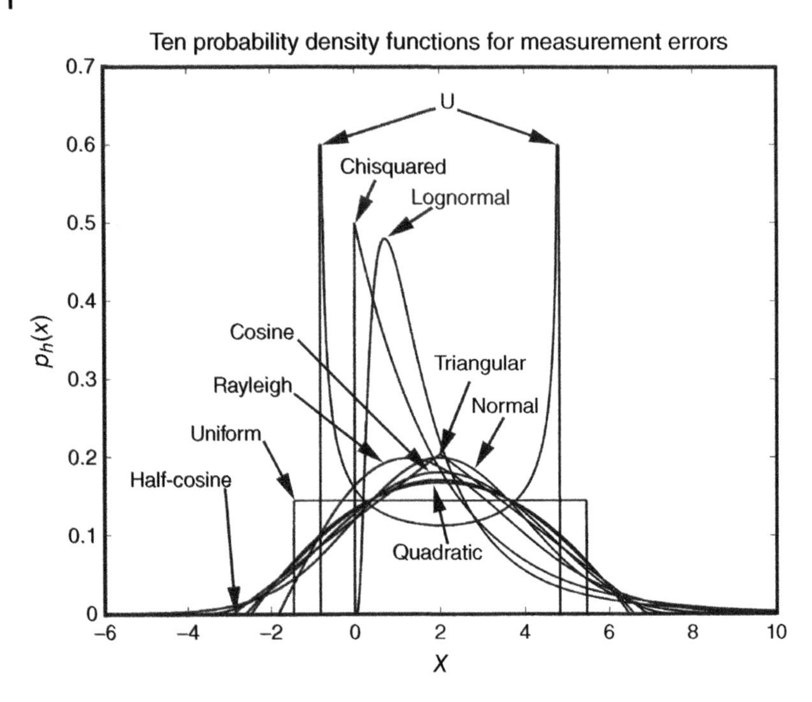

Figure C.1 Ten different univariate PDFs with common means $\mu_h = 2$ and variance $\sigma_h^2 = 4$.

and *kurtosis* (both described in the following list) of the distribution samples. The first few computed distribution statistics of all 10 distributions are listed in Table C.1, where the named m-files used for computing the respective PDFs are on www.wiley.com/go/grewal/gnss, and

$\sum$ is the numerically integrated PDF, included as a test for vetting the m-files.

μ_h is the computed mean of the PDF.

σ_h^2 is the computed variance of the PDF.

$\mathcal{S}_h$ is the computed *skewness,* a metric for PDF asymmetry.

$\mathcal{K}_h$ is the computed *kurtosis,* a metric for "tail weight" of the PDF.

Note, also, that this sampling includes no pathological PDFs (the Cauchy distribution, for example) which have no defined means or variances.

These statistics were computed using the MATLAB script `MSErrVsEst.m` on www.wiley.com/go/grewal/gnss, which also evaluates the test example described in the following text.

Example C.1 ***(Verifying PDF-independence of Eq. (10.20))***

The purpose of this example is to use Eq. (10.20) and the 10 PDFs from Table C.1 to demonstrate the approach, showing that – with no nonlinearities

Table C.1 Sample univariate PDF statistics

Distribution	m-file	Σ	μ_h	σ_h^2	$\mathcal{S}_h$	$\mathcal{K}_h$
Uniform	`pUniform.m`	1	2	4	0	2
Triangle	`pTriangle.m`	1	2	4	0	2
Normal[a]	`pNormal.m`	1	2	4	0	3
LogNormal	`pLogNormal.m`	1	2	4	4	38
Quadratic	`pQuadratic.m`	1	2	4	0	2
Cosine	`pCosine.m`	1	2	4	0	2
HalfCosine	`pHalfCosine.m`	1	2	4	0	2
U	`pU.m`	1	2	4	0	2
ChiSquared	`pChiSquared.m`	1	2	4	2	9
Rayleigh	`pRayleigh.m`	1	2	4	1	3

a) Gaussian.

involved – Eq. (10.20) does not depend on the error PDF beyond its mean and variance.

This equation expresses the mean squared estimation error as a function of the estimate $\hat{\mathbf{x}}$ as

$$E\langle|\hat{\mathbf{x}} - \mathbf{x}|^2\rangle = |\hat{\mathbf{x}} - \boldsymbol{\mu}|^2 + \mathrm{Tr}(\mathbf{P}) \tag{C.5}$$

where $\boldsymbol{\mu}$ is the mean of a PDF of $\mathbf{x}$ and P is its variance about the mean.

The right-hand side of Eq. (C.5) is a quadratic polynomial in the estimate $\hat{\mathbf{x}}$ which reaches its minimum value at $\hat{\mathbf{x}} = \boldsymbol{\mu}_h$, its mean, and that minimum value is the trace of $\mathbf{P}$, which is the sum of the diagonal elements of $\mathbf{P}$.

The left-hand side of Eq. (C.5) is an expected value, which – as an exercise for vetting the methodology – will be evaluated numerically for each of the 10 sample PDFs. In this example with univariate PDFs, all means will be scalars and all covariances will be variances. But, if everything works, this should show that the resulting quadratic curve on the right-hand side of the equation is the same for all 10 PDFs.

The MATLAB script `MSErrVsEst.m` on www.wiley.com/go/grewal/gnss performs numerical integrations of the mean squared estimation errors as a function of the estimated value $\hat{x}$ for each of the 10 PDFs. This is accomplished by numerical integration of $|\hat{x} - x|^2 p_h(x)$ over the interval $-60 \leq x \leq 100$ for each PDF p_h to demonstrate that the results are – as predicted by Gauss [4] – independent of the shapes of the PDFs so long as they have the same mean and variance about the mean.

Results are plotted in Figure C.2, which is actually 10 plots – one for each PDF. Note that the minimum value is at the mean of all the distributions ($\mu_h = 2$) and

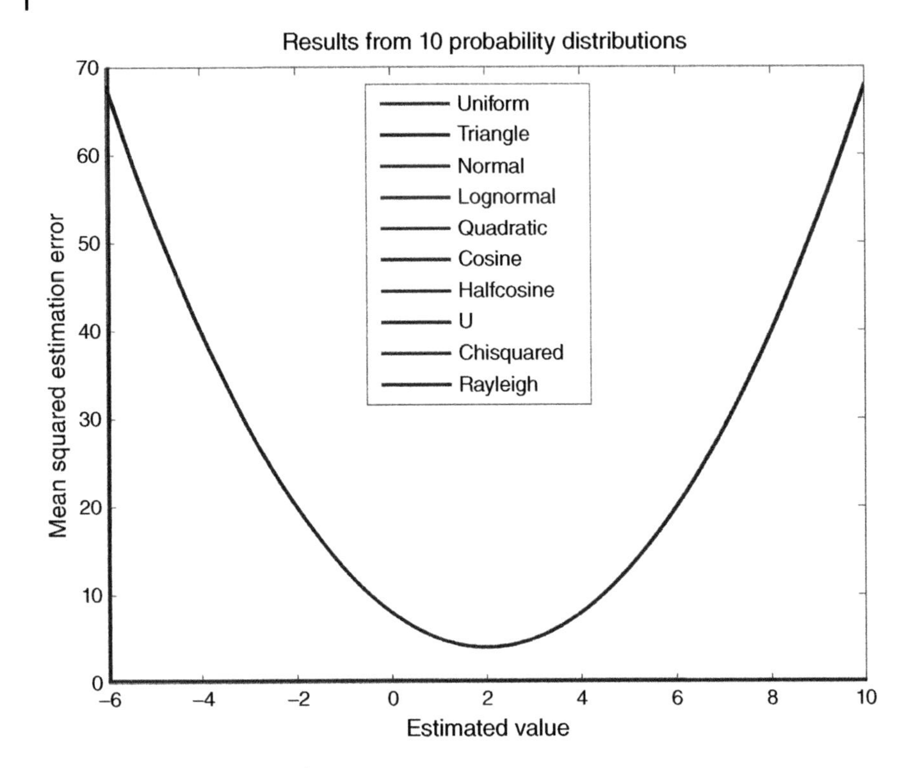

Figure C.2 Identical results from all 10 PDFs.

that minimum value is the variance (trace of the covariance matrix for multivariate distributions), which equals 4 in all cases.

This demonstrates that – at least for the 10 test PDFs used – the quadratic function of Eq. (C.5) defines the mean-squared estimation error as a function of the estimate and the mean and variance of the measurement errors – independent of the shapes of the PDFs assumed. That is the magic of Gauss's linear least mean squares estimator.

C.2.3 Parametric Class of Nonlinear Transformations Used

The analysis uses a parameterized family of nonlinear variate transformations $x \to y$ of the form

$$y_{ij}(x) \stackrel{\text{def}}{=} \begin{cases} x + a_i \ \cos(\omega_j(x - \mu_h)) & \text{(symmetric)} \\ x + a_i \ \sin(\omega_j(x - \mu_h)) & \text{(antisymmetric)} \end{cases} \tag{C.6}$$

These are nonlinear harmonic deviations from the linear transformation $y(x) = x$ with parameters

$0 \leq a_i \leq 1$, an amplitude parameter

$0 \leq \omega_j \leq 2$, a frequency parameter normalized to equal the number of wavelengths in the interval

$$\mu_h - \sigma_h \leq x \leq \mu_h + \sigma_h.$$

Equation (C.6) divides the nonlinearities into symmetric (cos) and antisymmetric (sin) nonlinearities about the mean as deviations from a linear transformation ($y = x$). The parameter a_i varies the magnitude of the nonlinearity relative to the linear part, and the parameter ω_j varies the relative rate of change of the nonlinearity. The frequency units in this case are normalized to the scale of the PDFs as "wave numbers," equal to the number of sinusoidal wavelengths in the interval between 1σ below the mean and 1σ above the mean, where $\sigma = 2$ is the common standard deviation of all PDFs used. Plots of some of the 882 nonlinear functions used in this analysis are shown in Figure C.3.

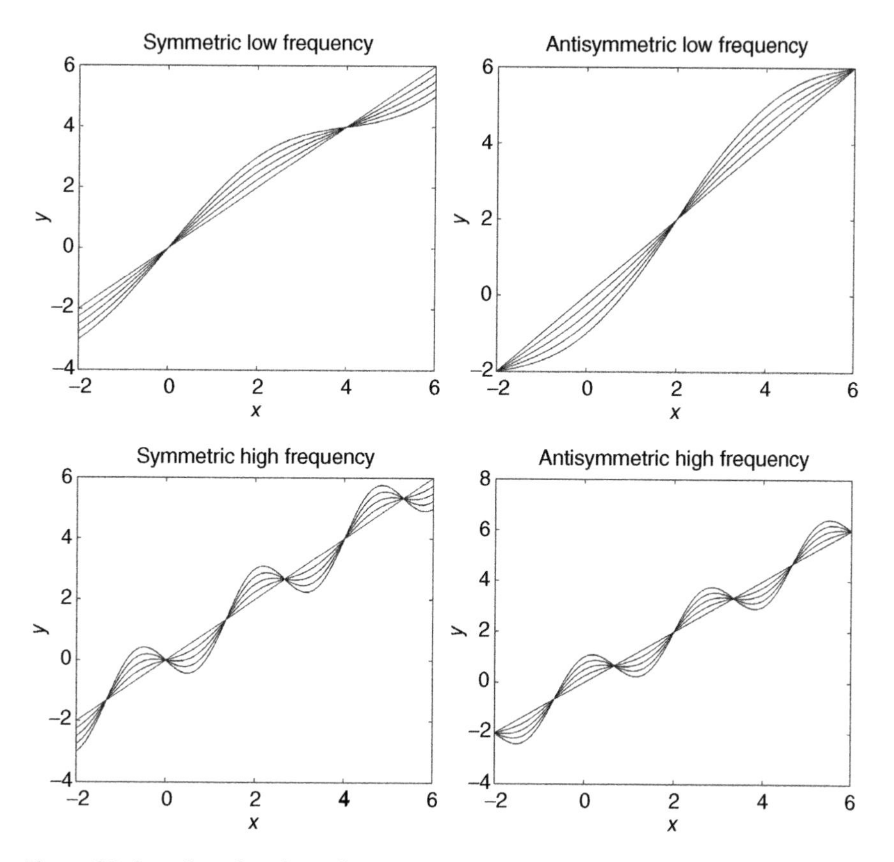

Figure C.3 Sampling of nonlinear functions used.

C.2.4 Ambiguity Errors in Nonlinearly Transformed Means and Variances

The objective here is to quantify the influence of error PDFs on predicted means and variances after nonlinear transformations of the variate. Because no probabilities have been assigned to the individual PDFs, this metric will not be a statistical one. The metric used is simply the total variation of the mean or variance as a percentage of the average. The term "average" is used here because the alternative term "mean" might imply some underlying probability distribution of probability distributions.

Percent variation The ith row $\mathbf{m}_{i,\cdot}$ of the matrix M from Eq. (C.4) would be a row vector containing the set of means (for example) resulting from the nonlinear transformation represented by that row. In that case, the percent total variation in the PDF-dependent means from that particular nonlinear transformation would be

$$\%_i = 100 \times \frac{\max\limits_h(m_{ih}) - \min\limits_h(m_{ih})}{\operatorname*{average}\limits_h(m_{ih})} \tag{C.7}$$

C.3 Results

C.3.1 Nonlinearly Transformed Means

The means μ_{hij} of the nonlinearly transformed variates y_{ij} with respect to the PDF $p_h(x)$ are defined by

$$\begin{aligned}
\mu_{hij} &= \mathop{\mathrm{E}}_{p_h}\langle y_{ij}\rangle \\
&= \int_x y_{ij}(x)\, p_h(x)\, dx \\
&= \int_x [x + a_i \text{ trig } (\omega_j(x-2))]\, p_h(x)\, dx \\
&= 2 + a_i \int_x \text{trig } (\omega_j(x-2))\, p_h(x)\, dx \\
&= 2 + a_i \begin{cases} c_{hj}, \text{ or} \\ s_{hj} \end{cases}
\end{aligned} \tag{C.8}$$

$$c_{hj} \stackrel{\text{def}}{=} \int_x \cos(\omega_j(x-2))\, p_h(x)\, dx \tag{C.9}$$

$$s_{hj} \stackrel{\text{def}}{=} \int_x \sin(\omega_j(x-2))\, p_h(x)\, dx \tag{C.10}$$

Rectangular numerical integrations of c_{hj} and s_{hj}, indexed over a range of values x_k with uniform spacing δ_x, can be implemented as dot products of the vectors

$$\{\mathbf{c}_j\}_k \stackrel{\text{def}}{=} \cos(\omega_j(x_k - 2)) \tag{C.11}$$

$$\{\mathbf{s}_j\}_k \stackrel{\text{def}}{=} \sin(\omega_j(x_k - 2)) \tag{C.12}$$

$$\{\mathbf{p}_h\}_k \stackrel{\text{def}}{=} p_h(x_k) \tag{C.13}$$

so that the numerically approximated means for the antisymmetric and symmetric nonlinearities become

$$\mu_{hij} \approx \begin{cases} 2 + \delta_x\, a_i \mathbf{c}_j^T \mathbf{p}_h & \text{(symmetric)} \\ 2 + \delta_x\, a_i \mathbf{s}_j^T \mathbf{p}_h & \text{(antisymmetric)} \end{cases} \tag{C.14}$$

where δ_x is the interval between adjacent x_k. The relative ranges of the nonlinearly transformed means due to differences in the assumed PDF can then be calculated as the percent differences

$$\%_{\mu ij} = 100 \times \frac{\max\limits_h(\mu_{hij}) - \min\limits_h(\mu_{hij})}{\operatorname{average}\limits_h(\mu_{hij})}$$

for the symmetric and antisymmetric cases.

Results are plotted in terms of the range of the computed means for the various PDFs, converted to percentages of the average mean and plotted as contours as functions of nonlinearity amplitude and frequency in Figures C.4 (symmetric nonlinearities) and C.5 (antisymmetric nonlinearities). The contour at 1% is plotted in addition to those at 5% intervals, to indicate where the nonlinear effects begin to take effect.

The results show a gradual increase in PDF sensitivity with amplitude, but vary by more than a factor of 2 with frequency.

Keep in mind that, because the Kalman estimate is essentially the mean of its PDF, unmodeled errors in propagating it as big as a few percent could be a very serious problem. However, the estimate is also in a feedback corrector loop that might tend to reduce the impact of unmodeled errors. The same is not true of unmodeled errors in the estimates of propagated variances, however.

C.3.2 Nonlinearly Transformed Variances

For these univariate PDFs, the variances can also be approximated in terms of vectors as

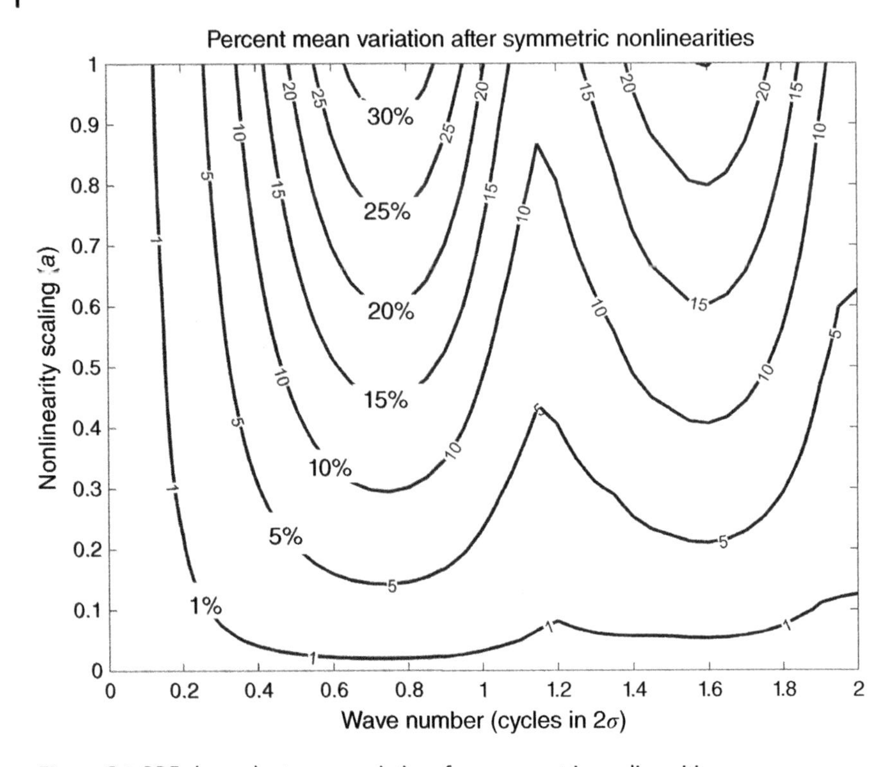

Figure C.4 PDF-dependent mean variations from symmetric nonlinearities.

$$\begin{aligned}\sigma_{hij}^2 &= \mathrm{E}\langle [y_{hij} - \mu_{hij}]^2 \rangle \\ &= \int_x [y_{hij}(x) - \mu_{hij}]^2\, p_h(x)\, dx \\ &= \int_x \Bigg[\underbrace{x + a_i \ \mathrm{trig}\ (\omega_j(x-2)) - \mu_{hij}}_{\psi_{hij}^{[\mathrm{trig}]}} \Bigg]^2 p_h(x)\, dx \end{aligned} \tag{C.15}$$

$$\psi_{hij}^{[\cos]} = x + a_i \ \cos(\omega_j(x-2)) - \mu_{hij} \tag{C.16}$$

$$\psi_{hij}^{[\sin]} = x + a_i \ \sin(\omega_j(x-2)) - \mu_{hij} \tag{C.17}$$

Consequently, if one lets the elements of the vectors ξ_{hij} be the squares of the elements of the ψ_{hij} (using the MATLAB `.^2` operator), then the dot-products

$$\sigma_{hij}^2 \approx \begin{cases} \delta_x\, \xi_{hij}^{[\cos]T} \mathbf{p}_h & \text{(symmetric)} \\ \delta_x\, \xi_{hij}^{[\cos]T} \mathbf{p}_h & \text{(antisymmetric)} \end{cases} \tag{C.18}$$

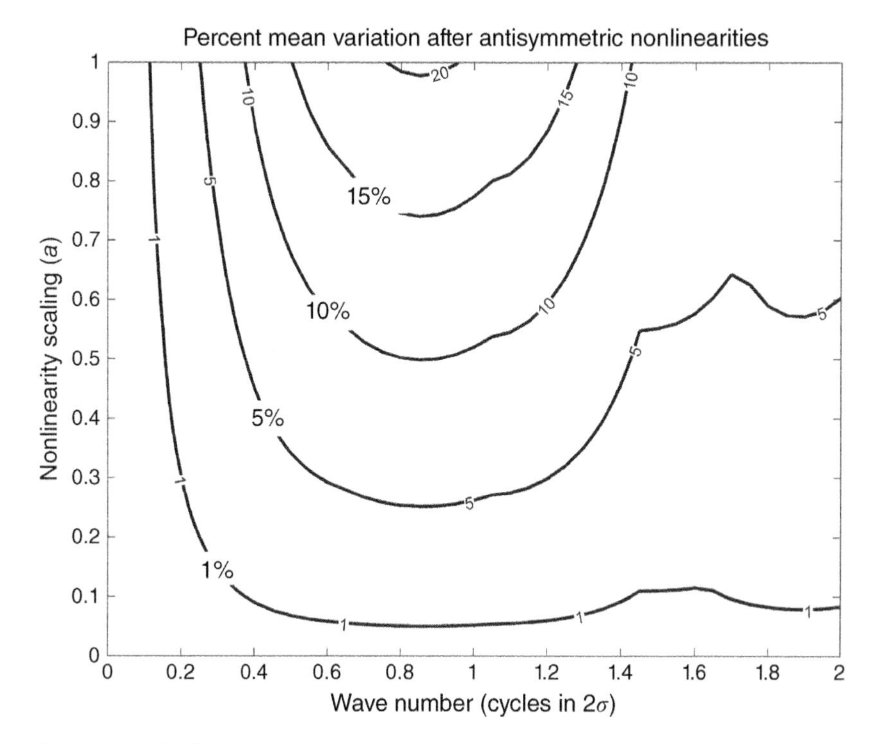

Figure C.5 PDF-dependent mean variations from antisymmetric nonlinearities.

The relative ranges of the nonlinearly transformed variances due to differences in the assumed PDF can then be calculated as the percent differences

$$\%_{\sigma ij} = 100 \times \frac{\max\limits_{h}(\sigma^2_{hij}) - \min\limits_{h}(\sigma^2_{hij})}{\operatorname*{average}\limits_{h}(\sigma^2_{hij})}$$

for the symmetric and antisymmetric cases.

Results from both cases (symmetric and antisymmetric) are calculated and contour plotted by the m-file `NonLinErr.m` on www.wiley.com/go/grewal/gnss.

The sensitivity of nonlinearly transformed means and variances to PDFs could be a serious issue in nonlinear approximations for Kalman filtering – especially for approximations focused on the nonlinear transformations and neglecting the influence of the likely PDFs involved.

Contour plots of the percent variation with PDFs of means and variances are shown in Figures C.4–C.7. Effects in the neighborhood of 1% come on at very low levels of nonlinearities in all cases, and nonlinear transformations cause serious PDF-dependent variations on the transformed mean – which is essentially the transformation of the estimate. Therefore, nonlinear approximations

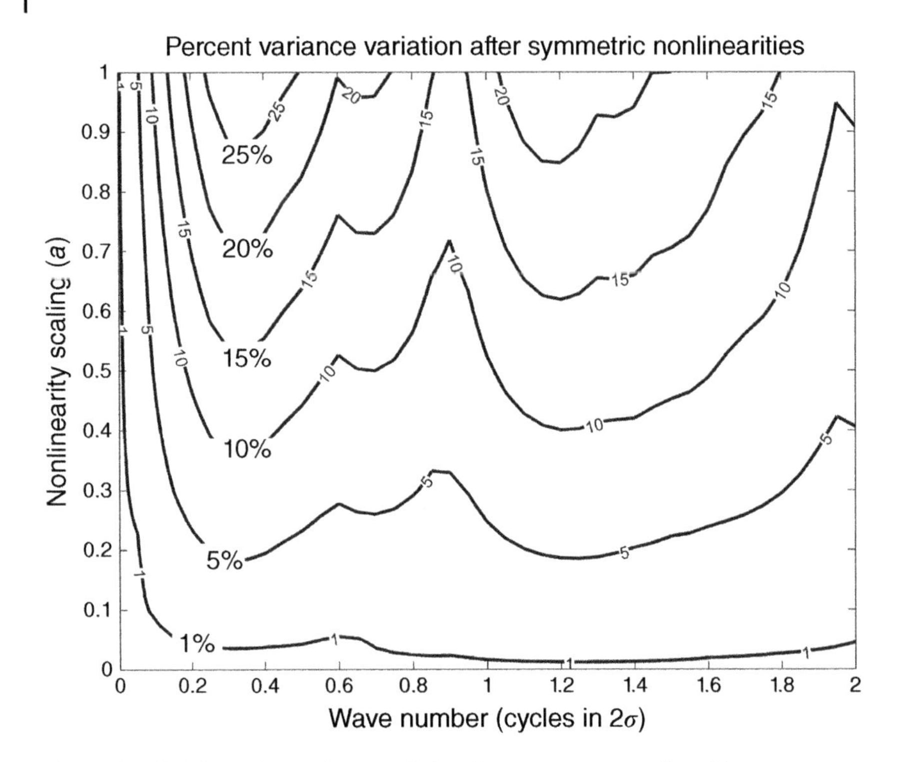

Figure C.6 PDF-dependent variance variations from symmetric nonlinearities.

focused on the nonlinear functions involved are neglecting errors incurred by the effects of the PDF on the outcome. As demonstrated in Chapter 10, this is not an issue with strictly linear applications of the Kalman filter.

The output differences in means and variances due to PDF differences varied with PDF as much as 60% for the test sets of nonlinearities and PDFs used. The resulting approximation errors are due to lack of adequate knowledge about the PDFs involved, but these might possibly be taken into account within the Kalman filter implementation by increasing the associated covariances of uncertainty in the system state estimate (for nonlinear dynamics) or measurement noise (for nonlinear sensors).

Although it is evident in Figures C.4–C.7 that there are wave numbers at which the percent variations of means or variances are relatively low for the given nonlinearity in amplitude, these are not at the same wave numbers in the different plots. The reason(s) for this were not studied.

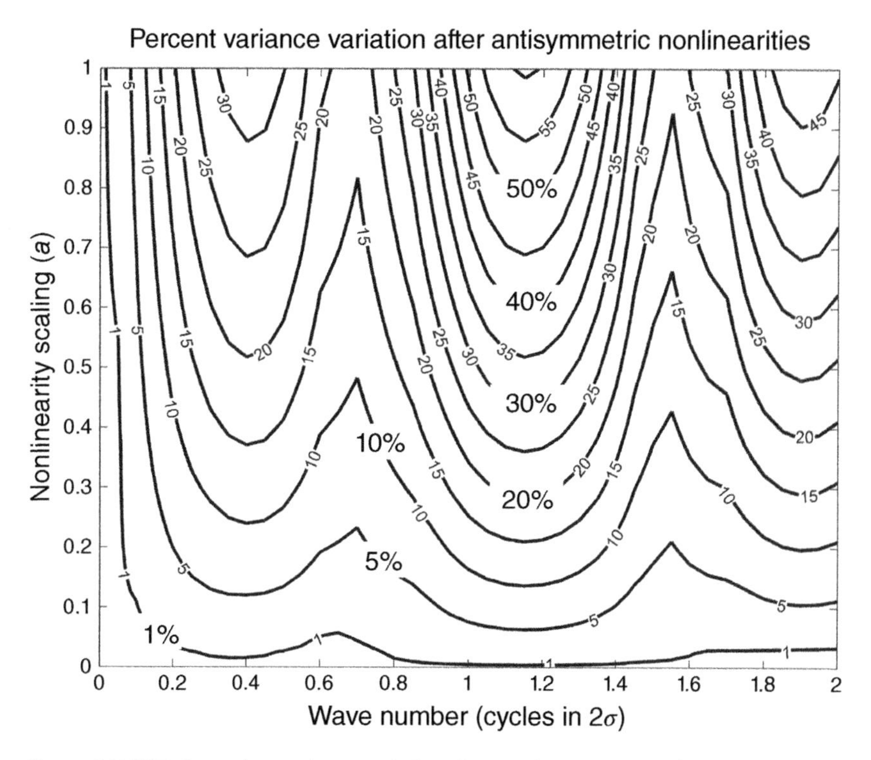

Figure C.7 PDF-dependent variance variations from antisymmetric nonlinearities.

C.4 Mitigating Application-specific Ambiguity Errors

If we cannot prevent ambiguity errors, then perhaps we can at least take them into account in the nonlinear Kalman filter implementations for specific applications, using statistical methods for estimating the PDFs of measurement errors and dynamic disturbance noise to better characterize and mitigate the approximation errors.

In the case of approximation errors in nonlinear propagation of the estimate, this might be mitigated to some degree by adding corrections to the covariance matrix of a priori estimation uncertainty, $\mathbf{P}_{k(-)}$. The problem is in determining how to assess those errors, and this might involve studies of their likely PDFs. It is unlikely that one can assume Gaussian distributions in this case, because Gaussianity is not preserved by nonlinear transformations. It could work in the case of nonlinear sensors, however.

In the case of approximation errors in nonlinear transformations of measurement noise, there is the possibility of adding a "nonlinear approximation" correction to the nonlinearly approximated measurement noise covariance.

Establishing suitable mitigation models for ambiguity errors for specific applications could come to involve off-line studies of source error distributions and possibly end-to-end Monte Carlo simulations using representative error distributions.

The vectorized numerical integration methods used here for univariate PDFs become higher-dimensioned data structures when applied to multivariate PDFs, so their computational requirements must be addressed before such an undertaking.

References

1 H. Castrup, "Distributions for Uncertainty Analysis," International Dimensional Workshop, Knoxville, Tennessee, 2001.

2 C. F. Gauss, *Theoria Combinationis Observationum Erribus Minimus Obnoxiae*, translated by G. W. Stewart, SIAM, Philadelphia, 1995.

3 Joint Committee for Guides in Metrology, Working Group 1, *Evaluation of measurement data - Guide to the expression of uncertainty in measurement*, Bureau International des Poids et Mesures, September 2008.

4 R. E. Kalman, "A New Approach to Linear Filtering and Prediction Problems," *ASME Transactions, Series D: Journal of Basic Engineering*, Vol. 82, pp. 35–45, 1960.

Index

Global Navigation Satellite Systems, Inertial Navigation, and Integration,
Fourth Edition. Mohinder S. Grewal, Angus P. Andrews, and Chris G. Bartone.

Companion website: www.wiley.com/go/grewal/gnss

Printed and bound by CPI Group (UK) Ltd, Croydon, CR0 4YY

16/04/2025

14658593-0001